Edited by
Jörg Grunenberg

Computational Spectroscopy

Related Titles

Feig, M. (ed.)

Modeling Solvent Environments

Applications to Simulations of Biomolecules

2010.
ISBN: 978-3-527-32421-7

Matta, Chérif F. (ed.)

Quantum Biochemistry

2010.
ISBN: 978-3-527-32322-7

Heine, Thomas / Joswig, Jan-Ole / Gelessus, Achim

Computational Chemistry Workbook

Learning Through Examples

2009
ISBN: 978-3-527-32442-2

Meyer, Hans-Dieter / Gatti, Fabien / Worth, Graham A. (eds.)

Multidimensional Quantum Dynamics

MCTDH Theory and Applications

2009.
ISBN: 978-3-527-32018-9

Reiher, Markus / Wolf, Alexander

Relativistic Quantum Chemistry

The Fundamental Theory of Molecular Science

2009
ISBN: 978-3-527-31292-4

van Santen, Rutger A. / Sautet, Philippe (eds.)

Computational Methods in Catalysis and Materials Science

An Introduction for Scientists and Engineers

2009.
ISBN: 978-3-527-32032-5

Willock, D.

Molecular Symmetry

2009
ISBN: 978-0-470-85347-4

Rode, B.M., Hofer, T., Kugler, M.

The Basics of Theoretical and Computational Chemistry

2007
ISBN: 978-3-527-31773-8

Edited by
Jörg Grunenberg

Computational Spectroscopy

Methods, Experiments and Applications

WILEY-VCH Verlag GmbH & Co. KGaA

The Editor

Dr. Jörg Grunenberg
TU Braunschweig
Institut für Organische Chemie
Hagenring 30
38106 Braunschweig
Germany

All books published by **Wiley-VCH** are carefully produced. Nevertheless, authors, editors, and publisher do not warrant the information contained in these books, including this book, to be free of errors. Readers are advised to keep in mind that statements, data, illustrations, procedural details or other items may inadvertently be inaccurate.

Library of Congress Card No.: applied for

British Library Cataloguing-in-Publication Data
A catalogue record for this book is available from the British Library.

Bibliographic information published by the Deutsche Nationalbibliothek
The Deutsche Nationalbibliothek lists this publication in the Deutsche Nationalbibliografie; detailed bibliographic data are available on the Internet at http://dnb.d-nb.de.

© 2010 WILEY-VCH Verlag GmbH & Co. KGaA, Boschstraße 12, 69469 Weinheim

All rights reserved (including those of translation into other languages). No part of this book may be reproduced in any form – by photoprinting, microfilm, or any other means – nor transmitted or translated into a machine language without written permission from the publishers. Registered names, trademarks, etc. used in this book, even when not specifically marked as such, are not to be considered unprotected by law.

Cover Design Grafik-Design Schulz, Fußgönheim
Typesetting Thomson Digital, Noida, India
Printing and Binding Fabulous Printers Pte Ltd

Printed in Singapore
Printed on acid-free paper

ISBN: 978-3-527-32649-5

Contents

Preface *XI*
List of Contributors *XIII*

1 **Concepts in Computational Spectrometry: the Quantum and Chemistry** *1*
J.F. Ogilvie
1.1 Introduction *1*
1.2 Quantum Laws, or the Laws of Discreteness *3*
1.3 Quantum Theories of a Harmonic Oscillator *5*
1.3.1 Matrix Mechanics *6*
1.3.2 Wave Mechanics *9*
1.3.3 Dirac's Operators for Creation and Destruction *15*
1.3.4 Discussion of Quantum Theories in Relation to a Harmonic Oscillator *17*
1.4 Diatomic Molecule as Anharmonic Oscillator *20*
1.5 Quantum Mechanics and Molecular Structure *23*
1.6 Conclusions *33*
References *35*

2 **Computational NMR Spectroscopy** *37*
Ibon Alkorta and José Elguero
2.1 Introduction *37*
2.2 NMR Properties *37*
2.3 Chemical Shifts *37*
2.4 NICS and Aromaticity *41*
2.5 Spin–Spin Coupling Constants *45*
2.6 Solvent Effects *53*
2.7 Conclusions *54*
2.8 The Problem of the Error in Theoretical Calculations of Chemical Shifts and Coupling Constants *55*
References *56*

3	**Calculation of Magnetic Tensors and EPR Spectra for Free Radicals in Different Environments** 63
	Paola Cimino, Frank Neese, and Vincenco Barone
3.1	Introduction 63
3.2	The General Model 64
3.3	Spin Hamiltonian, g-Tensor, Hyperfine Coupling Constants, and Zero-Field Splitting 66
3.3.1	The Spin Hamiltonian 66
3.3.2	Electronic Structure Theory 67
3.3.3	Additional Terms in the Hamiltonian 69
3.3.4	Linear Response Theory 72
3.3.5	Linear Response Equations for Spin Hamiltonian Parameters 76
3.3.6	Computational Aspects: Functionals and Basis Sets 82
3.4	Stereoelectronic, Environmental, and Dynamical Effects 84
3.4.1	Structures and Magnetic Parameters 84
3.4.2	Environmental Effects 86
3.4.3	Short-Time Dynamical Effects 89
3.5	Line Shapes 98
3.6	Concluding Remarks 101
	References 102
4	**Generalization of the Badger Rule Based on the Use of Adiabatic Vibrational Modes** 105
	Elfi Kraka, John Andreas Larsson, and Dieter Cremer
4.1	Introduction 105
4.2	Applicability of Badger-Type Relationships in the Case of Diatomic Molecules 112
4.3	Dissection of a Polyatomic Molecule into a Collection of Quasi-Diatomic Molecules: Local Vibrational Modes 118
4.3.1	Localized Vibrational Modes 122
4.3.2	The Adiabatic Internal Coordinate Modes 124
4.3.3	Properties of Adiabatic Internal Coordinate Modes 126
4.3.4	Characterization of Normal Modes in Terms of AICoMs 127
4.3.5	Advantages of AICoMs 129
4.4	Local Mode Properties Obtained from Experiment 132
4.4.1	Isolated Stretching Modes 132
4.4.2	Local Mode Frequencies from Overtone Spectroscopy 134
4.4.3	Local Mode Information via an Averaging of Frequencies: Intrinsic Frequencies 135
4.4.4	Compliance Force Constants 139
4.5	Badger-type Relationships for Polyatomic Molecules 140
4.6	Conclusions 143
	References 144

5 The Simulation of UV-Vis Spectroscopy with Computational Methods *151*
Benedetta Mennucci
5.1 Introduction *151*
5.2 Quantum Mechanical Methods *152*
5.3 Modeling Solvent Effects *157*
5.4 Toward the Simulation of UV-Vis Spectra *161*
5.5 Some Numerical Examples *162*
5.6 Conclusions and Perspectives *167*
References *168*

6 Nonadiabatic Calculation of Dipole Moments *173*
Francisco M. Fernández and Julián Echave
6.1 Introduction *173*
6.2 The Molecular Hamiltonian *174*
6.3 Symmetry *178*
6.4 The Hellmann–Feynman Theorem *179*
6.5 The Born–Oppenheimer Approximation *180*
6.6 Interaction between a Molecule and an External Field *182*
6.7 Experimental Measurements of Dipole Moments *184*
6.8 The Born–Oppenheimer Calculations of Dipole Moments *185*
6.9 Nonadiabatic Calculations of Dipole Moments *186*
6.10 Molecule-Fixed Coordinate System *192*
6.11 Perturbation Theory for the Stark Shift *195*
6.12 Conclusions *196*
References *197*

7 The Search for Parity Violation in Chiral Molecules *201*
Peter Schwerdtfeger
7.1 Introduction *201*
7.2 Experimental Attempts *205*
7.2.1 Vibration–Rotation Spectroscopy *206*
7.2.2 Mössbauer Spectroscopy *207*
7.2.3 NMR Spectroscopy *208*
7.2.4 Electronic Spectroscopy *209*
7.2.5 Other Experiments *209*
7.3 Theoretical Predictions *211*
7.4 Conclusions *215*
References *216*

8 Vibrational Circular Dichroism: Time-Domain Approaches *223*
Hanju Rhee, Seongeun Yang, and Minhaeng Cho
8.1 Introduction *223*
8.2 Time-Correlation Function Theory *224*

8.3	Direct Time-Domain Calculation with QM/MM MD Simulation Methods 227
8.4	Direct Time-Domain Measurement of VOA Free Induction Decay Field 231
8.4.1	Conventional Differential Measurement Method 231
8.4.2	Femtosecond Spectral Interferometric Approach 232
8.4.2.1	Cross-Polarization Detection Configuration 232
8.4.2.2	Fourier Transform Spectral Interferometry 234
8.4.2.3	Vibrational OA-FID Measurement 237
8.5	Summary and a Few Concluding Remarks 238
	References 239
9	**Electronic Circular Dichroism** 241
	Lorenzo Di Bari and Gennaro Pescitelli
9.1	Introduction 241
9.2	Molecular Anatomy 243
9.3	Conformational Manifolds and Molecular Structure 246
9.4	Hybrid Approaches 247
9.4.1	Coupled Oscillators and the DeVoe Method 248
9.4.2	The Matrix Method 251
9.4.3	Applications 252
9.5	The QM Approach 256
9.5.1	Assignments of Absolute Configurations 261
9.5.1.1	The Solid-State ECD TDDFT Method 266
9.5.2	Interpretations of ECD Spectra 268
9.5.3	Other Applications 270
9.6	Conclusions and Perspectives 271
	References 272
10	**Computational Dielectric Spectroscopy of Charged, Dipolar Systems** 279
	Christian Schröder and Othmar Steinhauser
10.1	Methods 279
10.1.1	Dielectric Field Equation 279
10.1.2	Molecular Resolution of the Total Collective Dipole Moment 282
10.1.3	Computing the Generalized Dielectric Constant in Equilibrium 286
10.1.4	Finite System Electrostatics 294
10.2	Applications and Experiments 299
10.2.1	Solvated Biomolecules 303
10.2.1.1	Peptides 304
10.2.1.2	Proteins 305
10.2.1.3	DNA 309
10.2.1.4	Biological Cells 310
10.2.2	Molecular Ionic Liquids 311
10.2.2.1	Conductivity and Dielectric Conductivity 312

10.2.2.2	Dielectric Permittivity	*314*
10.2.2.3	Generalized Dielectric Constant	*315*
10.3	Summary and Outlook	*317*
	References	*318*

11 Computational Spectroscopy in Environmental Chemistry *323*
James D. Kubicki and Karl T. Mueller

11.1	Introduction	*323*
11.1.1	Need for Computational Spectroscopy	*323*
11.1.1.1	Speciation	*323*
11.1.1.2	Surface Reactions	*324*
11.1.2	Types of Spectra Calculated	*325*
11.1.2.1	IR/Raman	*325*
11.1.2.2	NMR	*328*
11.1.2.3	EXAFS + CTR + XSW	*329*
11.1.2.4	QENS and INS	*330*
11.2	Methods	*331*
11.2.1	Model Building	*331*
11.2.2	Selecting a Methodology	*333*
11.3	Examples	*334*
11.3.1	IR/Raman Phosphate on Goethite	*334*
11.3.2	Solution-State NMR of Al–Organic Complexes	*337*
11.3.3	Solid-State NMR of Phosphate Binding on Alumina	*339*
11.3.4	Solid-State NMR of Aluminum Species at Mineral and Glass Surfaces	*341*
11.3.5	Water and Zn(II) on TiO_2	*341*
11.3.6	Water Dynamics on TiO_2 and SnO_2	*343*
11.4	Summary and Future	*345*
	References	*346*

12 Comparison of Calculated and Observed Vibrational Frequencies of New Molecules from an Experimental Perspective *353*
Lester Andrews

12.1	Introduction	*353*
12.2	Experimental and Theoretical Methods	*353*
12.2.1	The LiO_2 Ionic Molecule	*354*
12.3	Aluminum and Hydrogen: First Preparation of Dibridged Dialane, Al_2H_6	*356*
12.4	Titanium and Boron Trifluoride Give the Borylene $FB=TiF_2$	*359*
12.5	Ti and CH_3F Form the Agostic Methylidene Product $CH_2=TiHF$	*360*
12.6	Zr and CH_4 Form the Agostic Methylidene Product $CH_2=ZrH_2$	*362*
12.7	Mo and $CHCl_3$ Form the Methylidyne $CH\equiv MoCl_3$	*364*
12.8	Tungsten and Hydrogen Produce the $WH_4(H_2)_4$ Supercomplex	*366*
12.9	Pt and CCl_4 Form the Carbene $CCl_2=PtCl_2$	*367*

12.10	Th and CH_4 Yield the Agostic Methylidene Product $CH_2=ThH_2$	*371*
12.11	U and CHF_3 Produce the Methylidyne $CH\equiv UF_3$	*371*
	References	*374*

13 Astronomical Molecular Spectroscopy *377*
Timothy W. Schmidt

13.1	The Giants' Shoulders	*377*
13.2	The First Spectroscopists and Seeds of Quantum Theory	*379*
13.3	Small Molecules	*383*
13.3.1	CH, CN, CO, CO^+	*383*
13.3.2	Dicarbon: C_2	*385*
13.3.3	The Carbon Trimer: C_3	*387*
13.3.4	Radioastronomy	*389*
13.4	The Diffuse Interstellar Bands	*390*
13.4.1	The Hump	*392*
13.5	The Red Rectangle, HD44179	*392*
13.6	The Aromatic Infrared Bands	*394*
13.7	The Holy Grail	*394*
	References	*395*

Index *399*

Preface

...with its help, debates can be resolved forever, if they can be settled on the basis of some data; and if one took the pen it would be enough for the two disputing men to say to one another: Let's calculate.
Leibniz in a letter to P. J. Spener, 1687

Computational chemistry has reached a high degree of maturity and comprehension making it one of the vivid research areas in modern chemical and physical research in general. This is true because an accurate simulation of spectroscopic properties is one of the major challenges and – at the same time – a precious benefit of modern theoretical chemistry. Predictions concerning single molecules, molecular clusters, or even the solid state in combination with detailed information from apparatus-based experiments are therefore providing ingredients to an auspicious revolution in the borderland between theory and experiment: computational spectroscopy. At first sight, the term seems to contradict itself: from the traditional point of view, spectroscopy (or spectrometry) belongs to the realm of the experimentalists, while computational chemistry is allocated to the domain of theory. The frantic developments in both areas during the last years have nevertheless helped build new bridges between both worlds.

This is important because until the end of the last millennium theoretical and experimental chemistry were separated by respectable gaps. Studying chemistry in the 1990s was yet sometimes accompanied by dialectical training: equipped with the sanguine knowledge that molecular orbitals are artifacts (learned from an exciting theoretical chemistry course), one stumbled into an organic chemistry exam being forced to explain the formation of a covalent bond in terms of those very orbitals. Those days are history now for several – in part ambivalent – reasons. The main cause nevertheless is a simple one: modern computational chemistry deals with observable properties and this positivistic shift does not leave too much room for "overinterpretations." One can always try to find an experiment, which allows either falsification or confirmation of the computer simulation. This is in sharp contrast to the second major application area of computational chemistry: the underpinning of chemical concepts. Led by Coulson's famous request, "... give us insights, not numbers ...," more and more chemical perceptions as well as new molecular categories were introduced. It is, however, still unclear whether the addition of those ad hoc concepts is always helpful in characterizing the huge variety of chemical phenomena. On the contrary, many of these early chemical concepts resembled Leibniz's *voces metaphysicae*, that means phrases, which we use believing that we understand entities just by pinning names on them. Or, to quote Wolfgang Pauli, many of the earlier concepts deduced from approximate quantum chemistry were so fuzzy that they were not even wrong.

In order to keep pace with new developments in terms of more rigorous solutions for Schrödinger equation, we anyhow may not demand that ideas from the early days of numerical theoretical chemistry persist permanently. The decade-long debate and struggle for a unique definition of aromaticity is only one of the many examples of those fruitless endeavors. The (in part humoristic) suggestion by Heilbronner as early as 1971 at the famous Jerusalem symposium on "Aromaticity, Pseudo-Aromaticity, Anti-Aromaticity" to introduce the term "schizo-aromaticity" for molecules, which are aromatic by one definition and nonaromatic by another, illuminates this dilemma quite graphically.

The situation changed dramatically during the last 20 years. Reliable first-principle electronic structure calculations on the one hand and sophisticated molecular dynamic simulations for complex systems on the other hand are nowadays well-established instruments in the toolbox of theoretical chemists, and these rapid developments are paving the way for the study of increasingly large and chemically complex systems. At the same time, experimental molecular spectroscopy is also an extremely active and fast-developing field, which is evolving toward the possibility of performing precise measurements for single molecules and, even more intriguing, for the hub of chemistry itself, the individual covalent bond. The title of this book *Computational Spectroscopy* states its aim: From basic research to commercial applications in the area of environment relevance, we will compile the major developments during the past 5–10 years. A multitude of apparatus-driven technologies will be covered. Nevertheless, the selection of topics is of course a subjective one. Summarizing the results of so many different disciplines, I hope that this book will on the one hand attract the attention of newcomers and on the other hand inform the experts about developments in scientific areas adjacent to their own expertise.

At Wiley, I would especially like to thank Dr. Elke Maase and Dr. Martin Graf for their guidance through all phases (from the first concept of the book to the final cover design) of this challenging and fascinating project.

July 2010 *Jörg Grunenberg*

List of Contributors

Ibon Alkorta
Instituto de Química Médica
CSIC
Juan de la Cierva 3
E-28006
Madrid
Spain

Lester Andrews
University of Virginia
Department of Chemistry
Charlottesville
VA 22904-4319
USA

Lorenzo Di Bari
Università degli Studi di Pisa
Dipartimento di Chimica e Chimica
Industriale
Via Risorgimento 35
I-56516 Pisa
Italy

Vincenco Barone
Scuola Normale Superiore
Piazza dei Cavalieri 7
56126 Pisa
Italy

Minhaeng Cho
Korea University
Department of Chemistry and Center
for Multidimensional Spectroscopy
5-1 Anam-dong
Songbuk-ku
Seoul 136-701
Korea

and

Korea Basic Science Institute
Multidimensional Spectroscopy
Laboratory
Seoul 136-713
Korea

Paola Cimino
University of Salerno
Department of Pharmaceutical Science
Via Ponte don Melillo
84084 Fisciano
Italy

Dieter Cremer
Southern Methodist University
Department of Chemistry
3215 Daniel Ave
Dallas
TX 75275-0314
USA

Julián Echave
INIFTA (UNLP, CCTLa Plata-
CONICET)
Diag. 113 y 64 (S/N)
Sucursal 4
Casilla de Correo 16
1900 La Plata
Argentina

José Elguero
Instituto de Química Médica
CSIC
Juan de la Cierva
3, E-28006
Madrid
Spain

Francisco M. Fernández
INIFTA (UNLP CCTLa Plata-CONICET)
Diag. 113 y 64 (S/N)
Sucursal 4
Casilla de Correo 16
1900 La Plata
Argentina

Elfi Kraka
Southern Methodist University
Department of Chemistry
3215 Daniel Ave
Dallas
TX 75275-0314
USA

James D. Kubicki
The Pennsylvania State University
Department of Geosciences
University Park
PA 16802
USA

J. Andreas Larsson
Southern Methodist University
Department of Chemistry
3215 Daniel Ave
Dallas
TX 75275-0314
USA

Benedetta Mennucci
University of Pisa
Department of Chemistry
Via Risorgimento 35
56126 Pisa
Italy

Karl T. Mueller
The Pennsylvania State University
Department of Chemistry
University Park
PA 16802
USA

Frank Neese
University of Bonn
Institute for Physical and Theoretical
Chemistry
53115 Bonn
Germany

List of Contributors

J.F. Ogilvie
Universidad de Costa Rica
Ciudad Universitaria Rodrigo Facio
Escuela de Quimica
San Pedro de Montes de Oca
San Jose 2060
Costa Rica

and

Simon Fraser University
Centre for Experimental and
Constructive Mathematics
Department of Mathematics
8888 University Drive
Burnaby
British Columbia V5A 1S6
Canada

Gennaro Pescitelli
Università degli Studi di Pisa
Dipartimento di Chimica e Chimica
Industriale
Via Risorgimento 35
I-56516 Pisa
Italy

Hanju Rhee
Korea University
Department of Chemistry and Center
for Multidimensional Spectroscopy
5-1 Anam-dong
Songbuk-ku
Seoul 136-701
Korea

Timothy W. Schmidt
University of Sydney
School of Chemistry
NSW 2006
Australia

Christian Schröder
University of Vienna
Department of Computational
Biological Chemistry
Währinger Str. 17
1090 Vienna
Austria

Peter Schwerdtfeger
Massey University
New Zealand Institute for Advanced
Study
Centre for Theoretical Chemistry and
Physics
Auckland Campus
Private Bag 102904
North Shore City
0745 Auckland
New Zealand

Othmar Steinhauser
University of Vienna
Department of Computational
Biological Chemistry
Währinger Str. 17
1090 Vienna
Austria

Seongeun Yang
Korea University
Department of Chemistry and Center
for Multidimensional Spectroscopy 5-1
Anam-dong
Songbuk-ku
Seoul 136-701
Korea

1
Concepts in Computational Spectrometry: the Quantum and Chemistry

J. F. Ogilvie

1.1
Introduction

During the nineteenth century and most of the first half of the twentieth century, after Dalton's recognition of the atomic nature of chemical matter, which is everything tangible, that matter was regarded by most chemists as a material. Even though chemists, following Couper, Kekule, van't Hoff, and others, drew structural formulae in terms of atoms connected by bonds represented as lines, chemical samples were generally regarded as materials or "stuff". When, after 1955, molecular spectra, particularly of organic compounds, began to be recorded routinely in the mid-infrared region and with nuclear magnetic resonance, the outlook of chemists shifted from macroscopic properties, such as density, melting point, and refractive index, to purportedly molecular properties, such as the effect of adjacent moieties on the characteristic infrared absorption associated with a carbonyl group or on the chemical shift of a proton. The first "quantum-chemical" calculations, on H_2^+ by Burrau and on H_2 by Heitler and London, all physicists, had as subjects chemical species remote from common laboratory experience, but Pauling's brilliant insight and evangelical manner stimulated great qualitative interest in a theoretical interpretation of chemical properties, even though a large gap existed between the primitive calculations on methane and other prototypical molecules and molecules of substances of practical interest. This gap was bridged largely through the efforts of Pople and his collaborators during the second half of the twentieth century in developing computer programs that enabled efficient calculation of observable molecular properties; not coincidentally, Pople was also an early exponent of the application of nuclear-magnetic-resonance spectra in the publication in 1959 of an authoritative monograph [1] that was seminally influential in the general application of this spectral method [2].

Chemists concerned with quantitative analysis have always understood the distinction between spectroscopy and spectrometry: spectroscopy implies the use of a human eye as a visual detector with a dispersive optical instrument and hence necessarily qualitative and imprecise observations, whereas spectrometry pertains to

Computational Spectroscopy: Methods, Experiments and Applications. Edited by Jörg Grunenberg
Copyright © 2010 WILEY-VCH Verlag GmbH & Co. KGaA, Weinheim
ISBN: 978-3-527-32649-5

an instrument with an electrical detector amenable to quantitative measurement of both frequency and intensity. For spectra throughout the entire accessible range of frequencies from 10^6 Hz, characteristic of nuclear quadrupole or nuclear magnetic resonance, to radiation in the X-ray region sufficiently energetic to cause ionization, a significant use of the numerical results of computations based nominally on quantum mechanics, such as of molecular electronic structure and properties, is to assist that spectral analysis. Pople's programs were based, to an increasing extent over the years, on selected quantum-mechanical principles that arose from quantum theories. During the past century, the practice of chemistry has thus evolved much, from being a largely empirical science essentially involving operations in a laboratory and their discussion, to having – allegedly – an underpinning based on quantum *theories*.

During the nineteenth century, a standard paradigm for most chemical operations was that both matter and energy are continuous; following a philosophical point of view of Greek savants and concrete ideas of Bacon and Newton, Dalton's contention that matter is particulate provided a basis to explain chemical composition, but Ostwald remained skeptical of the existence of atoms until 1909 [3]. The essence of the quantum concept is that both energy and matter ultimately comprise small packets, or chunks, not further divisible retaining the same properties. In Latin, quantum means *how much?*. A descriptor more enlightening than quantum is discrete, so we refer to the ultimate prospective discreteness of matter and energy. (In a mathematical context, integers take discrete values, even though they number uncountably, and have a constant unit increment, whereas real numbers 1.1, 1.11, 1.111, ... vary continuously, with an increment between adjacent representatives as small as desired.) One accordingly distinguishes between the laws of discreteness, based on experiment, and various theories that have been devised to encompass or to reproduce those discrete properties. The distinctions between physical laws and theories or mathematical treatments are poorly appreciated by chemists; our objective is thus to clarify the nature of both quantum laws and quantum theories, thereby to propose an improved understanding of the purported mathematical and physical basis of chemistry and the application of computational spectrometry. After distinguishing between quantum laws and quantum theories, we apply to a prototypical problem three distinct quantum-mechanical methods that nevertheless conform to the fundamental postulate of quantum mechanics; we then consider molecular structure in relation to quantum-mechanical principles and their implications for the practice of chemistry aided by computational spectrometry.

For many chemists, the problem so called the *particle in a box* is the only purportedly quantum-mechanical calculation that they are ever required to undertake as a manual exercise, but its conventional solution is at least problematic. Any or all treatments of a harmonic oscillator in Section 1.3 serve as a viable alternative to that deficient model. The connection between quantum mechanics and chemistry might be based on a notion that "quantum mechanics governs the behavior of electrons and atoms in molecules," which is merely supposition. While Dirac and Einstein had, to the ends of their lives, grave misgivings about fundamental aspects of quantum mechanics [4], and even Born was never satisfied with a separate – and thereby inconsistent – treatment of the motions of electrons and atomic nuclei that underpins common quantum-chemical calculations, almost all chemists accept, as recipes, these highly mathematical theories, in a mostly qualitative manner embodied in orbitals – "for fools rush in where angels fear to tread" (Pope). For those chemists who undertake calculations, typically with standard computer programs developed by mathematically knowledgeable specialists who have no qualms about producing more or less efficient coding but who might refrain from questioning the underlying fundamental aspects, the emphasis is placed on the credibility of the results. For the molecular structures of stable species that have been established by essentially experimental methods, although a theoretical component is invariably present, the empirical nature of the computer coding – its parameters are invariably set to reproduce, approximately, various selected properties of selected calibration species – reduces its effect to a sophisticated interpolation scheme; for the molecular structures of such fabulous species as *transition states*, as these are inherently impossible to verify, the results of the calculations merely reinforce preconceived notions of those undertaking such calculations. We trust that reconsideration of the current paradigm in chemistry that abides such questionable content will motivate an improved understanding of the mathematical and physical bases of chemistry and a reorientation of chemistry as an experimental and logical science of both molecules and materials. For this purpose, computational spectrometry has a substantial role to play in a fertile production of information about the structure and properties of molecules and materials.

1.2
Quantum Laws, or the Laws of Discreteness

The universe comprises matter and energy; as chemists, we might ignore the possibility of their interconversion. With regard to matter, we classify anything on or above an atomic scale and that displays a rest mass as either material or molecule. Molecules exist only in a gaseous state of aggregation under conditions in which intermolecular interactions are negligible, thus describable as constituting an ideal gas; an isolated atom is simply a monatomic molecule. Molecules hence exist most purely in interstellar clouds, but even appropriate gaseous samples in a terrestrial laboratory that exhibit properties nearly characteristic of a free molecule might, to a sufficient approximation, be deemed to contain molecular entities. A material is found in a condensed phase or in a compressed gas, under which conditions rotational degrees of freedom are much hampered. A few condensed samples, such as liquid dihydrogen, have properties, such as spectra in the infrared region, that might resemble those of free molecules. In contrast, a single macroscopic crystal of diamond or sodium chloride or a sheet of "polyethylene", for instance, might on structural grounds be regarded as constituting a single molecule. For H_2O, the smallest internuclear distances in the gaseous phase and in liquid water or solid ice differ by more than their experimental uncertainties, and their infrared spectra concomitantly differ. How can we consider H_2O to be the same species in the

Table 1.1 Physical properties of molecules and photons.

Property	Value for molecule	Value for photon
Charge	$Z = 0, \pm 1e, \pm 2e, \ldots$	0
(Rest) mass	$M > 0$	0
Energy	$W \cong W_{tr} + W_{rot} + W_{vib} + W_{el} + \cdots$	$W = h\nu$
Linear momentum	$|\vec{p}| > 0$	$|\vec{p}| = h/\lambda$
Angular momentum	$|\vec{\Omega}| = [J(J+1)]^{1/2}\hbar$	$|\vec{\Omega}| = \hbar$

molecular vapor and the material solid? For energy, we limit attention to its radiant form as constituting a free state like an interstellar molecule; a discrete unit of radiation is called a photon.

We summarize in Table 1.1 some fundamental properties of molecules and photons [5].

The experiment best known to demonstrate the discreteness of a property of a particle with rest mass is Millikan's measurement of the charge of an electron on a drop of oil suspended in a vertical electric field. Assuming that discreteness of electric charge, one readily observes the discreteness of mass of molecular ions with a conventional mass spectrometer. The unit of charge is that on the proton or the magnitude of charge that an electron carries. Masses of individual atoms are classified with exact integers, but actual atomic masses, or masses of not too large molecules of particular isotopic composition, assume nearly integer values in terms of unified atomic mass unit or dalton. Under appropriate conditions, the total energy W of a molecule confined to a finite spatial volume might take discrete values, but even for an unconfined molecule the rotational, vibrational, and electronic contributions to total energy might be observed to alter in discrete increments. The linear momentum $\vec{p}$ of a confined molecule is supposed to assume discrete values. Not only the total angular momentum $\vec{\Omega}$ of a molecule but also the separate electronic, nuclear, and rotational contributions thereto appear to assume, in a particular discrete state, integer or half-integer values in terms of Dirac constant $\hbar$ (Planck constant h divided by 2π); these angular momenta likewise alter in unit increments in various processes.

A photon possesses neither net electric charge nor rest mass, but its energy and linear momentum are directly related to its wave attributes – frequency $\nu = E/h$ and wavelength $\lambda = h/|p|$; their product $\nu\lambda$ equals the speed c of light. Any photon carries, independent of frequency, intrinsic angular momentum $\vec{\Omega}$ to the extent of one unit in terms of $\hbar$, in the direction of propagation if it be circularly polarized in one sense or opposite the direction of propagation for circular polarization in the other sense. As a limiting case of elliptical polarization, linearly polarized light as a coherent superposition of these two circular polarizations lacks net angular momentum and so imparts no total angular momentum to an absorbing target.

Even when neglecting a distinction between molecules and materials, these laws of discreteness, or quantum laws, of molecules and photons provide an ample basis for the conduct and explanation of chemical operations for almost all practical purposes beyond which a continuum of properties suffices.

1.3
Quantum Theories of a Harmonic Oscillator

In attempts to explain or to encompass various experimental data interpreted in terms of discrete properties at a molecular level, scientists have devised various quantum theories. Planck proposed reluctantly in 1900 the first theory to involve a discrete quantity, for which he invoked harmonic oscillators; he attempted to explain the distribution of energy, as a function of wavelength, radiated by a *black body* [3]. As that distribution is continuous, the requirement for a discrete quantity in a theoretical derivation is not obvious. That distribution has been derived alternatively with classical statistical thermodynamics [6], although the incorporation therein of the Planck constant remains enigmatic. The second application of a quantum condition appeared in Einstein's treatment of the photoelectric effect in 1905, but in retrospect a recognition of the quantum laws makes that derivation almost trivial. In relation to infrared spectra of gaseous hydrogen halides, in 1911 Bjerrum, a Danish chemist, sought to develop an explicit quantum theory of molecules for vibrational and rotational motions; as this treatment preceded Rutherford's revelation of the structure of the nuclear atom, this endeavor was bound to fail. Equally incorrect but far better known is Bohr's theory, in 1913, of an atom with one electron for which some enhancements by Sommerfeld and Wilson failed to remedy the fundamental deficiencies [3].

In 1924, Born and Heisenberg recognized that a proper description of an atomic particle must be concerned with its mechanics and dynamics, hence with equations of motion in terms of position, momentum, and time. In Heisenberg's development of the first enduring quantum theory in 1925 [3], the crucial particular in his paper is expressible as

$$p_j q_k - q_k p_j = -i\hbar\,\delta_{j,k} \qquad (1.1)$$

Therein appear symbols to denote a component of momentum p or of position q, $i = \sqrt{-1}$, Dirac constant $\hbar$, and Kronecker delta function $\delta_{j,k}$ that equals unity if $j=k$ or zero otherwise; the left side of this equation contains a commutator, printed as $[p_j, q_k]$. In one dimension, this equation becomes

$$pq - qp \equiv [p,q] = -i\hbar \qquad (1.2)$$

From this relation are derivable both de Broglie's relation, $\lambda = h/p$, and Heisenberg's principle of indeterminacy [5], $\Delta q\,\Delta p \geq 1/2\,\hbar$, whereas the reverse derivations are less obvious. One may thus regard this equation, as Dirac recognized directly in 1925, to constitute the *fundamental postulate of quantum mechanics*. A parallel postulate in the form of a commutation relation involving energy and time is less relevant here. Among quantities that naturally fail to commute are matrices, and a variable with its differential operator; such quantities to represent p and q might hence form a basis of quantum-mechanical calculations.

To illustrate and to contrast three methods of quantum mechanics in a nonrelativistic approximation, we apply this commutator to a canonical linear harmonic oscillator in one spatial dimension. According to classical mechanics, the frequency of its oscillation is independent of its amplitude, whereas according to quantum mechanics a harmonic oscillator has states of discrete energies with equal increments between adjacent states, as we derive below. Because the latter oscillator possesses no angular momentum, it behaves as a boson. The classical potential energy V associated with this canonical form is expressed as

$$V(q) = 1/2\, k_e\, q^2 \tag{1.3}$$

in which V exhibits a parabolic dependence on displacement coordinate q; coefficient k_e is also the factor of proportionality in Hooke's law, $F(q) = -k_e q$, relating a restoring force to that displacement.

1.3.1
Matrix Mechanics

According to matrix mechanics, each physical quantity has a representative matrix [7]. For coordinate matrix $\mathbf{Q}$, we accordingly define its elements $q_{n,m}$. Combining the relation for the restoring force of an oscillator of mass μ with Newton's second law in nonrelativistic form, we obtain

$$F(q) = -k_e\, q = \mu\, d^2 q/dt^2 \tag{1.4}$$

Expressing a ratio k_e/μ of parameters as a square of a radial frequency ω_0, for which the units are appropriate, we rewrite this differential equation as

$$d^2 q(t)/dt^2 = -\omega_0^2\, q(t) \tag{1.5}$$

For this equation to be applicable to a system described by means of matrix mechanics, each element of matrix $\mathbf{Q}$ must separately obey this equation; we express this solution in exponential form as

$$q_{n,m}(t) = q_{n,m}^0 \exp(-i\,\omega_{n,m}\, t) \tag{1.6}$$

in which appear two arbitrary constants $q_{n,m}^0$ and $\omega_{n,m}$, appropriate to an ordinary differential equation of second order. Substitution of this solution into that differential equation yields the following condition:

$$\left(\omega_0^2 - \omega_{n,m}^2\right) q_{n,m}^0 = 0 \tag{1.7}$$

Hence, either $q_{n,m}^0 = 0$ or $\omega_{n,m} = \pm\omega_0$. Because numbering of matrix elements is arbitrary, we apply a convention that a condition $\omega_{n,m} = +\omega_0$ corresponds to emission of a photon as the oscillator passes from a state of energy with index n to another state with energy with index $n-1$, whereas a condition $\omega_{n,m} = -\omega_0$ corresponds to absorption of a photon as the oscillator passes from a state of energy with index n to another state with energy with index $n+1$. With numbering of elements

1.3 Quantum Theories of a Harmonic Oscillator

beginning at zero, the coordinate matrix thus assumes this form,

$$Q = \begin{pmatrix} 0 & q^0_{0,1} & 0 & 0 & \cdots \\ q^0_{1,0} & 0 & q^0_{1,2} & 0 & \cdots \\ 0 & q^0_{2,1} & 0 & q^0_{2,3} & \cdots \\ \cdots & \cdots & \cdots & \cdots & \cdots \end{pmatrix} \quad (1.8)$$

in which nonzero elements accordingly appear only on the first diagonals above and below the principal diagonal. As momentum, in a nonrelativistic approximation, is defined as a product of a constant mass and the temporal derivative of coordinate q, so that $p = \mu \, dq/dt$, we have for each element of the momentum matrix $p_{n,m} = i\,\mu\,\omega_{n,m} q_{n,m}$; with $\omega_{n,m} = \pm\omega_0$ and $m = n \pm 1$, we obtain

$$P = i\mu\omega_0 \begin{pmatrix} 0 & q^0_{0,1} & 0 & 0 & \cdots \\ q^0_{1,0} & 0 & q^0_{1,2} & 0 & \cdots \\ 0 & q^0_{2,1} & 0 & q^0_{2,3} & \cdots \\ \cdots & \cdots & \cdots & \cdots & \cdots \end{pmatrix} \quad (1.9)$$

which has nonzero elements along diagonals only directly above and below the principal diagonal.

The total energy W of a state of the oscillator is a sum of kinetic T and potential V contributions, which together constitute the Hamiltonian $\hat{H}$ applicable to this problem,

$$W = \hat{H} = 1/2 \, p^2/\mu + 1/2 \, k_e q^2 = 1/2 \, p^2/\mu + 1/2 \, \mu \, \omega_0^2 q^2 \quad (1.10)$$

We form accordingly an energy matrix W as a sum of squares of matrices P for momentum and Q for coordinate with their indicated multiplicands, $1/2\,\mu^{-1}$ and $1/2\,\mu\,\omega_0^2$, respectively, which yields

$$W = \mu\omega_0^2 \begin{pmatrix} q^0_{0,1} q^0_{1,0} & 0 & 0 & 0 & \cdots \\ 0 & q^0_{0,1} q^0_{1,0} + q^0_{1,2} q^0_{2,1} & 0 & 0 & \cdots \\ 0 & 0 & q^0_{1,2} q^0_{2,1} + q^0_{2,3} q^0_{3,2} & \cdots \\ \cdots & \cdots & \cdots & \cdots \end{pmatrix} \quad (1.11)$$

Nonzero elements appear therein only along the principal diagonal. Moreover, all factors dependent on time have vanished, which signifies that the energies of states are independent of time, thus corresponding to *stationary states*. Amplitude coefficients $q^0_{n,m}$, which originate as constants of integration, remain to be evaluated; for this purpose, we apply directly the commutation law, which here contains a unit matrix on the right side.

$$[pq-qp] = -2i\mu\omega_0 \begin{pmatrix} q^0_{1,0}q^0_{0,1} & 0 & 0 & \cdots \\ 0 & q^0_{2,1}q^0_{1,2}-q^0_{1,0}q^0_{0,1} & 0 & \cdots \\ 0 & 0 & q^0_{3,2}q^0_{2,3}-q^0_{2,1}q^0_{1,2} & \cdots \\ \cdots & \cdots & \cdots & \cdots \end{pmatrix}$$

$$= -i\hbar \begin{pmatrix} 1 & 0 & 0 & 0 & \cdots \\ 0 & 1 & 0 & 0 & \cdots \\ 0 & 0 & 1 & 0 & \cdots \\ \cdots & \cdots & \cdots & \cdots & \cdots \end{pmatrix}$$

(1.12)

Therefore,

$$q^0_{1,0}q^0_{0,1} = -i\hbar/(-2i\mu\omega_0) = \hbar/(2\mu\omega_0)$$
$$q^0_{2,1}q^0_{1,2} - q^0_{1,0}q^0_{0,1} = \hbar/(2\mu\omega_0), \ldots$$

(1.13)

Solving successively these equations and consistent with microscopic reversibility, we obtain

$$q^0_{n+1,n}q^0_{n,n+1} = q^0_{n,n+1}q^0_{n+1,n} = |q^0_{n,n+1}|^2 = (n+1)\hbar/(2\mu\omega_0)$$

(1.14)

We substitute this general relation into the energy matrix. The corresponding elements $q_{n,m}$ of coordinate matrix $\mathbf{Q}$ increase along each diagonal according to $[1/2\,(n+1)]^{1/2}$. When we replace radial frequency ω_0 by circular frequency $v_0 = \omega_0/(2\pi)$, we derive a general result

$$W_n \equiv W_{n,n} = (n+1/2)\hbar\omega_0 = (n+1/2)h\,v_0$$

(1.15)

This result signifies that the interval of energy between states characterized with adjacent integers is constant, equal to $h\,v_0$, and that the state of least energy, characterized with $n=0$, has a residual, or *zero-point*, energy equal to $1/2\,h\,v_0$. Transitions in absorption or emission, according to type electric dipole of form charge times distance, $e\,q$, are thus governed by the nonzero elements of coordinate matrix $\mathbf{Q}$; these transitions are hence possible only between states of adjacent energies.

In principle, the rows and columns of all matrices here number infinitely, but to form each matrix with a dozen rows and columns suffices for any practical purpose. Although these calculations by hand with matrices of even such an order are tedious, calculation with mathematical software [8] such as Maple is readily effected; according to contemporary methods of teaching mathematics, many students are introduced to such software in calculus courses, so there is no major impediment to such use for chemical applications. In the same way, one shows directly that the numbering of matrix elements is arbitrary, so that nonzero elements $q^0_{n,m}$ of matrix $\mathbf{Q}$ might occur for $n = m \pm k$, for instance, with $k = 2$ or 3 rather than 1 as above; in that case, the energies still have values $(n + 1/2)\, h\, \nu_0$ with nonnegative integer n, and two or three states have the same energy. Transitions of type electric dipolar still occur only between states of adjacent distinct energies.

1.3.2
Wave Mechanics

According to wave mechanics, an observable quantity might be represented with a differential operator. To conform to the fundamental postulate of quantum mechanics, either coordinate q or momentum p, but not both, might be selected to be a differential operator. According to a coordinate representation, we choose momentum p to become $-i\hbar\, d/dq$, whereas according to a momentum representation we choose coordinate q to become $i\hbar\, d/dp$; the reason for such choices is simply to impose conformity with that fundamental postulate. A differential operator requires an operand, called an amplitude function or wavefunction; for operator d/dq, we choose ψ to denote its operand, whereas χ for operand of d/dp. Among properties that $\psi(q)$ and $\chi(p)$ must obey are that these functions must be continuous, remain everywhere finite and singly valued, and satisfy appropriate boundary conditions; the first derivatives of $\psi(q)$ and $\chi(p)$ with respect to their specified arguments must likewise be well behaved except possibly at infinite discontinuities of potential energy. Amplified discussion of various properties of $\psi(q)$ is available elsewhere [9].

Also according to wave mechanics, the possible energies W of a system in a stationary state are obtained upon solution of Schrodinger's equation independent of time. For such a system, the coordinate representation is generally preferable to the momentum representation, because the potential energy is typically expressible more readily in terms of coordinate than in terms of momentum. For a particle of mass μ subject to displacement q, the kinetic energy according to classical formula $T(q) = 1/2\, p^2/\mu$ becomes operator $-1/2\, (\hbar^2/\mu)\, d^2/dq^2$ in the wave-mechanical coordinate representation. For a canonical linear harmonic oscillator, the potential energy in terms of coordinate q remains $V(q) = 1/2\, k_e q^2$, as in matrix mechanics. Inserting these quantities into Schrodinger's equation, we obtain

$$\hat{H}(q)\, \psi(q) = [-1/2\, (\hbar^2/\mu)\, d^2/dq^2 + 1/2\, k_e\, q^2]\, \psi(q) = W\psi(q) \tag{1.16}$$

in which the terms between brackets constitute the Hamiltonian operator $\hat{H}(q)$ that is applicable to this particular problem. To solve this differential equation, of type second order with linear symmetries, we best invoke mathematical software [8], as for matrix mechanics above: with Maple the direct solution, again with ω_0 substituted for $\sqrt{(k_e/\mu)}$, is directly expressed as

$$\psi(q) = c_1\, W_M(\pi\, W/(h\, w_0),\, 1/4,\, 2\pi\, \omega_0\, \mu\, q^2/h)/\sqrt{q}$$
$$+ c_2\, W_W(\pi\, W/(h\, \omega_0),\, 1/4,\, 2\pi\, \omega_0\, \mu\, q^2/h)/\sqrt{q} \tag{1.17}$$

With coefficients c_1 and c_2 as *constants of integration*, two independent solutions contain Whittaker M, as W_M, and Whittaker W, as W_W, functions, each with three arguments. For amplitude functions to be well behaved according to a condition specified above, namely, that $\psi_n(q) \to 0$ as $q \to \infty$, the difference between the first and second arguments must be equal to half a nonnegative integer: so $\pi\, W/(h\, \omega_0) - 1/4 = 1/2\, n$. Replacing radial frequency ω_0 by circular frequency ν_0, we hence obtain

$$W_n = (n+1/2)\, h\, \nu_0 \tag{1.18}$$

as in the solution according to matrix mechanics.

Plotting the part of the solution above containing the Whittaker M functions shows that, for even values of integer n, the curves diverge for positive and negative values of q; for this reason, we set c_1 equal to zero. In terms of Whittaker W functions, the amplitude function $\psi(q)$ thus becomes

$$\psi_n(q) = c_2\, W_W(n/2+1/4,\ 1/4,\ 2\pi\,(k\mu)^{1/2}\, q^2/h)/q^{1/2} \tag{1.19}$$

Integration constant c_2 remains to be evaluated; for this purpose, because Maple is unable to perform a general integration for symbolic integer n, we integrate $\psi_n(q)^*\psi_n(q)$ over q from $-\infty$ to ∞ for n from 0 to 5, with $\psi_n(q)^*$ as complex conjugate of $\psi_n(q)$; as $\psi_n(q)$ here has no imaginary part, $\psi_n(q)^*\psi_n(q) = \psi_n(q)^2$. On inspection of those results of integration, we discern that

$$c_2 = (2^n/n!)^{1/2}/\pi^{1/4} \tag{1.20}$$

causes each integral to become equal to unity, corresponding to normalization of amplitude function $\psi_n(q)$. After we test this result by integrating $\psi_n(q)^2$ from $-\infty$ to ∞ for further values of n to verify our deduction, $\psi_n(q)$ becomes thereby completely defined for n of arbitrary value. In Figure 1.1, with each function displaced upward n units for clarity, we plot $\psi_n(q)$ in terms of Whittaker W functions for $n = 0 \ldots 3$ and with q in a domain $-2.5 \ldots 2.5$; for the purpose of these plots, we take $h = k_e = \mu = 1$, but such values affect only the scales on the axes, not the shapes of the curves.

In Figure 1.2, we plot similarly a product $\psi_n(q)^2$. According to Born's interpretation, a product $\psi_n(q)^*\psi_n(q)\, dq$ represents a probability of a displacement of an oscillator having a value between q and $q + dq$; the unit integral for normalization is consistent with this concept.

We test two properties of these amplitude functions. When we integrate over q between $-\infty$ and ∞ a product of the first two amplitude functions,

$$\int_{-\infty}^{\infty} \psi_1(q)^*\psi_0(q)\, dq = 0 \tag{1.21}$$

or any other two distinct functions, we obtain zero; this result verifies that these amplitude functions are orthogonal. When we integrate likewise the same product with a further multiplicand q within the integrand, for the purpose of calculating

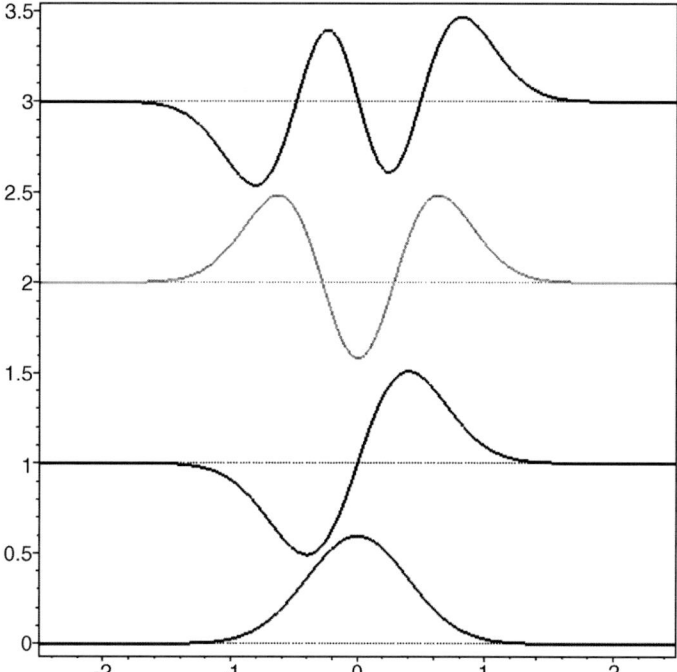

Figure 1.1 For a canonical linear harmonic oscillator according to wave mechanics and either coordinate or momentum representation, amplitude functions $\psi_n(q)$ or $\chi_n(p)$ in terms of Whittaker W functions on the ordinate axis versus displacement q or momentum p, respectively, on the abscissal axis for $n = 0, 1, 2, 3$, calculated with $h = k_e = \mu = 1$; each curve is displaced n units.

a transition probability or the intensity of a transition between the two states with which these amplitude functions are associated, we obtain a real quantity.

$$\int_{-\infty}^{\infty} \psi_1(q)^* \, q \, \psi_0(q) \, dq = 1/2 \, (h/\pi)^{1/2} / (\mu \, k_e)^{1/4} \qquad (1.22)$$

According to further integrals of this type, for two amplitude functions $\psi_n(q)$ and $\psi_m(q)$, only when $m = n \pm 1$ does this integral differ from zero, and the values of integrals of $\psi_{n+1}(q)^* q \psi_n(q)$ increase with n as $[1/2 \, (n+1)]^{1/2}$, in accordance with the result from matrix mechanics.

With Mathematica software, the solution of this Schrodinger equation is expressed directly in terms of parabolic cylinder functions. A conventional approach to this solution yields a product of an Hermite polynomial and an exponential function of Gaussian form, to which these Whittaker W functions are equivalent. The parabolic cylinder functions are in turn related closely to Whittaker functions; both are related to confluent hypergeometric functions in a product with an exponential term. Conversion from Whittaker or parabolic cylinder functions to Hermite or hypergeometric functions is unnecessary, because the former satisfy the pertinent

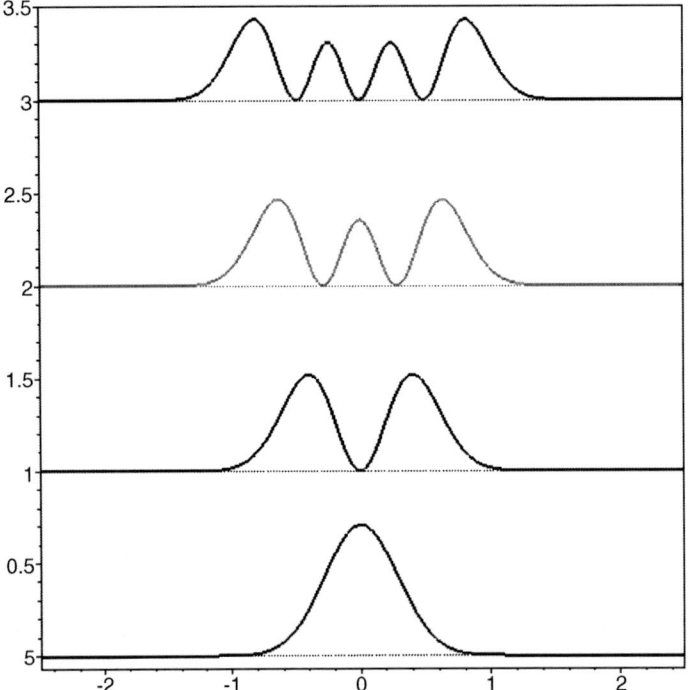

Figure 1.2 For a canonical linear harmonic oscillator according to wave mechanics and either a coordinate or a momentum representation, probability functions $\psi_n(q)^2$ or $\chi_n(p)^2$ in terms of Whittaker W functions on the ordinate axis versus displacement q or momentum p, respectively, on the abscissal axis for $n = 0, 1, 2, 3$, calculated with $h = k_e = \mu = 1$; each curve is displaced n units.

differential equation as effectively as the latter functions multiplied by an exponential term; avoiding that conversion makes this direct solution practicable without prior knowledge of its form. All these special functions arise in solutions of various differential equations, and their use in one or other chosen form has comparable convenience.

We repeat this calculation within the momentum representation, with amplitude function $\chi(p)$; this representation is useful for scattering conditions such as chemical reactions. Converting coordinate q to $i\hbar\, d/dp$, we obtain Schrodinger's equation accordingly in this form:

$$\hat{H}(p)\chi(p) = [1/2\, p^2/\mu - 1/2\, k_e\, \hbar^2\, d^2/dp^2]\chi(p) = W\chi(p) \tag{1.23}$$

the terms between brackets constitute the Hamiltonian $\hat{H}(p)$ applicable to this formulation. With Maple, the solution of this equation appears again in terms of Whittaker functions,

$$\chi(p) = c_1\, W_M[\pi\, W/(h\,\omega_0),\, 1/4,\, 2\pi\, p^2/(h\,\omega_0\,\mu)]/\sqrt{p}$$
$$+ c_2\, W_W[\pi\, W/(h\,\omega_0),\, 1/4,\, 2\pi\, p^2/(h\,\omega_0\,\mu)]/\sqrt{p} \tag{1.24}$$

With the difference of the first two arguments of these Whittaker functions set equal to half a nonnegative integer, for the same reason as before, the resulting functions satisfy requisites of an amplitude function $\chi(p)$; with n denoting that integer, solving for W and replacing the radial frequency ω_0 by circular frequency ν_0 as before, we obtain

$$W_n = (n+1/2)\, h\, \nu_0 \tag{1.25}$$

as in the solutions according to both matrix mechanics and wave mechanics in its alternative coordinate representation. With Mathematica, the solution of Schrodinger's equation is again expressed in terms of parabolic cylinder functions.

Plots show again that the Whittaker M functions in $\chi(p)$ above must be eliminated because they diverge for $|p| \gg 0$ for even n; with the other, Whittaker W, functions we proceed exactly as above. The amplitude function $\chi(p)$ in terms of momentum thus becomes

$$\chi_n(p) = c_2 W_W(n/2 + 1/4,\ 1/4,\ 2\pi\, p^2/(hk^{1/2}\, \mu^{1/2}))/p^{1/2} \tag{1.26}$$

which has no imaginary part. To evaluate integration constant c_2, we integrate $\chi_n(p)^2$ over p from $-\infty$ to ∞ for particular values of n from 0 to 5, and discern that

$$c_2 = (2^n/n!)^{1/2}/\pi^{1/4} \tag{1.27}$$

is required for each integral to become equal to unity, corresponding to normalization of amplitude function $\chi_n(p)$; after we test this result by integrating $\chi_n(p)^2$ from $-\infty$ to ∞ for further values of n to verify our deduction, $\chi_n(p)$ becomes completely defined. Taking again $h = k_e = \mu = 1$ for convenience, we plot $\chi_n(p)$ for $n = 0\ldots 3$ for p in a domain $-2.5\ldots 2.5$; the resulting plot appears in Figure 1.1, and the corresponding functions $\chi_n(p)^2$, for which a probabilistic interpretation is applicable analogously to $\psi_n(q)^2$, yield a plot in Figure 1.2.

The fact that, with $h = k_e = \mu = 1$, these plots of both $\chi_n(p)$ and $\chi_n(p)^2$ for these functions of momentum p are identical with the corresponding plots of $\psi_n(q)$ and $\psi_n(q)^2$ for the functions of coordinate q merely reflects the symmetry of the Hamiltonian: before conversion of either p or q to a differential operator according to a coordinate or momentum representation, respectively, the Hamiltonian contains p^2 and q^2, with their respective coefficients $1/2\,\mu^{-1}$ and $1/2\, k_e$. If we absorb these factors into variables p' and q', the Hamiltonian becomes expressed simply as $\hat{H} = p'^2 + q'^2$, thus exhibiting an entirely symmetric form. All these curves involving $\chi_n(p)$ that contain Whittaker W functions are likewise identical to the corresponding curves of products of Hermite polynomials and Gaussian functions according to a common representation of these amplitude functions for the canonical linear harmonic oscillator.

To test some properties of these momentum functions, we integrate a product of two amplitude functions for the first two states, with $n = 0$ and $n = 1$, between $-\infty$ and ∞:

$$\int_{-\infty}^{\infty} \chi_1(p)^* \chi_0(p)\, dp = 0 \tag{1.28}$$

As for any other two distinct functions, the result zero verifies that these amplitude functions $\chi_n(p)$ are orthogonal. To calculate a transition probability or intensity of a transition between these two states, before we integrate the same product with a further multiplicand q within the integrand, we convert q to its corresponding operator, $i\hbar \, (d/dp)$, appropriate for the momentum representation.

$$\int_{-\infty}^{\infty} \chi_1(p)^* \, i\,\hbar (d\chi_0(p)/dp) \, dp = i\,1/2(h/\pi)^{1/2}/(\mu k_e)^{1/4} \tag{1.29}$$

This integration yields a purely imaginary quantity. Apart from that additional factor i, this result is equivalent to that for the corresponding integral in the coordinate representation above, as expected because a transition probability is proportional to the modulus of this integral; for that reason, the presence of factor i is immaterial. According to further integrals of this type, for two amplitude functions $\chi_n(p)$ and $\chi_m(p)$, only when $m = n \pm 1$ does this integral differ from zero; the values of this integral of $\chi_{n+1}(p) \, p \, \chi_n(p)$ increase as $[1/2 \, (n+1)]^{1/2}$, in accordance with the results from both matrix mechanics and the coordinate representation.

According to some authors, an amplitude or wavefunction in a momentum representation is merely a Fourier transform of the corresponding function in the coordinate representation; this assertion is incorrect because coordinate and momentum are not mutually reciprocal quantities. On incorporating a factor $h^{-1/2}$, we, however, proceed with such a transform,

$$\chi_n(p)' = \int_{-\infty}^{\infty} \psi_n(q) \exp(-2\pi \, i \, q \, p/h) \, h^{-1/2} dq \tag{1.30}$$

and its inverse Fourier transform,

$$\psi_n(q)' = \int_{-\infty}^{\infty} \chi_n(p) \exp(2\pi \, i \, pq/h) h^{-1/2} dp \tag{1.31}$$

We distinguish the results of these transforms as primed quantities because the calculation in the former case shows that, with $\chi_n(p)$ defined above, the ratio $\chi_n(p)'/\chi_n(p) = i^n$. Such a result is expected because a Fourier transform of a real even function, such as $\psi_0(q)$, must yield a real function, $\chi_0(p)'$, whereas a Fourier transform of a real odd function, such as $\psi_1(q)$, must yield a purely imaginary function, $\chi_1(p)'$. This factor i is again immaterial because only the product of an amplitude function with its complex conjugate might have physical significance: any solution $\psi(q)$ or $\chi(p)$ of Schrodinger's equation independent of time, multiplied by $i\,a$ or $\exp(i\,a)$, remains a solution because that equation is homogeneous; if the magnitude of constant a differ from unity, the normalization might suffer.

In summary of this application of wave mechanics to a canonical linear harmonic oscillator, with this approach we find the energies of discrete states and the nonzero values of integrals over amplitude functions for two adjacent states; expressions for amplitude function $\psi(q)$ or $\chi(p)$ according to either a coordinate or a momentum representation, respectively, arise almost incidentally. With regard to the plots of these functions, with the mass of the oscillator taken to be that, m_e, of an electron at rest and with frequency ν_0 taken to be that of blue light, force coefficient k_e assumes a value approximately 13 N m^{-1}; the width at half maximum value of $\psi_0(q)$ in Figure 1.1 is then approximately 3×10^{-9} m and the corresponding width of $\chi_0(p)$ in Figure 1.2 is approximately $m_e c/180$, with speed of light c. The latter value justifies a nonrelativistic approximation for the purpose of these calculations; the ratio of the two widths is approximately $h^{-1/2}$, as expected.

1.3.3
Dirac's Operators for Creation and Destruction

As both matrix multiplication and the product of a quantity and its differential operator naturally exhibit in general a failure to commute, the fundamental postulate of quantum mechanics is readily implemented in their terms. Other quantities might be defined to conform to this postulate. In 1928, Dirac adapted, for use in a space of occupation number, two noncommuting operators, a for destruction and $a^\dagger$ for creation, that so conform and that he had previously applied to treat a radiation field. The state of a system, denoted with index n, is represented in a dual space of complex vectors of kinds *ket*, denoted $|n\rangle$, and *bra*, denoted $\langle n|$; such a ket or bra is supposed to carry complete information about that state. An observable property of a system is represented by an operator, such as A, that operates on a ket, as in $A|n\rangle$, to yield another ket. The creation and destruction operators act on a state through its ket to alter the index, denoting the occupation number, of that state in favor of an adjacent integer value, either a decrement for a or an increment for $a^\dagger$. An inner product $\langle m|n\rangle$ of two kets, defined with one ket and the bra corresponding to the other ket, is in general a complex number, and $\langle m|n\rangle = \langle n|m\rangle^*$, in which an asterisk denotes, as before, a complex conjugate. Two kets are orthogonal if an inner product of one ket with the bra corresponding to the other ket evaluate to zero, so $\langle m|n\rangle = 0$. A ket is normalized if the corresponding inner product with itself equal unity: $\langle n|n\rangle = 1$.

To apply this formalism to a canonical linear harmonic oscillator [10], we define destruction operator a to have a property such that its operation on a ket,

$$a|n\rangle = \sqrt{n}|n-1\rangle \qquad (1.32)$$

yields a resulting ket with its index decreased by unity and a scalar multiplicand $\sqrt{n}$, with a special case $a|0\rangle = 0$. A creation operator $a^\dagger$ has analogously a property,

$$a^\dagger|n\rangle = \sqrt{(n+1)}|n+1\rangle \qquad (1.33)$$

The resulting ket has its index increased by unity and its scalar multiplicand is $\sqrt{(n+1)}$. An auxiliary operator, N, defined as $a^\dagger a$, is called a number operator because, consistent with relations above, it operates on a ket to yield the value of its index like an eigenvalue, as in

$$N|n\rangle = a^\dagger a|n\rangle = n|n\rangle \qquad (1.34)$$

For application to a linear harmonic oscillator, we relate these creation and destruction operators to coordinate and momentum, as follows, in conformity with the fundamental postulate above:

$$q = (a^\dagger + a)\sqrt{(h/4\pi\mu\omega_0)}$$
$$p = i(a^\dagger - a)\sqrt{(\mu\omega_0 h/4\pi)} \quad (1.35)$$

in which the symbols imply the same quantities as in preceding sections. With such definitions, $a^\dagger$ and a are precisely complex conjugates of one another:

$$a = \pi q\sqrt{(2\mu\nu_0/h)} + ip/\sqrt{(2\mu\nu_0 h)}$$
$$a^\dagger = \pi q\sqrt{(2\mu\nu_0/h)} - ip/\sqrt{(2\mu\nu_0 h)} \quad (1.36)$$

We test the commutator of operators p and q, acting on a ket, as follows, by substituting their definitions above in terms of $a^\dagger$ and a,

$$[p,q]|n\rangle = (pq - qp)|n\rangle = -i\hbar|n\rangle \quad (1.37)$$

which yields the expected result.

The nonrelativistic Hamiltonian, $\hat{H} = 1/2\, p^2/\mu + 1/2\, k_e q^2$, in which p and q become expressed simply in terms of operators $a^\dagger$ and a according to the above relations, acts on a ket to generate the energy of its state, either directly,

$$\hat{H}|n\rangle = (n+1/2)\, h\nu_0|n\rangle \quad (1.38)$$

or in bracket notation,

$$\langle n|\hat{H}|n\rangle = (n+1/2)\, h\nu_0 \quad (1.39)$$

Another quantity of interest is the bracket for q that is evaluated as

$$\langle n+1|q|n\rangle = \sqrt{[(n+1)\, h/(2\mu\nu_0)]/2\pi} \quad (1.40)$$

Other quantities are generated analogously.

Those who have been shown, in an unbiased manner, these three approaches in manual form to solve for the energies of a canonical linear harmonic oscillator according to quantum mechanics have found it easiest to use Dirac's operators, and wave mechanics (with Hermite functions) the most difficult and complicated. The use of Dirac's operators for this problem has a distinct advantage in yielding general results in terms of the symbolic designation of states, whereas, with current software in application of both matrix and wave mechanics, one must at present solve for particular values of n and attempt to perceive a general relation. With appropriate general software for computer algebra [8], including operations involving noncommutative algebra, all three approaches are readily implemented; tedious manipulations are thereby eliminated in favor of both enhanced understanding of the underlying mathematical operations and greatly extended results.

1.3.4
Discussion of Quantum Theories in Relation to an Harmonic Oscillator

What do these results of calculations on a linear harmonic oscillator signify? Matrix mechanics of Heisenberg, Born, and Jordan and wave mechanics of Schrodinger together constitute *pioneer quantum mechanics* [11]. That these two approaches are equivalent was proved first by Eckart and Schrodinger, then by Pauli, Dirac, and von Neumann in increasingly sophisticated treatments. The major significance of that equivalence is that amplitude functions, $\psi(q)$ and $\chi(p)$ for instance, remain artifacts of wave mechanics: they are absent from, and play no part in, our calculations with either matrix mechanics or Dirac's operators. After one generates a Hamiltonian matrix according to matrix-mechanical methods, a standard procedure to obtain the eigenvalues converts that matrix to a diagonal form. In the preceding derivations, all three approaches to quantum mechanics, each based on the fundamental postulate of quantum mechanics, yield precisely the same results for the *observable* properties of that system of the harmonic oscillator, specifically the energy differences of its discrete states and the intensities of transitions between those states. The latter are proportional to the squares of a matrix element $q_{n,m}$ in matrix mechanics, of an integral $\int \psi_m(q)^* q \psi_n(q) dq$ over all space in wave mechanics in the coordinate representation, and of a bracket $\langle m|q|n \rangle$ evaluated with Dirac's operators.

Apart from these three approaches to nonrelativistic quantum mechanics that we illustrate here, there are relativistic quantum mechanics in various formulations with which one takes into account the dependence of electronic mass on velocity, quantum electrodynamics, quantum chromodynamics pertaining to a strong nuclear force, and higher ω algebras [11], among others. With the wave mechanics of Schrodinger, the only quantum theory with which most chemists become acquainted, one might calculate the discrete energies and angular momenta of molecules in a nonrelativistic approximation, and possibly also the discrete linear momentum of a confined molecule – only three of ten properties in Table 1.1. In relation to that entire table, wave mechanics evidently constitutes a grossly incomplete theory, with thus circumscribed validity and applicability, which Dirac and Einstein appreciated. The discrete mass and net charge of a particle or molecule must be treated as parameters wholly experimental in origin. As matrix mechanics is formally equivalent to wave mechanics, it provides no improved explanation of these laws of discreteness. To encompass an interaction of photons with matter such as a treatment of the emission of a photon by an atom or molecule, one must apply quantum electrodynamics [12], according to which the radiation field becomes quantized; this theory is applicable to interactions between molecules [13]. Despite chemists being almost totally ignorant of quantum mechanics in a form other than nonrelativistic wave mechanics in a coordinate representation, Schrodinger's approach applied in quantum electrodynamics leads to intractable infinities, whereas Heisenberg's approach is practicable [14]. On that basis, one might consider *matrix mechanics* [15, 16], despite little or no development since 1926 [17], to be *more fundamental than wave mechanics*.

As an inevitable conclusion from the preceding calculations, quantum mechanics is not a physical theory, certainly not a chemical theory, but merely a mathematical algorithm – or rather a *collection of mathematical algorithms* – applicable to problems on microscopic systems. The unifying property of these disparate approaches is their conformity to the fundamental postulate of quantum mechanics – $p\,q - q\,p = -i\,\hbar$ in one dimension – so that dynamical variables denoting position and momentum fail to commute with one another. Another and practical formulation of quantum mechanics has been constructed on the basis of quaternions [18, 19]; these quantities, which are hypercomplex numbers that commute according to addition but not multiplication, have not only useful chemical applications in describing, for instance, the orientational motions of molecular entities in lattices, but also common practical applications in the manipulation of animated figures in computer games. That commutator relation is mysterious as quantities coordinate q and momentum p, for which symbols on the left side appear to be nominally real, consistent with our experience of classical mechanics, whereas the right side is purely imaginary. Such a condition is clearly consistent with a mathematical formalism, not with a physical theory. One might object to this conclusion on the basis that the quantities involved are operators: yes, Dirac's $a^\dagger$ and a are undoubtedly operators, and $p \to -i\,\hbar\,d/dq$ in wave mechanics in its coordinate representation and $q \to i\,\hbar\,d/dp$ in its momentum representation undeniably become differential operators as indicated, but their conjugate variables, q in a coordinate representation and p in a momentum representation, remain in wave mechanics simply scalar or vectorial quantities applied with multiplication sign, $\times$, as operator within those respective calculations. A matrix can certainly act as an operator, for instance, in operating on an eigenvector to produce an eigenvalue, but one cannot consider energy matrix **W**, in the above derivation according to matrix mechanics, to operate on anything: it is a meaningful quantity that stands on its own. The various quantities through which this fundamental postulate is applied in the various methods, implicitly or explicitly, possess, nevertheless, a complex nature sufficient for that postulate to be properly obeyed. Quantum mechanics is thus a collection of mathematical methods that one might apply to calculate a desired quantity. If the latter quantity be a classical property, such as a molecular structure, a classical method such as *molecular mechanics* can yield equivalent results: in either case, apart from tests on the simplest molecules such as H_2 and HeH^+, the accuracy of those calculations of equilibrium internuclear distances falls short of an accuracy derived from measurements of optical spectra according to the experimental standard errors, cf. HCl for instance [20]. One might labor under a delusion that a typical calculation of a molecular structure with one or other common computer programs for quantum chemistry is more fundamental than the unquestionably empirical approach in molecular mechanics; first of all, in those programs the separate treatment of nuclear and electronic motions is an empirical concession, and even then, apart from the use of experimental masses and charges of particles, there is typically no optimization of orbital exponents that is at least essential for a description *ab initio*. If and when one undertakes a calculation, one should simply choose a convenient approach consistent with the computational resources at hand, and with any approximations or assumptions that one is prepared to accept to obtain some answer.

Another query that likely springs to a reader's mind is our application here to a system described as a *canonical linear harmonic oscillator* according to potential energy expressed as $V(q) = 1/2\, k_e q^2$ in each case. This description is consistent with the fact that, for the property of discrete energies at equal intervals that defines a quantum-mechanical linear harmonic oscillator, there exist uncountably many harmonic functions of potential energy [21]. A Davidson function for potential energy is expressible as $V(q) \propto (q/q_0 - q_0/q)^2$, with q_0 as a reference coordinate [20, 22]; such an oscillator might be taken also to constitute a *rigid rotor* because an integral $\int \psi_n(q)^* q^{-2} \psi_n(q)\, dq$ over all space, which enters a calculation of a rotational parameter, has a constant value for all states n. In contrast, for a linear harmonic oscillator with potential energy in its canonical form $V(q) = 1/2\, k_e q^2$, the value of that integral *increases* for states with increasing n or energy, contrary to our experience of *decreasing* values for rotational parameters B_n of real diatomic molecules [20].

An harmonic oscillator, canonical or otherwise, is a convenient model for exercises in either classical or quantum-mechanical methods, convenient for the three derivations above, for instance, but what is its relevance to the distribution of radiant energy from a black body? The experimental measurements on which the empirical curve was based were best performed with a *hohlraum* – a heated cavity with a small orifice through which light escapes toward a detector; this device presumably consists of solid walls of refractory material, which might be heated electrically, and which enclose air. Where are the harmonic oscillators? What is the actual source of the radiation with that quintessential distribution? The best that one might state is that a field of radiation appears dynamically equivalent, in some sense, to harmonic oscillators in some set [12, 23]. An harmonic oscillator is clearly a fictitious concept because no real entity has equally spaced states of infinite number. Moreover, a canonical linear harmonic oscillator is a boson, possessing no intrinsic angular momentum. According to Table 1.1, each photon carries angular momentum to the extent of one unit in terms of $\hbar$ if it be circularly polarized. Like energy, angular momentum is a rigorously conserved quantity; for that reason, no purely vibrational transition of a free diatomic molecule in electronic state $^1\Sigma$, such as CO, occurs with absorption or emission of a single photon involving circularly polarized light: an altered rotational state invariably accompanies that vibrational transition under those conditions. For an hypothetical system such as a linear harmonic oscillator in one spatial dimension, angular momentum is irrelevant: the emitted radiation is linearly polarized [12]. For a cavity in three dimensions, the situation is less clear because of the uncertain nature of the emitters, but in any case the connection between that hohlraum and an harmonic oscillator is tenuous, other than a pragmatic success of the model.

Among many facets of quantum mechanics that become discernible through our preceding calculations on the canonical linear harmonic oscillator, one might pose a question about the choice of a vehicle for this demonstration. Why have we not devoted attention, instead, to a particle confined to a segment of a line, sometimes inelegantly called *the particle in a one-dimensional box*? The answer is that despite its popular use in demonstrating wave mechanics, even though its quantized energies are derivable without wave mechanics, the solution fails signally to conform to the properties of such a system: explicitly, although the amplitude or wavefunction varies continuously everywhere, its first derivative has major discontinuities at the boundaries of the confining segment. Such an infinite, positive, potential energy except for a small domain is a highly artificial condition, and one for which the corresponding solution in momentum space is all but impracticable. To avoid discontinuities of $d\psi_n(q)/dq$ for a particle confined to a line segment that arise from those infinite discontinuities of potential energy at the boundaries, relaxing that infinity to a finite difference of $V(q)$ causes the particle to become no longer confined to the line segment – the particle is no longer completely *in the box*. Even worse, taking the boundary conditions into account with a finite barrier as the difference between $V(q)$ within the segment and $V(q)$ outside deprives the solution of its contrived simplicity. An appropriate solution of this dilemma is to apply a finite barrier at the boundaries of the segment, and then to suggest that the particle becomes increasingly confined as the barrier increases, but this stratagem merely shirks the fundamental deficiency of this artificial problem. A particle confined to the circumference of a circle evokes no such internal inconsistency of the model, but that example entails other disadvantages. In relation to quantum mechanics, this model of a particle confined to a line segment with an infinite potential energy becomes thus totally misleading, but in applications in statistical mechanics this model is less objectionable. A major problem with quantum theories is that their mathematical structure is poorly understood by chemists, despite their application of recipes to chemical systems for which they are obviously inappropriate. For the canonical linear harmonic oscillator according to wave mechanics, amplitude functions $\psi(q)$ and $\chi(p)$ and their first and second derivatives are everywhere continuous and well behaved.

1.4
Diatomic Molecule as Anharmonic Oscillator

A frequency of light absorbed or emitted by any real molecule of a particular species depends on the amount of energy that that molecule has; its dissociation requires an energy of only a finite amount: for these reasons, the properties of that species bear only a slight resemblance to the properties of a quantum-mechanical harmonic oscillator. A function described as potential energy that is central to a harmonic oscillator arises, moreover, only indirectly for a real molecular species, as we proceed to explain. Assuming an assembly comprising two atomic nuclei and N electrons, $N > 1$, we consider that those nuclei remain fixed in space with the electrons free to move in their vicinity. To calculate the possible energies of this system, we postulate this Hamiltonian operator that takes into account the kinetic energy of the electrons and their electrostatic interactions with each other and with the nuclei, plus an electrostatic repulsion between the two nuclei.

$$\hat{H}(\underline{r}_i, \underline{R}) = T(\underline{r}_i, \underline{R}) + V(\underline{r}_i, \underline{R})$$
$$= \sum_{i=1}^{N} 1/2 p_i^2/m_e + \left\{ \sum_{i=1}^{N}\sum_{j>i}^{N} 1/r_{ij} - \sum_{i=1}^{N}(Z_a/\varrho_{ia} + Z_b/\varrho_{ib}) \right.$$
$$\left. + Z_a Z_b/R \right\} e^2/(4\pi\varepsilon_0) \qquad (1.41)$$

Here appear symbols denoting distance r_{ij} between two electrons, distance R between two nuclei of atomic numbers Z_a and Z_b, distance ϱ_i between an electron and one or other nucleus, and electronic linear momentum p_i; the sums run accordingly. The energies $E(R)$ possessed by this system that depend on the internuclear distance become a potential energy $V(R)$ for the motion of the nuclei in the field of the electrons. If the two atomic centers could be considered to have no internal structure, their energies would become calculated from this Hamiltonian [20],

$$\hat{H}(R) = \hat{p}^2/(2\mu) + \hbar^2 J(J+1)/(2\mu R^2) + V(R) \tag{1.42}$$

which contains terms for the kinetic energy of vibration of those atomic centers along a line joining them, the kinetic energy of those atomic centers perpendicular to that line that corresponds to a rotation of the diatomic species about the center of mass, and the potential energy resulting from $E(R)$ calculated with the preceding Hamiltonian; reduced mass μ is defined as the product of the atomic masses divided by their sum, $M_a M_b/(M_a + M_b)$. We assume a bound electronic ground state for this diatomic system for which the vibration–rotational energies [20] become expressible for a particular isotopic variant i as

$$E_{vJ}^i = \sum_{k=0}^{\infty} \sum_{l=0}^{\infty} Y_{kl} (v+1/2)^k [J(J+1)]^l \tag{1.43}$$

for states with vibrational v and rotational J quantum numbers sufficiently small. Because each atomic center possesses an internal structure comprising an atomic nucleus and multiple associated electrons, we postulate instead an *effective Hamiltonian* for the nuclear motion of this form [20],

$$\hat{H}_{\text{eff}}(R) = \hat{p}\,[1 + g_v(R)m_e/m_p]\,\hat{p}/(2\mu)$$
$$+ [1 + g_r(R)m_e/m_p]\,\hbar^2 J(J+1)/(2\mu R^2) + V(R) + \Delta V(R) \tag{1.44}$$

in which $\mu/(1 + g_v(R)\,m_e/m_p)$ serves as an effective reduced mass $\mu_v(R)$ for vibration and $\mu/(1 + g_r(R)\,m_e/m_p)$ analogously as $\mu_r(R)$ for rotation; both are functions of internuclear distance. These two g factors and $\Delta V(R)$ take into account that the atomic centers are composite; $g_v(R)$ and $g_r(R)$ are called vibrational and rotational g factors, respectively, and conventionally have as coefficient a ratio of electronic and proton rest masses so that their magnitudes assume order unity. Each g factor is assumed to have an electronic and a nuclear contribution,

$$g_{r,v} = g_{r,v}^{\text{el}} + g_{r,v}^{\text{nu}} \tag{1.45}$$

The electronic contributions take into account that the vibrational and rotational motions of the nuclei are interpreted formally to induce interactions between the electronic ground state of interest and electronically excited states. Each g factor is assumed also to be expressible as a sum of contributions due to each atomic center separately,

$$g_{r,v} = g_{r,v}^a + g_{r,v}^b \tag{1.46}$$

The other term $\Delta V(R)$ is also divisible into contributions from atomic centers A and B and contains two parts: one takes into account that the internuclear potential energy $V(R)$ depends slightly on the nuclear masses and includes that dependence as an adiabatic correction that reflects only the electronic ground state; the other part takes into account further interactions between the electronic ground and excited states. As a result of the terms in this effective Hamiltonian additional to that in the preceding simplified Hamiltonian, there arise further terms as coefficients of vibrational and rotational quantum numbers in the vibration-rotational energies for particular isotopic variant i,

$$E_{vJ}^i = \sum_{k=0} \sum_{l=0} \left(Y_{kl} + Z_{kl}^{a,v} + Z_{kl}^{a,r} + Z_{kl}^{b,v} + Z_{kl}^{b,r} \right) (v+1/2)^k [J(J+1)]^l \quad (1.47)$$

Mechanical term coefficients Y_{kl} contain molecular reduced mass μ and parameters in some assumed algebraic form of $V(R)$. Extra-mechanical term coefficients depend also on parameters in some assumed algebraic forms of other terms in the effective Hamiltonian as follows: $Z_{kl}^{a,v}$ contains parameters in some assumed algebraic forms of $g_v^a(R)$ and $\Delta V^a(R)$ for atomic center A, whereas $Z_{kl}^{a,r}$ contains parameters from $g_v^a(R)$ and $g_r^a(R)$ for atomic center A, and analogously for $Z_{kl}^{b,v} + Z_{kl}^{b,r}$ for atomic center B. There hence occur two observable quantities for each set k,l that might in principle be evaluated from isotopic variants of each atomic type, but three contributors to them arise in general, specifically radial functions $g_v^a(R)$, $g_r^a(R)$, and $\Delta V^a(R)$ for atomic center of type A and analogous quantities for atomic center of type B. As spectral lines split upon the imposition of an appropriate magnetic field according to the Zeeman effect, one can evaluate the rotational g factor separately from the other contributors, but for no molecular species have such splittings been measured for many states of vibrational and rotational energies. No comparable magnetic, or other, property provides direct experimental information about the vibrational g factor, and the adiabatic corrections, like $V(R)$, are purely artifacts of a separate treatment of electronic and nuclear motions. Any radial function is fundamentally an artifact of that separate treatment, but for the rotational and vibrational g factors in principle, and for the rotational g factor in practice [24], an expectation value of that radial function in a particular molecular state is an experimentally measurable quantity. The inversion of such data with both a vibrational and a rotational dependence might yield a unique representation of that radial function for internuclear distance within a particular domain; in that sense, the rotational g factor and even its radial function is a quantity determined from experiment.

A prototypical application of computational spectrometry [25] enables separate evaluation of parameters for potential energy, vibrational and rotational g factors, and adiabatic corrections as follows. Methods have been developed for accurate calculations of the rotational g factor that reproduce the few experimental values measured with various techniques; the results of those calculations are then applied as constraints in fitting the spectral data of diatomic molecules so as to enable evaluation of parameters, of number limited by the extent of spectral data according to the range of quantum numbers v and J for appropriate isotopic variants, of the other three radial functions.

We have applied this approach to a few molecular species – H$_2$ [26], HeH$^+$ [27], LiH [28], and NaCl [29] – of diverse types, but in each case not involving atomic centers of large atomic number Z because further complications would then arise from the finite and isotopically varying nuclear volume; for small Z, such effects on the spectra from finite nuclear volume are much smaller than those that might be considered to arise from finite nuclear mass [20]. This approach has been tested on comparison of radial functions for potential energy, vibrational g factor, and adiabatic corrections defined for a small domain of internuclear distance from spectral analysis with the corresponding functions from separate quantum-chemical calculations, which are not limited to a small domain. These tests have satisfactorily established the validity of the approach, which has in turn verified the algebraic basis of this method of spectral analysis [25]. To avoid extended discussion of technical details, we refer the reader to an application to H$_2$ [26] that effectively demonstrates the power of this computational spectrometry.

The sophistication of theoretical treatment or calculation of a molecular species is in inverse proportion to the number of atomic centers [30]. For this reason, diatomic molecules provide a proving ground for theories and algorithms. Like an hypothetical linear harmonic oscillator, a diatomic molecule is effectively a system of one spatial dimension in which vibrational motion occurs along the internuclear axis; although rotational motion occurs perpendicular to this axis, this motion is readily enveloped in a quantitative treatment because only kinetic, not potential, energy is associated with this motion, despite its inclusion within an effective potential energy.

1.5
Quantum Mechanics and Molecular Structure

Let us consider, instead of a hypothetical oscillator or even a diatomic molecule as an anharmonic oscillator, a real polyatomic molecule of which we might seek to calculate the structure and molecular properties. For this purpose, we choose first a chemist's perhaps favorite organic chemical compound, ethanol. On the basis of a chemical formula C_2H_5OH, we construct a molecular Hamiltonian for use in a conventional calculation; for this purpose, we take consistently into account the kinetic energy T of both nine atomic nuclei, each of mass M_k, and their 26 associated electrons, each of mass m_e, and the potential energy V of electrostatic interaction between each two particles.

$$\hat{H}(r_i, \underline{R}_k) = T(\underline{r}_i, \underline{R}_k) + V(\underline{r}_i, \underline{R}_k)$$

$$= \sum_{i=1}^{26} 1/2\, p_i^2/m_e + \sum_{k=1}^{9} 1/2\, p_k^2/M_k$$

$$+ \left\{ \sum_{i<j} 1/r_{ij} + \sum_{k<l} Z_k Z_l / R_{kl} - \sum_{i,k} Z_k/\varrho_{ik} \right\} e^2/(4\pi\varepsilon_0) \quad (1.48)$$

This formula exceeds that, 1.41, for a diatomic molecular species in containing a sum for the kinetic energy of the atomic nuclei and a further sum of terms for repulsion between nuclei. The additional symbols denote distance R_{kl} between two nuclei of atomic numbers Z_k and Z_l, distance ϱ_{ik} between an electron and a nucleus, and nuclear linear momentum p_k; the sums run accordingly. According to wave mechanics in its coordinate representation for which we seek to solve Schrodinger's equation independent of time, we replace those squared momenta, p^2, with operators, $-\hbar^2 \nabla^2$, in three dimensions; ignoring any magnetic effects, we form this expression:

$$\hat{H}(\underline{r}_i, \underline{R}_k) = -1/2\,\hbar^2 \sum_{i=1}^{26} \nabla_i^2/m_e - 1/2\,\hbar^2 \sum_{k=1}^{9} \nabla_k^2/M_k$$

$$+ \left\{ \sum_{i<j} 1/r_{ij} + \sum_{k<l} Z_k\,Z_l/R_{kl} - \sum_{i,k} Z_k/\varrho_{ik} \right\} e^2/(4\pi\varepsilon_0) \quad (1.49)$$

This Hamiltonian requires an operand for its differential operators in laplacian ∇^2; for this purpose, we assume that $\Psi(\underline{r}_i, \underline{R}_k)$ is an amplitude function of both electronic and nuclear coordinates with requisite properties. Because, for a system of multiple bodies, no exact algebraic solution of

$$\hat{H}(\underline{r}_i, \underline{R}_k)\,\Psi(\underline{r}_i, \underline{R}_k) = E\,\Psi(\underline{r}_i, \underline{R}_k) \quad (1.50)$$

is practicable, after eliminating the motion of the center of mass we have recourse to integration.

$$E = \iint \Psi(\underline{r}_i, \underline{R}_k)^*\, \hat{H}(\underline{r}_i, \underline{R}_k)\, \Psi(\underline{r}_i, \underline{R}_k)\, d\tau_i\, d\tau_k \Big/ \iint \Psi(\underline{r}_i, \underline{R}_k)^*\, \Psi(\underline{r}_i, \underline{R}_k)\, d\tau_i\, d\tau_k$$

$$(1.51)$$

Here, $d\tau_i$ and $d\tau_k$ are elements of volume for integration over coordinates of all electrons and all nuclei, respectively, minus three coordinates to define the center of mass; each integration is performed over all space, from $-\infty$ to ∞ for each Cartesian coordinate. This integration yields merely a sequence of values of energy E_n, some discrete and some continuous; the state of least energy is discrete because the motion of the center of mass of this stable molecular system is separated.

Let us contemplate another such calculation, this time on dimethyl ether, H_3COCH_3. For this purpose, we construct the appropriate Hamiltonian and again prepare to solve Schrodinger's equation – but this Hamiltonian $\hat{H}(\underline{r}_i, \underline{R}_k)$ is *exactly the same* as the preceding one! The energies E_n must perforce be the same, and likewise the amplitude functions $\Psi(\underline{r}_i, \underline{R}_k)$ that reflect merely the content of an Hamiltonian on the basis of which they are evaluated in a particular case. As a result of this exercise, either molecule in a quantum state has only indefinite extension in both space and time [31, 32] – no atoms, no association of electrons with particular atomic nuclei, no conformers, and no electric dipolar moment (with respect to laboratory coordinates); possessing spherical symmetry, each molecule acts like – so effectively constitutes – a point mass. That same calculation is applicable not only to a molecule of ethanol or dimethyl ether but also to methane and methanal together, or to ethene and water together – any combination of the same nine atomic nuclei of C, O, and H and 26 electrons, always avoiding generating magnetic couplings, such as for free radical species, for which additional terms in the Hamiltonian arise. For the same reason, we associate neither rotational nor vibrational nor electronic state, nor a combination of these, with a particular energy – the calculation yields just numerical values of energies, some discrete and some continuous. Selected differences between energies in that single sequence or manifold might be compared with energies proportional to the frequencies of spectral lines of any compounds fitting the stoichiometry – C_2H_5OH or H_3COCH_3 or $CH_4 + H_2CO$ or $H_2O + C_2H_4$. A calculation of this kind would not be a fruitful practice of computational spectrometry.

Such a systematic calculation on ethanol or comparable system is still beyond the capabilities of extant computational resources, but an almost rigorous calculation was performed on the trihydrogen cation, H_3^+, in a nonrelativistic approximation [33]. As the simplest polyatomic species, this molecular ion comprises three protons as separate atomic nuclei and two associated electrons; after separation of variables pertaining to the origin of coordinates defined at one nucleus, a solution of Schrodinger's equation, similar to above, yields energies of internal, or spectrally pertinent, states of this system. Quantum mechanics requires all identical particles to be treated as *indistinguishable*, thus not only indistinguishable electrons but also indistinguishable protons; because no operator representing distances between particles commutes with the Coulombic Hamiltonian, particular internuclear distances or *bond lengths* are immeasurable for this – or any other – molecule in a stationary or quantum state. Taking into account the indistinguishability of identical particles, one might, however, calculate a distribution of probability of a distance between particles of any two types. Cafiero and Adamowicz reported mean distances, and their dispersions, for the separations between particles in H_3^+ [33]; to decide on that basis whether H_3^+ is, in a conventional sense, a linear or a triangular molecule was impossible. Applying isotopic substitution according to a separate calculation on HDT^+, so that these nuclei become distinguishable, these authors deduced that the equilibrium geometric shape defined by the atomic nuclei of that species corresponds to a nearly equilateral triangle. An analogous rigorous calculation on benzene, C_6H_6, in its stationary state of least energy would yield only a single, mean, internuclear distance $C-C = 2.09 \times 10^{-10}$ m, a mean distance $C-H = 2.69 \times 10^{-10}$ m, a mean distance $H-H = 3.80 \times 10^{-10}$ m, and another mean distance between any two electrons of the 42 associated with each formula unit. According to such a calculation, there is no indication whatsoever of a molecular structure or conformation. With regard to isomerism and chirality, Sutcliffe and Woolley discussed classical molecular structure in relation to prospective quantum-mechanical calculations on HCN, C_3H_4, and C_8H_8 [34].

Several computer programs, such as those named Gaussian (originating from Pople's work) or Dalton, are available for calculating a molecular structure on the basis of a separate treatment of electronic and nuclear motions, proceeding iteratively from initially estimated internuclear distances and interbond angles until a local minimum of energy is attained. On the hypersurface in a hyperspace of 21 dimensions corresponding to 3 N -6 internal degrees of freedom for a noncollinear structure with the nine atomic nuclei of C_2H_6O in its electronic ground state, there would hence exist not only local minima corresponding to those four specified structures constituting ethanol and its structural isomers but also combinations of free radicals. One attractive feature of these computer programs is the provision of not only one or more desired structures, in the form of optimized internuclear distances and interbond angles in some set, but also wavenumbers of harmonic nuclear vibrations, at a modest additional computational cost. These wavenumbers might be based on algebraic second derivatives of energy with respect to displacements from the equilibrium structure according to normal coordinates – that is, according to a modeling of the internal vibrational degrees of freedom as if the molecule were a set of harmonic oscillators with some distinct and some identical vibrational frequencies. Even though this model is a gross approximation for a real molecule that has only too finite energies of dissociation into various fragments, the information is helpful for analytical purposes – associating a measured infrared spectrum with a prospective structure of an unknown compound. To compensate for the crudeness of the approximation, an empirical factor, such as 0.95 for bond-stretching modes, is typically applied to each calculated *harmonic wavenumber* to aid a comparison with experiment. In that sense, despite its purely fictitious nature, the harmonic oscillator is a valuable armament in a chemist's arsenal of mathematical methods that constitute computational spectrometry to attack a dearth of information about a particular molecular structure. An alternative and accurate calculation of wavenumbers for anharmonic vibrational modes is at present an operation far from routine, and computationally expensive; when distinguishing among similar structures, such intensive calculation is necessary. As an instance of this application, computational spectrometry involved the simulation of the infrared spectrum of HC_5 on the basis of extensive calculations of molecular electronic structure and with anharmonic vibrations; of two absorption lines recorded in infrared spectra of products of photolysis of methane with vacuum ultraviolet radiation, their assignment not only to HC_5 but also to a conformer of a particular geometry had to be confirmed with such calculations [35] because experimental data in the form of wavenumbers and relative intensities of lines of isotopic variants were insufficient to enable a positive identification in the absence of such calculations. *As an empirical procedure, a separate treatment of electronic and nuclear motions might pragmatically preserve molecular structure, but at a loss of quantum-mechanical justification of its application.*

With regard to general molecular structure and electronic structure in particular, much explanation of chemical phenomena in textbooks of general, inorganic, and organic chemistry invokes orbitals in an almost invariably qualitative manner. An atomic orbital is precisely a solution of Schrodinger's temporally independent equation for an atom comprising a single atomic nucleus and a single electron; that solution is a wavefunction or amplitude function of a nature in three spatial dimensions analogous to amplitude functions $\psi(q)$ or $\chi(p)$ that we derived in our analysis for the canonical linear harmonic oscillator. As we demonstrated above for that system, the wavefunction is indeterminate within a factor of a complex number of modulus unity, which contains i = $\sqrt{-1}$ as one case; *such a condition precludes any possible observation.* Despite the fact that chemists have a limited interest in the properties of isolated atoms, the atomic orbitals that pertain only to atomic hydrogen or an equivalent species, such as He$^+$ that has only one electron, are almost invariably cited to explain properties of molecular entities or materials with many electrons – incontestably a non sequitur! Moreover, their application is by no means limited to H atomic centers because such a H atom is much less interesting than other atomic centers for which multiple simultaneously directed chemical bonds are postulated. From our scrutiny of the canonical harmonic oscillator, we recall that a wavefunction of spatial coordinate, denoted $\psi(q)$, arose in only one of the three approaches that yielded identical solutions of the energies and matrix elements of transitions of that system, and even then that approach yielded also an alternative solution $\chi(p)$ of Schrodinger's equation in terms of a momentum variable; that either wavefunction is purely an artifact of that particular approach is hence incontestable, but the existence of multiple successful approaches might not imply that one particular approach or its artifact is less viable or fundamental. A Liouville transformation [36] enables a conversion of that differential equation into a form applicable to a nominally disparate problem; such a transformation converts that equation for a canonical linear harmonic oscillator into a differential equation applicable to a Morse oscillator, or to a Davidson oscillator, or even to a central Coulombic system – namely, an atom with one electron, H. The implication is that both matrix mechanics and Dirac's operators are applicable to the H atom; Pauli indeed solved the H atom according to matrix mechanics [17] before Schrodinger's solution according to wave mechanics, and quaternions have since been applied also to solve this one-electron atom [18]. The consequence of this recognition is that an atomic orbital is by no means an essential feature of quantum mechanics applied to the H atom.

Because Schrodinger failed to accommodate the then known requirements of general relativity in designing his differential equation, wave mechanics is merely an approximate treatment for the H atom. Dirac's equation in his relativistic version of quantum mechanics is also an approximation because it fails to cope, for instance, with the emission spectrum of H, which was the result of a simple experiment that eventually launched quantum mechanics [3]; for an acceptable account of that emission spectrum, quantum electrodynamics [9, 12] is required. Wave mechanics is clearly a deficient approach for even the simplest atom. An orbital is formally inapplicable to any other atomic or molecular system because the derivation of such a wavefunction begins with a Hamiltonian that lacks any term to take into account repulsion between electrons that comprise part of any such other system. For orbitals to be applied to describe the hydride anion H$^-$, for instance, would imply an Hamiltonian in which terms appear for attraction of both electrons to the proton as atomic nucleus but no term for repulsion between those electrons; as all terms, and distances between any two particles, have similar magnitudes, this neglect of repulsion is incontestably a gross approximation. For other atoms, or molecular systems, in which an attraction exists between an electron and one or other atomic nucleus of atomic number Z greater than unity, the analogous approximation might be less severe but remains appreciable. In any practical application of a quantum-chemical calculation, such as is effected with Dalton or Gaussian, the iterative process inherent in a self-consistent field and further mathematical contortions to take account of electron correlation accomplish numerically by brute force, to an extent satisfactory for many purposes such as for a treatment of HC_5 [35], what is algebraically unachievable for any system comprising independent particles numbering more than two. These calculations have traditionally been initiated with trial atomic wavefunctions that resemble those of H orbitals, even though these functions are almost invariably decomposed directly into sums of Gaussian functions for ease of computation of integrals over electronic coordinates. When density-functional theory was first developed, it required a basis set of orbitals, but even that impediment is now overcome [37], allowing a significant and usefully accurate calculation of molecular structure and molecular properties with no explicit invocation of wavefunctions. Despite this advance, it seems inevitable that unenlightened authors of textbooks and even research papers will blissfully continue to invoke unseen and unseeable artifacts – orbitals – in their qualitative discussion of molecular structure and chemical properties – "where ignorance is bliss, 'tis folly to be wise" (Thomas Gray).

As an atomic orbital is defined as a solution of Schrodinger's temporally independent equation according to nonrelativistic wave mechanics applied to an atom comprising only one electron, such an orbital is accordingly just an artifact of a particular calculational method; our three detailed treatments of the canonical linear harmonic oscillator had as motivation just such a proof. One might argue that even though alternative approaches yield, in the particular instance of that oscillator, identical results for its observable properties, namely, the frequency and intensity of its optical spectrum, this condition neither diminishes the validity of the wavefunction nor makes it less important in circumstances in which alternative computational methods are impracticable. Such a reluctance to accept the fallacy of an orbital ignores the fact that Schrodinger's equation is a nonrelativistic approximation that fails quantitatively even for the hydrogen atom: four quantum numbers, not just three that result from the solution of Schrodinger's equation, are required to specify the state of that atom and to provide a nearly correct account of the spectrum in gross details. For even an application to the hydrogen atom, this failure entirely undermines the theoretical credibility of an orbital as being other than an artifact of one particular approximate calculation.

For any atom with multiple electrons, the applicability of an orbital to explain any property of that atom is clearly lacking, apart from the indeterminacy because of factor i clearly illustrated above, because the total wavefunction expressed as a product, such as $1s^2 2s^2 2p^2$ for atomic C, of orbitals or functions for the hydrogen atom ignores the repulsion between electrons. Such a wavefunction or an orbital has at any point in space both a magnitude and a sign, plus or minus, depending on whether the numerical value of that wavefunction is greater or less than zero, but that sign is subject to an arbitrary phase convention; for that reason, such a sign is fundamentally unobservable even if one might suppose that there exists a "true" wavefunction of which a calculated result according to Schrodinger's wave mechanics is an approximation. No mathematical function is directly observable in a chemical or physical experiment. A molecule contains only atomic nuclei and their associated electrons; the total electronic density in a particular small volume of space containing tangible matter is an observable and measurable quantity, for instance, on a solid surface with an appropriate electron microscope, or through a Fourier transform of a diffraction pattern of X-rays for the interior of a small crystal. If one assume that such a density of electronic charge be proportional to the square of a total electronic wavefunction, with the proportionality factor again arbitrarily defined, the corresponding value of a product of that proportionality factor and the square of the wavefunction might be measurable in a particular small spatial volume. Whatever the value of that proportionality factor, the product decays exponentially with distance d away from a particular atomic nucleus, so of form e^{-d}, but becomes zero only at an infinite distance from that nucleus. According to wave mechanics, a single hydrogen atom occupies the entire universe, but not to the exclusion of other atomic centers! For comparison, the value of ψ^2 or χ^2 for the canonical linear harmonic oscillator in terms of distance or momentum decays more rapidly away from a maximum, namely, proportionally to e^{-d^2}; in that sense, the region of large magnitude of ψ^2 or χ^2 is more localized for that harmonic oscillator than for the corresponding property with which is associated a density of electronic charge near an atomic center. For this reason, the shape of that distribution of ψ^2 or χ^2 associated with electronic charge in space depends on some other arbitrary criterion. A partition of a total assumed wavefunction for the atomic or molecular system into components subject to the same total energy, like a partition of electronic density in space associated with multiple atomic centers, is arbitrary because Schrodinger's temporally independent wave equation is of homogeneous type, $\hat{H}\psi = E\psi$, as applied to a harmonic oscillator above.

In the light of this almost universally prevailing indoctrination of chemists about orbitals, some scientists have claimed even to have *observed* orbitals. Of two particular experiments, one involved spectrometry of a sort and the other diffractometry; the latter has stimulated greater discussion in the literature. On the basis of diffraction experiments on a crystal of cuprite, Cu_2O, authors Zuo et al. [38] – three physicists and a chemist – made several novel claims, not the least of which was to have "seen an orbital," specifically "an unusual d-orbital hole"; pertinent quotations from these authors have been collected elsewhere [39]. Scrutiny of the original text [38] reveals that not an orbital was observed but an orbital density, which implies a distribution of electronic charge, and not even that charge but a "hole" or deficit of that charge. Zuo et al. [38] plotted the difference between an electronic density in the crystalline unit cell, fitted rather than reconstructed from measurements independent of phase, and a model of electronic density of a metallic ion from theoretical atomic densities; the result of this rather arbitrary difference of electronic densities attained a shape that might be construed to resemble a d-orbital – this shape is itself a questionable quantity, as remarked above. For a few decades, crystallographers have produced such plots called deformation densities, but have generally refrained from their fallacious association with orbitals of one or other kind. A question that is impossible to suppress is whether such a deficit of electronic density might be expected logically to resemble an orbital. A deficit, or lack of anything, is clearly not directly observable, but must arise as a difference between two calculated quantities; in this case, one was loosely based on experimental measurements of X-ray and electron diffraction, after the requisite Fourier transformation – the calculations constitute computational diffractometry. The basis of the sensational claim by Zuo et al. [38] was an experimental measurement of the density of electronic charge in a crystallite about 150 nm thick selected to be "free from defects" [38]; this density resulted from an analysis of data obtained with the diffraction of electrons in a convergent beam, supplemented with measurements of the diffraction of X-rays on the same crystal. Despite the claim of a crystallite selected to be "free from defects," a significant consequence of the impact of either electrons or X-rays on a sample is the production of defects; as crystalline Cu_2O is a hard mineral and the amplitudes of its atomic vibrations are small, likely its strength resists such production, but an enduring lesson from quantum mechanics is that an observation on a microscopic system, or an interaction with any particle or wave, even if less energetic than electrons or X-rays in the case of these experiments, inevitably alters that system. In any case, the partition of electronic charge within a particular distribution into components described as orbital densities is undeniably an operation susceptible to ambiguity. Compounding the fatuous claim to have "seen an orbital," which even according to the further clarification by the authors [38] is a gross embellishment of a purported deficit of electronic density, is an invocation of *hybridization*, as some linear combination of atomic orbitals with desired directional properties, to explain the distribution of charge, or its lack, between particular atomic centers in the crystal [38, 39]: hybridization exists only in the mind of the beholder [5]. The density of other than spherical shape was represented as a multipole expansion of the experimental data subject to constraints of symmetry [38], as sites of Cu atomic centers lie on threefold axes. Could some purported deficit really resemble other than a d_{z^2} orbital, apart from an unavoidable ambiguity of direction z in a cubic crystal? Those calculations of electronic structure on a crystal inevitably involve assumptions and approximations beyond those practicable on a single small molecule, and in this case involve atomic center Cu or Cu^+ with 29 or 28 electrons per center. Further discussions [40, 41] of these diffraction experiments [38], which were intrinsically not a spectacular innovation but a significant improvement upon the preceding work elsewhere, contain pertinent references.

As another instance of an audacious claim regarding orbitals from experiments, Itatani et al. [42] published an incredibly regular and detailed diagram of a "molecular orbital" obtained through "tomographic imaging". Their claims are exaggerated and inaccurate in several ways. First, the authors depicted, and conceded to have "measured", only one molecular orbital, rather than the plural in their title. They asserted that "the electrons that make up molecules are organized by energy in orbitals", but electrons are fundamentally indistinguishable. What was claimed to be "measured" was the so-called *Dyson orbital*, which has physical dimensions different from those of an atomic (such as for H) or molecular (such as for H_2^+) orbital as generally defined; on those grounds, a Dyson orbital is not an orbital at all. A Dyson orbital represents an overlap between the total molecular electronic wavefunction of a neutral molecule with N electrons and the total electronic wavefunction of the corresponding molecular cation with $N-1$ electrons. In the particular case, the wavelength of radiation that effects ionization seems indefinite, being described as harmonics 17–51 of light at 800 nm; there is proffered no explanation of the effect of that uncertainty on the possible electronic and vibrational states of the molecular ion and on the concomitant ramifications for the shape of the Dyson orbital. That Dyson orbital is inconsistent with the definition by the same authors [42] of "orbitals" that "single-electron wavefunctions are the mathematical constructs ...". The particular Dyson orbital was described as the "HOMO" of N_2, that is, an acronym for highest occupied molecular orbital, which the authors mistakenly equate with "the highest electronic state." The deduced picture in their Figure 4 [42] that is "reconstructed" from a tomographic inversion of "high harmonic spectra" results from a Fourier transform of the measured squares of a transition dipolar moment, which is a complex tensor, calibrated on the basis of a supposed $2p_x$ orbital of Ar from computational spectrometry according to calculations of uncertain quality; according to a standard convention of labeling atomic orbitals, the latter, unlike $3d_{z^2}$ of Cu, arises necessarily from a real, linear combination of complex $2p_{+1}$ and $2p_{-1}$ wavefunctions of H. Moreover, the "highest occupied *atomic* orbital" of Ar is 3p, not 2p. The authors claim that the presence of positive and negative values of a quantity with symbol "$\psi(X)$" proves this quantity to be "a wavefunction, not the square of the wavefunction, up to an arbitrary phase", but the authors made assumptions about the sign of the magnitude of the matrix element and about the polarization of the emission. As in other spectral experiments, the measured quantity [42] is an intensity that is proportional to the square of that dipolar moment; this complex moment itself, rather than its measured square, is stated to be "a spatial Fourier transform of the [nonexistent] orbital in the x direction" (the relation between x and X was undefined), so taken as lying along the internuclear axis that is conventionally designated z. A Fourier transform of a complex quantity cannot yield a quantity symmetric, according to that figure, with respect to an inversion center between the two atomic nuclei.

To understand how the dependence of a molecular property on distance might be derived from experimental data, specifically the intensities of spectral transitions, we consider briefly the molecular electric dipolar moment of a heteronuclear diatomic molecule [20], which we assume to be expressible as a radial function, i.e., a function of internuclear distance. The integrated intensity of the pure rotational spectrum in absorption is proportional to the square of the total molecular electric dipolar moment with respect to the internuclear axis, thus producing an ambiguity of sign of that dipolar moment. The integrated intensities of vibration-rotational bands in absorption are proportional to the squares of transition moments that correspond to successive derivatives of that radial function with respect to internuclear distance, which fails to resolve the ambiguity of sign, but the variation of transition moment for individual lines within a vibration-rotational band varies linearly with that function; in this way, the ambiguity of sign is resolved, relative to a chosen sign for the permanent electric dipolar moment. The latter sign is in turn unambiguously determined from the isotopic dependence of the rotational g factor [20]. This molecular electric dipolar moment, moreover, is not only a physical attribute of a molecule but also has direct ramifications for a macroscopic sample: the extent to which the total molar (electric) polarization depends on the reciprocal of temperature is proportional to the square of that moment averaged over molecular vibration-rotational states occupied at the temperature of the experiment. This situation is in stark contrast to that of a particular orbital of any kind, which is a mathematical construct or a function that is based incontestably on an approximation of electrons being independent particles, that is, each electron not being subject to electrostatic or other interaction with other electrons even in the same molecule. In view of the multiple assumptions and approximations in the calculations of multiple stages between the measured intensities as a function of angle between the internuclear axis and the polarization axis in the experiment described as "tomographic imaging of an orbital" and the eventual purported picture of an orbital, this proof [42] of a wavefunction is clearly nonsensical. As an example of understatement by the same authors, there appears a remark "Yet single orbitals are difficult to observe experimentally" – in truth, not merely difficult but *impossible*, for manifest reasons expounded above. As their second reference, these authors cite a book, *The Nature of the Chemical Bond*, by an author named "C.P. Linus"! Although one is tempted to suspect that this paper [42] is a deliberate hoax, the authors merely delude themselves.

Several claims have been made to have observed or measured orbitals by means of electron spectroscopy [43], whereby an electron incident on a molecule effects ionization to yield two electrons plus a molecular cation; the orbitals allegedly observed are momentum functions related to $\chi(p)$ of the harmonic oscillator derived above, rather than to $\psi(q)$ that appears in – mistaken – typical discussions of a qualitative nature about molecular electronic structure. For a molecule, these quantities $\chi(p)^*\chi(p)$ representing the density in momentum space have, typically, small values corresponding to small kinetic energies of electrons in directions along "chemical bonds," and large magnitudes in directions between such internuclear vectors. Although the perpetrators of such experiments on electron or optical spectra have undoubtedly great expertise in vacuum plumbing, their understanding of fundamental mathematical and physical concepts that underlie their chemical interpretations is questionable. Further discussion of the physical and philosophical aspects of these experiments and their interpretations is available elsewhere [44]. A claimed "wavefunction for Be" [45] is likewise a result of calculations in several stages and with various assumptions and approximations, based on fitting 361 independent parameters from 58 experimental measurements!

An apparent triumph of an application of orbitals to explain a real chemical reaction was to a stereospecific electrocyclic process – the conversion of buta-1,3-diene to cyclobutene. Woodward and Hoffmann [46] postulated that the stereochemical course of closure to form a ring of four carbon atoms is set according to the symmetry of a particular orbital, that of greatest energy with which electrons are associated in the electronic ground state, in the acyclic precursor. This generalization was claimed to be supported by calculations based on an extension of Huckel theory, but Huckel theory is recognized to have a graph-theoretical basis, not quantum-mechanical whatsoever [47]. Longuet-Higgins and Abrahamson [48] developed an alternative and systematic procedure by considering the symmetries of electronic states, which are observable, thus generating correlations along the entire course between reactants and products, without involving numerical calculations. The so-called *conservation of orbital symmetry* lacks physical foundation because it relies on constituents of a basis set that one can in principle select arbitrarily, without regard for an ultimate energy that might result from a calculation as we have shown above. This *triumph* [46] of an application of orbitals is yet another rebuke of the mathematical ignorance of chemists who fail to distinguish between mathematical artifacts and chemical or physical reality [49].

1.6 Conclusions

Computational spectrometry is an established practice in at least two contexts – analyses of vibration–rotational spectra of diatomic molecules [25] and of vibrational spectra of polyatomic molecules, such as of C_5H [35]. The calculations involved might have a basis in classical or quantum mechanics, but their application invariably has as its objective an observable property such as wavenumber or intensity of spectral lines. As our derivations for the canonical linear harmonic oscillator demonstrate, such observable properties are amenable to calculations with varied methods; each method might have its particular artifacts that are hence of only internal significance. Even that harmonic oscillator is an artifact but a useful initial approximation upon which perturbation methods or other mechanisms might yield more realistic anharmonic oscillators for comparison with experimental data [20]. A particular artifact of calculations of atomic or molecular electronic structure is a wavefunction of an atom or molecule containing only one electron, which was given the name *orbital* of either atomic or molecular kind. Such artifacts, which are legitimate components of appropriate computational procedures, have long been accorded a far greater importance in the teaching and interpretation of chemical phenomena than their artifactual nature warrants. The reason for this unfortunate and unstable condition is that chemists lack mathematical understanding, so that these artifacts are not recognized as being mere components of prospective basis sets for calculations in a particular wave mechanical approach but have become the focus of attention – *"the medium is the message"* (McLuhan)! Orbitals are the cancers of chemistry.

As a remedy of this disease, chemists must improve their understanding of mathematical concepts and principles; one mechanism to achieve this objective that also greatly enhances the capabilities of chemists to implement mathematics for chemical applications involves the teaching and learning of mathematics with software for symbolic computation [8]. Mathematics is not difficult, at least at a level of the topics in calculus, linear algebra, differential equations, and statistics on which chemical applications are constructed, but most chemists find tedious or boring the traditional manipulations that manual operations entail. When powerful contemporary mathematical software serves as a vehicle both of teaching and learning of mathematics and of its implementation, a chemist can attain a sufficient mastery of the concepts and principles to undertake meaningful calculations that would never be routinely attempted by hand.

The quintessence of this essay is that a quantum-mechanical basis of molecular structure, and thus of chemistry itself, is absolutely fraudulent [5]. With irrational exuberance, Dirac proclaimed [50] "The underlying physical laws necessary for the mathematical theory of a large part of physics and the whole of chemistry are thus completely known, and the difficulty is only that the exact application of these laws leads to equations too complicated to be soluble", but the above discussion of their "exact application" to ethanol proves that the utmost central idea in chemistry since Couper's intuition about benzene in 1858 [51], namely, molecular structure, is fundamentally incompatible [31, 32] with "these laws". With his relativistic quantum mechanical wave equation, Dirac predicted the existence of the positron, thus antimatter, but that equation predicts no bound state for even a helium atom, or anything chemically more complicated or interesting. Which is more important for chemistry and chemists – quantum mechanics or molecular structure? They are incompatible.

Dirac recognized [50] that "approximate practical methods of applying quantum mechanics [needed to] be developed", although before the computer age he could scarcely have envisaged the success of present approximate practical methods that, on a partially empirical basis, enable the geometric structure and infrared absorption spectra of a molecular species to be predicted almost within experimental error of their measurement; Dirac alluded to an objective to be an "explanation of the main features of [complicated] atomic systems without too much computation". Molecular structure is a classical concept, which might be introduced into, or imposed upon, a quantum-mechanical calculation by means of a procedure due to Born and Oppenheimer or as subsequently refined by Born; this procedure has never been fully justified theoretically. Molecular structure is thus intrinsically foreign to quantum mechanics; because a molecule in a quantum state has a completely indefinite structure [31, 32], an attempt to explain a classical molecular structure with quantum mechanics is a logical fallacy [11]. For the calculation of a classical property such as molecular structure, one might apply a classical method, such as molecular mechanics, or one might choose, subject to the preceding proviso, a quantum-mechanical method. In the latter case, an association of details of the structure with artifacts of the particular method of calculation, such as atomic orbitals in a basis set for a wave-mechanical calculation within a coordinate representation, would constitute a logical fallacy.

Through the application of symbolic computation for both the teaching of mathematics and the implementation of mathematical operations, as we have illustrated here for the harmonic oscillator without the tortuous contortions of a tedious manual approach [52], both the learning and understanding of mathematics by students of chemistry and their capabilities of undertaking mathematical operations can be significantly enhanced without increased time and effort [8]. An enhanced mathematical knowledge is not only essential for chemical applications of quantum mechanics but also important for other aspects of chemistry in the undergraduate curriculum. How much content of core curricula is devoted to, for instance, mesophases [53] – applications of fine chemical materials with these properties in display devices have generated an enormous global industry, but many chemists are ignorant of these states of aggregation – or to atmospheric constituent molecular species and their reactions in relation to *global warming*? An improved understanding of mathematics and of its applications by chemists is applicable to real problems of our society, thus strongly justifiable, not merely for calculating energies of a linear harmonic oscillator or as a basis of the practice of computational spectrometry.

From an appreciation of the distinction between quantum laws and quantum theories, there is much that one can learn; there is equally much that one can learn on comparing quantum theories applied to a particular chemical system, but in almost all such applications quantum theories are much less important than the quantum laws. The latter, one applies directly to interpret atomic and molecular spectra for analytical and other purposes, and for other aspects of chemical reactions. Quantum mechanics might have profound philosophical implications [54], which are poorly understood and almost invariably ignored in textbooks of chemistry; there is, nevertheless, subtlety involved in a distinction between quantum laws and quantum theories that we extenuate here.

Acknowledgment

For helpful comments and discussion, I thank numerous colleagues, in particular Professors D.P. Craig F.R.S., F.M. Fernandez, W.H.E. Schwarz, and B.T. Sutcliffe.

References

1 Pople, J.A., Schneider, W.G., and Bernstein, H.J. (1959) *High-Resolution Nuclear Magnetic Resonance*, McGraw-Hill, New York.
2 Roberts, J.D. (2005) *Can. J. Chem.*, **83**, 1626–1628.
3 Laidler, K.J. (1993) *The World of Physical Chemistry*, Oxford University Press, Oxford, UK.
4 Pais, A., Jacob, M., Olive, D.I., and Atiyah, M.F. (1998) *Paul Dirac, The Man and His Work*, Cambridge University Press, Cambridge, UK.
5 Ogilvie, J.F. (1994) The nature of the chemical bond 1993, in *Conceptual Trends in Quantum Chemistry* (eds E.S. Kryachkoand J.L. Calais), Kluwer, Dordrecht, The Netherlands, p. 171, and references therein.
6 Boyer, T.H. (1969) *Phys. Rev.*, **186**, 1304.
7 Born, M. (1969) *Atomic Physics*, 8th edn, Blackie, Glasgow, UK.
8 Ogilvie, J.F. and Monagan, M.B. (2007) *J. Chem. Educ.*, **84**, 889.
9 Kemble, E.C. (1958) *Fundamental Principles of Quantum Mechanics*, Dover, New York.
10 Dirac, P.A.M. (1981) *Principles of Quantum Mechanics*, 4th revised edn, Oxford University Press, Oxford, UK.
11 Primas, H. (1983) *Chemistry, Quantum Mechanics and Reductionism*, 2nd edn, Springer-Verlag, Berlin.

12. Craig, D.P. and Thirunamachandran, T. (1984) *Molecular Quantum Electrodynamics: An Introduction to Radiation–Molecule Interactions*, Academic Press, London.
13. Craig, D.P. and Thirunamachandran, T. (1986) *Acc. Chem. Res.*, **19**, 10.
14. Dirac, P.A.M. (1964) *Nature*, **203**, 115.
15. Aitchison, I.J.R., MacManus, D.A., and Snyder, T.M. (2004) *Am. J. Phys.*, **72**, 1370–1379.
16. Fedak, W.A. and Prentis, J.J. (2009) *Am. J. Phys.*, **77**, 128–139.
17. Jordan, T.F. (2005) *Quantum Mechanics in Simple Matrix Form*, Dover, Mineola, USA.
18. Adler, S.L. (1995) *Quaternionic Quantum Mechanics and Quantum Fields*, Oxford University Press, New York.
19. Jiang, T. and Chen, L. (2008) *Comput. Phys. Commun.*, **178**, 795.
20. Ogilvie, J.F. (1998) *The Vibrational and Rotational Spectrometry of Diatomic Molecules*, Academic Press, London.
21. Nieto, M.M. (1981) *Phys. Rev.*, **D24**, 1030.
22. Davidson, P.M. (1932) *Proc. Roy. Soc. London*, **A135**, 459.
23. Green, H.S. (1965) *Matrix Mechanics*, Noordhoff, Groningen, The Netherlands.
24. Ogilvie, J.F., Oddershede, J., and Sauer, S.P.A. (2000) *Adv. Chem. Phys.*, **111**, 475–536.
25. Ogilvie, J.F. and Oddershede, J. (2005) *Adv. Quantum Chem.*, **48**, 254–317.
26. Bak, K.L., Sauer, S.P.A., Oddershede, J., and Ogilvie, J.F. (2005) *Phys. Chem. Chem. Phys.*, **7**, 1747–1758.
27. Sauer, S.P.A., Jensen, H.J.Aa., and Ogilvie, J.F. (2005) *Adv. Quantum Chem.*, **48**, 319–334.
28. Sauer, S.P.A., Paidarova, I., Oddershede, J., Bak, K.L., and Ogilvie, J.F. (2010) *Int. J. Quantum Chem.*, in press.
29. Ogilvie, J.F., Jensen, H.J.Aa., and Sauer, S.P.A. (2005) *J. Chin. Chem. Soc.*, **52**, 631–639.
30. Pople, J.A. (1964) *J. Chem. Phys.*, **43**, S229–S230.
31. Woolley, R.G. (1976) *Adv. Phys.*, **25**, 27.
32. Woolley, R.G. (1985) *J. Chem. Ed.*, **62**, 1082.
33. Cafiero, M. and Adamowicz, L. (2004) *Chem. Phys. Lett.*, **387**, 136.
34. Sutcliffe, B.T. and Woolley, R.G. (2005) *Chem. Phys. Lett.*, **408**, 445.
35. Wu, Y.-J., Chou, H.-F., Comacho, C., Witek, H.A., Hsu, S.-C., Lin, M.-Y., Chou, S.-L., Ogilvie, J.F., and Cheng, B.-M. (2009) *Astrophys. J.*, **701**, 8–11.
36. Fernandez, F.M. and Castro, E.A. (1996) *Algebraic Methods in Quantum Chemistry and Physics*, CRC Press, Boca Raton, FL.
37. Ho, G.S., Ligneres, V.L., and Carter, E.A. (2008) *Comput. Phys. Commun.*, **179**, 839–854.
38. Zuo, J.M., Kim, M., O'Keeffe, M., and Spence, J.C.H. (1999) *Nature*, **401**, 49–52.
39. Wang, S.G. and Schwarz, W.H.E. (2000) *Angew. Chem. Int. Ed.*, **39**, 1757–1762.
40. Zuo, J.M., Kim, M., O'Keeffe, M., and Spence, J.C.H. (2000) *Angew. Chem. Int. Ed.*, **39**, 3791–3794.
41. Wang, S.G. and Schwarz, W.H.E. (2000) *Angew. Chem. Int. Ed.*, **39**, 3794–3796.
42. Itatani, J., Levesque, J., Zeidler, D., Niikura, H., Pepin, H., Kieffer, J.C., Corkum, P.B., and Villeneuve, D.M. (2004) *Nature*, **432**, 867–871.
43. McCarthy, I.E. (2001) *Z. Phys. Chem.*, **215**, 1303–1313.
44. Schwarz, W.H.E. (2006) *Angew. Chem. Int. Ed.*, **45**, 1508–1517.
45. Jaytilaka, D. (2000) A wave function for Be from X-ray diffraction data, Chapter 21, in *Electron, Spin and Momentum Densities and Chemical Reactivity* (eds P.G. Mezey and B.E. Robertson), Springer, pp. 253–263.
46. Woodward, R.B. and Hoffmann, R. (1965) *J. Am. Chem. Soc.*, **87**, 395–396.
47. Trinajstic, N. (1992) *Chemical Graph Theory*, 2nd edn, CRC Press, Boca Raton, USA.
48. Longuet-Higgins, H.C. and Abrahamson, E.W. (1965) *J. Am. Chem. Soc.*, **87**, 2045–2046.
49. Ogilvie, J.F. (1997) Aspects of the chemical bond 1996, in *Conceptual Perspectives in Quantum Chemistry* (eds J.L. Calais and E.S. Kryachko), Kluwer, Dordrecht, The Netherlands, p. 127, and references therein.
50. Dirac, P.A.M. (1929) *Proc. Roy. Soc. London*, **A123**, 714.
51. Duff, D.G. (1987) *Chem. Brit.*, **23**, 354.
52. Dushman, S. (1935) *J. Chem. Educ.*, **12**, 381.
53. Ogilvie, J.F. (1989) *J. Chin. Chem. Soc.*, **36**, 375 and 501.
54. Selinger, B.T. (1982) *Chem. Australia*, **49**, 448.

2
Computational NMR Spectroscopy
Ibon Alkorta and José Elguero

2.1
Introduction

The topic of computational NMR spectroscopy was mostly covered by a multiauthor book published in 2004, *Calculation of NMR and EPR Parameters* [1]. With very few exceptions, the most recent references in this book are from 2003. Therefore, this chapter will cover essentially the 2004–2009 period. In 2009, an article appeared entitled "How aromaticity affects the chemical and physicochemical properties of heterocycles: a computational approach" in Vol. 19 of *Topics in Heterocyclic Chemistry* [2], in which chemical shifts and nucleus-independent chemical shifts (NICS) of aromatic heterocycles are discussed. This chapter is not intended to be exhaustive as far as examples are concerned but rather highlights the general significance of our contribution to this topic.

2.2
NMR Properties

The two main NMR properties that interest experimentalists are chemical shifts (relative to a reference and measured in parts per million) and spin–spin coupling constants (SSCC, not needing a reference and measured in hertz but depending on the magnetogyric ratios). Other properties such as nuclear quadrupole coupling constants (NQCC, not needing a reference and measured in megahertz [3]), relaxation times [4], isotope effects [1], and shielding tensors [1] are not included in this review, in part, because we have not contributed to them.

2.3
Chemical Shifts

Literature results on the use of theoretical methods to calculate chemical shifts (δ, ppm), via absolute shieldings (σ, ppm), are very abundant, mainly due to the

Computational Spectroscopy: Methods, Experiments and Applications. Edited by Jörg Grunenberg
Copyright © 2010 WILEY-VCH Verlag GmbH & Co. KGaA, Weinheim
ISBN: 978-3-527-32649-5

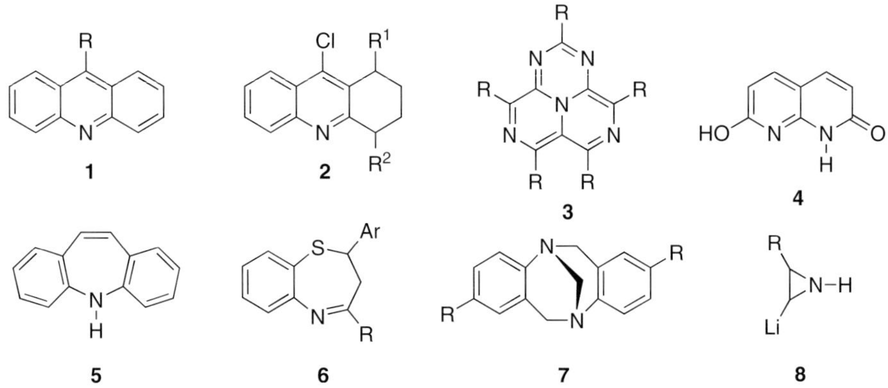

Scheme 2.1 Six-, seven-, eight- (Tröger's bases), and three-membered rings.

facility to carry out accurate high-level calculations. In 1999, Helgaker, Jaszunski, and Ruud published a review "*Ab initio* methods for the calculation of NMR shielding and indirect spin–spin coupling constants" that is still an excellent summary of the situation [3]. Some illustrative examples include papers devoted to buspirone analogues [5], five-membered aromatic heterocycles [6], indoloquinoline alkaloids [7], and rhodium(II) complexes with azoles (imidazoles and pyrazoles) [8].

At the beginning of our studies reporting GIAO calculations of absolute shieldings [1] [9], we used the B3LYP/6-31G(d) approximation. In recent years, we consider that more reliable calculations can be obtained at the GIAO/B3LYP/6-311++G(d,p) level [10]. For particular atoms, such as Xe, we have carried out calculations at the GIAO/DGDZVP level. Our publications concern mainly heterocycles (Schemes 2.1 and 2.2) and also some aliphatic and aromatic compounds (Scheme 2.3).

In general, the results are highly satisfactory; that is, absolute shieldings and chemical shifts are highly correlated (see Tables 2.1 and 2.2). To obtain the chemical shifts, some authors use the absolute shielding of the reference and subtract from it the calculated σ. We prefer to scale the σ values in a way similar to that Pople used for IR frequencies [11, 12]. Thus, we assume that σ and δ values are related by an equation of the form $\delta = \mathbf{a} + \mathbf{b} \times \sigma$, where **a**, the intercept, should be as close as possible to the σ of the reference and **b**, the slope, as close as possible to |1|, close but not necessarily identical (Table 2.1).

From Scheme 2.1, **1** [25], **2** [26], **3** [27], **4** [28], **5** [29], **6** [30], **7** [31], and **8** [32]. From Scheme 2.2, **9** [33–36], **10** [37–39], **11** [40–43], **12** [44–46], **13** [47, 48], **14** [24, 40, 49], **15** [40], **16** [50], **17** [49], **18** [51, 52], **19** [53], **20** [54], **21** [43, 55], **22** [56, 57], **23** [58], **24** [17], **25** [22], and also subazaporphyrins [59] and trindoles [60]. From Scheme 2.3, **26–30** [61], **31** [16, 62], **32** [63], **33** [64], **34** [23], **35** [15] as well as azolides (*N*-acyl azoles) [65], amino acids [66], plumbagin [67], and 2-fluorobenzamide [19].

---

1) GIAO acronym was initially used for "gauge invariant atomic orbital," but since it is generally agreed that this is an unfortunate name, today it means "gauge including atomic orbital" method.

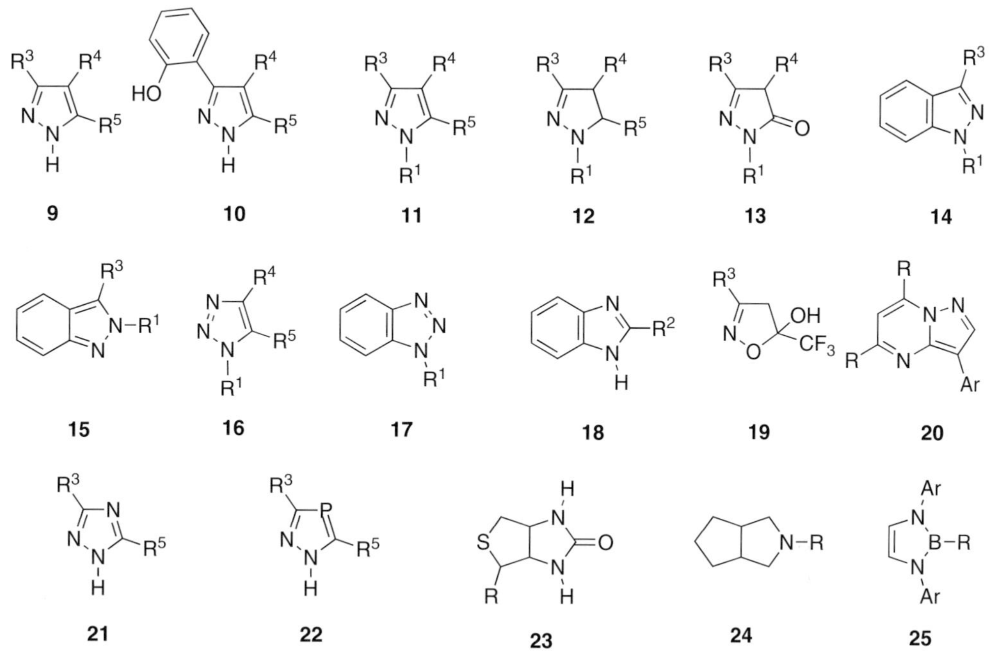

Scheme 2.2 Five-membered rings.

The equations reported in Table 2.1 cover a wide range of values; in some cases, it is better to consider a subset, for instance, only aromatic carbon atoms [21]. Although these equations work extremely well, we have noted some exceptions. The most notable are carbon atoms bearing halogen atoms (mainly Br but also Cl) [26, 30, 35, 43, 55] and also carbon atoms directly linked to sulfur atoms [17, 30, 58]. For these atoms, the calculated value overestimates the experimental one by 7–10 ppm (e.g., experimental value is 127.7 ppm and calculated value is 136.5 ppm for an aromatic carbon atom). Using methane monosubstituted derivatives CH_3X (X = H, CH_3, CN, NH_2, NO_2, OH, and F) as a training set, we have found that to reproduce the chemical

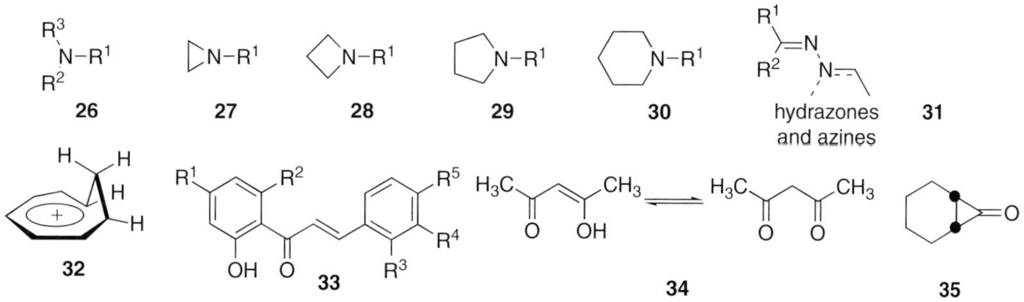

Scheme 2.3 Amines (including cyclic ones) and other structures.

Table 2.1 Best equations for B3LYP/6-311++G(d,p) calculations.

Equation	Nucleus	Conditions	Intercept	Slope	σ (reference) (ppm)	References
Robust (50 or more points)						
1	^{1}H	—	31.8	−1.000	TMS: 31.97	[13–15]
2	^{1}H	—	31.0	−0.970	TMS: 31.97	[13, 16–20]
3	^{13}C	sp, sp^2, sp^3 C atoms	175.7	−0.963	TMS: 184.75	[13, 17, 20, 21]
4	^{15}N	Without nitroso compounds	−152.0	−0.946	CH$_3$NO$_2$: −154.43	[13, 21]
5	^{15}N	With nitroso compounds	−154.0	−0.874	CH$_3$NO$_2$: −154.43	[13, 21]
Not robust (less than 50 points)						
6	^{11}B	—	106.5	−0.900	BF$_3$OEt$_3$: 101.95	[13, 22]
7	^{17}O	sp^2, sp^3	309.7	−0.926	H$_2$O: 322.27	[13, 23]
8	^{19}F	—	164.0	−0.970	CFCl$_3$: 153.70	[13, 24]
9	^{31}P	—	228.6	−0.785	PO$_4$H$_3$: 292.33	[13] [a]

a) Frideling, A., Faure, R., Galy, J.-P., Kenz, A., Alkorta, I., and Elguero, J. (2004) *Eur. J. Med. Chem.*, **39**, 37–48.

Table 2.2 Relationships between intercept, slope, and reference compound (all values in ppm).

Equation	Nucleus	Intercept	Slope	Reference	Calc.	Error
1	^{1}H	31.8	−1.000	31.97	31.80	0.17
2	^{1}H	31.0	−0.970	31.97	31.96	0.01
3	^{13}C	175.7	−0.963	184.75	182.45	2.30
4	^{15}N	−152.0	−0.946	−154.43	−160.68	6.25
5	^{15}N	−154.0	−0.874	−154.43	−176.20	21.77
6	^{11}B	106.5	−0.900	101.95	118.33	−16.38
7	^{17}O	309.7	−0.926	322.27	334.45	−12.18
8	^{19}F	164.0	−0.970	153.70	169.07	−15.37
9	^{31}P	228.6	−0.785	292.33	291.21	1.12

shifts of derivatives (X = Cl, Br, SH, PH$_2$, and SeH) it is necessary to increase the level of the calculations to MP2 [2].

In general, we have not carried out corrections for the solvent effects (for an exception, see Ref. [35]) because we are rather skeptical about continuum models while specific solvation effects (supermolecules) are costly (see Section 2.6). Heterocycles bearing nitro groups have been studied by Katritzky *et al.* at different theoretical levels including the B3LYP/6-311++G(d,p) [68]. They have carried out continuum model calculations to simulate the solvent.

2) Alkorta, I. and Elguero, J., unpublished results.

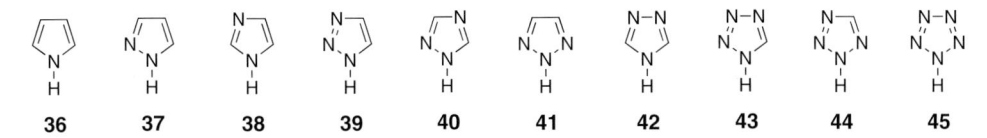

Scheme 2.4 Azoles.

2.4
NICS and Aromaticity

According to Bachrach [69], Schleyer became interested in NMR and its relationship with aromaticity in the 1980s. This led to the development of nucleus-independent chemical shift [70], where the virtual chemical shift can be computed at any point of space. Schleyer first advocated using the geometrical center of the ring, and then a point 1 Å above the ring center [71]. His more recent paper postulated using just the component of the chemical shift tensor perpendicular to the ring evaluated at the center of the ring [72] because it is only this component that is related to aromaticity. Nevertheless, NICS(0) and NICS(1) continue to be used by most authors.

We use the convention of the Gaussian package (aromatic compounds have positive NICS) that is the opposite to that of Schleyer (aromatic compounds have negative NICS).

NICS have been reviewed [73, 74], compared with other aromaticity criteria [75–77], used in relation with tautomeric equilibrium [78] and with hydrogen bonds [79], applied to a great variety of heterocycles: all parent azoles, 36–45 (Scheme 2.4), concluding that all of them are aromatic [80], phosphorus heterocycles [81], other five-membered heterocycles 46 (Scheme 2.5) [82], and substituted pyridines and pyridinium cations [83].

Of particular interest, in relation to NICS and aromaticity, is the work of Kleinpeter and Koch who introduced the "through-space" NMR shieldings (TSNMRS) visualized as isochemical shielding surfaces (ICSS) to study many structural problems. One of their latest papers reports the visualization of aromaticity in cations, neutral molecules, and anions by spatial magnetic properties (through-space NMR shieldings) [84], for instance, benzene (Figure 2.1).

Our contribution is mainly related to the application of NICS calculations to the structure and aromaticity of heterocycles including two reviews where this topic is analyzed [2, 10]. Besides heterocycles, we have discussed the cases of benzene versus cyclohexatriene [86], compounds related to the Mills–Nixon effect, dehydroannulenes, polyacenes, s-indacene, and Möbius rings [87], as well as the structure of homotropylium cation [63] and that of the cation obtained by protonation of 5 [29].

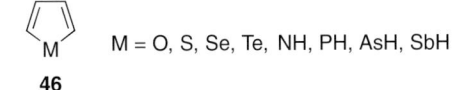

M = O, S, Se, Te, NH, PH, AsH, SbH

46

Scheme 2.5 Furan, thiophene, selenophene, tellurophene, pyrrole, phosphole, arsole, and stilbole.

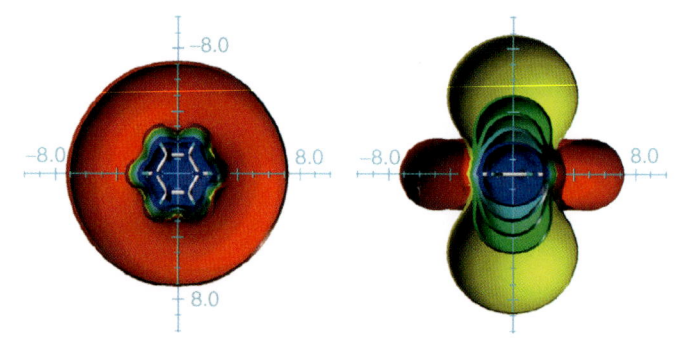

Figure 2.1 An ICSS view of benzene [85].

We have devoted several papers to the aromaticity of azoles (related to those of Scheme 2.4) and to the effect of their protonation (azolium salts) [88–91]. We published a paper discussing phosphole (**46**, M = PH), pyrrole (**36**), and their tetrahydro derivatives where NICS were used to assess their aromaticity.

The problem of the structure and aromaticity of heteropentalenes [92] was examined using NICS as an analytical tool, so we will use this series of compounds to illustrate our approach. These compounds result formally by replacing C atoms of the pentalene dianion **47** by heteroatoms X and Y (Scheme 2.6). Since the other peripheral atoms can be carbon or nitrogen, it results in hundreds of heteropentalenes that are called azapentalenes when X and Y = NR. There are three main classes of heteropentalenes: (i) no nitrogen atom in the junction (3a–6a bond) (**48** and **49**); (ii) one nitrogen atom in the junction (**50** and **51**); and (iii) two nitrogen atoms in the junction (**52**) (a zwitterion).

In the first paper, we studied the compounds reported in Scheme 2.7 (all of them belonging to the **48** and **49** classes) together with some model compounds (**36–39**, **41**, and **46**; X = O, S) [93]. Concerning NIST, both (0) and (1), they decrease (the aromaticity decreases) from the parent compounds to the azapentalenes. 1,2-DiH derivatives **53**, **54**, and **55** being nonplanar have NICS(1) above and below the molecular plane (Figure 2.2).

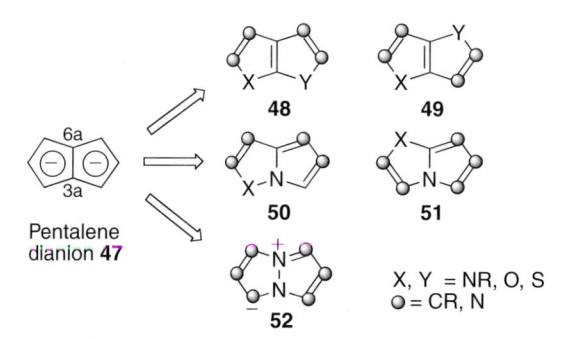

X, Y = NR, O, S
◯ = CR, N

Scheme 2.6 The different heteropentalenes.

Scheme 2.7 Heteropentalenes without N atoms in the junction.

The next publication deals with a compound of the **52** class, the fully nitrogenated N_8 molecule, pentazolo[1,2-*a*]pentazole (**56**) [94]. According to the NICS calculated values, this molecule is more aromatic than benzene and pyrrole (**36**). The aromaticity of **56** increases by complexation with anions (Figure 2.3).

The third paper [95] described the application of Free–Wilson matrices to the analysis of the tautomerism and aromaticity of azapentalenes of the classes **50** and **51**. A total of 44 neutral and 60 protonated azapentalenes were studied, some examples being reported in Scheme 2.8. We concluded that aromaticity, as defined by Schleyer's NICS(1) values, provides a coherent picture for azapentalenes but this picture is not consistent with other aromaticity criteria.

The fourth paper concerns the simplest example of class **52**, 1*H*-pyrazolo[1,2-*a*]pyrazol-4-ium hydroxide inner salt (**57**) (often represented as **57a**, Scheme 2.9) [96], NICS(1) = 9.94 ppm. In azapentalenes lacking N atoms in positions 3a and 6a,

Figure 2.2 NICS(1) above and below the molecular plane for compounds **53**, **54**, and **55**.

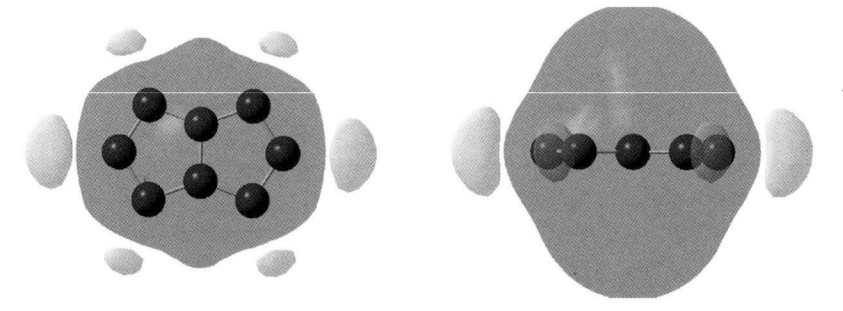

Figure 2.3 Two views of the molecular electrostatic potential of **56** at the ±0.02 au isosurface. Positive regions in dark and negative ones in light.

NICS(1) = 9–11 ppm [93], with only one N atom NICS(1) is about 10 ppm [95], and in the N_8 molecule (which has two N atoms in these positions), NICS (1) = 12.8 ppm [94].

The last paper on the heteropentalene series applies the NICS methodology to bimanes **58** and **59** and related compounds **60** and **61**, all of them of the class **52** (Scheme 2.10) [97]. Since the rings are folded, there are two NICS per ring, one above (the convex part) and the other below (the concave part, see Figure 2.4).

In summary, these five publications concerning the aromaticity of heteropentalenes as seen by NICS provide a coherent picture but different from other aromaticity criteria. A large collection of NICS(1) values is now available to the reader interested in heteroaromaticity.

A paper on the effect of perfluorination on the aromaticity of benzene and heterocyclic six-membered rings resulted from a collaboration with the Schleyer's group [98].

Scheme 2.8 Tautomerism of azapentalenes with one N atom in the junction.

Scheme 2.9 3a,6a-Diazapentalene.

syn-Bimane **58**
(1,7-dione)

anti-Bimane **59**
(1,5-dione)

1,3-Dione **60**

61

Scheme 2.10 Bimanes and related compounds.

2.5
Spin–Spin Coupling Constants

Many other authors and ourselves have carried out theoretical calculations of SSCC. Our studies mostly refer to a situation like X–H $\cdots$ Y where there are three coupling constants: a covalent $^1J_{XH}$, a hydrogen bond $^{1h}J_{HY}$, and a coupling of the heavy atoms through the hydrogen bond $^{2h}J_{XY}$. Some authors such as Limbach [99] prefer a phenomenological nomenclature not differentiating the covalent and the hydrogen-bond ^1J couplings because when the proton moves along X $\cdots$ Y the difference blurs and disappears; this can happen only when X and Y are identical, for instance, ^{15}N nuclei. When comparing SSCC involving different nuclei, it is useful to transform J into K, the so-called reduced CC [100], to avoid the magnetogyric complication.

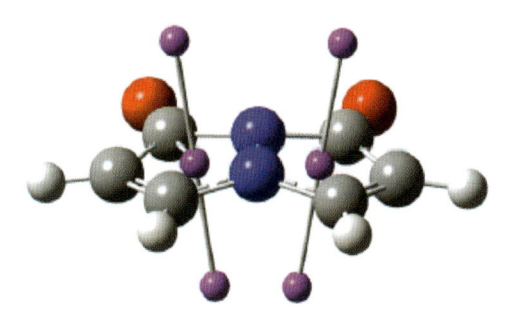

Figure 2.4 NICS(1) above and below the molecular plane for compound **58**.

At present, the two best methods for calculating SSCC are SOPPA (second-order polarization propagator) and EOM-CCSD (equation-of-motion/coupled cluster singles and doubles) [101–103]. There exists a SOPPA-CCSD version [104]. An acceptable, although less good, approach consists in using DFT-based methods implemented in Gaussian 03 [11].

In terms of the work we have done, the basis sets we have used are as follows:

Ahlrich qzp [105] :   C, N, O, F
Ahlrich qz2p [105] :   P, S, Cl, H

Del Bene has established two Ahlrich-like "hybrid" bases for boron (^{11}B) and lithium (^{7}Li) [106].

The only exception to the above scheme was using the Dunning VDZ basis on H. When the systems were too large, we replaced the qz2p basis on H by VDZ [107–109]. This reduces the number of basis functions on H from 10 to 5. However, when we did that, we did not report coupling constants for any H atom that had Dunning's VDZ basis set. Note that the presence of VDZ does not change coupling involving other atoms.

It has been documented in the literature that the use of smaller contracted basis sets (such as Pople's and Dunning's) does not in general perform well for coupling constants. The Ahlrich basis sets are contracted only for one s function, one p function, and so on. The remaining functions are primitives. That is part of the reason why they perform well.

There are results in the literature that suggest that a very large Dunning's or Pople's basis set would be needed [110–113], something perhaps better than cc-pVTZ or maybe aug-cc-pVTZ [107, 108]. Consider, for instance, cc-pVTZ. This basis is not considered very large, but it already has 30 basis functions for second-period elements such as C, and aug-cc-pVTZ has 46 basis functions per carbon. The Ahlrich basis set has 24 basis functions on C (used only as an example). Thus, using Dunning's basis set we would not have been able to carry out some of the calculations we are performing because the number of basis functions would become too large (these calculations scale something like n^7, where n is the number of basis functions). There are some other basis sets in the literature (e.g., Sadlej-J [114, 115]) that have been constructed for coupling constant calculations.

According to Malkina, the Dirac vector model (DVM, also called Dirac–van Vleck vector model) of SSCC was proposed by Duval and Koide in 1964 [1, 116, 117]; however, there are at least two earlier papers on the DVM: one by McConnell in 1955 "Dirac vector model for electron coupled nuclear spin interactions" [118] and another by Alexander in 1961 "Spin–spin interactions in nuclear magnetic resonance. Contact contribution" [119]. Subsequently, it has been used so many times that it has become a classical tool in NMR spectroscopy [120–122]. Nevertheless, it has long been recognized that the DVM is too simple for an adequate interpretation of the dependence of the SSCC signs on the number of bonds [123].

In 2003, Del Bene introduced the "nuclear magnetic resonance triplet wavefunction model" (NMRTWM) [124] considering only the Fermi contact (FC) contribution, and we will refer to the sign of J based on the sign of the Fermi contact term.

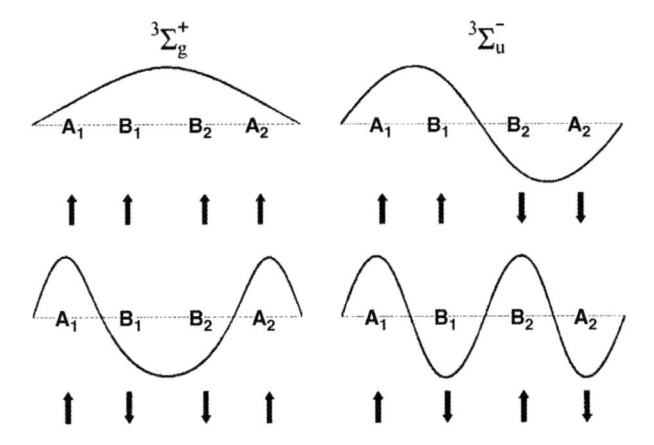

Figure 2.5 Nodal patterns for a four-atom system with $D_{\infty h}$ symmetry.

For a linear system with $D_{\infty h}$ symmetry, $A_1-B_1 \cdots B_2-A_2$, the only excited triplet states that can have nonzero contributions to the coupling constants are those with $^3\Sigma^+$ or $^3\Sigma^-$ symmetry, states that do not have nodes containing the $D_{\infty h}$ axis (Figure 2.5).

The model used for discussing the sign of coupling constants arose from the well-known expression for computing J, namely $J_{AB} = \partial^2 E/\partial\mu_A \cdot \partial\mu_B$, which indicates that the energy (E) depends on the nuclear magnetic moments (μ) of atoms A and B. This implies that the wavefunction also depends on these moments. Alternatively, the nuclear magnetic moments might be sensitive to the wavefunction. How would this sensitivity be manifested? The sum-over-states expression for J_{AB} extends over the entire manifold of triplet states that can interact with the ground state. What in the triplet wavefunctions might influence the alignment of magnetic nuclei? One possibility is the phases of these wavefunctions at atoms A and B. The NMRTWM was introduced as a tool to gain insight into what determines the sign of J_{AB} and employ it to consider the orientation of nuclear spins in $^3\Sigma^+$ and $^3\Sigma^-$ states by considering the nodal properties of wavefunctions for these states in a model linear $A_1-B_1 \cdots B_2-A_2$ system. In this model, it was assumed that the orientation of the nuclear magnetic moment vector responds to the phase of the excited triplet state wavefunction, and arbitrarily assigns nuclear spin up (↑) when the wavefunction is positive and down (↓) when it is negative. For a four-atom system with $D_{\infty h}$ symmetry, the nodal patterns for $^3\Sigma^+$ and $^3\Sigma^-$ states and the resulting nuclear alignments are shown in Figure 2.5. This assignment is consistent with the convention adopted for the sign of J, which states that a positive coupling constant corresponds to a reduction in the interaction energy when the nuclear spins are antiparallel. We have used NMRTWM several times [106, 125–132], but the impossibility to make predictions led us to stop using it in 2006.

In Table 2.3, we have summarized, in chronological order, our publications dealing with the calculations of SSCC both J and K.

Table 2.3 Calculations of SSCC.

System	Couplings	Method	Main conclusions and comments	Reference
X–H⋯Y [X = ^{13}C, ^{15}N, ^{17}O, ^{19}F]	$^1J_{XH}$	E-C	NMRTWM applies to Ks	[129]
(HF)$_2$ clusters	J_{FF}	E-C	The Fermi term dominates	[125]
H$_m$X–YH$_n$ [Y = ^{13}C, ^{15}N, ^{31}P]	$^1J_{XY}$	E-C	The X–Y bond may be a single, double, or triple bond	[126]
X–H⋯Y [Y = ^{15}N, ^{17}O, ^{19}F]	$^{1h}J_{HY}$	E-C	Determining the sign of $^1K_{XH}$ and $^{1h}K_{HY}$ should allow to determine if the HB is proton shared or not	[133]
X–H⋯Y	$^{2h}K_{XY}$	E-C	All $^{2h}K_{XY}$ are positive	[128]
Scheme 2.11	$^{2h}J_{OO}$, $^{2h}J_{NN}$	E-C	Neither the Js nor the δ^{1}H provide evidence of RAHB [134]	[127]
Scheme 2.12	$^2J_{CH}$	DFT	Tautomerism. $^2J_{CH}$ is always positive	[135]
H–X–Y–H, H–X⋯H⋯Y–H	$^3J_{HH}$, $^{4h}J_{HH}$	FPT	Karplus: molecular versus supramolecular	[136]
Scheme 2.11	$^{2h}J_{OO}$, $^{2h}J_{NN}$	E-C	RAHB: we ratify [127]	[137]
N–H⋯O=P	$^{3h}J_{NP}$, $^{2h}J_{HP}$	E-C	Urea/phosphoric and urea/phosphate	[138]
AH:XH:YH$_3$ [A, X = ^{19}F, ^{35}Cl; Y = ^{15}N, ^{31}P]	$^1J_{XH}$, $^{1h}J_{HY}$, $^{2h}J_{XY}$	E-C	The presence of a third molecule (AH) increases the proton-shared character of the X–H–Y HB	[130]
(HCN)$_n$ (HNC)$_n$	$^1J_{NC}$, $^{2h}J_{NC}$	SO	Complexes n = 1–6, cooperativity	[139]
N–H⋯F⋯H–N	$^1J_{NH}$, $^{1h}J_{HF}$, $^{2h}J_{FN}$	E-C	The HB in the gas phase has more proton-shared character relative to HBs in solution	[140]
(FH)$_2$⋯collidine (see Figure 2.6)	$^1J_{FH}$, $^{1h}J_{HN}$, $^{2h}J_{FN}$	E-C	The 2:1 FH:NH$_3$ complex as a model. Effect of the proton transfer on the Js	[141]
Compound 64	$^2J_{PP}$, $^2J_{FF}$	DFT	Error in one exp. J value: see Table 2.4	[142]

Compound/System	Couplings	Method	Comments	Ref.
Compound **11** ($R^3 = R^4 = R^5 = H$)	$^1J_{NN}$, $^1J_{NC}$, $^1J_{CC}$	DFT	131 exp. coupling constants versus 243 calcd. All agree well save $^1J_{CC}$	[143]
Compound **19**	$^2J_{HH}$, $^4J_{HFu}$	DFT	$^2J_{HH}$ (gem) well reproduced	[53]
N—H···N	$^1J_{NH}$, $^{1h}J_{HN}$, $^{2h}J_{NN}$	E-C	66 complexes, all kind of N atoms; effect of the NN distance on $^{1h}J_{HN}$ sign	[144]
Scheme 2.13	$^1J_{BN}$	E-C	F increases and Li decreases $^1J_{BN}$	[106]
H_mX-YH_n [X, Y = N, O, P, S]	$^1J_{XY}$	E-C	Karplus-type equations: (dihedral angle) orientation of lone pairs is determining	[145]
$X(CH_3)_nH_{(4-n)}$ [X = C, N]	J_{HH}, J_{CH}, J_{NH}, J_{CC}, J_{NC}	E-C	Importance of solvating ammonium salts such as NH_4^+ with H_2O molecules	[146]
C_6H_5X [X = CH (benzene), N (pyridine), P (phosphinine)]	J_{CC}, J_{NC}, J_{PC}	E-C	Protonation on the heteroatom produces dramatic changes in the Js	[147]
RBNR, RCCR	J_{NB}, J_{BH}, J_{NH}, J_{CC}	E-C	Effect of substituent R on the Js	[148]
$P(CH_3)_nH_{(4-n)}$	J_{PH}, J_{PC}, J_{CH}, J_{HH}	DFT	$N(Et)_4^+$; only the Me are coupled with ^{14}N	[149]
Diazaboroles **25**	J_{NB}, J_{NC}, J_{CC}, J_{BR}	E-C	$^1J_{NB}$ is always negative	[22]
H_mX-YH_n [X = N, O, P, S]	$^1J_{XY}$	E-C	Essential role of the number of lone pairs in X and Y	[150]
$[P-H\cdots P]^+$	$^1J_{PH}$, $^{1h}J_{HP}$, $^{2h}J_{PP}$	E-C	See Figure 2.7	[151]
X—H···Y	$^1J_{FH}$, $^{1h}J_{BLi}$	E-C	^1J couplings as probes of PA of bases	[152]
$[P-H\cdots P]^+$	$^{2h}J_{PP}$, $^{2h}J_{PC}$	E-C	Also $^{2h}J_{PP}$ and $^{2h}J_{PC}$ with C=P bonds	[153]
25, R = Li	$^1J_{BLi}$	E-C	$^1J(^{11}B^{-7}Li)$ is very sensitive to solvation	[154]
F—H···P	$^1J_{FH}$, $^{1h}J_{HP}$, $^{2h}J_{FP}$	E-C	Effect of the proton transfer on Js and Ks	[155]
Compound **13**	$^1J_{CN}$	DFT	CH, NH, OH tautomers of pyrazolinones	[47]
X—H···Y [X, Y = N, O, F]	$^1J_{XH}$, $^{1h}J_{HY}$, $^{2h}J_{XY}$	E-C versus SO	And monomers XY, X—H and H—Y protonated and deprotonated	[156]
Formamide [solvent X: NH_3, H_2O, FH]	$^1J_{NH}$, $^{1h}J_{HX}$, $^{2h}J_{NX}$	E-C	Effect of one or two solvent molecules on formamide Js	[157]
F—Cl···N	$^1J_{FCl}$, $^{1X}J_{ClN}$, $^{2X}J_{FN}$	E-C	SSCC across halogen bonds	[158]

(Continued)

Table 2.3 (Continued)

System	Couplings	Method	Main conclusions and comments	Reference
HXH [X from Li to Cl]	$^2J_{HH}$, $^1J_{XH}$	DFT and SO	Clearly, SOPPA/sad] yield better values than SOPPA/aug-cc-pVTZ	[159]
N–H$^+$···N	$^1J_{NH}$, $^{1h}J_{HN}$, $^{2h}J_{NN}$	E-C	Models of proton sponges	[160]
All systems	$^{nh}J_{XY}$	Review	Comprehensive data on experimental indirect scalar NMR spin–spin coupling constants across hydrogen bonds	[161]
Compound 31	$^1J_{NN}$, J_{NC}, J_{HH}	DFT	SSCC in azines are well reproduced	[16]
F–C≡C–F	$^3J_{FF}$, $^1J_{CC}$, J_{CF}	E-C versus SO	$^3J_{FF}$ strongly depends on the geometry	[162]
H$_2$O, H$_3$O$^+$	$^{2h}J_{OO}$	E-C	Water and hydronium clusters	[163]
Cyclic (FH)$_n$	$^1J_{FH}$, $^{1h}J_{HF}$, $^{2h}J_{FF}$	E-C	For $n = 2$–6, $^{1h}J_{HF}$ is always negative	[164]
Compound 8	$^1J_{LiC}$	DFT	$^1J_{LiC}$ is very sensitive to solvent effects	[32]
X=Y, X≡Y [X, Y = C, N, O]	J_{CC}, J_{CN}, J_{CO}, J_{CF}, J_{NN}, J_{NO}, J_{NF}, J_{FF}	E-C versus SO	And selected F-substituted derivatives	[165]
N–H$^+$···N	$^1J_{NH}$, $^{1h}J_{HN}$, $^{2h}J_{NN}$	DFT and SO	Statistical modeling (hybridization of both N atoms) and charge of the complex	[166]
Compound 66	$^{1h}J_{FH}$	DFT	Solvent effects	[19]
HLB=BLH [diboranes, 67 and 68; L = CO, NH$_3$, OH$_2$, PH$_3$, SH$_2$, ClH]	$^1J_{BB}$, $^1J_{BH}$, $^1J_{BL}$	E-C	$^1J_{BB}$ and $^1J_{BH}$ are always positive	[167]
Scheme 2.14		E-C	1,2-Dihydro-1,2-azaborine (69)	[168]
31 molecules	J_{CC}, J_{CN}, J_{NN}	E-C versus SO	E-C underestimates $^1J_{CH}$ by ~10 Hz	[169]

Methods: E-C = EOM-CCSD; SO = SOPPA; FTP = finite perturbation theory; DFT = B3LYP/6-311++G(d,p).

Scheme 2.11 Intramolecular hydrogen bonds in RAHB (resonance-assisted hydrogen bond) structures **62** and **63**.

Scheme 2.12 Geometry of the calculated and measured $^2J_{CH}$ coupling constants.

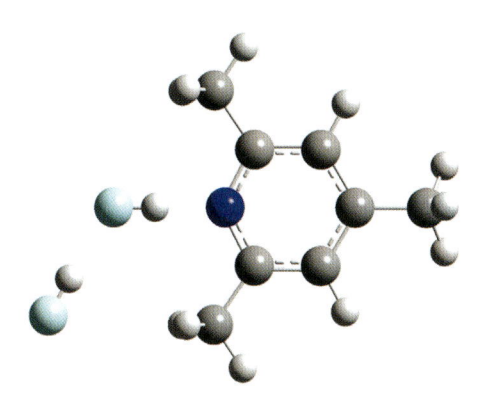

Figure 2.6 The FH···FH···collidine complex experimentally studied by Limbach and coworkers [170]. Deep blue: nitrogen atom; pale blue: fluorine atoms.

Scheme 2.13 Borazine (**65**) and F-substituted borazines.

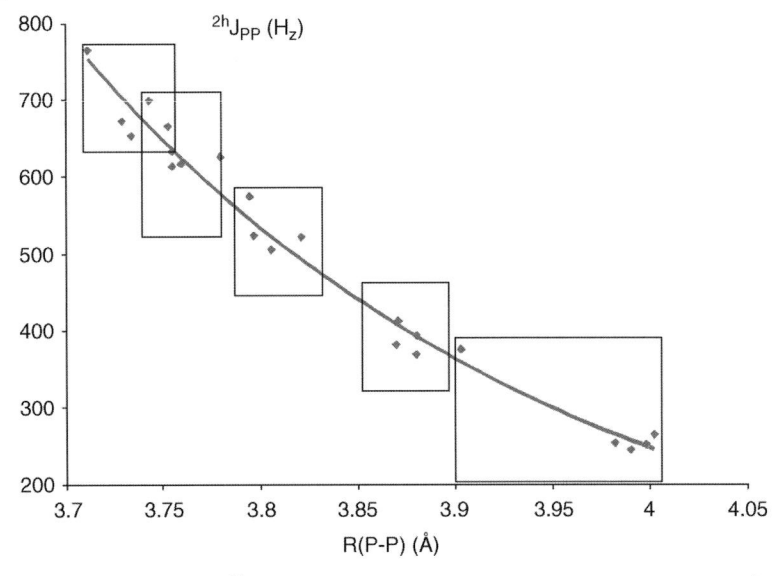

Figure 2.7 Variation of $^{2h}J_{PP}$ with P···P distance for 22 complexes with linear P–H$^+$···P hydrogen bonds. The blocks correspond to different proton donor ions.

Scheme 2.14 1,2-Dihydro-1,2-azaborine (**69**) and related molecules.

2.6
Solvent Effects

The perturbation of NMR parameters (δ and J) due to solvent effects can be modeled in computational studies in two ways: with continuum models to simulate dielectric solvent effects (DSE) [172] and with specific solute–solvent interactions (supermolecules).

a) *Continuum models: shielding constants (chemical shifts)*. Only in a few papers [14, 35, 36], we carried out GIAO/B3LYP/6-311++G(d,p) calculations on bromopyrazoles in DMSO using the PCM model. On the other hand, many authors have used continuum models to calculate shielding constants [173–176], especially important are those of Ruud and coworkers [177, 178]. A last paper by the same group reports solvent effects on nitrogen NMR shieldings in 1-methyltriazoles [179].

b) *Continuum models: spin–spin coupling constants*. We have not used PCM-type calculations for SSCC, a field where the most important contributions are those of Contreras and coworkers. Using $HN=CH_2$ as a model molecule, they studied both dielectric solvation effect and specific solute–solvent interactions (H_2O and DMSO); $^1J_{CH(anti)}$ increases monotonically with increasing dielectric constant ε [180]. These authors found that SSCC including PCM effects are in better agreement with experimental values than those without them for pyridinecarboxaldehydes [181]. A statistical analysis of their data indicates that this is true for J_{CC} (largest differences about 2 Hz) but not for J_{CH} (largest differences 9 Hz). They worked also on amides [182, 183]. Both Ruud [184] and Sauer [185] have considerably improved the performance of PCM models. For reviews on this important topic, see Refs [186, 187].

c) *Specific solute–solvent interactions: shielding constants (chemical shifts)*. The subject is very broad since it is difficult to distinguish normal molecules from solvent molecules in complexes. Others and we have calculated many supramolecules; when one of these is a classical NMR solvent (water, acetone, DMSO, etc.), they belong to this category. In the case of benzene as solvent, we have calculated the aromatic solvent-induced shifts (ASIS) with a considerable success [188]. In the case of fluorobenzamide (**66**), we have calculated the effect of water and acetone on the chemical shifts using the 1:1 and 1:2 complexes [19]. One can consider the solid state as a special case of solvation; only, instead of solvent there are other molecules in the unit cell. For comparing experimental CPMAS chemical shifts, we have reported calculations for dimers [10] and trimers [34].

d) *Specific solute–solvent interactions: spin–spin coupling constants*. When discussing quaternary ammonium salts, such as NH_4^+, it is fundamental to add water molecules forming $N-H^+ \cdots OH_2$ HBs to reproduce the experimental SSCC [146]. In the case of 2-fluorobenzamide (**66**), we have calculated its SSCC in complexes with one water molecule and one and two acetone molecules [19].

Coupling constants involving ^{7}Li are very sensitive to solvation, for instance, $^1J(^7\text{Li}-^{13}\text{C})$ [32]. To rationalize this observation, we carried out EOM-CCSD calculations using several H$_2$O and FLi molecules [154].

2.7
Conclusions

We hope that this summary of "computational NMR spectroscopy" will convince the still skeptical readers that computation of NMR properties is a powerful and reliable tool. This is so to the point that when there is a large disagreement between calculated and experimental data, the experiment should be questioned (see Table 2.4). At this moment, size and complexity of molecules are not a limitation for calculating absolute shieldings; on the other hand, some high-level calculations of coupling constants are limited by size, symmetry, and the nature of the atoms involved. For instance, a compound such as **64** with its 345 basis functions is today out of reach of EOM-CCSD calculations. In not too distant a future, all these limitations will be overcome and computational and experimental chemists will advance hand in hand. We recommend least squares fitting (regression) using GIAO/B3LYP/6-311++G(d,p) for the chemical shifts and SOPPA (without intercept) for the spin–spin coupling constants. Improving the level of the calculations (basis set, vibrational and thermal corrections, etc.) will not necessarily result in better correlations because the experimental data that interest the experimental chemist came from solution or solid state, conditions far from the ideal ones.

Table 2.4 Calculated and experimental coupling constants for hexafluorocyclotriphosphazene **64** in CDCl$_3$ at 300 K.

	J	Calcd.	Exp.
	$^2J_{PP}$	+178.5	+155.0
	$^1J_{PN}$	−3.5	−24.9
	$^3J_{PN}$	−0.8	—
	$^1J_{PF}$	−1252.0	−916.6
	$^3J_{PF}$	+6.3	+21.2
	$^2J_{NN}$	−0.8	—
	$^2J_{NF}$	+5.9	—
	$^4J_{NF}$	+0.2	—
	$^2J_{FF}$	−192.4	+50.0[a]
	$^4J_{FF(cis)}$	+0.1	−2.0
	$^4J_{FF(trans)}$	+12.9	+17.0

a) Excluding this value, Exp. = 0.735 × Calcd., n = 6, R2 = 0.998. This equation predicts a value of −137.0 Hz for $^2J_{FF}$. At 183 K in CD$_2$Cl$_2$, Kapicka found −71.4 Hz [171], in much better agreement with the calculations.

Progress in any area of physical science requires a combination of experimental and theoretical investigations. However, in some cases, a theoretical approach may assume increased importance, particularly if the experimental data are fragmentary or subject to large experimental errors. In such circumstances, if one wants to examine a molecular property with only an incomplete set of experimental data available, a reasonable strategy is to calculate that property in a systematic manner for an entire collection of related molecules and compare the available experimental values with the corresponding calculated results. If the comparison shows that theory and experiment agree, or at least are highly correlated, then the calculated values can be used as a basis for discussing the property of interest. In addition, theory often has the advantage of allowing an analysis based on partitioning of the property, which is not possible with the experimental data. Such is the situation for scalar coupling constants (spin–spin coupling constants) in NMR. In many cases, understanding the variations in related spin–spin coupling constants is difficult without knowing their components, given in the Ramsay model as the paramagnetic spin–orbit (PSO), diamagnetic spin–orbit (DSO), FC, and spin–dipole (SD) terms. It is only theory that can provide this information.

2.8
The Problem of the Error in Theoretical Calculations of Chemical Shifts and Coupling Constants

Although *sensu strictu* there is no error in a theoretical calculation, it is used to signify the difference between the calculated value and the experimental one. Concerning chemical shifts, the results of the calculations are absolute shieldings σ. Some authors calculate the reference (e.g., TMS in ^{13}C NMR) at the same level and transform the calculated σ to δ by subtraction. This is equivalent to assuming that $\delta = \sigma_{compd} - \sigma_{ref}$. We, and many others, prefer to assume that $\delta = a + b \times \sigma_{compd}$, a being close but not identical to σ_{ref} and b being close to 1 but not identical. At the GIAO/B3LYP/6-311++G(d,p) level and with the exception of some atoms (e.g., C atoms bearing Cl, Br, and S substituents), the RMS residuals of the regressions are about 3 ppm for ^{13}C and 9 ppm for ^{15}N [21] (see also Table 2.2). Concerning coupling constants, no transformation is necessary since the calculations afford directly the coupling in Hz. Naturally, calculated and experimental values should be identical for $J = 0$ Hz (or very small), thus allowing to impose that the trendline goes through the origin (no intercept). Concerning the more accessible SOPPA calculations, RMS is about 3–4 Hz [189], and introducing some dummy variables can reduce it to 2 Hz [190].

Acknowledgments

This work has been financed by the Spanish Ministerio de Ciencia e Innovación (Project No. CTQ-13129-C02-02; subprograma BQ) and Comunidad Autónoma de Madrid (Project MADRISOLAR, Ref. S-0505/PPQ/0225).

References

1 Kaupp, M., Bühl, M. and Malkin, V.G. (eds) (2004) *Calculation of NMR and EPR Parameters*, Wiley-VCH Verlag GmbH, Weinheim, p. 159.
2 Alkorta, I. and Elguero, J. (2009) *Top. Heterocycl. Chem.*, **19**, 155–202.
3 Helgaker, T., Jaszunski, M., and Ruud, K. (1999) *Chem. Rev.*, **99**, 293–352.
4 Kirchner, B., Ermakova, E., Steinebrunner, G., Dyson, A.J., and Huber, H. (1998) *Mol. Phys.*, **94**, 257–268; Dixon, A.M., Larive, C.K., Nantsis, E.A., and Carper, W.R. (1998) *J. Phys. Chem. A*, **102**, 10573–10578; Bagno, A., Casella, G., and Saielli, G. (2006) *J. Chem. Theor. Comput.*, **2**, 37–46.
5 Pisklak, M., Kossakowski, J., Perlinski, M., and Wawer, I. (2004) *J. Mol. Struct.*, **698**, 93–102.
6 Katritzky, A.R., Akhmedov, N.G., Doskocz, J., Mohapatra, P.P., Hall, C.D., and Güven, A. (2007) *Magn. Reson. Chem.*, **45**, 532–543.
7 Tousek, J., Van Miert, S., Pieters, L., Van Baelen, G., Hostyn, S., Maes, B.U.W., Lemière, G., Dommisse, R., and Marek, R. (2008) *Magn. Reson. Chem.*, **46**, 42–51.
8 Bocian, W., Jazwinski, J., and Sadlej, A. (2008) *Magn. Reson. Chem.*, **46**, 156–165.
9 Ditchfield, R. (1974) *Mol. Phys.*, **27**, 789–807; London, F. (1937) *J. Phys. Radium*, **8**, 397–409.
10 Alkorta, I. and Elguero, J. (2003) *Struct. Chem.*, **14**, 377–389.
11 Frisch, M.J., Trucks, G.W., Schlegel, H.B., Scuseria, G.E., Robb, M.A., Cheeseman, J.R., Montgomery, J.A., Jr., Vreven, T., Kudin, K.N., Burant, J.C., Millam, J.M., Iyengar, S.S., Tomasi, J., Barone, V., Mennucci, B., Cossi, M., Scalmani, G., Rega, N., Petersson, G.A., Nakatsuji, H., Hada, M., Ehara, M., Toyota, K., Fukuda, R., Hasegawa, J., Ishida, M., Nakajima, T., Honda, Y., Kitao, O., Nakai, H., Klene, M., Li, X., Knox, J.E., Hratchian, H.P., Cross, J.B., Adamo, C., Jaramillo, J., Gomperts, R., Stratmann, R.E., Yazyev, O., Austin, A.J., Cammi, R., Pomelli, C., Ochterski, J.W., Ayala, P.Y., Morokuma, K., Voth, G.A., Salvador, P., Dannenberg, J.J., Zakrzewski, V.G., Dapprich, S., Daniels, A.D., Strain, M.C., Farkas, O., Malick, D.K., Rabuck, A.D., Raghavachari, K., Foresman, J.B., Ortiz, J.V., Cui, Q., Baboul, A.G., Clifford, S., Cioslowski, J., Stefanov, B.B., Liu, G., Liashenko, A., Piskorz, P., Komaromi, I., Martin, R.L., Fox, D.J., Keith, T., Al-Laham, M.A., Peng, C.Y., Nanayakkara, A., Challacombe, M., Gill, P.M.W., Johnson, B., Chen, W., Wong, M.W., Gonzalez, C., and Pople, J.A. (2003) *Gaussian 03*, Gaussian, Inc., Pittsburgh PA.
12 Kim, H.-W., Chechla, A.A., and Kim, B. (2007) *J. Mol. Struct. THEOCHEM*, **802**, 105–110.
13 Alkorta, I. and Elguero, J. (1998) *Struct. Chem.*, **9**, 187–202.
14 Cavero, E., Giménez, R., Uriel, S., Beltrán, E., Serrano, J.L., Alkorta, I., and Elguero, J. (2008) *Cryst. Growth Des.*, **8**, 838–847.
15 Alkorta, I., Blanco, F., and Elguero, J. (2009) *J. Mol. Struct. THEOCHEM*, **896**, 92–95.
16 Silva, A.M.S., Sousa, R.M.S., Jimeno, M.L., Blanco, F., Alkorta, I., and Elguero, J. (2008) *Magn. Reson. Chem.*, **46**, 859–864.
17 Santa María, M.D., Claramunt, R.M., Herranz, F., Alkorta, I., and Elguero, J. (2009) *J. Mol. Struct.*, **920**, 323–326.
18 Santa María, M.D., Claramunt, R.M., Alkorta, I., and Elguero, J. (2009) *Magn. Reson. Chem.*, **47**, 472–477.
19 Alkorta, I., Elguero, J., Limbach, H.-H., Shenderovich, I.G., and Winkler, T. (2009) *Magn. Reson. Chem.*, **47**, 585–592.
20 Claramunt, R.M., Sanz, D., López, C., Pinilla, E., Torres, M.R., Elguero, J., Nioche, P., and Raman, C.S. (2009) *Helv. Chim. Acta*, **92**, 1952–1962.
21 Blanco, F., Alkorta, I., and Elguero, J. (2007) *Magn. Reson. Chem.*, **45**, 797–800.
22 Del Bene, J.E., Elguero, J., Alkorta, I., Yáñez, M., and Mó, O. (2007) *J. Phys. Chem. A*, **111**, 419–421.
23 Claramunt, R.M., López, C., Lott, S., Santa María, M.D., Alkorta, I., and Elguero, J. (2005) *Helv. Chim. Acta*, **88**, 1931–1942.

24 Teichert, J., Oulié, P., Jacob, K., Vendier, L., Etienne, M., Claramunt, R.M., López, C., Pérez Medina, C., Alkorta, I., and Elguero, J. (2007) *New J. Chem.*, **31**, 936–946.

25 Dardonville, C., Jimeno, M.L., Alkorta, I., and Elguero, J. (2004) *ARKIVOC*, **ii**, 206–212.

26 Frideling, A., Faure, R., Galy, J.-P., Kenz, A., Alkorta, I., and Elguero, J. (2004) *Eur. J. Med. Chem.*, **39**, 37–48.

27 Alkorta, I., Jagerovic, N., and Elguero, J. (2004) *ARKIVOC*, **iv**, 130–136.

28 Alvarez-Rua, C., García-Granda, S., Goswami, S., Mukherjee, R., Dey, S., Claramunt, R.M., Santa María, M.D., Rozas, I., Jagerovic, N., Alkorta, I., and Elguero, J. (2004) *New J. Chem.*, **28**, 700–707.

29 Dardonville, C., Jimeno, M.L., Alkorta, I., and Elguero, J. (2004) *Org. Biomol. Chem.*, **2**, 1587–1591.

30 Prakash, O., Kumar, A., Sadana, A., Prakash, R., Singh, S.P., Claramunt, R.M., Sanz, D., Alkorta, I., and Elguero, J. (2005) *Tetrahedron*, **61**, 6642–6651.

31 Pardo, C., Alkorta, I., and Elguero, J. (2006) *Tetrahedron Asymm.*, **17**, 191–198.

32 Capriati, V., Florio, S., Luisi, R., Musio, B., Alkorta, I., Blanco, F., and Elguero, J. (2008) *Struct. Chem.*, **19**, 785–792.

33 Claramunt, R.M., García, M.A., López, C., Trofimenko, S., Yap, G.P.A., Alkorta, I., and Elguero, J. (2005) *Magn. Reson. Chem.*, **43**, 89–91.

34 López, C., Claramunt, R.M., García, M.A., Pinilla, E., Torres, M.R., Alkorta, I., and Elguero, J. (2007) *Cryst. Growth Des.*, **7**, 1176–1184.

35 Trofimenko, S., Yap, G.P.A., Jove, F.A., Claramunt, R.M., García, M.A., Santa María, M.D., Alkorta, I., and Elguero, J. (2007) *Tetrahedron*, **63**, 8104–8111.

36 Sanz, D., Claramunt, R.M., Alkorta, I., Elguero, J., Thiel, W.R., and Rüffer, T. (2008) *New J. Chem.*, **32**, 2225–2232.

37 Lévai, A., Silva, A.M.S., Pinto, D.C.G.A., Cavaleiro, J.A.S., Alkorta, I., Elguero, J., and Jekö, J. (2004) *Eur. J. Org. Chem.*, 4672–4679.

38 Lévai, A., Silva, A.M.S., Cavaleiro, J.A.S., Alkorta, I., Elguero, J., and Jekö, J. (2006) *Eur. J. Org. Chem.*, 2825–2832.

39 Lévai, A., Silva, A.M.S., Cavaleiro, J.A.S., Alkorta, I., Elguero, J., and Jekö, J. (2007) *Aust. J. Chem.*, **60**, 905–914.

40 Claramunt, R.M., Santa María, M.D., Sanz, D., Alkorta, I., and Elguero, J. (2006) *Magn. Reson. Chem.*, **44**, 566–570.

41 Santa María, M.D., Claramunt, R.M., Alkorta, I., and Elguero, J. (2007) *Dalton Trans.*, 3995–3999.

42 Alkorta, I., Alvarado, M., Elguero, J., García-Granda, S., Goya, P., Jimeno, M.L., and Menéndez-Taboada, L. (2009) *Eur. J. Med. Chem.*, **44**, 1864–1869.

43 Alkorta, I., Alvarado, M., Elguero, J., García-Granda, S., Goya, P., Torre-Fernández, L., and Menéndez-Taboada, L. (2009) *J. Mol. Struct.*, **920**, 82–89.

44 Yap, G.P.A., Alkorta, I., Jagerovic, N., and Elguero, J. (2004) *Aust. J. Chem.*, **57**, 1103–1108.

45 Alkorta, I., Elguero, J., Fruchier, A., Jagerovic, N., and Yap, G.P.A. (2004) *J. Mol. Struct.*, **689**, 251–254.

46 Yap, G.P.A., Alkorta, I., Elguero, J., and Jagerovic, N. (2004) *Spectroscopy*, **18**, 605–611.

47 Sanz, D., Claramunt, R.M., Alkorta, I., and Elguero, J. (2007) *Struct. Chem.*, **18**, 703–708.

48 Holzer, W., Claramunt, R.M., López, C., Alkorta, I., and Elguero, J. (2008) *Solid State Nucl. Magn. Reson.*, **34**, 68–76.

49 Alkorta, I., Elguero, J., Jagerovic, N., Fruchier, A., and Yap, G.P.A. (2004) *J. Heterocycl. Chem.*, **41**, 285–289.

50 Yap, G.P.A., Jové, F.A., Claramunt, R.M., Sanz, D., Alkorta, I., and Elguero, J. (2005) *Aust. J. Chem.*, **58**, 817–822.

51 Claramunt, R.M., López, C., Alkorta, I., Elguero, J., Yang, R., and Schulman, S. (2004) *Magn. Reson. Chem.*, **42**, 712–714.

52 García, M.A., Claramunt, R.M., Solcan, T., Milata, V., Alkorta, I., and Elguero, J. (2009) *Magn. Reson. Chem.*, **47**, 100–104.

53 Sanz, D., Claramunt, R.M., Singh, S.P., Kumar, V., Aggarwal, R., Elguero, J., and Alkorta, I. (2005) *Magn. Reson. Chem.*, **43**, 1040–1043.

54 Sanz, D., Claramunt, R.M., Saini, A., Kumar, V., Aggarwal, R., Singh, S.P.,

Alkorta, I., and Elguero, J. (2007) *Magn. Reson. Chem.*, **45**, 513–517.

55 Claramunt, R.M., López, C., García, M.A., Otero, M.D., Torres, M.R., Pinilla, E., Alarcón, S.H., Alkorta, I., and Elguero, J. (2001) *New J. Chem.*, **25**, 1061–1068.

56 Wan, L., Alkorta, I., Elguero, J., Sun, J., and Zheng, W. (2007) *Tetrahedron*, **63**, 9129–9133.

57 Pi, C., Elguero, J., Wan, L., Alkorta, I., Zheng, W., Weng, L., Chen, Z., and Wu, L. (2009) *Chem. Eur. J.*, **15**, 6581–6585.

58 Santa María, D., Claramunt, R.M., Alkorta, I., and Elguero, J. (2009) *Magn. Reson. Chem.*, **47**, 472–477.

59 Rodríguez-Morgade, M.S., Claessens, C.G., Medina, A., González-Rodríguez, D., Gutiérrez-Pueblam, E., Monge, A., Alkorta, I., Elguero, J., and Torres, T. (2008) *Chem. Eur. J.*, **14**, 1342–1350.

60 García-Frutos, E.M., Gómez-Lor, B., Monge, A., Gutiérrez-Puebla, E., Alkorta, I., and Elguero, J. (2008) *Chem. Eur. J.*, **14**, 8555–8561.

61 Alkorta, I. and Elguero, J. (2004) *Magn. Reson. Chem.*, **42**, 955–961.

62 Langa, F., de la Cruz, P., Delgado, J.L., Haley, M.M., Shirtcliff, L., Alkorta, I., and Elguero, J. (2004) *J. Mol. Struct.*, **699**, 17–21.

63 Alkorta, I., Elguero, J., Eckert-Maksic, M., and Maksic, Z.B. (2004) *Tetrahedron*, **60**, 2259–2265.

64 Barros, A.I.R.N.A., Silva, A.M.S., Alkorta, I., and Elguero, J. (2004) *Tetrahedron*, **60**, 6513–6521.

65 Claramunt, R.M., Sanz, D., Alkorta, I., Elguero, J., Foces-Foces, C., and Llamas-Saiz, A.L. (2001) *J. Heterocycl. Chem.*, **38**, 443–450.

66 Ballano, G., Jiménez, A.I., Cativiela, C., Claramunt, R.M., Sanz, D., Alkorta, I., and Elguero, J. (2008) *J. Org. Chem.*, **73**, 8575–8578.

67 Reviriego, F., Alkorta, I., and Elguero, J. (2008) *J. Mol. Struct.*, **891**, 325–328.

68 Katritzky, A.R., Akhmedov, N.G., Doskocz, J., Hall, C.D., Akhmedova, R.G., and Majumder, S. (2007) *Magn. Reson. Chem.*, **45**, 5–23.

69 Bachrach, S.M. (2007) *Computational Organic Chemistry*, Wiley–Interscience, Hoboken, NJ.

70 Schleyer, P.v.R., Maerker, C., Dransfeld, A., Jiao, H., and Hommes, N.J.R.v.E. (1996) *J. Am. Chem. Soc.*, **118**, 6317–6318.

71 Schleyer, P.v.R., Jiao, H., Hommes, N.J.R.v.E., Malkin, V.G., and Malkina, O.L. (1997) *J. Am. Chem. Soc.*, **119**, 12669.

72 Corminboeuf, C., Heine, T., Seifert, G., Schleyer, P.v.R., and Weber, J. (2004) *Phys. Chem. Chem. Phys.*, **6**, 273–276.

73 Lazzeretti, P. (2000) *Prog. NMR Spectrosc.*, **36**, 1–88.

74 Chen, Z., Wannere, C.S., Corminboeuf, C., Puchta, R., and Schleyer, P.v.R. (2005) *Chem. Rev.*, **105**, 3842–3888.

75 Cyranski, M.K., Krygowski, T.M., Katritzky, A.R., and Schleyer, P.v.R. (2002) *J. Org. Chem.*, **67**, 1333–1338.

76 Balaban, A.T., Oniciu, D.C., and Katritzky, A.R. (2004) *Chem. Rev.*, **104**, 2777–2812.

77 Alonso, M. and Herradón, B. (2007) *Chem. Eur. J.*, **13**, 3913–3923.

78 Raczynska, E.D., Kosinska, W., Osmialowski, B., and Gawinecki, R. (2005) *Chem. Rev.*, **105**, 3561–3612.

79 Sobczyk, L., Grabowski, S.J., and Krygowski, T.M. (2005) *Chem. Rev.*, **105**, 3513–3560.

80 Vianello, R. and Maksic, Z.B. (2005) *Mol. Phys.*, **103**, 209–219.

81 Nyulászi, L. and Benkö, Z. (2009) *Top. Heterocycl. Chem.*, **19**, 27–81;Krygowski, T.M. and Cyranski, M.K. (eds) (2009) *Aromaticity in Heterocyclic Compounds*, Springer, Berlin.

82 Vessally, E. (2008) *J. Struct. Chem.*, **49**, 979–985.

83 Blanco, F., O'Donovan, D.H., Alkorta, I., and Elguero, J. (2008) *Struct. Chem.*, **19**, 339–352.

84 Kleinpeter, E. and Koch, A. (2009) *Tetrahedron*, **65**, 5350–5360 and references therein.

85 Kleinpeter, E. and Koch, A. (2008) *J. Mol. Struct. THEOCHEM*, **857**, 89–94.

86 Alkorta, I. and Elguero, J. (1999) *New J. Chem.*, **23**, 951–954.

87 Alkorta, I., Rozas, I., and Elguero, J. (2001) *Tetrahedron*, **57**, 6043–6049.

88 Claramunt, R.M., López, C., García, M.Á., Denisov, G.S., Alkorta, I., and Elguero, J. (2003) *New J. Chem.*, **27**, 734–742.

89 Trifonov, R.E., Alkorta, I., Ostrovskii, V.A., and Elguero, J. (2004) *J. Mol. Struct. THEOCHEM*, **668**, 123–132.

90 Alkorta, I., Elguero, J., and Liebman, J.F. (2006) *Struct. Chem.*, **17**, 439–444.

91 Alkorta, I. and Elguero, J. (2006) *Tetrahedron*, **62**, 8683–8686.

92 Elguero, J., Claramunt, R.M., and Summers, A.J.H. (1978) *Adv. Heterocycl. Chem.*, **22**, 183–320.

93 Alkorta, I., Blanco, F., and Elguero, J. (2008) *J. Mol. Struct. THEOCHEM*, **851**, 75–83.

94 Alkorta, I., Blanco, F., and Elguero, J. (2008) *J. Phys. Chem. A*, **112**, 1817–1822.

95 Alkorta, I., Blanco, F., and Elguero, J. (2008) *Tetrahedron*, **64**, 3826–3836.

96 Alkorta, I., Blanco, F., and Elguero, J. (2009) *Tetrahedron*, **65**, 5760–5766.

97 Blanco, F., Alkorta, I., and Elguero, J. (2009) *Tetrahedron*, **65**, 6244–6250.

98 Wu, J.I., Pühlhofer, F.G., von, P., Schleyer, R., Puchta, R., Kiran, B., Mauksch, M., van Eikema Hommes, N.J.R., Alkorta, I., and Elguero, J. (2009) *J. Phys. Chem. A*, **113**, 6789–6794.

99 Pietrzak, M., Wehling, J.P., Kong, S., Tolstoy, P.M., Shenderovich, I.G., López, C., Claramunt, R.M., Elguero, J., Denisov, G.S., and Limbach, H.-H. (2010) *Chem. Eur. J.*, **16**, 1679–1690.

100 Dickson, R.M. and Ziegler, T. (1996) *J. Phys. Chem.*, **100**, 5286–5290.

101 Perera, S.A. and Bartlett, R.J. (2000) *J. Am. Chem. Soc.*, **122**, 1231–1232.

102 Del Bene, J.E., Perera, S.A., Bartlett, R.J., Alkorta, I., and Elguero, J. (2000) *J. Phys. Chem. A*, **104**, 7165–7166.

103 Del Bene, J.E., Jordan, M.J.T., Perera, S.A., and Bartlett, R.J. (2001) *J. Phys. Chem. A*, **105**, 8399–8402.

104 Kupka, T. (2009) *Magn. Reson. Chem.*, **47**, 210–221.

105 Schafer, A., Horn, H., and Alhrichs, R. (1992) *J. Chem. Phys.*, **97**, 2571–2577.

106 Del Bene, J.E., Elguero, J., Alkorta, I., Yáñez, M., and Mó, O. (2006) *J. Phys. Chem. A*, **110**, 9959–9966.

107 Dunning, T.H., Jr. (1989) *J. Chem. Phys.*, **90**, 1007–1023.

108 Kendall, R.A., Dunning, T.H., Jr., and Harrison, R. (1992) *J. Chem. Phys.*, **96**, 6796–6806.

109 Woon, D.E. and Dunning, T.H., Jr. (1995) *J. Chem. Phys.*, **103**, 4572–4585.

110 Pecul, M., Dodziuk, H., Jaszunski, M., Lukin, O., and Leszczynski, J. (2001) *Phys. Chem. Chem. Phys.*, **3**, 1986–1991.

111 Wu, A., Cremer, D., Auer, A.A., and Gauss, J. (2002) *J. Phys. Chem. A*, **106**, 657–667.

112 Barone, V., Peralta, J.E., Contreras, R.H., and Snyder, J.P. (2002) *J. Phys. Chem. A*, **106**, 5607–5612.

113 Kupka, T. (2009) *Magn. Reson. Chem.*, **47**, 210–221.

114 Sadlej, A.J. (1988) *Collect. Czech. Chem. Commun.*, **53**, 1995–2015.

115 Peralta, J.E., Scuseria, G.E., Cheeseman, J.R., and Frisch, M.J. (2003) *Chem. Phys. Lett.*, **375**, 452–458.

116 Duval, E. and Koide, S. (1964) *Phys. Lett.*, **8**, 314–315.

117 Soncini, A. and Lazzeretti, P. (2005) *Chem. Phys. Lett.*, **409**, 177–186.

118 McConnell, H.M. (1955) *J. Chem. Phys.*, **23**, 2454–2455.

119 Alexander, S. (1961) *J. Chem. Phys.*, **34**, 106–117.

120 Bothner-By, A.A. and Harris, R.K. (1965) *J. Am. Chem. Soc.*, **87**, 3451–3455.

121 Hecht, H.G. (1967) *J. Phys. Chem.*, **71**, 1761–1764.

122 Barfield, M. and Karplus, M. (1969) *J. Am. Chem. Soc.*, **91**, 1–10.

123 Malkina, O.L. (2004) Chapter 19: Interpretation of indirect nuclear spin–spin coupling constants, in *Calculation of NMR and EPR Parameters* (eds M. Kaupp, M. Bühl, and V.G. Malkin), Wiley-VCH Verlag GmbH, Weinheim, p. 307.

124 Del Bene, J.E. and Elguero, J. (2003) *Chem. Phys. Lett.*, **382**, 100–105.

125 Del Bene, J.E., Elguero, J., Alkorta, I., Yáñez, M., and Mó, O. (2004) *J. Chem. Phys.*, **120**, 3237–3243.

126 Del Bene, J.E., Elguero, J., and Alkorta, I. (2004) *J. Phys. Chem. A*, **108**, 3662–3667.

127 Alkorta, I., Elguero, J., Mó, O., Yáñez, M., and Del Bene, J.E. (2004) *Mol. Phys.*, **102**, 2563–2574.

128 Del Bene, J.E. and Elguero, J. (2004) *Magn. Reson. Chem.*, **42**, 421–423.

129 Del Bene, J.E. and Elguero, J. (2004) *J. Am. Chem. Soc.*, **126**, 15624–15631.

130 Del Bene, J.E., Elguero, J., Alkorta, I., Mó, O., and Yáñez, M. (2005) *J. Phys. Chem. A*, **109**, 2350–2355.

131 Del Bene, J.E. and Elguero, J. (2006) Chapter 5: Predicting and understanding the signs of one- and two-bond spin–spin coupling constants across X–H–Y hydrogen bonds, in *Computational Chemistry: Reviews of Current Trends* (ed. J. Leszczynski), World Scientific, Singapore, pp. 229–264.

132 Krivdin, L.B. and Contreras, R.H. (2007) *Annu. Rep. NMR Spectrosc.*, **61**, 133–245.

133 Del Bene, J.E. and Elguero, J. (2004) *J. Phys. Chem. A*, **108**, 11762–11767.

134 Gilli, G., Bellucci, F., Ferretti, V., and Bertolasi, V. (1989) *J. Am. Chem. Soc.*, **111**, 1023–1028.

135 Holzer, W., Kautsch, C., Laggner, C., Claramunt, R.M., Pérez-Torralba, M., Alkorta, I., and Elguero, J. (2004) *Tetrahedron*, **60**, 6791–6805.

136 Alkorta, I. and Elguero, J. (2004) *Theor. Chem. Acc.*, **111**, 31–35.

137 Alkorta, I., Elguero, J., Mó, O., Yáñez, M., and Del Bene, J.E. (2005) *Chem. Phys. Lett.*, **411**, 411–415.

138 Alkorta, I., Elguero, J., and Del Bene, J.E. (2005) *Chem. Phys. Lett.*, **412**, 97–100.

139 Provasi, P.F., Aucar, G.A., Sánchez, M., Alkorta, I., Elguero, J., and Sauer, S.P.A. (2005) *J. Phys. Chem. A*, **109**, 6555–6564.

140 Del Bene, J.E. and Elguero, J. (2005) *J. Phys. Chem. A*, **109**, 10753–10758.

141 Del Bene, J.E. and Elguero, J. (2005) *J. Phys. Chem. A*, **109**, 10759–10769.

142 Fruchier, A., Vicente, V., Alkorta, I., and Elguero, J. (2005) *Magn. Reson. Chem.*, **43**, 471–474.

143 Claramunt, R.M., Sanz, D., Alkorta, I., and Elguero, J. (2005) *Magn. Reson. Chem.*, **43**, 985–991.

144 Del Bene, J.E. and Elguero, J. (2006) *J. Phys. Chem. A*, **110**, 7496–7502.

145 Del Bene, J.E. and Elguero, J. (2006) *J. Phys. Chem. A*, **110**, 12543–12545.

146 Jimeno, M.-L., Alkorta, I., Elguero, J., and Del Bene, J.E. (2006) *Magn. Reson. Chem.*, **44**, 698–707.

147 Del Bene, J.E. and Elguero, J. (2006) *Magn. Reson. Chem.*, **44**, 784–789.

148 Del Bene, J.E., Elguero, J., Alkorta, I., Yáñez, M., and Mó, O. (2007) *J. Chem. Theor. Comput.*, **3**, 549–556.

149 Jimeno, M.-L., Alkorta, I., and Elguero, J. (2007) *J. Mol. Struct.*, **837**, 147–152.

150 Del Bene, J.E. and Elguero, J. (2007) *J. Phys. Chem. A*, **111**, 2517–2526.

151 Del Bene, J.E., Elguero, J., and Alkorta, I. (2007) *J. Phys. Chem. A*, **111**, 3416–3422.

152 Del Bene, J.E. and Elguero, J. (2007) *J. Phys. Chem. A*, **111**, 6443–6448.

153 Alkorta, I., Del Bene, J.E., and Elguero, J. (2007) *J. Phys. Chem. A*, **111**, 9924–9930.

154 Del Bene, J.E., and Elguero, J. (2007) *Magn. Reson. Chem.*, **45**, 484–487.

155 Del Bene, J.E. and Elguero, J. (2007) *Magn. Reson. Chem.*, **45**, 714–719.

156 Del Bene, J.E., Alkorta, I., and Elguero, J. (2008) *J. Chem. Theor. Comput.*, **4**, 967–973.

157 Del Bene, J.E., Alkorta, I., and Elguero, J. (2008) *J. Phys. Chem. A*, **112**, 6338–6343.

158 Del Bene, J.E., Alkorta, I., and Elguero, J. (2008) *J. Phys. Chem. A*, **112**, 7925–7929.

159 Alkorta, I., Provasi, P.F., Aucar, G.A., and Elguero, J. (2008) *Magn. Reson. Chem.*, **46**, 356–361.

160 Del Bene, J.E., Alkorta, I., and Elguero, J. (2008) *Magn. Reson. Chem.*, **46**, 457–463.

161 Alkorta, I., Elguero, J., and Denisov, G.S. (2008) *Magn. Reson. Chem.*, **46**, 599–624.

162 Del Bene, J.E., Provasi, P.F., Alkorta, I., and Elguero, J. (2008) *Magn. Reson. Chem.*, **46**, 1003–1006.

163 Del Bene, J.E. and Elguero, J. (2008) *Mol. Phys.*, **106**, 1461–1471.

164 Del Bene, J.E. and Elguero, J. (2008) *Solid State NMR*, **34**, 86–92.

165 Del Bene, J.E., Alkorta, I., and Elguero, J. (2009) *J. Chem. Theor. Comput.*, **5**, 208–216.

166 Alkorta, I., Blanco, F., and Elguero, J. (2009) *Magn. Reson. Chem.*, **47**, 249–256.

167 Alkorta, I., Del Bene, J.E., Elguero, J., Mó, O., and Yáñez, M. (2009) *Theor. Chem. Acc.*, **124**, 187–195.

168 Del Bene, J.E., Yáñez, M., Alkorta, I., and Elguero, J. (2009) *J. Chem. Theor. Comput.*, **5**, 2239–2247.

169 Del Bene, J.E., Alkorta, I., and Elguero, J. (2009) *J. Phys. Chem. A*, **113**, 12411–12420.

170 Shenderovich, I.G., Tolstoy, P.M., Golubev, N.S., Smirnov, N.S., Denisov, G.S., and Limbach, H.-H. (2003) *J. Am. Chem. Soc.*, **125**, 11710–11720.

171 Kapicka, L., Dastych, D., Richterová, V., Alberti, M., and Kubácek, P. (2005) *Magn. Reson. Chem.*, **43**, 294–301.

172 Mennucci, B. and Cammi, R. (eds) (2007) *Continuum Solvation Models in Chemical Physics: From Theory to Applications*, John Wiley & Sons, Inc., Chichester.

173 Jaszunski, M., Mikkelsen, K.V., Rizzo, A., and Witanowski, M. (2000) *J. Phys. Chem. A*, **104**, 1466–1473.

174 Manalo, M.N., de Dios, A.C., and Cammi, R. (2000) *J. Phys. Chem. A*, **104**, 9600–9604.

175 Mennucci, B., Martínez, J.M., and Tomasi, J. (2001) *J. Phys. Chem. A*, **105**, 7287–7296.

176 Kleinpeter, E., Koch, A., and Shainyan, B.A. (2008) *J. Mol. Struct. THEOCHEM*, **863**, 117–122.

177 Aidas, K., Møgelhøj, A., Kjaer, H., Nielsen, C.B., and Mikkelsen, K.V. (2007) *J. Phys. Chem. A*, **111**, 4199–4210.

178 Kongsted, J. and Ruud, K. (2008) *Chem. Phys. Lett.*, **451**, 226–232.

179 Møgelhøj, A., Aidas, K., Mikkelsen, K.V., and Kongsted, J. (2008) *Chem. Phys. Lett.*, **460**, 129–136.

180 Zaccari, D.G., Snyder, J.P., Peralta, J.E., Taurian, O.E., Contreras, R.H., and Barone, V. (2002) *Mol. Phys.*, **100**, 705–715.

181 Taurian, O.E., De Kowalewski, D.G., Pérez, J.E., and Contreras, R.H. (2005) *J. Mol. Struct.*, **754**, 1–9.

182 Pedersoli, S., Tormena, C.E., dos Santos, F.P., Contreras, R.H., and Rittner, R. (2008) *J. Mol. Struct.*, **891**, 508–513.

183 Pedersoli, S., dos Santos, F.P., Rittner, R., Contreras, R.H., and Tormena, C.E. (2008) *Magn. Reson. Chem.*, **46**, 202–205.

184 Ruud, K., Frediani, L., Cammi, R., and Mennucci, B. (2003) *Int. J. Mol. Sci.*, **4**, 119–134.

185 Møgelhøj, A., Aidas, K., Mikkelsen, K.V., Sauer, S.P.A., and Kongsted, J. (2009) *J. Chem. Phys.*, **130**, 134508 1–12.

186 Krivdin, L.B. and Contreras, R.H. (2007) *Annu. Rep. NMR Spectrosc.*, **61**, 133–245.

187 Helgaker, T., Jaszunski, M., and Pecul, M. (2008) *Prog. NMR Spectrosc.*, **53**, 249–268.

188 Alkorta, I. and Elguero, J. (1998) *New J. Chem.*, 381–385.

189 Alkorta, I., Blanco, F., Del Bene, J.E., Elguero, J., Hernández-Folgado, L., and Jimeno, M.L. (2010) *Magn. Reson. Chem.*, **48**, 68–73.

190 Alkorta, I., Blanco, F., and Elguero, J. (2010) *J. Mol. Struct.*, **964**, 119–125.

3
Calculation of Magnetic Tensors and EPR Spectra for Free Radicals in Different Environments

Paola Cimino, Frank Neese, and Vincenco Barone

3.1
Introduction

The tools needed by EPR spectroscopists are from the world of quantum mechanics (QMs), as far as the parameters of the spin Hamiltonian are concerned, and molecular dynamics (MDs) and statistical thermodynamics for the simulation of spectral line shapes. As a matter of fact, data reduction in experimental EPR spectroscopy is achieved using a powerful device: the phenomenological spin Hamiltonian (SH). Virtually every EPR spectroscopist is familiar with fitting the relatively small number of parameters that enter the SH through least square or eyeball procedures to the experimental spectra. The outcome of the analysis is a compact description of the information content of the EPR spectrum: numerical values for the SH parameters. These are the elements (principal values) of the g-matrix, the hyperfine couplings of various magnetic nuclei, and perhaps the nuclear quadrupole couplings, or – if applicable – the zero-field splitting tensor. Usually only the principal values of the SH parameters are obtained. However, single-crystal and modern pulse experiments often yield more precise information with respect to the orientation of the various magnetic coupling tensors in the molecular coordinate frame. The other challenging experimental–theoretical match, EPR spectral shape versus probe dynamics, has a long history too. The two limits of essentially fixed molecular orientation as in a crystal, and of rapidly rotating probes in solutions of low viscosity (Redfield limit) [1], have been overcome by methods based on the stochastic Liouville equation (SLE), allowing the simulation of spectra in any régime of motion and in any type of orienting potential [2]. The ongoing integration of the above two aspects, namely, improved QM methods for the calculation of magnetic tensors and effective implementations of SLE approaches for increasing numbers of degrees of freedom, paves the route toward quantitative evaluations of EPR spectra in different phases and large temperature intervals starting from the chemical formula of the radical and the physical parameters of the solvent.

In the following sections, we will try to sketch the building blocks of an integrated computational approach [3] to the EPR spectra of organic free radicals in solution and

Computational Spectroscopy: Methods, Experiments and Applications. Edited by Jörg Grunenberg
Copyright © 2010 WILEY-VCH Verlag GmbH & Co. KGaA, Weinheim
ISBN: 978-3-527-32649-5

to illustrate the key issues of its application. Besides presenting the main framework of the proposed general model, special attention will be paid to the computation of magnetic parameters, whereas the problem of line shapes will be only briefly illustrated in the last part of the chapter. The selected examples will show that last-generation models rooted in the density functional theory (DFT) provide an accurate description of molecular structure and values of the magnetic parameters in quantitative agreement with experiments. Next, we will see that a suitable theoretical treatment of solvent effects on the magnetic parameters is able to give full account of the bulk and specific interaction. In particular, the last-generation continuum models perform a remarkable job in reproducing nonspecific solvent effects, whereas in the presence of specific interactions (e.g., solute–solvent H-bonds), they have to be integrated by explicit inclusion of some solvent molecules strongly and specifically interacting with the solute. The resulting discrete/continuum description represents a very versatile tool that can be adapted to different structural and spectroscopic situations. It is noteworthy that recent developments of classical and *ab initio* dynamics approaches enforcing proper boundary conditions permit extension of the same general approach from static to dynamic situations, thus allowing to take into proper account the averaging effects issuing from solute vibrations and solvent fluctuations. As mentioned above, longer timescale dynamical effects determining line shapes require a different approach, whose integration in a consistent general framework is under active development.

Clearly, the task at hand is a very large one and the chapter cannot serve as a substitute for a textbook in theoretical chemistry. Therefore, some familiarity with the concepts of molecular quantum mechanics is assumed. Many books on the theoretical background are recommended for further reading [4]. We will also not try to give an extensive coverage of the literature as this has been done several times earlier [5]. Rather, we will try to provide a clear description of the necessary theoretical apparatus paving the route from reliable SH parameters to complete spectral shapes and hint at how to employ the methodology in practical applications. Most of the discussion as well as the results makes reference to actual implementations in modern electronic structure program packages, such as ORCA [6] and Gaussian 03 [7], that offer advanced tools for the prediction of SH parameters.

3.2
The General Model

The calculation of ESR observables can be in principle based on a "complete" Hamiltonian $\hat{H}(\{\mathbf{r}_i\}, \{\mathbf{R}_k\}, \{\mathbf{q}_\alpha\})$, including electronic $\{\mathbf{r}_i\}$ and nuclear $\{\mathbf{R}_k\}$ coordinates of the paramagnetic probe together with solvent coordinates $\{\mathbf{q}_\alpha\}$:

$$\hat{H}(\{\mathbf{r}_i\}, \{\mathbf{R}_k\}, \{\mathbf{q}_\alpha\}) = \hat{H}_{\text{probe}}(\{\mathbf{r}_i\}, \{\mathbf{R}_k\})$$
$$+ \hat{H}_{\text{probe-solvent}}(\{\mathbf{r}_i\}, \{\mathbf{R}_k\}, \{\mathbf{q}_\alpha\}) + \hat{H}_{\text{solvent}}(\{\mathbf{q}_\alpha\})$$

(3.1)

3.2 The General Model

Any spectroscopic observable can then be linked to the density matrix $\hat{\varrho}(\{\mathbf{r}_i\}, \{\mathbf{R}_k\}, \{\mathbf{q}_\alpha\}, t)$ governed by the Liouville equation

$$\frac{\partial}{\partial t}\hat{\varrho}\{\mathbf{r}_i\}, \{\mathbf{R}_k\}, \{\mathbf{q}_\alpha\}, t) = -i[\hat{H}(\{\mathbf{r}_i\}, \{\mathbf{R}_k\}, \{\mathbf{q}_\alpha\}), \hat{\varrho}(\{\mathbf{r}_i\}, \{\mathbf{R}_k\}, \{\mathbf{q}_\alpha\}, t)]$$

$$= -\hat{\mathcal{L}}(\{\mathbf{r}_i\}, \{\mathbf{R}_k\}, \{\mathbf{q}_\alpha\})\hat{\varrho}(\{\mathbf{r}_i\}, \{\mathbf{R}_k\}, \{\mathbf{q}_\alpha\}, t) \quad (3.2)$$

Solving Equation 3.2 as a function of time would allow, in principle, a direct evaluation of $\hat{\varrho}(\{\mathbf{r}_i\}, \{\mathbf{R}_k\}, \{\mathbf{q}_\alpha\}, t)$ and hence calculation of any molecular property. However, the diverse timescales characterizing different sets of coordinates allow the introduction of a number of generalized adiabatic approximations. In particular, the nuclear coordinates $\mathbf{R} \equiv \{\mathbf{R}_k\}$ can be separated into fast vibrational coordinates $\mathbf{R}_{\text{fast}}$ and slow probe coordinates (e.g., overall probe rotations and, if required, large amplitude intramolecular degrees of freedom) $\mathbf{R}_{\text{slow}}$, relaxing at least in a picosecond timescale. Then the probe Hamiltonian is averaged on (i) femtosecond and subpicosecond dynamics, pertaining to probe electronic coordinates, and (ii) picosecond dynamics, pertaining to fast intraprobe degrees of freedom. The averaging on the electron coordinates is the usual implicit procedure for obtaining a spin Hamiltonian from the complete electronic Hamiltonian of the probe. In the frame of Born–Oppenheimer approximation, the averaging on the picosecond dynamics of nuclear coordinates allows to introduce in the calculation of magnetic parameters the effect of the vibrational motions, which can be very relevant in some cases [8]. We end up with an averaged magnetic Hamiltonian $\hat{H}(\mathbf{R}_{\text{slow}}, \{\mathbf{q}_\alpha\})$.

$$\hat{H}(\mathbf{R}_{\text{slow}}, \{\mathbf{q}_\alpha\}) = \hat{H}_{\text{probe}}(\mathbf{R}_{\text{slow}}, \{\mathbf{q}_\alpha\}) + \hat{H}_{\text{probe-solvent}}(\mathbf{R}_{\text{slow}}, \{\mathbf{q}_\alpha\})$$
$$+ \hat{H}_{\text{solvent}}(\{\mathbf{q}_\alpha\}) \quad (3.3)$$

The last two terms do not affect directly the magnetic properties and account for probe–solvent [$\hat{H}_{\text{probe-solvent}}(\mathbf{R}_{\text{slow}}, \{\mathbf{q}_\alpha\})$] and solvent–solvent [$\hat{H}_{\text{solvent}}(\{\mathbf{q}_\alpha\})$] interactions. An explicit dependence is left in the magnetic tensor definition from slow probe coordinates (e.g., geometrical dependence upon rotation) and solvent coordinates. The averaged density matrix becomes $\hat{\varrho}(\mathbf{R}_{\text{slow}}, \{\mathbf{q}_\alpha\}, t) = \langle \hat{\varrho}(\{\mathbf{r}_i\}, \{\mathbf{R}_k\}, \{\mathbf{q}_\alpha\}, t)\rangle_{\{\mathbf{r}_i\}, \mathbf{R}_{\text{fast}}}$ and the corresponding Liouville equation, in the hypothesis of no residual dynamic effect of averaging with respect to subpicosecond processes, can be simply written as in Equation 3.2 with $\hat{H}(\mathbf{R}_{\text{slow}}, \{\mathbf{q}_\alpha\})$ instead of $\hat{H}(\{\mathbf{r}_i\}, \{\mathbf{R}_k\}, \{\mathbf{q}_\alpha\})$.

Finally, the dependence upon solvent or bath coordinates can be treated at a classical mechanical level, either by solving explicitly the Newtonian dynamics of the explicit set $\{\mathbf{q}_\alpha\}$ or by adopting the standard statistical thermodynamics arguments leading to an effective averaging of the density matrix with respect to solvent variables $\hat{\varrho}(\mathbf{R}_{\text{slow}}, t) = \langle\hat{\varrho}(\mathbf{R}_{\text{slow}}, \{\mathbf{q}_\alpha\}, t)\rangle_{\{\mathbf{q}_\alpha\}}$. One of the most effective ways of dealing with the modified time evolution equation for $\hat{\varrho}(\mathbf{R}_{\text{slow}}, t)$ is represented by the SLE, that is, by the direct inclusion of motional dynamics in the form of stochastic (Fokker–Planck/diffusive) operators in the Liouvillean governing the time evolution of the system:

$$\frac{\partial}{\partial t}\hat{\varrho}(\mathbf{R}_{\text{slow}},t) = -i\left[\hat{H}(\mathbf{R}_{\text{slow}}),\hat{\varrho}(\mathbf{R}_{\text{slow}},t)\right] - \hat{\Gamma}\hat{\varrho}(\mathbf{R}_{\text{slow}},t) = -\hat{\mathcal{L}}(\mathbf{R}_{\text{slow}})\hat{\varrho}(\mathbf{R}_{\text{slow}},t) \quad (3.4)$$

The effective Hamiltonian, averaged with respect to the solvent coordinates, is made up of probe magnetic tensors averaged with respect to fast intramolecular motions and solvent coordinates, while $\hat{\Gamma}$ is the stochastic operator modeling the dependence of the reduced density matrix on relaxation processes described by stochastic coordinates $\mathbf{R}_{\text{slow}}$.

This is a general scheme that allows additional considerations and further approximations. First, the average with respect to picosecond dynamic processes is carried on, in practice, together with the average with respect to solvent coordinates to allow the QM evaluation of magnetic tensors corrected for solvent effects and for fast vibrational and solvent librational motions. The effective treatment of these aspects is the core of this chapter.

Dynamics on longer timescales determines spectral line shapes and requires more "coarse-grained" models rooted into a stochastic approach. For semirigid systems, the relevant set of stochastic coordinates can be restricted to the set of orientational coordinates $\mathbf{R}_{\text{slow}} \equiv \Omega$, which can be described, in turn, in terms of a simple formulation for a diffusive rotator, characterized by a diffusion tensor $\mathbf{D}$ [9], that is,

$$\hat{\Gamma} = \hat{\mathbf{J}}(\Omega) \cdot \mathbf{D} \cdot \hat{\mathbf{J}}(\Omega) \quad (3.5)$$

where $\hat{\mathbf{J}}(\Omega)$ is the angular momentum operator for body rotation.

Once the effective Liouvillean is defined, the direct calculation of the cw-ESR signal is possible without resorting to a complete solution of the SLE by evaluating the spectral density from the expression [2, 10]

$$I(\omega-\omega_0) = \frac{1}{\pi}\text{Re}\langle v|[i(\omega-\omega_0) + i\hat{\mathcal{L}}]^{-1}|vP_{\text{eq}}\rangle \quad (3.6)$$

where the Liouvillean $\hat{\mathcal{L}}$ acts on a starting vector, which is defined as proportional to the x component of the electron spin operator $\hat{S}_x$.

In the last part of the chapter, we will give a sketch of how this approach can be used for obtaining line shapes from first principles, a topic that is still under active development.

3.3
Spin Hamiltonian, g-Tensor, Hyperfine Coupling Constants, and Zero-Field Splitting

3.3.1
The Spin Hamiltonian

In order to more precisely state the nature of the problem, the leading spin Hamiltonian parameters are briefly introduced. The SH is an effective Hamiltonian and contains only spin variables of a "fictitious" electron spin S and the nuclear spins I_A, I_B, and so on. All references to the spatial part of the many-electron wavefunction and therefore to detailed molecular electronic and geometric structure are implicitly contained in the well-known SH parameters **D** (zero-field splitting), **g** (g-tensor), **A** (hyperfine coupling), **Q** (quadrupole coupling), **σ** (chemical shift), and **J** (spin–spin coupling), which are considered as adjustable parameters in the analysis of experiments and which will be explained in detail below. The SH that includes the interactions covered above is

$$\hat{H}_{\text{spin}} = \hat{\mathbf{S}} \mathbf{D} \hat{\mathbf{S}} + \beta \mathbf{B} \mathbf{g} \hat{\mathbf{S}} + \sum_A \left[\hat{\mathbf{S}} \mathbf{A}^{(A)} \hat{\mathbf{I}}^{(A)} + \beta_N \mathbf{B} \mathbf{g}_N^{(A)} \hat{\mathbf{I}}^{(A)} + \hat{\mathbf{I}}^{(A)} \mathbf{Q}^{(A)} \hat{\mathbf{I}}^{(A)} \right]$$
$$+ \sum_{A<B} \left[\hat{\mathbf{I}}^{(A)} \mathbf{J}^{(AB)} \hat{\mathbf{I}}^{(B)} \right]$$
(3.7)

where the sum over A refers to the magnetic nuclei, **B** is the magnetic flux density, and β and $β_N$ are the electronic and nuclear Bohr magnetons, respectively. The SH acts on a basis of product functions $|SM_S\rangle \otimes \left|I^{(A)} M_I^{(A)}\right\rangle \otimes \cdots \otimes \left|I^{(N)} M_I^{(N)}\right\rangle$. For not too many spins, this basis is often small enough to allow exact diagonalization of the SH and therefore exact QM treatments of the spin physics in the SH framework.

3.3.2
Electronic Structure Theory

On a fundamental level, the leading interactions between the positively charged nuclei and the negatively charged electrons are summarized in the (nonrelativistic) Born–Oppenheimer Hamiltonian:

$$\hat{H}_{\text{BO}} = -\frac{1}{2} \sum_i \vec{\nabla}^2 - \sum_{i,A} \frac{Z_A}{|\mathbf{r}_i - \mathbf{R}_A|} + \frac{1}{2} \sum_{i \neq j} \frac{1}{|\mathbf{r}_i - \mathbf{r}_j|} + \frac{1}{2} \sum_{A \neq B} \frac{Z_A Z_B}{|\mathbf{R}_A - \mathbf{R}_B|}$$
$$= \hat{T} + \hat{V}_{\text{eN}} + \hat{V}_{\text{ee}} + \hat{V}_{\text{NN}}$$
$$= \hat{h} + \hat{V}_{\text{ee}} + \hat{V}_{\text{NN}}$$
(3.8)

The terms describe the kinetic energy of the electrons, the electron–nuclear attraction, the electron–electron repulsion, and the nuclear–nuclear repulsion, respectively. In Equation 3.8, i and j sum over electrons at positions $\mathbf{r}_i$, A, and B over nuclei with charge Z_A at positions $\mathbf{R}_A$. The nuclear positions are assumed to be fixed and the electrons are supposed to readjust immediately to the positions of these (classical) nuclei. The BO Hamiltonian accounts for the vast amount of the molecular total energy and most problems of chemical structure and energetics can be satisfactorily discussed in terms of these comparatively simple interactions. Yet, the BO operator contains the coupled motion of N electrons and to find the exact eigenfunctions and eigenvalues of the (time-independent) BO–Schrödinger equation

$$\hat{H}_{\text{BO}} \Psi(\mathbf{x}_1, \ldots, \mathbf{x}_N | \mathbf{R}) = E(\mathbf{R}) \Psi(\mathbf{x}_1, \ldots, \mathbf{x}_N | \mathbf{R})$$
(3.9)

is a hopelessly complicated task. In Equation 3.9, the many-electron wavefunction $\Psi(\mathbf{x}_1, \ldots, \mathbf{x}_N | \mathbf{R})$ has been introduced that depends on the space (**r**) and spin (σ) variables of the N electrons ($\mathbf{x}_i \equiv (\mathbf{r}_i, \sigma_i)$). **R** collectively denotes the positions of the nuclei on which the many-electron wavefunction and the eigenvalues $E(\mathbf{R})$ depend parametrically. According to the basic quantum theory, all that can be known about the molecular system in the time-independent case are contained in $\Psi(\mathbf{x}_1, \ldots, \mathbf{x}_N | \mathbf{R})$. It is also important to note that *all* measurements always probe the N electron system. Molecular orbitals (MOs) to be introduced below are *never* observable and, in fact, the entire theory of molecular electronic structure can be exactly formulated without any recourse to orbitals. Yet, MOs are very convenient building blocks in the majority of approximate methods that have been developed to date.

In the density functional theory approach to the molecular Born–Oppenheimer Schrödinger equation, one does not attempt to approximate the many-particle wavefunction. Rather, one attempts to obtain the correct energy (or at least an energy that is sufficiently parallel to the correct energy) as a functional of the electron density. Owing to the celebrated Hohenberg–Kohn theorems, it is known that, *in principle*, the knowledge of $\varrho(\mathbf{r})$ is sufficient to deduce the exact ground-state energy. This comes at the price of introducing an unknown exchange–correlation functional $E_{xc}[\varrho]$. Since a systematic procedure to approach the exact $E_{xc}[\varrho]$ appears to be unknown, physically motivated guesses have to be introduced. Over the years, many such approximations have been suggested and new functionals appear in the literature almost on a weekly basis. Unfortunately, each functional has its own strengths and weaknesses that need to be assessed through extensive series of test calculations.

Without going into much detail, it is noted that the so-called Kohn–Sham procedure allows one to solve a set of pseudo-single particle equations that would provide the exact ground-state energy if the exact $E_{xc}[\varrho]$ is known. This procedure introduces the so-called "noninteracting reference system" that is described by a single Slater determinant and that shares with the physical system the electron density calculated through Equation 3.10a:

$$\varrho(\mathbf{r}) = \Sigma_i^N |\phi_i(r)|^2 \tag{3.10a}$$

In order to expand the single-particle wavefunctions (orbitals) that occur in the Slater determinant of the noninteracting reference determinant, one commonly introduces a set of auxiliary one-electron functions $\{\varphi(\mathbf{x})\}$ (basis functions) that are used to expand the orbitals as

$$\psi_i(\mathbf{x}) = \sum_\mu c_{\mu i} \varphi_\mu(\mathbf{x}) \tag{3.10b}$$

The minimization of the energy is then performed with respect to the coefficient $c_{\mu i}$, while the basis functions are held fixed. The expansion is exact only in the limit of a mathematically complete basis set $\{\varphi(\mathbf{x})\}$, which is impossible to obtain in practice. Thus, the results depend on the size and nature of the employed basis functions, but there is a well-defined *basis set limit*. Since the BO operator is spin free, it is customary to let the orbitals be eigenfunctions of the single-electron spin operator by choosing them to be either spin-up or spin-down orbitals ψ_i^σ ($\sigma = \alpha, \beta$). The spin-unrestricted Kohn–Sham equations in a finite basis set take on the form (upon dividing the electron density into its spin components $\varrho^\alpha(\mathbf{r})$ and $\varrho^\beta(\mathbf{r})$)

$$F_{\mu\nu}^{\sigma} = h_{\mu\nu} + \sum_{\varkappa\tau} P_{\varkappa\tau}(\mu\nu|\varkappa\tau) + \int \varphi_{\mu}(\mathbf{r})\varphi_{\nu}(\mathbf{r}) V_{XC}^{\sigma}[\varrho^{\alpha}, \varrho^{\beta}](\mathbf{r})d\mathbf{r} \quad (3.11)$$

Thus, in place of the exchange term that is familiar from Hartree–Fock theory, there now appears a *local* exchange–correlation potential, which is defined as the functional derivative of $E_{xc}[\varrho^{\alpha}, \varrho^{\beta}]$ with respect to $\varrho(\mathbf{r})$:

$$V_{XC}^{\sigma}[\varrho^{\alpha}, \varrho^{\beta}](\mathbf{r}) = \frac{\delta E_{xc}[\varrho]}{\delta \varrho^{\sigma}(\mathbf{r})} \quad (\sigma = \alpha, \beta) \quad (3.12)$$

The total Kohn–Sham energy is

$$E_{UKS} = \sum_{\mu\nu} P_{\mu\nu} h_{\mu\nu} + \frac{1}{2} \sum_{\varkappa\tau} P_{\mu\nu} P_{\varkappa\tau}(\mu\nu|\varkappa\tau) + E_{XC}[\varrho^{\alpha}, \varrho^{\beta}] + V_{NN} \quad (3.13)$$

The second term consists of the Coulombic self-interaction of the electron cloud and it can be written in a perhaps somewhat more illuminating way as

$$E_J = \frac{1}{2} \int \int \varrho(\mathbf{r}_1) \frac{1}{|\mathbf{r}_1 - \mathbf{r}_2|} \varrho(\mathbf{r}_2) d\mathbf{r}_1 d\mathbf{r}_2 \quad (3.14)$$

Likewise, the Coulomb contribution to the Kohn–Sham matrix is

$$J_{\mu\nu} = \int \int \varphi_{\mu}(r_1)\varphi_{\nu}(r_1) \frac{1}{|\mathbf{r}_1 - \mathbf{r}_2|} \varrho(\mathbf{r}_2) d\mathbf{r}_1 d\mathbf{r}_2 \quad (3.15)$$

which emphasizes the local nature of the Coulomb potential $V_C(\mathbf{r})$. Since this potential is of long range, its calculation usually dominates the computational effort of a Hartree–Fock (HF) or Kohn–Sham (KS) calculation. The precise functional forms of the various approximations to $E_{xc}[\varrho^{\alpha}, \varrho^{\beta}]$ are complicated and involve odd powers of $\varrho(\mathbf{r})$ such as $\varrho(\mathbf{r})^{4/3}$. If the functional also depends on the gradient of ϱ ($\vec{\nabla}\varrho(\mathbf{r})$), one obtains functionals from the "generalized gradient approximation" (GGA) family. Modern functionals may also depend on the Laplacian of the density ($\vec{\nabla}^2 \varrho(\mathbf{r})$) and the kinetic energy density ($\tau(\mathbf{r})$), which leads to the family of "meta-GGA" functionals. In recent years, the so-called "hybrid functionals" have become very popular, which involve a fraction of the nonlocal Hartree–Fock exchange and this was found to improve the results for total energies as well as many molecular properties.

There are many important conceptual and practical subtleties in DFT that cannot be discussed in the framework of this chapter and the interested reader is referred to a recent review that also provides pointers to the specialist literature.

3.3.3
Additional Terms in the Hamiltonian

Given an approximation to the ground-state energy of the BO Hamiltonian by some method, one needs to introduce the smaller field- and spin-dependent terms in the Hamiltonian that give rise to the interactions one actually probes by EPR spectroscopy. These terms can be derived through relativistic quantum chemistry, which is outside the scope of this chapter. Among the many terms that arise, we will mainly need the following interactions:

a) **The spin–orbit coupling.** Unlike found in many textbooks, this term in the Hamiltonian is of a *two*-electron nature and reads within the Breit–Pauli approximation:

$$\hat{H}_{SOC} = \hat{H}_{SOC}^{(1)} + \hat{H}_{SOC}^{(2)} \tag{3.16}$$

$$\hat{H}_{SOC}^{(1)} = \sum_i \hat{h}_i^{\text{1el-SOC}} = \sum_i \hat{\mathbf{h}}_i^{\text{1el-SOC}} \hat{\mathbf{s}}_i = \frac{\alpha^2}{2} \sum_i \sum_A Z_A r_{iA}^{-3} \hat{\mathbf{l}}_{iA} \hat{\mathbf{s}}_i \tag{3.17}$$

$$\hat{H}_{SOC}^{(2)} = \hat{H}_{SSO}^{(2)} + \hat{H}_{SOO}^{(2)} = \sum_i \sum_{j \neq i} \hat{g}_{i,j}^{\text{2el-SOC}} = -\frac{\alpha^2}{2} \sum_i \sum_{j \neq i} r_{ij}^{-3} \hat{\mathbf{l}}_{ij} (\hat{\mathbf{s}}_i + 2\hat{\mathbf{s}}_j) \tag{3.18}$$

Here, $\alpha = c^{-1}$ in atomic units is the fine structure constant ($\sim 1/137$), $\hat{\mathbf{r}}_i, \hat{\mathbf{p}}_i$, and $\hat{\mathbf{s}}_i$ are the position, momentum, and spin operators, respectively, of the ith electron, and $\hat{\mathbf{l}}_{iA} = (\hat{\mathbf{r}}_i - \mathbf{R}_A) \times \hat{\mathbf{p}}_i$ is the angular momentum of the ith electron relative to nucleus A. The vector $\hat{\mathbf{r}}_{iA} = \hat{\mathbf{r}}_i - \mathbf{R}_A$ of magnitude r_{iA} is the position of the ith electron relative to atom A. Likewise, the vector $\hat{\mathbf{r}}_{ij} = \hat{\mathbf{r}}_i - \hat{\mathbf{r}}_j$ of magnitude r_{ij} is the position of the ith electron relative to electron j and $\hat{\mathbf{l}}_{ij} = (\hat{\mathbf{r}}_i - \hat{\mathbf{r}}_j) \times \hat{\mathbf{p}}_i$ is its angular momentum relative to this electron. The one-electron term is familiar from many phenomenological treatments, for example, in atomic spectroscopy and ligand field theory. The two-electron term has contributions from the spin-same-orbit (SSO) and spin-other-orbit (SOO) terms, which are both important for a quantitatively correct treatment of SOC. They essentially provide a screening of the one-electron term in much the same way as the nuclear–electron attraction and electron–electron repulsion contributions counteract each other in the Born–Oppenheimer Hamiltonian. Since the full SOC operator is difficult to handle in large-scale molecular applications, it is desirable to approximate it as accurately as possible. This is possible through the accurate spin–orbit mean field approximation (SOMF) developed by Hess *et al*. Without going into the details of the derivation, we merely state the form of this operator:

$$\hat{h}^{\text{SOMF}} = \sum_i \hat{\mathbf{z}}^{\text{SOMF}}(i) \hat{\mathbf{s}}(i) \tag{3.19}$$

With the matrix elements of the kth component of the SOMF operator given by

$$\langle \varphi_\mu | \hat{z}_k^{\text{SOMF}} | \varphi_\nu \rangle = \langle \varphi_\mu | \hat{h}_k^{\text{1el-SO}} | \varphi_\nu \rangle$$

$$+ \sum_{\varkappa\tau} P_{\varkappa\tau} \left[(\varphi_\mu \varphi_\nu | \hat{g}_k^{\text{SO}} | \varphi_\varkappa \varphi_\tau) - \frac{3}{2} (\varphi_\mu \varphi_\varkappa | \hat{g}_k^{\text{SO}} | \varphi_\tau \varphi_\nu) - \frac{3}{2} (\varphi_\tau \varphi_\nu | \hat{g}_k^{\text{SO}} | \varphi_\mu \varphi_\varkappa) \right]$$

$$\tag{3.20}$$

and

$$\hat{h}_k^{\text{1el-SO}}(\mathbf{r}_i) = \frac{\alpha^2}{2} \sum_i \sum_A Z_A r_{iA}^{-3} \hat{l}_{iA;k} \quad (3.21)$$

$$\hat{g}_k^{\text{SO}}(\mathbf{r}_i, \mathbf{r}_j) = -\frac{\alpha^2}{2} \hat{l}_{ij;k} r_{ij}^{-3} \quad (3.22)$$

Here **P** is the total charge density matrix calculated by some theoretical method. Essentially like the HF approximation gives 99% of the total molecular energy, the SOMF operator covers around 99% of the two-electron SOC operator. It will be exclusively used below in order to approximate the SOC terms that will arise in the equations for the SH parameters.

b) **The direct magnetic dipolar spin–spin interaction.** This interaction is described by a genuine two-electron operator of the form

$$\hat{H}_{\text{SS}} = \frac{g_e^2 \alpha^2}{8} \sum_{i \neq j} \left(\frac{\hat{s}_i \hat{s}_j}{r_{ij}^3} - 3 \frac{(\hat{s}_i \mathbf{r}_{ij})(\hat{s}_j \mathbf{r}_{ij})}{r_{ij}^5} \right) \quad (3.23)$$

where the free-electron g-value $g_e = 2.002319\ldots$ appears.

c) **The hyperfine coupling.** This term describes the well-known dipolar interaction between the electron and the nuclear spins:

$$\hat{H}_{\text{SI}} = \frac{\alpha}{2} g_e \beta_N \sum_A g_N^{(A)} \sum_i \left(\frac{\hat{s}_i \hat{\mathbf{I}}^{(A)}}{r_{iA}^3} - 3 \frac{(\hat{s}_i \mathbf{r}_{iA})(\hat{\mathbf{I}}^{(A)} \mathbf{r}_{iA})}{r_{iA}^5} \right) \quad (3.24)$$

Here β_N is the nuclear magneton, $g_N^{(A)}$ is the g-value of the A'th nucleus, and $\hat{\mathbf{I}}^{(A)}$ is the spin operator for the nuclear spin of the A'th nucleus. While the isotropic Fermi contact term is frequently introduced as a separate operator, it arises naturally as a boundary term in the partial integration of the singular operator in Equation 3.24.

d) **The nuclear–orbit interaction.** The interaction of the nuclear spin with the orbital angular momentum of the electrons leads to the following term in the Hamiltonian:

$$\hat{H}_{\text{LI}} = \frac{\alpha}{2} \beta_N \sum_A g_N^{(A)} \sum_i \frac{\mathbf{l}_i^A \hat{\mathbf{I}}^{(A)}}{r_{iA}^3} \quad (3.25)$$

e) **The quadrupole coupling.** The quadrupole coupling describes the interaction of the electric field gradient (EFG) at a given nucleus with the quadrupole moment of that nucleus (only present for nuclei with spin $I > 1/2$). The electronic quantity of interest is the field gradient operator. The quadrupole interaction may be written as an operator of the following form:

$$\hat{H}_Q = e^2 \sum_{A,i} Q^{(A)} \hat{\mathbf{I}}^{(A)} \hat{\mathbf{F}}^{(A)}(i) \hat{\mathbf{I}}^{(A)} \quad (3.26)$$

where $Q^{(A)}$ is the quadrupole moment of the A'th nucleus, e is the elementary charge, and the field gradient operator is given by

$$\hat{F}^{(A)}_{\mu\nu}(i) = \frac{r^2_{iA}\delta_{\mu\nu} - 3r_{iA;\mu}r_{iA;\nu}}{r^5_{iA}} \tag{3.27}$$

f) **The electronic Zeeman interaction.** The interaction of the electrons with a static external magnetic field is described by

$$\hat{H}_{LB} = \frac{\alpha}{2}\sum_i \mathbf{B}(\mathbf{l}_i + g_e\hat{\mathbf{s}}_i) \tag{3.28}$$

From the fully relativistic treatment, there arises a "kinetic energy correction" (relativistic mass correction) to the spin–Zeeman energy that is given by

$$\hat{H}^{RMC}_{SB} = \frac{\alpha^3 g_e}{2}\sum_i \nabla^2_i \mathbf{B}\mathbf{s}_i \tag{3.29}$$

3.3.4
Linear Response Theory

The equations given in the above sections are important from a conceptual point of view as they show most clearly the basic physics that is involved in SH parameters. However, as a basis of actual calculations, they are unfortunately much less useful owing to the presence of the second-order terms. The evaluation of these terms would require an infinite sum over excited many-electron states. In practice, at most a few dozen many-electron states can be calculated. Although quite useful results have been obtained with this approach, the convergence of the perturbation sum is uncertain and can, in the general case, hardly be guaranteed. In the case of DFT, the excited states cannot be obtained explicitly since the Hohenberg–Kohn theorems apply only to the electronic ground state. Hence, it is important to look for an alternative definition of the various SH parameters. An approach of substantial generality and elegance is provided by the so-called linear response theory (LRT). In our view, LRT is just one realization of a family of methods that are all formulated in a similar spirit. If time does not explicitly occur in the equations (which is not necessary for the formulation of EPR and NMR parameters), these methods can also be called "analytic derivative approaches." In the framework of HF and DFT methods, they are known as "coupled-perturbed self-consistent field" (CP-SCF) methods and "double perturbation theory" (DPT), respectively. All these acronyms stand for computational methods that provide identical results and it is a matter of taste which framework one prefers. These methods have developed to high sophistication in quantum chemistry and have been proven to be extremely useful in many contexts, including geometry optimization and frequency calculations (geometric derivatives), as well as in the precise prediction of many molecular properties. In fact, the LRT approaches can be shown to implicitly involve an untruncated sum over excited states of the system that are (implicitly) described at the same level of sophistication as the ground state. The key quantities of interest in LRT are the derivatives of the (approximate) total ground-state energy with respect to external perturbation parameters $\lambda, \varkappa, \ldots$, where λ and $\varkappa$ may denote components of an external field or a nuclear magnetic moment of an electronic magnetic moment. Formally, we could take the derivative of the perturbation sum and then make the connection to the response formalism by it with the appropriate derivative of the approximate total energy calculated with the theoretical method of choice. Since all SH parameters are bilinear in external perturbations, the desired quantity is the second partial derivative of the total energy.

In order to appreciate the general concepts that are involved, the linear response equations for a self-consistent field (SCF) ground state will be sketched below. This description is appropriate if the state of interest is well described by a HF or DFT single determinant. The ground-state energy is written here as

$$E = V_{NN} + \sum_i h_{ii} + \frac{1}{2}\sum_{i,j}(ii|jj) - c_{HF}(ij|ij) + c_{DF}E_{XC}[\varrho] \tag{3.30}$$

The parameters c_{HF} and c_{DF} are scaling parameters for the HF exchange energy and the XC energy, respectively. Thus, HF theory corresponds to $c_{HF} = 1$; $c_{DF} = 0$, "pure" DFT corresponds to $c_{HF} = 0$; $c_{DF} = 1$, while hybrid DFT methods choose $0 < c_{HF} < 1$. The energy has been written here in terms of the occupied orbitals $\psi_i(\mathbf{x})$. They are determined self-consistently from the SCF (HF or KS) equations:

$$\left\{\hat{h} + \int \frac{\varrho(\mathbf{x}')}{|\mathbf{x}-\mathbf{x}'|}d\mathbf{x}' - c_{HF}\sum_i \hat{K}^{ii} + c_{DF}\frac{\delta E_{XC}[\varrho]}{\delta\varrho(\mathbf{x})}\right\}\psi_i(\mathbf{x}) = \varepsilon_i\psi_i(\mathbf{x}) \tag{3.31}$$

For illustrating the concepts, it is sufficient to consider the case where the basis functions are chosen to be independent of the external perturbations. To include such dependence (as is necessary, for example, for geometric or magnetic field perturbations) is straightforward, but would lead to more lengthy equations that are not of interest for the purpose of this chapter. Since the MO coefficients c are determined in a variational procedure, one has

$$\frac{\partial E}{\partial c_{\mu i}}\frac{\partial c_{\mu i}}{\partial \lambda} = 0 \tag{3.32}$$

And, therefore, the first derivative of the energy with respect to a perturbation λ is

$$\left.\frac{\partial E}{\partial \lambda}\right|_{\lambda=0} = \sum_{\mu\nu}P_{\mu\nu}\langle\varphi_\mu|\hat{h}|\varphi_\nu\rangle_\lambda \tag{3.33}$$

where

$$\langle\varphi_\mu|\hat{h}|\varphi_\nu\rangle_\lambda = \left\langle\varphi_\mu\left|\frac{\partial\hat{h}}{\partial\lambda}\right|\varphi_\nu\right\rangle \tag{3.34}$$

if λ is a one-electron perturbation and if the basis functions are independent of λ.

Through an additional differentiation, the second partial derivative becomes

$$\frac{\partial^2 E}{\partial \lambda \partial \varkappa}\bigg|_{\lambda=0;\varkappa=0} = \sum_{\mu\nu} P_{\mu\nu} \langle \varphi_\mu | \hat{h} | \varphi_\nu \rangle_{\lambda\varkappa} + \sum_{\mu\nu} \frac{\partial P_{\mu\nu}}{\partial \varkappa} \langle \varphi_\mu | \hat{h} | \varphi_\nu \rangle_\lambda \quad (3.35)$$

This important equation contains two contributions: the first term is referred to as a first-order contribution since it depends only on the ground-state density. The second term is a second-order contribution since it requires the knowledge of the first derivative of the density matrix with respect to an external perturbation and the first derivative of the Hamiltonian. These two terms substitute the two first- and second-order terms in the perturbation sum discussed above. It remains to be shown how the perturbed density matrix can be calculated.

A simple approach will be followed for the calculation of the perturbed density matrix. To this end, the perturbed orbitals $\psi_i^{(\lambda)} \equiv |i^{(\lambda)}\rangle$ are calculated in terms of the zeroth-order orbitals $\psi_i^{(0)} \equiv |i^{(0)}\rangle$. Differentiation of the SCF equations yields

$$\{\hat{F}^{(0)} - \varepsilon_i^{(0)}\}|i^{(\lambda)}\rangle + \{\hat{F}^{(\lambda)} - \varepsilon_i^{(\lambda)}\}|i^{(0)}\rangle = 0 \quad (3.36)$$

The perturbed orbitals $|i^{(\lambda)}\rangle$ are expanded as

$$|i^{(\lambda)}\rangle = \sum_a U_{ai}^{(\lambda)} |a^{(0)}\rangle \quad (3.37)$$

(here and below, the labels i, j, k, and l refer to occupied orbitals and a, b, c, and d to unoccupied ones). The unitary matrix **U** has only occupied/virtual blocks in the case that the basis functions do not depend on the perturbation. In order to determine the unique elements of **U**, one uses the perturbed SCF equations:

$$\sum_b U_{bi}^{(\nu)} \langle a^{(0)} | \hat{F}^{(0)} - \varepsilon_i^{(0)} | b^{(0)} \rangle + \langle a^{(0)} | \hat{F}^{(\lambda)} - \varepsilon_i^{(\lambda)} | i^{(0)} \rangle = 0 \quad (3.38)$$

$$= U_{ai}^{(\lambda)} (\varepsilon_a^{(0)} - \varepsilon_i^{(0)}) + \langle a^{(0)} | \hat{F}^{(\lambda)} | i^{(0)} \rangle = 0 \quad (3.39)$$

However, $\hat{F}^{(\lambda)}$ depends on the perturbed orbitals. Therefore, ones needs to take the derivative of the SCF operator carefully.

$$\hat{F}^{(\lambda)} = \hat{h}^{(\lambda)} + \int \frac{\varrho^{(\lambda)}(\mathbf{x}')}{|\mathbf{x}-\mathbf{x}'|} - c_{HF} \sum_{jb} U_{bj}^{(\lambda)*} \hat{K}^{bj} + U_{bj}^{(\lambda)} \hat{K}^{jb}$$

$$+ c_{DF} \int\int \frac{\delta^2 E_{XC}[\varrho]}{\delta\varrho(\mathbf{x})\delta\varrho(\mathbf{x}')} \varrho^{(\lambda)}(\mathbf{x}') d\mathbf{x} d\mathbf{x}'$$

$$= \hat{h}^{(\lambda)} + \sum_{jb} U_{bj}^{(\lambda)*} \hat{J}^{bj} + U_{bj}^{(\lambda)} \hat{J}^{jb} - c_{HF} U_{bj}^{(\lambda)*} \hat{K}^{bj} - c_{HF} U_{bj}^{(\lambda)} \hat{K}^{jb}$$

$$+ c_{DF} \int f_{xc}[\varrho] \varrho^{(\lambda)}(\mathbf{x}) d\mathbf{x}$$

3.3 Spin Hamiltonian, g-Tensor, Hyperfine Coupling Constants, and Zero-Field Splitting

$$= \hat{h}^{(\lambda)} + \sum_{jb} U_{bj}^{(\lambda)*} \left[\hat{J}^{bj} - c_{HF} \hat{K}^{bj} + c_{DF} \int f_{xc}[\varrho] \psi_b^{(0)*}(\mathbf{x}) \psi_j^{(0)}(\mathbf{x}) d\mathbf{x} \right]$$

$$+ U_{bj}^{(\lambda)} \left[\hat{J}^{jb} - c_{HF} \hat{K}^{jb} + c_{DF} \int f_{xc}[\varrho] \psi_j^{(0)*}(\mathbf{x}) \psi_b^{(0)}(\mathbf{x}) d\mathbf{x} \right] \quad (3.40)$$

Here, the "XC kernel" $f_{xc}[\varrho]$ has been defined as the second functional derivative with respect to ϱ and it has been tacitly assumed that for all functionals in use, this yields a factor $\delta(\mathbf{x}-\mathbf{x}')$ that reduces the double integral to a single integral. Taken together, this results in the first-order equations:

$$U_{ai}^{(\lambda)} \left(\varepsilon_a^{(0)} - \varepsilon_i^{(0)} \right) + \langle a^{(0)} | \hat{h}^{(\lambda)} | i^{(0)} \rangle$$

$$+ \sum_{jb} U_{bj}^{(\lambda)*} [(bj|ai) - c_{HF}(ba|ji) + (ai|f_{xc}|jb)] \quad (3.41)$$

$$+ U_{bj}^{(\lambda)} [(jb|ai) - c_{HF}(ja|bi) + (ai|f_{xc}|jb)] = 0$$

At this point it is useful to distinguish two different types of perturbation: First, the "electric field-like perturbations" yield purely real $\langle a^{(0)} | \hat{h}^{(\lambda)} | i^{(0)} \rangle$ and, consequently, purely real and symmetric **U** matrices. In this case, one has

$$\mathbf{A}^{(E)} \mathbf{U}^{(\lambda)} = -\mathbf{V}^{(\lambda)} \quad (3.42)$$

with

$$A_{ia,jb}^{(E)} = \delta_{ij} \delta_{ab} \left(\varepsilon_a^{(0)} - \varepsilon_i^{(0)} \right) + 2(jb|ia) + 2c_{DF}(ai|f_{xc}|jb) - c_{HF} \{(ba|ji) + (ja|bi)\} \quad (3.43)$$

$$V_{ai}^{(\lambda)} = \langle a^{(0)} | \hat{h}^{(\lambda)} | i^{(0)} \rangle \quad (3.44)$$

Note that the **A** matrix (the "electric Hessian") is independent of the nature of the perturbation and that the **U** and **V** matrices have been written as vectors with a compound index (ai).

Second, the "magnetic field-like perturbations" yield purely Hermitian imaginary $\langle a^{(0)} | \hat{h}^{(\lambda)} | i^{(0)} \rangle$ and, consequently, purely imaginary and Hermitian **U** matrices. This leads to

$$\mathbf{A}^{(M)} \mathbf{U} = -\mathbf{V}^{(\lambda)} \quad (3.45)$$

with

$$A_{ia,jb}^{(M)} = \delta_{ij} \delta_{ab} \left(\varepsilon_a^{(0)} - \varepsilon_i^{(0)} \right) + c_{HF} \{(ib|ja) - (ba|ij)\} \quad (3.46)$$

Thus, the magnetic field-like perturbations yield much easier response (or "coupled perturbed") equations in which the contributions from any local potential vanish. In fact, in the absence of HF exchange, the **A** matrix becomes diagonal and the linear equation system is trivially solved. This then leads to a "sum-over-orbital" like equation for the second derivative, which resembles in some way a "sum-over-states" equation. One should, however, carefully distinguish the sum-over-states picture from linear response or analytic derivative techniques since they have a very different origin. For the electric or magnetic field-like perturbations in the presence of HF exchange, one thus has to solve a linear equation system of the size N (occupied) × N (virtual) that may amount to dimensions of several hundred thousand coefficients in large-scale applications. However, there are efficient iterative techniques to solve such large equation systems without ever explicitly constructing the full **A** matrix. Once the perturbed orbitals have been determined, the perturbed density is found from

$$P_{\mu\nu}^{(\lambda)} = \sum_i U_{ai}^{(\lambda)*} c_{\mu a}^{(0)} c_{\nu i}^{(0)} + U_{ai}^{(\lambda)} c_{\mu i}^{(0)} c_{\nu a}^{(0)} \qquad (3.47)$$

3.3.5
Linear Response Equations for Spin Hamiltonian Parameters

Using the results of the preceding sections, it is now possible to provide explicit expressions for all SH parameters.

a) **Zero-field splitting.** The ZFS is the least well-developed SH parameter in EPR spectroscopy. It is also the most complicated one since the SS contribution is a genuine two-electron property. For this contribution, McWeeny and Mizuno have shown [11]

$$D_{kl}^{(SS)} = -\frac{g_e^2}{16} \frac{\alpha^2}{S(2S-1)} \sum_{\mu\nu} \sum_{\varkappa\tau} \{P_{\mu\nu}^{\alpha-\beta} P_{\varkappa\tau}^{\alpha-\beta} - P_{\mu\varkappa}^{\alpha-\beta} P_{\nu\tau}^{\alpha-\beta}\} \langle \mu\nu|r_{12}^{-5}\{3r_{12,k}r_{12,l}-\delta_{kl}r_{12}^2\}|\varkappa\tau\rangle \qquad (3.48)$$

The integrals appearing in Equation 3.48 look complicated at first glance but are readily calculated and owing to the factorization of the two-particle spin density matrix, Equation 3.48 can be implemented for the large-scale application without creating storage of computation time bottlenecks.

It is very interesting to look at the physical content of Equation 3.48 in a little more detail. From the form of the operator and the appearance of the spin density matrix, it is obvious that it describes the (traceless) direct electron–electron magnetic dipole–dipole interaction between unpaired electrons. Such a term is widely used in modeling the EPR spectra of interacting electron spins within the "point dipole" approximation. Equation 3.48 consists of two parts. In analogy to HF theory, the first part should be recognized as a "Coulomb" contribution, while the second one as an "exchange" contribution. Thus, even the direct dipolar spin–spin interaction contains an exchange contribution that is of fundamentally different origin than that of the "genuine" exchange interaction used in the modeling of interacting spins. This point does not appear to be widely recognized. Nevertheless, assuming an exponential decay of the basis functions, it becomes evident that the exchange term is expected to fall off much more quickly with interspin distance than the Coulomb contribution. The "distributed point dipole" like equations can be recovered from Equation 3.48 by (a) neglecting the exchange contribution, (b) assuming that the spin density matrix is diagonal in the chosen basis, and (c) "compressing" the basis functions to δ-functions centered at the atomic positions. One then obtains

$$D_{kl} \approx -\frac{g_e^2}{16} \frac{\alpha^2}{S(2S-1)} \sum_{AB} P_A^{\alpha-\beta} P_B^{\alpha-\beta} R_{AB}^{-5} [3 R_{AB;k} R_{AB;l} - \delta_{kl} R_{AB}^2] \qquad (3.49)$$

where A and B sum over nuclei and $P_A^{\alpha-\beta} = \sum_{\mu \in A} P_{\mu\mu}^{\alpha-\beta}$ is the "gross" spin population on atom A. Equation 3.49 describes the interaction of point dipoles centered at atomic positions where each atom pair is weighted by the product of the spin populations that reside on this atom. Note that this gross atomic spin population differs from the usual numbers that are predicted by Mulliken or Löwdin analysis and that are part of the output of many electronic structure programs since the latter contain terms that depend on the basis function overlap while $P_A^{\alpha-\beta}$ does not. However, since the approximation leading to the point dipole formula (Equation 3.49) has been rather crude, large additional errors may not be expected if Mulliken or Löwdin spin populations are inserted into Equation 3.49. If the distance between two spin-carrying fragments is large enough, it may even be possible to reduce Equation 3.49 to a single term where R_{AB} must then refer to an "effective" distance. We propose to calculate it as follows: First the center of gravity of the spin density of fragments $F1$ and $F2$ are defined as

$$\mathbf{R}^{(F1)} = \sum_{A \in F1} \bar{P}_A^{\alpha-\beta} \mathbf{R}_A \qquad (3.50)$$

$$\mathbf{R}^{(F2)} = \sum_{B \in F2} \bar{P}_B^{\alpha-\beta} \mathbf{R}_B \qquad (3.51)$$

And then replace the vector $\mathbf{R}_{AB}$ in Equation 3.49 by $\mathbf{R}_{12} = \mathbf{R}^{(F1)} - \mathbf{R}^{(F2)}$. The barred quantity $\bar{P}_A^{\alpha-\beta}$ refers to a normalized spin population such that the sum $\sum_{A \in F1} \bar{P}_A^{\alpha-\beta} = 1$. This appears to be a slightly more rigorous approach than the commonly used approach in which the intercenter distance is fixed by subjective plausible choices that may, however, differ between different workers. Note also that the full g-tensor does not enter the above equations. We have already criticized the use of the g-tensor in the point dipole approximations for the hyperfine couplings [12] and a similar situation also applies to the case of the dipolar ZFS tensor.

The SOC contribution to the ZFS involves the response of the orbitals to the SOC. The formalism to achieve an analytic derivative formulation of this part of the SH has been worked out recently by the present author and is somewhat more involved than the analogous methodology for the g-tensor [13]. Without going into much detail, the three contributions to the **D**-tensor may be rewritten as follows:

$$D_{kl}^{(0)} = -\frac{1}{4S^2} \sum_{\mu\nu} \langle \mu | h_k^{SOC} | \nu \rangle \frac{\partial P_{\mu\nu}^{(0)}}{\partial S_l^{(0)}} \tag{3.52}$$

$$D_{kl}^{(+1)} = \frac{1}{2(S+1)(2S+1)} \sum_{\mu\nu} \langle \mu | h_k^{SOC} | \nu \rangle \frac{\partial P_{\mu\nu}^{(-1)}}{\partial S_l^{(+1)}} \tag{3.53}$$

$$D_{kl}^{(-1)} = \frac{1}{2S(2S-1)} \sum_{\mu\nu} \langle \mu | h_k^{SOC} | \nu \rangle \frac{\partial P_{\mu\nu}^{(+1)}}{\partial S_l^{(-1)}} \tag{3.54}$$

The components $S_l^{(m)}$ ($m = 0, \pm 1$) are the vector operator components of the total spin. The spin densities $\mathbf{P}^{(m)}$ are the response densities with respect to a SOC perturbation. They are calculated from a nonstandard set of coupled perturbed equations analogous to the ones described above for the g-tensor as

$$\frac{\partial P_{\mu\nu}^{(0)}}{\partial S_l^{(0)}} = \sum_{i_\alpha a_\alpha} U_{a_\alpha i_\alpha}^{(0);l} c_{\mu i}^\alpha c_{\nu a}^\alpha + \sum_{i_\beta a_\beta} U_{a_\beta i_\beta}^{(0);l} c_{\mu i}^\beta c_{\nu a}^\beta \tag{3.55}$$

$$\frac{\partial P_{\mu\nu}^{(+1)}}{\partial S_l^{(-1)}} = \sum_{i_\alpha a_\beta} U_{a_\beta i_\alpha}^{(-1);l} c_{\mu i}^\alpha c_{\nu a}^\beta - \sum_{i_\beta a_\alpha} U_{a_\alpha i_\beta}^{(-1);l} c_{\mu a}^\alpha c_{\nu i}^\beta \tag{3.56}$$

$$\frac{\partial P_{\mu\nu}^{(-1)}}{\partial S_l^{(+1)}} = -\sum_{i_\alpha a_\beta} U_{a_\beta i_\alpha}^{(+1);l} c_{\mu a}^\beta c_{\nu i}^\alpha + \sum_{i_\beta a_\alpha} U_{a_\alpha i_\beta}^{(+1);l} c_{\mu i}^\beta c_{\nu a}^\alpha \tag{3.57}$$

With the U-coefficients calculated from

$m = 0$:

$$\left(\varepsilon_{a_\alpha}^{(0)} - \varepsilon_{i_\alpha}^{(0)}\right) U_{a_\alpha i_\alpha}^{k(0)} + c_{HF} \sum_{j_\alpha b_\alpha} U_{b_\alpha j_\alpha}^{k(0)} \{(b_\alpha i_\alpha | a_\alpha j_\alpha) - (j_\alpha i_\alpha | a_\alpha b_\alpha)\} = -\left(a_\alpha | h_k^{SOC} | i_\alpha\right) \tag{3.58}$$

$$\left(\varepsilon_{a_\beta}^{(0)} - \varepsilon_{i_\beta}^{(0)}\right) U_{a_\beta i_\beta}^{k(0)} + c_{HF} \sum_{j_\beta b_\beta} U_{b_\beta j_\beta}^{k(0)} \{(b_\beta i_\beta | a_\beta j_\beta) - (j_\beta i_\beta | a_\beta b_\beta)\} = -\left(a_\beta | h_k^{SOC} | i_\beta\right) \tag{3.59}$$

$m = +1$:

$$\left(\varepsilon_{a_\alpha}^{(0)} - \varepsilon_{i_\beta}^{(0)}\right) U_{a_\alpha i_\beta}^{k(+1)} + c_{HF} \sum_{j_\alpha b_\alpha} U_{b_\beta j_\alpha}^{k(+1)} (b_\beta i_\beta | a_\alpha j_\alpha) - c_{HF} \sum_{b_\alpha j_\beta} U_{b_\beta j_\alpha}^{k(+1)} (j_\beta i_\beta | a_\alpha b_\alpha)$$
$$= -\left(a_\alpha | h_k^{SOC} | i_\beta\right) \tag{3.60}$$

3.3 Spin Hamiltonian, g-Tensor, Hyperfine Coupling Constants, and Zero-Field Splitting

$$\left(\varepsilon_{a_\beta}^{(0)}-\varepsilon_{i_\alpha}^{(0)}\right)U_{a_\beta i_\alpha}^{k(+1)}+c_{HF}\sum_{j_\beta b_\alpha}U_{b_\alpha j_\beta}^{k(+1)}(b_\alpha i_\alpha|a_b j_\beta)$$
$$-c_{HF}\sum_{b_\beta j_\alpha}U_{b_\beta j_\alpha}^{k(+1)}(j_\alpha i_\alpha|a_\beta b_\beta)=0 \qquad (3.61)$$

$m = -1$:

$$\left(\varepsilon_{a_\beta}^{(0)}-\varepsilon_{i_\alpha}^{(0)}\right)U_{a_\beta i_\alpha}^{k(-1)}+c_{HF}\sum_{j_\beta b_\alpha}U_{b_\alpha j_\beta}^{k(-1)}(b_\alpha i_\alpha|a_b j_\beta)$$
$$-c_{HF}\sum_{b_\beta j_\alpha}U_{b_\beta j_\alpha}^{k(-1)}(j_\alpha i_\alpha|a_\beta b_\beta)=-\left(a_\beta|h_k^{SOC}|i_\alpha\right) \qquad (3.62)$$

$$\left(\varepsilon_{a_\alpha}^{(0)}-\varepsilon_{i_\beta}^{(0)}\right)U_{a_\alpha i_\beta}^{k(-1)}+c_{HF}\sum_{j_\alpha b_\alpha}U_{b_\beta j_\alpha}^{k(-1)}(b_\beta i_\beta|a_\alpha j_\alpha)-c_{HF}\sum_{b_\alpha j_\beta}U_{b_\beta j_\alpha}^{k(-1)}(j_\beta i_\beta|a_\alpha b_\alpha)=0$$
$$(3.63)$$

This formalism is the exact analogue of the ones used to compute the g-tensor and the SOC contribution to the HFC tensor and directly follows from the general equations given above. It is available in the ORCA package and has been shown to correct some deficiencies of earlier formulation of the ZFS tensor in the DFT framework. Since the latter procedures are more commonly met in the literature, they are briefly described. Pederson and Khanna have suggested the equation [14]

$$D_{kl}^{(SOC)} = -\frac{1}{4S^2}\sum_{i_\beta,a_\beta}\frac{\langle\psi_i^\beta|h_k^{SOC}|\psi_a^\beta\rangle\langle\psi_a^\beta|h_l^{SOC}|\psi_i^\beta\rangle}{\varepsilon_a^\beta-\varepsilon_i^\beta}$$
$$-\frac{1}{4S^2}\sum_{i_\alpha,a_\alpha}\frac{\langle\psi_i^\alpha|h_k^{SOC}|\psi_a^\alpha\rangle\langle\psi_a^\alpha|h_l^{SOC}|\psi_i^\alpha\rangle}{\varepsilon_a^\alpha-\varepsilon_i^\alpha}$$
$$+\frac{1}{4S^2}\sum_{i_\alpha,a_\beta}\frac{\langle\psi_i^\alpha|h_k^{SOC}|\psi_a^\beta\rangle\langle\psi_a^\beta|h_l^{SOC}|\psi_i^\alpha\rangle}{\varepsilon_a^\beta-\varepsilon_i^\alpha}$$
$$+\frac{1}{4S^2}\sum_{i_\beta,a_\alpha}\frac{\langle\psi_i^\alpha|h_k^{SOC}|\psi_a^\alpha\rangle\langle\psi_a^\alpha|h_l^{SOC}|\psi_i^\alpha\rangle}{\varepsilon_a^\alpha-\varepsilon_i^\beta} \qquad (3.64)$$

which is valid in the case of DFT functionals that do not contain the HF exchange ($c_{HF}=0$). It can be shown to be a special case of the more general treatment outlined above if the prefactors in front of the individual terms in Equation 3.64 are all set to $1/4S^2$. We had previously implemented this equation and compared it to the following equation that had also been motivated from the general equations:

$$D_{kl}^{(SOC)} = -\frac{1}{4S^2} \sum_{i,p} \frac{\langle \psi_i | h_k^{SOC} | \psi_p \rangle \langle \psi_p | h_l^{SOC} | \psi_i \rangle}{\varepsilon_p^\beta - \varepsilon_i^\beta}$$

$$-\frac{1}{4S^2} \sum_{p,a} \frac{\langle \psi_p | h_k^{SOC} | \psi_a \rangle \langle \psi_a | h_l^{SOC} | \psi_p \rangle}{\varepsilon_a^\alpha - \varepsilon_p^\alpha}$$

$$+\frac{1}{2} \frac{1}{S(2S-1)} \sum_{p \neq q} \frac{\langle \psi_p | h_k^{SOC} | \psi_q \rangle \langle \psi_q | h_l^{SOC} | \psi_p \rangle}{\varepsilon_q^\beta - \varepsilon_p^\alpha}$$

$$+\frac{1}{2} \frac{1}{(S+1)(2S+1)} \sum_{i,a} \frac{\langle \psi_i | h_k^{SOC} | \psi_a \rangle \langle \psi_a | h_l^{SOC} | \psi_i \rangle}{\varepsilon_a^\alpha - \varepsilon_i^\beta} \quad (3.65)$$

In this equation, there enters a set of "quasi-restricted" orbitals (QROs), which are explained in detail in Ref. [13]. It may be appreciated that both formulations involve the terms that are already apparent in the general treatment. Namely, the first two terms correspond to the contributions from the spin-conserving excitations, while the third and fourth term correspond to the contributions from the excited states of lower and higher multiplicity than the ground state, respectively. However, this QRO formalism is now superseded by the more general development in Equations 3.52–3.54.

g-Tensor. The g-tensor is well studied by now with many implementations and applications available. One obtains the following expressions for the four contributions:

$$g_{kl} = g_e \delta_{kl} + \Delta g^{RMC} \delta_{kl} + \Delta g_{kl}^{GC} + \Delta g_{kl}^{OZ/SOC} \quad (3.66)$$

$$\Delta g^{RMC} = -\frac{\alpha^2}{S} \sum_{\mu,\nu} P_{\mu\nu}^{\alpha-\beta} \langle \varphi_\mu | \hat{T} | \varphi_\nu \rangle \quad (3.67)$$

$$\Delta g_{kl}^{GC} = \frac{1}{2S} \sum_{\mu,\nu} P_{\mu\nu}^{\alpha-\beta} \left\langle \varphi_\mu \left| \sum_A \xi(r_A)(\mathbf{r}_A \mathbf{r} - \mathbf{r}_{A,k} \mathbf{r}_l) \right| \varphi_\nu \right\rangle \quad (3.68)$$

$$\Delta g_{kl}^{(OZ/SOC)} = \frac{1}{2S} \sum_{\mu\nu} \frac{\partial P_{\mu\nu}^{(\alpha-\beta)}}{\partial B_k} \langle \varphi_\mu | \hat{z}_l^{SOMF} | \varphi_\nu \rangle \quad (3.69)$$

Here $\hat{T}$ is the kinetic energy operator. It is noted that the g-tensor expressions make, through the operators $\mathbf{r}$ in Equation 3.68 and l (implicit in Equation 3.69), reference to the global origin of the coordinate system. This would seem to imply the unphysical and unfortunate situation that the results of the computations depend on the choice of origin. This is indeed so in g-tensor calculations and would only disappear in the basis set limit that is, in practice, never reached. The way around this artifact is to employ magnetic field-dependent basis functions (gauge including atomic orbitals (GIAOs)). The GIAOs are an elegant way to solve the gauge problem but require some additional computational effort. They have been very successful in the prediction of NMR chemical shifts where it is essential to remove the gauge dependence entirely. The alternative "independent gauge for localized orbitals" (IGLO) that is also popular in chemical shift calculations is not successful in EPR spectroscopy since the separate localization of spin-up and spin-down orbitals introduces artifacts into the results. Fortunately, the gauge problem in EPR spectroscopy is not large and one obtains meaningful results even if slight origin dependence persists. In order to make results comparable, a reasonable choice of origin is still required and is conveniently provided by the center of electronic charge. The error made by this approximation is much smaller than the other remaining errors due to the functional, the basis set, and the molecular model or the treatment of environmental effects.

b) **Hyperfine coupling.** One finds for the three parts of the HFC the following expressions:

$$A_{kl}^{(A;c)} = \delta_{kl} \frac{8\pi}{3} \frac{P_A}{2S} \varrho^{\alpha-\beta}(\mathbf{R}_A) \tag{3.70}$$

$$A_{kl}^{(A;d)} = \frac{P_A}{2S} \sum_{\mu\nu} P_{\mu\nu}^{\alpha-\beta} \langle \varphi_\mu | r_A^{-5} (r_A^2 \delta_{kl} - 3 r_{A;k} r_{A;l}) | \varphi_\nu \rangle \tag{3.71}$$

$$A_{kl}^{(A;SO)} = -\frac{P_A}{S} \sum_{\mu\nu} \frac{\partial P_{\mu\nu}^{\alpha-\beta}}{\partial \hat{I}_k^{(A)}} \langle \varphi_\mu | z_l^{SOMF} | \varphi_\nu \rangle \tag{3.72}$$

with $P_A = g_e g_N \beta_e \beta_N$. Thus, the first two terms are straightforward expectation values, while the SOC contribution is a response property. In this case, one has to solve a set of coupled perturbed equations with the nucleus–orbit interaction taken as the perturbing operator. Since the solution of the coupled perturbed equations becomes time consuming for larger molecules, this should only be done for a few selected heavier nuclei. For light nuclei, the SOC correction is usually negligible.

c) **Electric field gradient.** The EFG tensor is straightforwardly calculated from

$$V_{\mu\nu}^{(A)} = \sum_{\varkappa,\tau} P_{\varkappa\tau} \langle \phi_\varkappa | r_A^{-5} (r_A^2 \delta_{\mu\nu} - 3 r_{A;\mu} r_{A;\nu}) | \phi_\tau \rangle \tag{3.73}$$

Once available, the EFG tensor can be diagonalized. The numerically largest element V_{max} (in atomic units) defines the value of q that is in turn used to calculate the quadrupole splitting parameter as $e^2 qQ = 235.28 V_{max} Q$, where Q is the quadrupole moment of the nucleus in barn. Transformed to its eigensystem, the quadrupole splitting enters the SH in the following form:

$$\hat{H}_Q = \hat{\mathbf{I}} \mathbf{Q} \hat{\mathbf{I}} = \frac{e^2 qQ}{4I(2I-1)} \hat{\mathbf{I}} \begin{pmatrix} -(1-\eta) & 0 & 0 \\ 0 & -(1+\eta) & 0 \\ 0 & 0 & 2 \end{pmatrix} \hat{\mathbf{I}} \tag{3.74}$$

The asymmetry parameter η is defined as

$$\eta = \frac{|V_{mid} - V_{min}|}{V_{max}} \tag{3.75}$$

It is to be noted that this is the only term that involves the total electron density rather than the spin density. The field gradient tensor is consequently of a nature quite different from the hyperfine coupling that depends on the same dipolar interaction integrals, but in the case of the HFC they are contracted with the spin density instead of the electron density.

It is important to realize that the equations of this section are not only valid in the case of a SCF ground-state description but are also of much wider applicability. In the case of correlated *ab initio* methods, the equations to be solved in order to determine the effective density and its response merely become much more complicated than the relatively simple CP-SCF equations sketched above. The general line of thought is, however, identical and merely the "mechanics" of the calculation become more involved.

3.3.6
Computational Aspects: Functionals and Basis Sets

The introduction of methods rooted into the density functional theory represents a turning point for the calculations of spin-dependent properties [15]. Before DFT, QM calculations of magnetic tensors were either prohibitively expensive for medium-sized radicals [16] or not sufficiently reliable for predictive and interpretative purposes. Today, the last-generation functionals coupled to purposely tailored basis sets allow to compute magnetic tensors in remarkable agreement with their experimental counterparts [17, 18]: computed data can take into proper account both average environmental effects and short-time dynamical contributions, for example, vibrational averaging from intramolecular vibrations and/or solvent librations [19–21], therefore providing a set of tailored parameters that can be confidently used for further calculations. On the basis of recent studies, the combination of the new basis set (N07D) [22–24] with PBE0 [25] and B3LYP [26] functionals seems to be the best choice for the calculation of several properties: geometrical parameters, dipole moments, and magnetic properties (hyperfine coupling constants and g-tensors). The N07D is a new polarized split-valence basis set for the second- and third-row atoms, which add a reduced number of polarization and diffuse functions to the 6-31G set, leading to an optimum compromise between reliability and computer time.

Hydrogen atoms (Table 3.1) require some specific considerations in view of the lack of inner shells and the overwhelming role of small hcc's in an unbiased statistics. In general, all DFT methods yield A_H values close to the experimental ones, and the best results are consistently delivered by the N07D basis set.

The performances of different basis sets for a large set of radicals containing second- and third-row atoms [22–24] are compared in Figure 3.1. The N07D results for all atoms are much better than those delivered by other (even significantly larger)

3.3 Spin Hamiltonian, g-Tensor, Hyperfine Coupling Constants, and Zero-Field Splitting

Table 3.1 Data analysis for hydrogen nuclei obtained with B3LYP functional.

	6-31G(d)	EPR-II	EPR-III	NO7D	Experimental
Hydrogen: $N=29$					
MAD	2.1	1.8	1.9	1.8	
Max Abs. Err.	11.3	6.0	6.0	9.0	
Average E%	25.3%	25.4%	25.4%	25.9%	
Max E%	113.1%	108.4%	104.6%	113.9%	
R_2	0.9946	0.9938	0.9934	0.9943	
Intercept	1.4936	0.8670	0.5675	1.0527	
Slope	0.9097	0.9666	0.9691	0.9051	
Max	121.5	128.2	128.7	120.0	132.7
Min	0.1	0.1	0.1	0.1	1.8

MAD (mean absolute deviation in Gauss) $= \Sigma |A_{calc} - A_{exp}|/N$; E% (percent error) $= |A_{calc} - A_{xp}|/A_{exp}$. The structures of the selected radicals are in Ref. [23].

basis sets both in terms of MADs and closeness of the slope of the linear regression to the theoretical value of 1.0.

The N07D basis set has also been assessed by comparison with EPR-III basis set for the g-tensors. The comparison with experimental results shown in Table 3.2 points

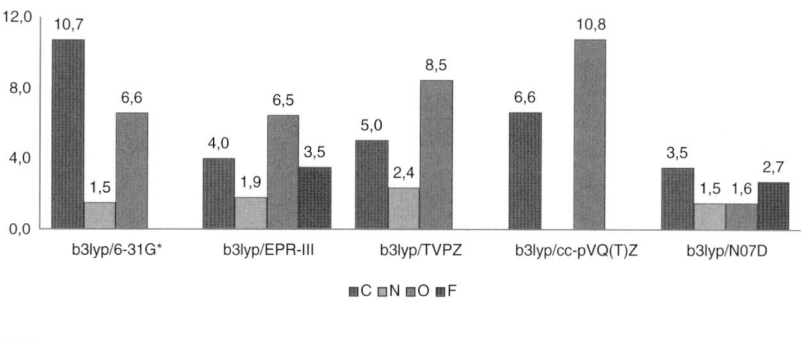

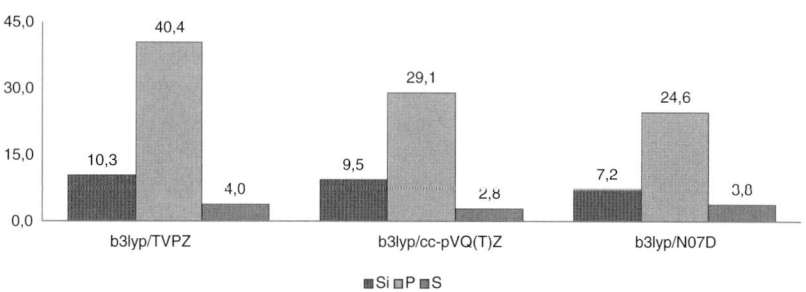

Figure 3.1 MAD (mean absolute deviation in Gauss) $= \Sigma |A_{calc} - A_{exp}|/N_{(total\ nuclei)}$ for the second- and third-row nuclei. The structures of the selected radicals and all statistical data are in Ref. [23].

Table 3.2 Data analysis for structures with second-row atoms.

	PBE0[a]	B3LYP[a]	PBE0/N07D	B3LYP/N07D	Experimental
60 Molecules[b]					
MAD	0.0006	0.0006	0.0005	0.0005	
Max Abs. Err.	0.0050	0.0050	0.0044	0.0048	
R_2	0.7303	0.7447	0.7403	0.7505	
Intercept	0.1046	0.0440	0.2439	0.1470	
Slope	0.9478	0.9781	0.8782	0.9266	
Max	2.0091	2.0091	2.0085	2.0089	2.0093
Min	1.9993	1.9993	1.9995	1.9994	2.0000

MAD (mean absolute deviation in Gauss) = $\Sigma |g_{calc} - g_{exp}|/N|$.
a) Single-point EPR-III calculations on geometries optimized the PBE0/N07D level.
b) The selected radicals are in Ref. [24].

out the accuracy of the computational approach that seems very promising for quantitative studies of the large systems.

3.4
Stereoelectronic, Environmental, and Dynamical Effects

In the following, we will make explicit reference only to organic free radicals so that only one unpaired electron will be considered and relativistic effects can be neglected. As a consequence, we will consider only hyperfine and gyromagnetic tensors without spin–orbit contributions.

3.4.1
Structures and Magnetic Parameters

The magnetic properties of a radical are tuned by small modifications of its geometry. Indeed, small geometry changes can be interpreted in terms of shape of the SOMO (single occupied molecular orbital), the orbital that directly affects the hyperfine coupling constants. For example, the deviation from planarity is the most influential geometric effect for π-like radicals (e.g., the methyl radical); the bond distance between specific pair of atoms in the radical is another one. In fact, the variation of the bond distance contributing to the π-system can modify the shape of the SOMO and thus the magnetic properties. In the same way, in a σ-radical (vinyl), the distance between two atoms modifies the localization of the SOMO.

As an illustration of the effect of the geometry changes on the EPR parameters (A and g-tensors), we have selected the class of nitroxide derivatives. In this case, there are two critical geometrical parameters of the molecular backbone, namely, the improper dihedral angle corresponding to the out-of-plane motion of the NO moiety and the nitroxide bond length. In order to gain further insight into the dependence of different molecular properties on these parameters, we have performed a molecular

3.4 Stereoelectronic, Environmental, and Dynamical Effects

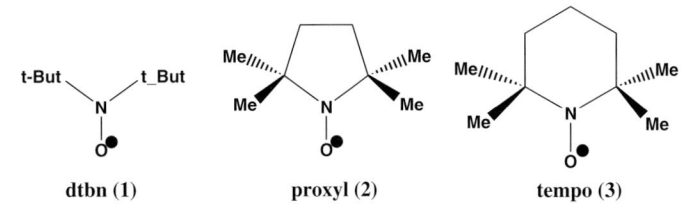

Figure 3.2 Structures of dtbn, proxyl, and tempo radicals.

dynamics run for proxyl (**2** in Figure 3.2) in the gas phase and computed the magnetic parameters for a significant number of snapshots [15].

As shown in Figure 3.3, the isotropic g-tensor shift (Δg_{iso}) is almost linearly dependent on the NO bond length, whereas it does not display any regular trend with respect to out-of-plane motion.

In summary, a good geometry is necessary to evaluate in the right way different effects, for example, the direct and spin polarization contributions to the hyperfine coupling constant. In this connection, the performance of the N07D basis set coupled with PBE0 and B3LYP functionals is comparable to those of much larger basis sets with increased computational efficiency.

While structural parameters are generally satisfactory, irrespective of the presence of diffuse functions on O, F, and Cl atoms, dipole moments are significantly

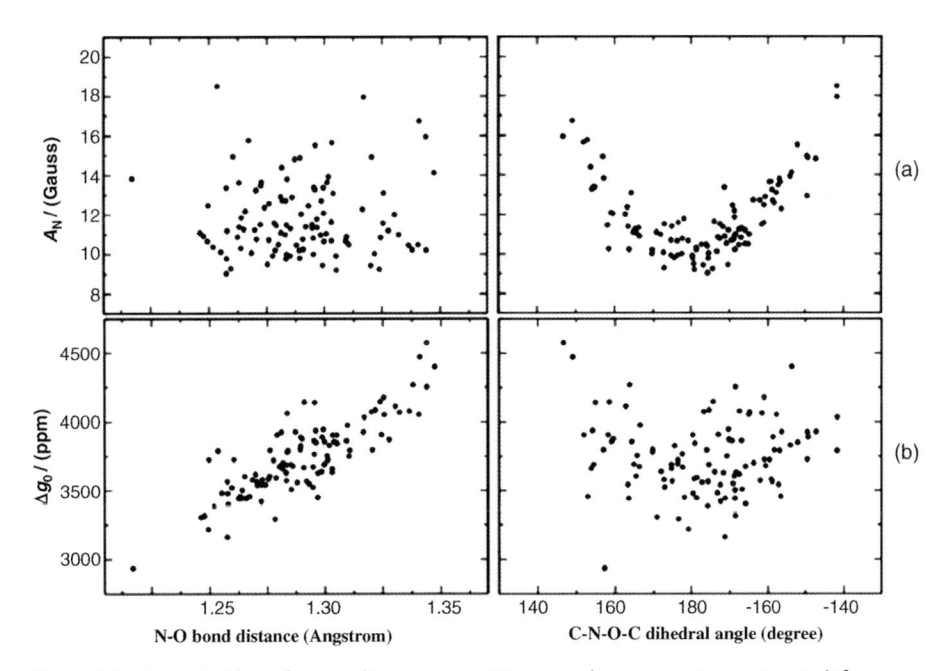

Figure 3.3 Computed hyperfine coupling constants (Gauss) and isotropic g-tensor (ppm) shift along Car–Parrinello molecular dynamic trajectory of proxyl in the gas phase.

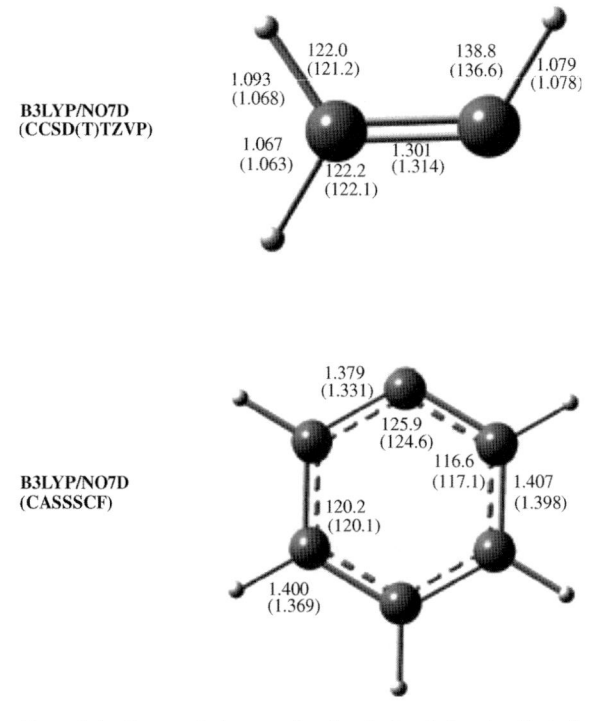

Figure 3.4 Geometrical parameters for vinyl and phenyl radicals (bond in Angstrom and angles in degrees) calculated at B3LYP/N07D and CCSD(T) theory levels.

improved by the addition of diffuse polarization functions, reaching quantitative agreement with experiment. Thus, the range of application of the B3LYP/N07D model is significantly enlarged by addition of diffuse polarization functions on electronegative atoms. As an illustration, we report in Figure 3.4 some significant parameters of vinyl and phenyl radicals.

It is quite apparent that B3LYP/N07D results are in remarkable agreement with the highly accurate computational studies at CCSD(T) [27, 28] and multireference [25] levels with extended basis sets. For C—C bond lengths, maximum deviations do not exceed 0.01 Å, and in many cases are as small as 0.002 Å. For C—H bond lengths, even better agreement has been found with a maximum discrepancy of 0.05 Å. The same stands for angles that are predicted by B3LYP/N07D model with accuracy of 2°. These findings are particularly encouraging having in mind larger systems for which expensive coupled cluster studies are still unfeasible.

3.4.2
Environmental Effects

The most promising general approach to the problem of environmental (e.g., solvent) effects can be based on a system–bath decomposition. The system includes the part of the solute where the essential of the process to be investigated is localized together with, possibly, the few solvent molecules strongly (and specifically) interacting with it. This part is treated at the electronic level of resolution and is immersed in a polarizable continuum, mimicking the macroscopic properties of the solvent. The solution process can then be dissected into the creation of a cavity in the solute (spending energy E_{cav}) and the successive switching on of dispersion–repulsion (with energy $E_{dis-rep}$) and electrostatic (with energy E_{el}) interactions with surrounding solvent molecules.

The so-called polarizable continuum model (PCM) [29] offers a unified and sound framework for the evaluation of all these contributions both for isotropic and anisotropic solutions. In the PCM, the solute molecule (possibly supplemented by some strongly bound solvent molecules to include short-range effects like hydrogen bonds) is embedded in a cavity formed by the envelope of spheres centered on the solute atoms. The procedures to assign the atomic radii [30] and to form the cavity have been described in detail together with effective classical approaches for evaluating E_{cav} and $E_{dis-rep}$ [31]. Here we recall that the cavity surface is finely subdivided into small tiles (tesserae) and that the solvent reaction field determining the electrostatic contribution is described in terms of apparent point charges appearing in tesserae and self-consistently adjusted with the solute electron density [32]. The solvation charges (q) depend, in turn, on the electrostatic potential (V) on tesserae through a geometrical matrix Q ($q = QV$) related to the position and size of the surface tesserae, so that the free energy in solution G can be written:

$$G = E[\varrho] + V_{NN} + \frac{1}{2}\mathbf{V}^\dagger \mathbf{Q}\mathbf{V} \tag{3.76}$$

where $E[\varrho]$ is the free solute energy, but with the electron density polarized by the solvent, and V_{NN} is the repulsion between the solute nuclei. The core of the model is then the definition of the Q matrix, which in the most recent implementations of PCM depends only on the electrostatic potentials, takes into the proper account the part of the solute electron density outside the molecular cavity, and allows the treatment of conventional, isotropic solutions, ionic strengths, and anisotropic media like liquid crystals. Furthermore, the analytical first and second derivatives with respect to geometrical, electric, and magnetic parameters have been coded, thus giving access to proper evaluation of structural, thermodynamic, kinetic, and spectroscopic solvent shifts.

Solvent can affect the electronic structure of the solute and, hence, its magnetic properties either directly (e.g., favoring more polar resonance forms) or indirectly through geometry changes. Furthermore, it can influence the dynamical behavior of the molecule: for example, viscous and/or oriented solvents (such as liquid crystals) can strongly dump the rotational and vibrational motions of the radical. Static aspects will be treated in the following, whereas the last one will be tackled in the section devoted to all the dynamical effects.

As an example, we illustrate the role of solvent effects on the EPR parameters of 2,2,6,6-tetramethylpiperidine-N-oxyl tempo (3 in Figure 3.2). The nitrogen isotropic hyperfine coupling constant (A_N) is tuned by the polarity of the medium in which the

R₁–N(–O•)–R₂ ⇌ R₁–N⁺(•)(–O⁻)–R₂

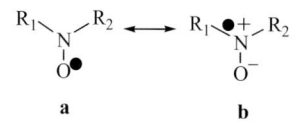

    a   b

Figure 3.5 Main resonance structures of nitroxide radicals.

nitroxide is embedded, as well as by the formation of specific hydrogen bonds to the oxygen radical center. Both factors contribute to a selective stabilization of the charge-separated resonance form of the NO functional group (Figure 3.5) with a consequent increase of A_N. Indeed, form b entails a larger spin density on nitrogen, which has a smaller spin–orbit coupling constant than oxygen.

The continuum solvent models (PCM) reproduce satisfactorily solvent effects on the A_N parameter only for aprotic solvents (bulk effects) [33], whereas there is a noticeable underestimation of solvent shifts for protic solvents (methanol and water), as showed in Figure 3.6. In these media, specific solute–solvent interactions also have to be taken into account [34].

In other words, since for solvents with H-bonding ability (methanol and water) the A_N of the nitroxide radical is shifted to higher values due to the influence of one or more hydrogen bonds between the solute and the solvent, it becomes necessary to build a model in which nonspecific effects are described in terms of continuum polarizable medium with a dielectric constant typical of the protic solvent under study, whereas specific effects are taken into account through an explicit hydrogen-bonded complex between radical and some solvent molecules. Figure 3.7 reports the A_N values for the complexes formed by tempo with methanol, and water measured experimentally at room temperature, and computed in the gas phase and in solution.

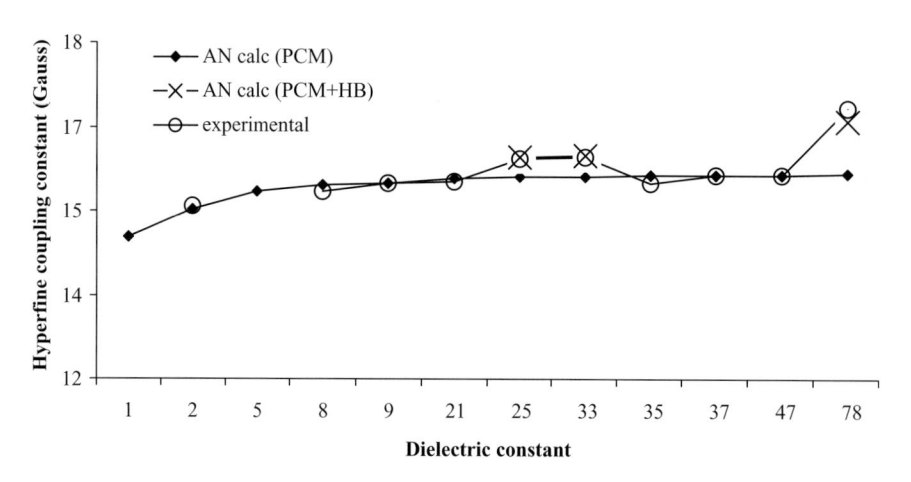

Figure 3.6 Experimental and calculated A_N values of tempo-choline [4-(N,N-dimethyl-N-(2-hydroxyethyl))ammonium-2,2,6,6-tetramethylpiperidine-1-oxyl chloride] as a function of the solvent dielectric constant.

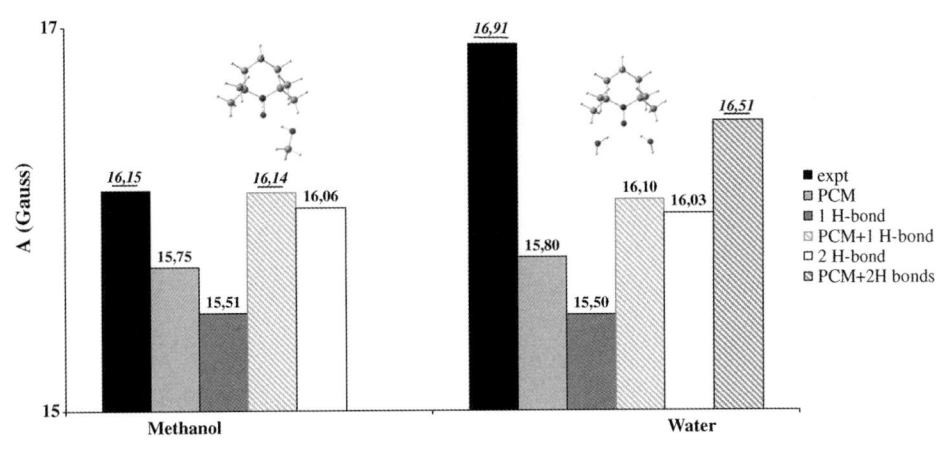

Figure 3.7 Computed and corresponding experimental A_N values (in Gauss) for the tempo–alcohol complexes in gas and in condensed phase. See text for details.

It is quite apparent that the values computed in solution fit well the experimental data.

From a more general perspective, the example at hand highlights a situation when PCM alone is unable to fully account for the solvent effects on spectroscopic properties (e.g., the A_N values in solution computed with PCM are 15.75 and 15.80 Gauss versus experimental values of 16.15 and 16.91 Gauss for methanol and water, respectively): this is typically related to the presence of strong, specific H-bond interactions. As shown in Figure 3.7, the inclusion of specific hydrogen bond effects results in a further increase of the computed A_N values, with final results close to their experimental counterparts (16.15 and 16.51 Gauss).

The accuracy of the cluster/PCM approach is so high that the computed EPR properties provide valuable indirect information on the nature of the H-bond network around the NO group. In the case of water, computed results in good agreement with experiment are obtained only when two explicit solvent molecules H-bonded to the nitroxyl moiety are introduced; in contrast, a single explicit solvent molecule is required for alcohols.

The same approach is able to reproduce the lowering of the isotropic g-value observed experimentally when going from nonprotic to protic solvents in terms of the reduced spin density on the oxygen atom: as a matter of fact, formation of intermolecular hydrogen bonds leads to a transfer of spin density from the oxygen to the nitrogen atom.

3.4.3
Short-Time Dynamical Effects

In the framework of the Born–Oppenheimer approximation, we can speak of a potential energy surface (PES) and of a "property surface" (PS), which can be obtained from quantum mechanical computations at different nuclear configurations. In this scheme, expectation values of observables (e.g., isotropic hcc's) are obtained by averaging the different properties on the nuclear wavefunctions. Semirigid molecules are well described in terms of a harmonic model, but a second-order perturbative inclusion of principal anharmonicities provides much improved results at a reasonable cost [35]. In perturbative model, the vibrational

$$E_n = \xi_0 + \Sigma_i \omega_i \left(n_i + \frac{1}{2}\right) + \Sigma_i \Sigma_{j<i} \xi_{ij} \left(n_i + \frac{1}{2}\right)\left(n_j + \frac{1}{2}\right) \quad (3.77)$$

where ω are the harmonic wave numbers and ξ are the simple functions of third (F_{ijk}) and semidiagonal fourth (F_{iijj}) energy derivatives with respect to normal modes Q [33]. Next, the fundamental vibrational frequencies (ν_i) and the zero-point vibrational energy (E_0) are given by

$$\nu_i = \omega_i + 2\xi_i + \frac{1}{2}\Sigma_{j\neq i}\xi_{ij} \quad (3.78)$$

$$E_0 = \xi_0 + \frac{1}{2}\Sigma_i\left(\omega_i + \frac{1}{2}\xi_{ii} + \Sigma_{j>i}\frac{1}{2}\xi_{ij}\right) \quad (3.79)$$

At the same level, the vibrationally averaged value of a property Ω is given by

$$\langle\Omega\rangle_n = \Omega_e + \Sigma_i A_i\left(n_i + \frac{1}{2}\right) \quad (3.80)$$

where Ω_e is the value at the equilibrium geometry and

$$A_I = \frac{\beta_{ii}}{\omega_i} - \Sigma_j\left(\frac{\alpha_j F_{iij}}{\omega_i \omega_j^2}\right) \quad (3.81)$$

where α_i and β_{ii} are the first and second derivatives of the property with respect to the ith normal mode. The first term on the rhs of Equation 3.81 will be referred to in the following as harmonic and the second one as anharmonic. Also, in this case, methods rooted in the density functional theory perform well provided that numerical problems are properly taken into account [36].

Here we just discuss vibrational averaging effects related to inversion at the radical center of typical π (methyl)- and σ (vinyl)- radicals.

The results shown in Figure 3.8 for vinyl radical point out the effect of harmonic and anharmonic terms in Equation 3.81 on the final results. The harmonic contribution is the most important (although anharmonic terms are not negligible) except for H^α and, especially, C^α. As a matter of fact, for this latter atom, there is a nearly exact compensation between harmonic contributions by in-plane and out-of-plane bendings, which have opposite signs. On the other hand, anharmonic terms are particularly large for C^α and H^α since they are not negligible for in-plane bending at the radical center, whereas they vanish (due to symmetry) for out-of-plane bendings.

The methyl radical has a planar equilibrium structure with a low-frequency out-of-plane motion. The behavior of hcc's as a function of the out-of-plane angle (θ) is the

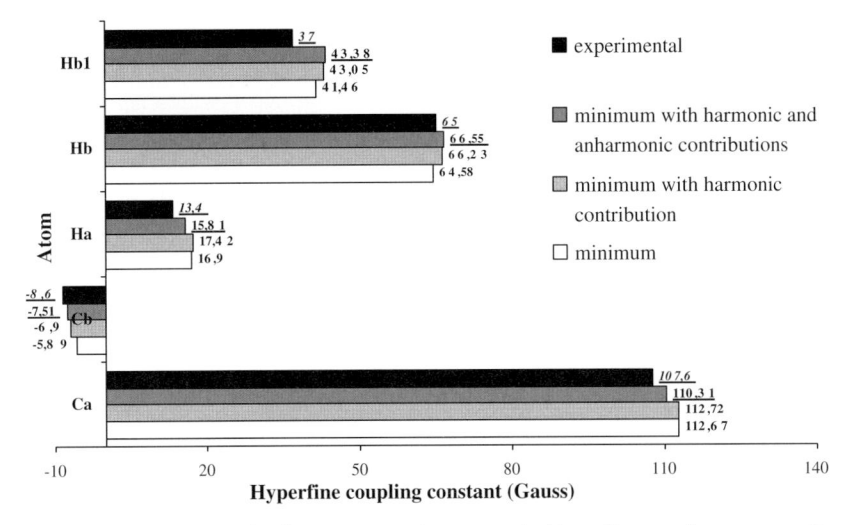

Figure 3.8 Harmonic and anharmonic contributions to vinyl hyperfine coupling constants from PBE0/N07D computations are compared to experimental results.

following: A_C is always positive and increases with θ due to the progressive contribution of carbon s orbitals to the SOMO. The effect is similar for A_H, but since A_H is negative for the planar structure (due to spin polarization), the absolute value of A_H decreases up to $\theta = 10°$ and then increases due to the direct contribution. The ground vibrational function is peaked at the planar structure. Vibrational averaging then changes the coupling constant toward values that would have been obtained for pyramidal structures in a static description. The wavefunction of the ground vibrational state being symmetrically spread around the planar reference configuration introduces contributions of pyramidal configurations. The effect is even more pronounced in the first excited vibrational state, whose wavefunction has a node at the planar structure and is more delocalized than the fundamental one, thus giving increased weight to pyramidal structures.

As a first approximation, the vibrationally averaged value of a property can be written as

$$\langle \Omega \rangle = \Omega_{\text{ref}} + \left(\frac{\partial \Omega}{\partial s} \right)_{\text{ref}} \langle s \rangle + \frac{1}{2} \left(\frac{\partial^2 \Omega}{\partial s^2} \right)_{\text{ref}} \langle s^2 \rangle \tag{3.82}$$

For the planar reference structures, the linear term is absent for symmetry reasons and the key role is played by mean square amplitudes, which, however, can be quite large and badly described at the harmonic level. From a quantitative point of view, vibrational averaging changes the equilibrium value of A_H by about 2 G (10%) and that of A_C by about 10 G (30%): thus, quantitative (and even semiquantitative) agreement with experiment cannot be obtained by static models, irrespective of the quality of the electronic model.

The low-frequency motions of vinyl radical correspond to out-of-plane vibrations (wagging and torsion) and in-plane inversion at the radical center. The out-of-plane motions have the same effect as the methyl inversion, albeit with a significantly smaller strength. On the other hand, in-plane inversion is characterized by a double-well potential with a significant barrier. Vibrational averaging now acts in an opposite direction, bringing the coupling constants to values that would have been obtained for less bent structures in a static description. The ground-state vibrational wavefunction is more localized inside the potential well, even under the barrier, than outside. So it introduces more contributions of "nearly linear" structures. Vibrational effects, while still operative, are less apparent in this case since high-energy barriers imply high vibrational frequencies with the consequent negligible population of excited vibrational states and smaller displacements around the equilibrium positions. Unless Boltzmann averaging gives significant weight to states above the barrier, this kind of vibration is effectively governed by a single-well potential unsymmetrically rising on the two sides of the reference configuration. Now $\partial\Omega/\partial s$ and $\langle s \rangle$ do not vanish and usually have opposite signs, thus counterbalancing the positive harmonic term. The resulting correction to Ω_e is small and can be treated by perturbative methods. The other large amplitude motions that have a significant effect on hcc's are internal rotations. However, in most cases, these can be treated by a simple average of the hcc's of different substituents.

Large amplitude motions and solvent librations cannot be described by the perturbative approach sketched above, but a classical treatment is usually sufficient. Then, the computational strategy involves two independent steps: first, MD simulations are run for sampling with one or more trajectories the general features of the solute–solvent configurational space; then, observables are computed exploiting the discrete/continuum approach for supramolecular clusters, made of the solute and its closest solvent molecules, as averages over a suitable number of snapshots. It is customary to carry out the same steps also for the molecule in the gas phase just to have a comparison term for quantifying solvent effects. The *a posteriori* calculation of spectroscopic properties, compared to other on-the-fly approaches, allows us to exploit different electronic structure methods for the MD simulations and the calculation of physical–chemical properties. In this way, a more accurate treatment for the more demanding molecular parameters, of both first (hyperfine coupling constants) and second (electronic g-tensor shifts) orders, could be achieved independent of structural sampling methods: first-principles, semiempirical force fields, as well as combined quantum mechanics/molecular mechanics approaches could be all exploited to the same extent once the accuracy in reproducing reliable structures and statistics is proven.

The extension of the discrete/continuum model described above to dynamical treatment has been recently formulated and validated [37]. In the framework of formally monoelectronic QM methods (e.g., Hartree–Fock or Kohn–Sham model), if $E_{QM/MM}(P_0, x)$ is the QM/MM (molecular mechanics) gas-phase energy of the explicit system expressed as a function of the nuclear coordinates, x, and the unpolarized (no solvent effects) one-electron density matrix, P_0, then the solvation free energy, $\Delta A_{sol}(x)$, at a specific molecular configuration can be written as the sum of the internal energy plus the so-called "mean field" (or potential of mean force) contribution that accounts for the interactions with the environment (solvent) minus the gas-phase energy:

$$\Delta A_{sol}(x) = (E_{QM/MM}(P, x) + W(P, x)) - E_{QM/MM}(P_0, x) \quad (3.83)$$

where $\Delta A_{sol}(x)$ is the free energy of the system at a given molecular configuration and $W(P, x)$ is the mean field term. Note that P is explicitly present in the first two terms on the rhs to remark that they are mutually polarized, that is, the mean field response is always considered at equilibrium and the electronic charge distribution is determined by a self-consistent calculations. In particular, we have integrated the mean field contributions as a modification of the ONIOM scheme for the isolated systems. The mean field W is the potential experienced by the explicitly treated molecules in a given configuration $\{x\}$ due to the average interactions with the environment. A number of discrete/continuum models have been proposed in the literature that differ in the way W is approximated. Here, according to the Ben-Naim's definition of the solvation process, we can conveniently assume that the mean field potential is composed of conceptually simple terms: a long-range electrostatic contribution, due to the linear response of the polarizable dielectric continuum, and a short-range dispersion–repulsion contribution, which accounts effectively for the interactions in proximity to the cavity boundary, $W = W_{elec} + W_{disp-rep}$. In the following, we describe the essential features of the GLOB model, a sophisticated and integrated method recently developed in our group that allows studying efficiently the solvent effects on generic solute molecules along with a few explicit solvent molecules and using different levels of theory, from computationally inexpensive, but less accurate, MM methods to more realistic hybrid QM/MM or full QM methods.

According to the GLOB model, the explicit system (solute + solvent) is embedded into a suitable cavity of a dielectric continuum possibly with a regular and smooth shape, such as a sphere, an ellipsoid, or a spherocylinder. In combination with molecular dynamics techniques, such a cavity could be kept fixed, corresponding to NVT ensemble conditions, or allowed to change volume, according to NPT ensemble simulations (see below). The long-range electrostatic interactions between the system and the dielectric continuum are modeled by means of the conductor-like polarizable continuum model (CPCM) [30b]. The continuum medium, which mimics the response of liquid bulk, is completely specified by a few parameters, for example, the dielectric permittivity (ε_r), and does depend on the nature of the solvent and the physical conditions, such as density and temperature. To be specific, the reaction field Φ_{RF}, that is, the electrostatic potential due to the induced polarization of the dielectric, is described in terms of apparent surface charges (q) centered on small tiles, called tesserae, which are the results of a fine subdivision of the cavity surface into triangular area elements of about equal size, and computed by a self-consistent calculation with respect to the solute electronic density. The computation of q requires the solution of a system of N_{tes} linear equations, with N_{tes} being the number of tesserae:

$$\mathbf{D} \cdot \mathbf{q} = -\Phi_I \quad (3.84)$$

where **q** is the array of the "apparent surface charges," Φ_I is the electrostatic potential evaluated at the center of each tessera due only to the charge distribution of the system, and **D** is a matrix that depends only on the surface topology and on the dielectric constant.

Hence, for a given molecular configuration of the explicit system, x, q are determined from Equation 3.84 and the corresponding free energy, W_{elec}, is given by

$$W_{elec} = -\frac{1}{2}\Phi^+ \Phi^{-1} \Phi \tag{3.85}$$

The dispersion–repulsion contribution $W_{disp-rep}$, which is related to the short-range solvent (explicit)–solvent (implicit) interactions, has been introduced to remove any possible source of physical anisotropy in proximity to the cavity surface, that is, deviation from bulk behavior. According to the several other methodologies developed in the framework of QM continuum models, we have also treated $W_{disp-rep}$ as a classical mean force potential not perturbing the system electronic density. In particular, $W_{disp-rep}$ is obtained from an effective empirical procedure parameterized on structural and thermodynamic properties. Briefly, we have assumed that $W_{disp-rep}$ can be represented by an effective potential acting on each explicit solvent molecule irrespective of the others, depending only on the molecule distance and, possibly orientation, with respect to the cavity surface. Further, $W_{disp-rep}$ is expanded in a series of terms corresponding to increasing levels of approximation, as

$$W_{disp-rep} = W^0_{disp-rep} + W^1_{disp-rep} + \cdots \tag{3.86}$$

As an example, the first term $W^0_{disp-rep}$, which depends only on the distance of the center of mass of the solvent molecule from the cavity surface, does ensure an isotropic density distribution of the liquid at the interface with the continuum, thus avoiding artifacts in the simulations due to the presence of a physical boundary as observed in other continuum-based methodologies. Analogously, higher order terms are introduced, if needed, to prevent other possible physical deviations from liquid bulk as the solvent polarization effect that may appear by using discrete/continuum models. Hence, $W_{disp-rep}$ can be expressed in a simple general form as

$$W_{disp-rep} = \Sigma_i \lambda(r_i) \tag{3.87}$$

where $\lambda(r_i)$ is the potential acting on the ith molecule and the sum is extended over the total number of explicit solvent molecules. The basic idea that has been followed to derive the dispersion–repulsion free energy term consists in building up such a potential "on the fly" from a test simulation of a neat liquid by discretizing the distance from the cavity boundary with a set of equally spaced Gaussian functions, whose heights are adjusted after a certain time interval on the basis of the local density. It is worth noting that the so obtained $W_{disp-rep}$ term is parametrized for a given solvent at specific physical conditions (e.g., density and temperature), but we can reasonably assume that it is constant for any solution of the same solvent irrespective of the cavity size and shape, provided that the boundary surface is smooth and the number of explicit solvent molecules sufficiently large.

For purposes of illustration, we consider a prototypical nitroxide spin probe molecule di-*tert*-butyl nitroxide (dtbn in Figure 3.2) analyzing the first-principle MD simulations in aqueous solution and, for comparison, in the gas phase. The results can be summarized in three main points: the effect of the solvent on the internal dynamics of the solute, the very flexible structure of the dtbn–water H-bonding network, and the rationalization of the solvent effects on the magnetic parameters. Magnetic parameters are quite sensitive to the configuration of the nitroxide backbone, and in the particular case of dtbn [38, 39], the out-of-plane motion of the nitroxide moiety is strongly affected by the solvent medium. While the average structure in the gas phase is pyramidal, the behavior of dtbn in solution presents a maximum probability of finding a planar configuration: this does not mean that the minimum in solution is planar, but that there is a significant flattening of the potential energy governing the out-of-plane motion and that the solute undergoes repeatedly an interconversion among pyramidal positions. The vibrational averaging effects of this large amplitude internal motion have been taken into account by computing the EPR parameters along the trajectories. The H-bonding network embedding the nitroxide moiety in aqueous solution presents a very interesting result: the dynamics of the system points out the presence of a variable number of H-bonds, from zero to two, with the highest probability of only one genuine H-bond. Such a feature of the dtbn–water interaction is actually system dependent, the high flexibility of the NO moiety and the steric repulsion of the *tert*-butyl groups decrease the energetically accessible space around the nitroxide oxygen.

As a matter of fact, simulations carried out in the same conditions and with the same level of theory, for a more rigid five-ring nitroxide (proxyl in Figure 3.2), in aqueous solution provided a different picture with an average of two nitroxide water H-bonds. In this case, the substituents embedding the NO moiety are constrained in a configuration where methyl groups are never close to the nitroxide oxygen, and also the backbone of the nitroxide presents an average value of the CNC angle that is lower than in the case of the dtbn, thus evidencing a better exposition of the NO moiety to the solvent molecules in the case of the proxyl radical. Nevertheless, the behavior of the closed-ring nitroxide in water could not be generalized to all the protic solvents: a similar simulation of the proxyl molecule in methanol solutions presents, on average, only one genuine solute–solvent H-bond, possibly because the more crammed H-bonded methanol molecule prevents an easy access to the NO moiety for other solvent molecules. The H-bonding picture arising from these simulations is depicted in Figure 3.9. Once again the reliable description of solvent dynamics plays a crucial role for an accurate prediction of spectroscopic data. Eventually, the discrete–continuum approach allowed the decoupling of the different contributions and also the quantification of their effect on each of the molecular parameters: the H-bonding interaction and the dielectric contribution of the solvent.

Comparison between experimental and computed EPR properties of "nonconjugated" nitroxides has been the subject of several studies; on the other hand, comparatively much less work has been devoted to the corresponding aromatic

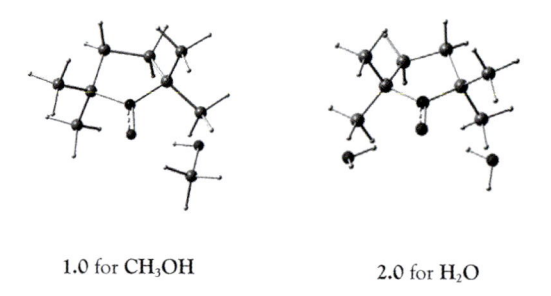

96 | *3 Calculation of Magnetic Tensors and EPR Spectra for Free Radicals in Different Environments*

1.0 for CH$_3$OH   2.0 for H$_2$O

Figure 3.9 Average number of solute–solvent hydrogen bonds.

systems. This is quite surprising in view of the remarkable thermal and chemical stability of aromatic nitroxides, in which the −NO spin density is delocalized over the aromatic ring system. These radicals are studied in several fields, as potential DNA intercalators, in the polymers to protect the material by retarding oxidative damage as well as generating fluorescent indicator for the degree of degradation to which the material has been exposed, in the synthesis of aromatic polyradical, or as the building block for the organic magnetic materials.

The performances of QM methods validated for several classes of organic free radicals proved till now quite disappointing for aromatic nitroxides. This prompted us to hypothesize that other effects (in particular, vibrational averaging) could play a significant role in determining some EPR parameters (especially the nitrogen isotropic hyperfine coupling). We have thus undertaken a systematic study of 1,2-dihydro-2methyl-2-phenyl-3H-indole-3-oxo-N-oxyl molecule, hereafter referred to as INDCO (Figure 3.10) [40, 41], a radical often used in organic synthesis in which a flexible five-ring nitroxide is conjugated with an aromatic system responsible for the spin delocalization. Therefore, accurate results for such a system demands for the correct evaluation of both factors (flexibility and delocalization), issuing from well-defined physical–chemical effects and not by simple parameter optimization.

Let us start from static computations at optimized geometries. The INDCO nitroxide with R-stereochemical configuration is constituted by three rings, labeled A, B1, and B2, respectively. The A ring is a cyclic five-member nitroxide, whereas B1 and B2 are aromatic systems. Geometry optimization (PBE0/N07D) of INDCO gives an almost flat structure, characterized by an improper CNOC angle of 176.6° and a

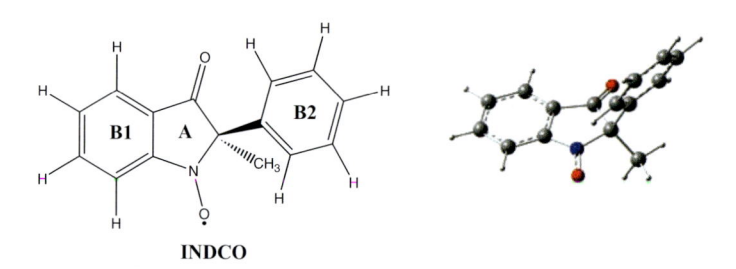

Figure 3.10 Structure of INDCO.

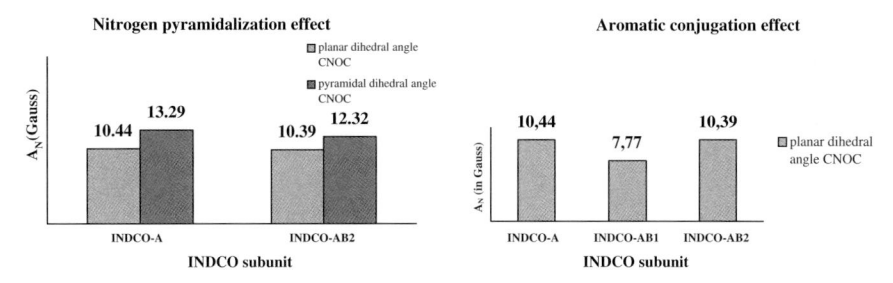

Figure 3.11 Calculated (PBE0/N07D) hyperfine coupling constant of nitrogen atom (A_N in Gauss) in INDCO subunits.

nitrogen isotropic hyperfine coupling constant (A_N) of 7.90 Gauss. In order to analyze the role of delocalization and other stereoelectronic effects, we have analyzed several substructures of INDCO: (1) the ring A with planar and pyramidal nitrogen; (2) the ring A supplemented by ring B1 or B2; (3) the above moieties in which methyl groups are replaced by hydrogen atoms. The results collected in Figure 3.11 show that two effects play a dominant role: out-of-plane deformation of the nitrogen environment shifts the A_N value by 2.85 Gauss (10.44 and 13.29 G for planar and pyramidal structures of ring A, respectively) and the presence of the ring B1 shifts A_N by 2.67 Gauss (10.44 versus 7.77 G for A and AB1 moieties, respectively).

The computed A_N for the planar conformation of INDCO is lower than the value obtained for the pyramidal structure because a nearly planar environment of nitrogen leads to the lack of any contribution of nitrogen s orbitals to the SOMO with the consequent strong reduction of A_N. The second trend points out that the conjugated aromatic ring is responsible for a significant spin delocalization, with the resulting decrease of the hyperfine coupling constant. Indeed, the second aromatic ring that is not conjugated with A does not have significant effects (0.05 Gauss). The effect of the methyl groups close to the –NO moiety is negligible (for ring A, 10.26 versus 10.52 G with four –H and four –CH$_3$, respectively).

The magnetic parameters of INDCO collected in Figure 3.12 show a fair agreement with experiment concerning hydrogen atoms and point out the role of delocalization in determining the nitrogen hyperfine coupling in aromatic nitroxides, whose final value is, however, lower than its experimental counterpart by more than 1G.

In order to take vibrational averaging effects into the proper account, we then resorted to a classical treatment. The analysis of the classical molecular dynamics simulations reported in Figure 3.12 shows that the average structure of INDCO is slightly pyramidal with an average CNOC improper dihedral angle of 168.4°. An analysis of the relationships between structural fluctuations measured along the MD trajectory and the computed EPR observables shows that the nitrogen hyperfine coupling constant mainly depends on the out-of-plane motion of the nitroxide group, but is apparently uncorrelated to the NO bond length. Conversely, the isotropic g shift is mainly affected by changes in the nitroxide bond length, but is quite insensitive to the C-N-O-C improper dihedral. Furthermore, the behavior of the g-tensor is

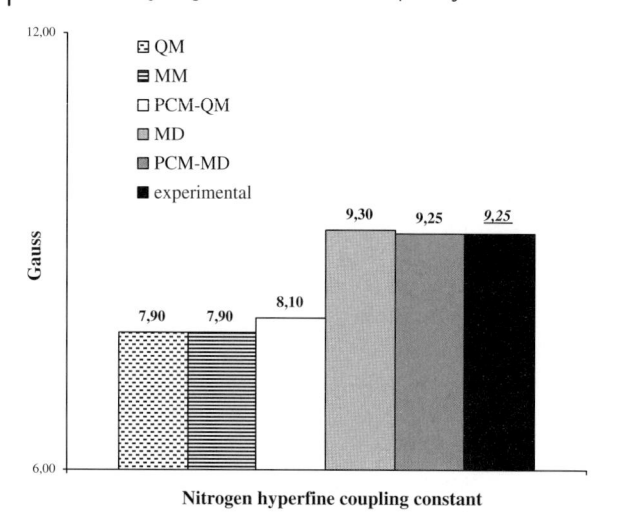

Figure 3.12 Hyperfine coupling constant (in Gauss) of the nitrogen in the INDCO molecule calculated on the minimum structure in gas phase (PBE0/N07D: A_N calculated on the QM minimum; PBE0/N07D: A_N calculated on the MM minimum) and in benzene with the PCM approach (PCM/PBE0/N07D) as well as the average upon the molecular dynamics trajectory in vacuum (PBE/N07D/MD) and in benzene (PCM/PBE0/N07D/MD).

dominated by its largest component g_{xx}, so that Δg_{iso} and g_{xx} show parallel trends. In all the calculations, the principal axes of the tensor are aligned with the NO bond (taken by convention x-axis) and with the average direction of π-orbitals (z-axis).

The overall vibrational correction of $\mathbf{A}_N$ amounts to 1.2 G ($\mathbf{A}_N$ = 9.1 Gauss in gas phase and 9.3 Gauss in benzene). The final result (whose contributions are summarized in Figure 3.12) is in remarkable agreement with experiment (9.25 G).

Our results show that a correct evaluation of dynamical and solvent effects is mandatory for a correct reproduction of the experimental hyperfine coupling constant of nitrogen. This is the reason why previous attempts have not been able to correctly describe the EPR features of such kind of radicals. On the other hand, dynamical effects do not influence the g-tensor since the computed values for the energy minimum and as average upon the MD trajectories are identical (2.00593) and very close to the experimental value of 2.00598 reported by Stipa [40].

3.5
Line Shapes

The radical species relaxation processes can be modeled as diffusive stochastic processes, so an important parameter needed is the (generalized) diffusion tensor of the molecule, which includes translational, rotational, and internal interaction terms, and depends on the molecular geometry. A reasonable estimate of the diffusion tensor can be achieved via a mesoscopic hydrodynamic approach [42]. The molecule is described as a set of N spheres immersed in a fluid, and it is partitioned into a number of rigid fragments connected by mobile bonds (torsional angles). While in the relatively trivial case of a rigid molecule there are only six degrees of freedom, position of the center of mass, and global orientation to be considered, in the case of flexible molecules, the torsional (internal) angles also need to be considered. Using geometrical arguments, it is possible to show that starting from the complete friction tensor of the unconstrained system of spheres ξ, which is represented by a $3N \times 3N$ matrix, and from the "constraints matrix" $\mathbf{B}$, which is a $3N \times n$ matrix, n being the number of degrees of freedom, the friction tensor for the constrained system (molecule) is given by

$$\Xi = \mathbf{B}^+ \xi \mathbf{B} \tag{3.88}$$

and the diffusion tensor is obtained from the Einstein's relation $\mathbf{D} = kT\Xi^{-1}$. This is the total diffusion tensor that is represented by a square symmetric positive definite matrix having dimension $n = 6 + N_{\text{tors}}$, where N_{tors} is the number of internal torsional angles. The tensor contains all the translational, rotational, and internal parts and coupling terms. This methodology showed to have good predictivity for the evaluation of the diffusion tensor of molecules of any size.

Once magnetic, structural, and dissipative parameters have been estimated, it is necessary to solve the SLE following, for instance, a standard variational approach, by spanning the Liouville operator over a proper basis set, obtained as the direct product of the space of spin transitions and the rotational space defined by stochastic variables. Details on the definition of the basis can be found elsewhere [43]. The spectrum is evaluated by standard algebraic methods. Matrix dimensions in a typical cw-ESR simulation can be quite large. As an example, the simplest case of a rigid nitroxide probe with one electron interacting with a nitrogen nucleus and subject to free fast rotation needs roughly 1260 basis functions, while adding a second equal probe the dimension makes this number grow to 45 360, that is, simply adding another electron and another coupled nitrogen nucleus, the dimension of the basis grows one order of magnitude. This makes it impossible to use the simple techniques for solving algebraic problems, even if the matrix associated with the Liouvillean is sparse. Symmetry arguments can be taken into account in order to reduce the dimensions, but in many cases the number of basis functions required remains high. A powerful numerical tool to deal with such big sparse matrices is the Lanczos algorithm [1], which is a well-known iterative method that at every step builds a tridiagonal approximation of the Liouvillean matrix. It has been shown that Lanczos and related algorithms are particularly apt to evaluate efficiently cw-ESR since convergence is usually achieved in a number of steps much smaller than the dimension of the basis (about 10%) with sufficient accuracy to extract the most important eigenvalues. Moreover, the diagonal α and subdiagonal β elements of the generated tridiagonal symmetric matrix 8 can be used to directly calculate the spectral density (i.e., the line shape) as a continued fraction. Handling big matrices does not imply only large calculation times, a problem that is in part solved by using iterative

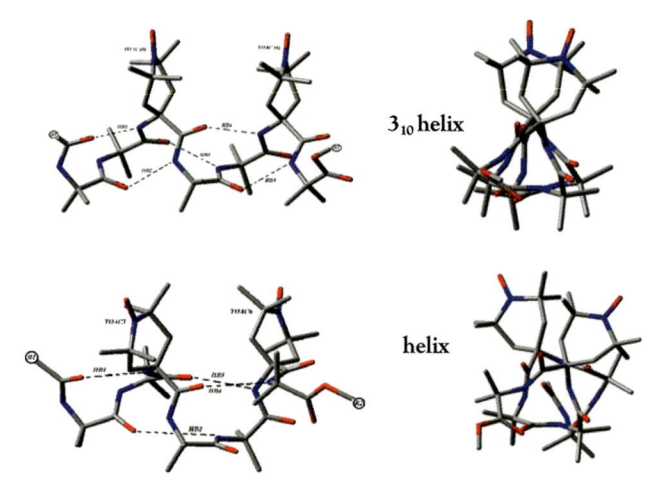

Figure 3.13 3_{10}- and α-helix conformations of the peptide Fmoc-(Aib-Aib-TOAC)2-Aib-Ome.

algorithms. It also carries problems with storage. Both time and memory problems can be solved by proper computational approaches.

Let us consider, for purposes of illustration, a small peptide with structure Fmoc-(Aib-Aib-TOAC)$_2$-Aib-OMe (Fmoc, fluorenyl-9-methoxycarbonyl; Aib, R-aminoisobutyric acid; TOAC, 2,2,6,6-tetramethylpiperidine-1-oxyl-4-amino, 4-carboxylic acid; and OMe, methoxy), which is characterized by the presence of two TOAC nitroxide free radicals at relative positions $i, i + 3$ (Figure 3.13). Spectra were collected at several temperatures in four different solvents: acetonitrile, chloroform, methanol, and toluene. Aib and TOAC are two strongly helicogenic, C^α-tetrasubstituted α-amino acids. The cw-ESR spectra have been compared with their theoretical counterparts pertaining to the deepest energy minima obtained by QM computations (3_{10}- and α-helices) [44]. We found that in specific solvents, the experimental spectra agree well with those expected for the 3_{10}-helix, in other solvents with those predicted for the α-helix, while for a final set of solvents with those associated with a mixture of α- and 3_{10}-helices with temperature-dependent relative percentages.

Modeling was based on the construction of the spin Hamiltonian and on the definition of the diffusive operator for the motion. We examined the dependence on both solvent and temperature of the spectra. The effect of temperature on the experimental observations is substantially reflected in the variation of the diffusion tensor, through the temperature dependence of the viscosity of the solvents. Note that a common assumption for all solvents is the presence of a monoradical impurity that might arise from the reduction of one of the nitroxide functions. The estimated amount of the impurity is below 4%, a low but still appreciable percentage. Figure 3.14 collects the theoretical and experimental spectra calculated for the four solvents. From a chemical point of view, our results provide evidence on the property of Aib-rich peptides changing their conformation from 3_{10}-helix to α-helix as a function of increasing polarity and hydrogen-bond donor capability of

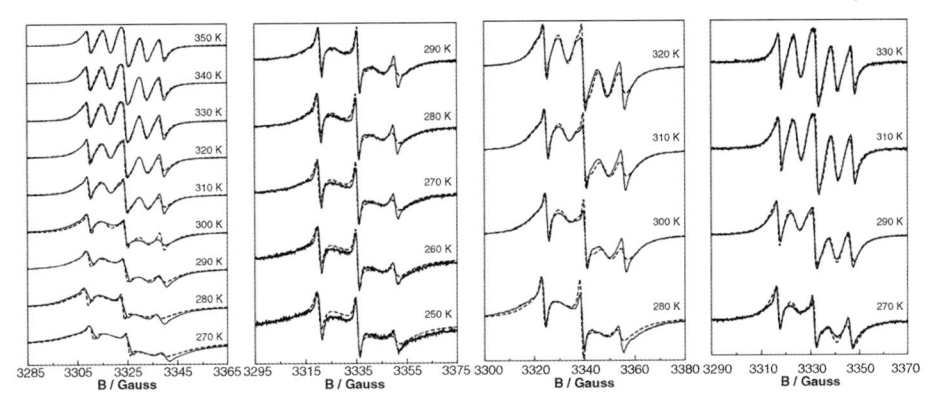

Figure 3.14 Spectra of the peptide in different solvents: toluene, chloroform, methanol, and acetonitrile, respectively. Solid lines: experimental spectra; dashed lines: simulated spectra.

the solvent: α-helix in protic solvents and at low temperature, whereas 3_{10}-helix in aprotic solvents.

In summary, feeding of magnetic and diffusion tensors derived from atomistic modeling in a general computational protocol based on the stochastic Liouville equation allowed us to reproduce in a remarkable way the ESR spectra in different solvents and at different temperatures without any adjustable parameter. The favorable scaling of the computational protocol with the dimensions of the system and its remarkable performances for both structural and magnetic properties might pave the route for systematic studies of spin-labeled peptides and proteins.

3.6
Concluding Remarks

In the present chapter, we have presented an integrated computational strategy for the study of structures and magnetic properties of organic free radicals. The first step has been the development of a new effective basis set (N07D), which coupled with hybrid density functionals (like B3LYP or PBE0) delivers remarkably accurate structures and physical–chemical properties of open-shell species, including magnetic parameters. Next, bulk solvent effects can be taken into account by continuum models (COSMO, PCM, etc.), whereas the treatment of the cybotactic region can be improved, when needed, by explicit consideration of specific first-shell solvent molecules. Finally, solute vibrations and, possibly, solvent libration effects can be described by classical molecular dynamics employing nonperiodic boundary conditions. A number of case studies show that only contemporary consideration of stereoelectronic, environmental, and dynamical effects allows to obtain magnetic properties of different classes of organic free radicals in diverse environments, without the need of any empirical parameter. Next we turned to line shapes and showed that inclusion of the above magnetic tensors together with diffusion tensors computed by a coarse-graining approach into a formalism rooted into the stochastic

Liouville equation allows to predict accurate spectra in different motional regimes without resorting to fitting procedures.

Together with its specific interest, our results point out, in our opinion, the importance of developing and validating computational approaches able to switch on and off different effects, including environmental and dynamical ones, in order to evaluate their specific role in determining the overall observable.

References

1. Moro, G. and Freed, J.H. (1986) in *Large-Scale Eigenvalue Problems,* Mathematical Studies Series, vol. **127** (eds J. Cullum and R. Willoughby), Elsevier, New York; Schneider, D.J. and Freed, J.H. (1989) *Adv. Chem. Phys.,* **73**, 487.
2. Osiecki, J.H. and Ullman, E.F. (1968) *J. Am. Chem. Soc.,* **90**, 1078.
3. (a) Barone, V. and Polimeno, A. (2006) *Phys. Chem. Chem. Phys.,* **8**, 4609; (b) Brustolon, M. and Giamello, E. (2009) *Electron Paramagnetic Resonance: A Practitioner's Toolkit,* John Wiley & Sons, Inc.
4. (a) Szabo, A. and Ostlund, N.S. (1982) *Modern Theoretical Chemistry,* Macmillan, New York; (b) Koch, W. and Holthausen, M.C. (2000) *A Chemist's Guide to Density Functional Theory,* Wiley-VCH Verlag GmbH, Weinheim; (c) Kaupp, M., Malkin, V., and Buhl, M. (2004) *The Quantum Chemical Calculation of NMR and EPR Properties,* Wiley-VCH Verlag GmbH, Heidelberg.
5. (a) Neese, F. (2007) *Electron Paramagn. Reson.,* **20**, 73; (b) Sinnecker, S. and Neese, F. (2007) *Top. Curr. Chem.,* **268**, 47; (c) Barone, V. and Polimeno, A. (2007) *Chem. Soc. Rev.,* **36**, 11.
6. Neese, F. (2006) ORCA: an ab initio density functional and semiempirical program package, Version 2.6.0, University of Bonn, Germany. Available at http://www.thch.uni-bonn.de/tc/orca/.
7. Frisch, M.J., Trucks, G.W., Schlegel, H.B., Scuseria, G.E., Robb, M.A., Cheeseman, J.R., Montgomery, J.J.A., Vreven, T., Kudin, K.N., Burant, J.C., Millam, J.M., Iyengar, S.S., Tomasi, J., Barone, V., Mennucci, B., Cossi, M., Scalmani, G., Rega, N., Petersson, G.A., Nakatsuji, H., Hada, M., Ehara, M., Toyota, K., Fukuda, R., Hasegawa, J., Ishida, M., Nakajima, T., Honda, Y., Kitao, O., Nakai, H., Klene, M., Li, X., Knox, J.E., Hratchian, H.P., Cross, J.B., Bakken, V., Adamo, C., Jaramillo, J., Gomperts, R., Stratmann, R.E., Yazyev, O., Austin, A.J., Cammi, R., Pomelli, C., Ochterski, J.W., Ayala, P.Y., Morokuma, K., Voth, G.A., Salvador, P., Dannenberg, J.J., Zakrzewski, V.G., Dapprich, S., Daniels, A.D., Strain, M.C., Farkas, O., Malick, D.K., Rabuck, A.D., Raghavachari, K., Foresman, J.B., Ortiz, J.V., Cui, Q., Baboul, A.G., Clifford, S., Cioslowski, J., Stefanov, B.B., Liu, G., Liashenko, A., Piskorz, P., Komaromi, I., Martin, R.L., Fox, D.J., Keith, T., Al-Laham, M.A., Peng, C.Y., Nanayakkara, A., Challacombe, M., Gill, P.M.W., Johnson, B., Chen, W., Wong, M.W., Gonzalez, C., and Pople F J.A. (2004) Gaussian 03, revision C.02, Wallingford, CT. http://www.gaussian.com/
8. (a) Nillson, J.A., Eriksson, L.A., and Laaksonen, A. (2001) *Mol. Phys.,* **99**, 247; (b) Nonella, M., Mathias, G., and Tavan, P. (2003) *J. Phys. Chem. A,* **107**, 8638; (c) Asher, J.R., Doltsinis, N.L., and Kaupp, M. (2005) *Magn. Reson. Chem.,* **43**, S237.
9. Pillet, S., Souhassou, M., Pontillon, Y., Caneschi, A., Gatteschi, D., and Lecomte, C. (2001) *New J. Chem.,* **25**, 131.
10. (a) Awaga, K., Inabe, T., Okayama, T., and Maruyama, Y. (1993) *Mol. Cryst. Liq. Cryst.,* **232**, 79; (b) Caneschi, A., Chiesi, P., David, L., Ferraro, F., Gatteschi, D., and Sessoli, R. (1993) *Inorg. Chem.,* **32**, 1445; (c) Caneschi, A., Ferraro, F., Gatteschi, D., Le Lirzin, A., Novak, M., Rentschler, E., and Sessoli, R. (1995) *Adv. Mater.,* **7**, 476;

(d) Gorini, L., Caneschi, A., and Menichetti, S. (2006) *Synlett*, **6**, 948; (e) Caneschi, A., David, L., Ferraro, F., Gatteschi, D., and Fabretti, A.C. (1995) *Inorganica Chim Acta*, **235**, 159.

11 McWeeny, R. and Mizuno, Y. (1961) *Proc. R. Soc. Lond. A*, **259**, 554.

12 Neese, F. (2003) *Curr. Opin. Chem. Biol.*, **7**, 125.

13 Neese, F. (2007) *Mol. Phys.*, **105**, 19.

14 Pederson, M.R. and Khanna, S.N. (1999) *Chem. Phys. Lett.*, **303**, 373.

15 (a) Barone, V.J. (1994) *Chem. Phys.*, **101**, 6834; (b) Barone, V.J. (1994) *Chem. Phys.*, **101**, 10666; (c) Barone, V. (1995) *Theor. Chem. Acc.*, **91**, 113; (d) Barone, V. (1995) *Advances in Density Functional Theory, Part I* (ed. D.P. Chong), World Scientific Publishing Co., p. 287; (e) Improta, R. and Barone, V. (2004) *Chem. Rev.*, **104**, 1231.

16 (a) Feller, D. and Davidson, E.R. (1988) *J. Chem. Phys.*, **88**, 5770; (b) Engels, B., Eriksson, L.A., and Lunell, S. (1996) *Advances in Quantum Chemistry*, vol. **27**, Academic Press, San Diego, CA, p. 297; (c) Perera, S.A., Salemi, L.M., and Bartlett, R.J. (1997) *J. Chem. Phys.*, **106**, 4061; (d) Al Derzi, A.R., Fan, S., and Bartlett, R.J. (2003) *J. Phys. Chem.*, **A107**, 6656.

17 Neese, F.J. (2001) *Chem. Phys.*, **115**, 11080.

18 Ciofini, I., Adamo, C., and Barone, V. (2004) *J. Chem. Phys.*, **121**, 6710.

19 (a) Barone, V. and Subra, R. (1996) *J. Chem. Phys.*, **104**, 2630; (b) Jolibois, F., Cadet, J., Grand, A., Subra, R., Barone, V., and Rega, N. (1998) *J. Am. Chem. Soc.*, **120**, 1864; (c) Barone, V. (2005) *J. Chem. Phys.*, **122**, 014108; (d) Barone, V. and Carbonniere, P. (2005) *J. Chem. Phys.*, **122**, 224308.

20 (a) Nillson, J.A., Eriksson, L.A., and Laaksonen, A. (2001) *Mol. Phys.*, **99**, 247; (b) Nonella, M., Mathias, G., and Tavan, P. (2003) *J. Phys. Chem.*, **A107**, 8638; (c) Asher, J.R., Doltsinis, N.L., and Kaupp, M.J. (2005) *Magn. Reson. Chem.*, **43**, S237.

21 (a) Pavone, M., Benzi, C., De Angelis, F., and Barone, V. (2004) *Chem. Phys. Lett.*, **395**, 120; (b) Pavone, M., Cimino, P., De Angelis, F., and Barone, V. (2006) *J. Am. Chem. Soc.*, **128**, 4338; (c) Pavone, M., Sillampa, A., Cimino, P., Crescenzi, O., and Barone, V. (2006) *J. Phys. Chem. B*, **110**, 16189.

22 Barone, V. and Cimino, P. (2008) *Chem. Phys. Lett.*, **454**, 139.

23 Barone, V., Cimino, P., and Stendardo, E. (2008) *J. Chem. Theory Comput.*, **4**, 751.

24 Barone, V. and Cimino, P. (2009) *J. Chem. Theory Comput.*, **5**, 192.

25 Adamo, C. and Barone, V. (1996) *J. Chem. Phys.*, **110**, 158.

26 Hariharan, P.C. and Pople, J.A. (1973) *Theor. Chim Acta*, **28**, 213.

27 Koziol, L., Levchenko, S.V., and Krylov, A.I. (2006) *J. Phys. Chem. A*, **110**, 2746.

28 Mebel, A.M., Chen, Y.-T., and Lin, S.-H. (1997) *Chem. Phys. Lett.*, **275**, 19.

29 Tomasi, J., Mennucci, B., and Cammi, R. (2005) *Chem. Rev.*, **105**, 2999.

30 Barone, V., Cossi, M., and Tomasi, J. (1997) *J. Chem. Phys.*, **107**, 3210.

31 Benzi, C., Cossi, M., Improta, R., and Barone, V. (2005) *J. Comput. Chem.*, **26**, 1096.

32 (a) Cossi, M., Scalmani, G., Rega, N., and Barone, V. (2002) *J. Chem. Phys.*, **117**, 43; (b) Cossi, M., Rega, N., Scalmani, G., and Barone, V. (2003) *J. Comput. Chem.*, **24**, 669.

33 Tedeschi, A.M., D'Errico, G., Busi, E., Basosi, R., and Barone, V. (2002) *Phys. Chem. Chem. Phys.*, **4**, 2180.

34 (a) Barone, V. (1996) *Chem. Phys. Lett.*, **262**, 201; (b) Rega, N., Cossi, M., and Barone, V. (1996) *J. Chem. Phys.*, **105**, 11060; (c) di Matteo, A., Adamo, C., Cossi, M., Rey, P., and Barone, V. (1999) *Chem. Phys. Lett.*, **310**, 159; (d) Adamo, C., di Matteo, A., Rey, P., and Barone, V. (1999) *J. Phys. Chem. A*, **103**, 3481; (e) Cimino, P., Pavone, M., and Barone, V. (2005) *Chem. Phys. Lett.*, **409**, 106.

35 (a) Clabo, D.A., Jr., Allen, W.D., Remington, R.B., Yamaguchi, Y., and Schaefer, H.F., III (1988) *Chem. Phys.*, **123**, 187; (b) Barone, V. and Minichino, C. (1995) *Theochem*, **330**, 325; (c) Barone, V. (2005) *J. Chem. Phys.*, **122**, 014108; (d) Ruud, K., Astrand, P.-O., and Taylor, P.R. (2000) *J. Chem. Phys.*, **112**, 2668; (e) Christiansen, O. (2007) *Phys. Chem. Chem. Phys.*, **9**, 2942.

36 (a) Barone, V. (1994) *J. Chem. Phys.*, **101**, 10666; (b) Dressler, S. and Thiel, W. (1977) *Chem. Phys. Lett.*, **273**, 71.
37 Brancato, G., Rega, N., and Barone, V. (2008) *Chem. Phys.*, **128**, 144501.
38 Rega, N., Brancato, G., and Barone, V. (2006) *Chem. Phys. Lett.*, **422**, 367.
39 Pavone, M., Cimino, P., De Angelis, F., and Barone, V. (2006) *J. Am. Chem. Soc.*, **128**, 4338.
40 Stipa, P. (2006) *Chem. Phys.*, **323**, 501.
41 Cimino, P., Pedone, A., Stendardo, E., and Barone, V. (2010) *Phys. Chem. Chem. Phys.*, **12**, 3741.
42 (a) Moro, G. (1987) *Chem. Phys.*, **118**, 167; (b) Moro, G. (1987) *Chem. Phys.*, **118**, 181.
43 Meirovitch, E., Igner, D., Igner, D., Moro, G., and Freed, J.H. (1982) *J. Chem. Phys.*, **77**, 3915.
44 Carlotto, S., Cimino, P., Zerbetto, M., Franco, L., Corvaja, C., Crisma, M., Formaggio, F., Toniolo, C., Polimeno, A., and Barone, V. (2007) *J. Am. Chem. Soc.*, **129**, 11248.

4
Generalization of the Badger Rule Based on the Use of Adiabatic Vibrational Modes

Elfi Kraka, John Andreas Larsson, and Dieter Cremer

4.1
Introduction

Empirical relationships relating bond lengths to the corresponding bond stretching frequencies or bond stretching force constants were first derived in 1920s (see Table 4.1 for a summary) and have ever since been a topic of research on the nature of the chemical bond [1–67]. It is remarkable that in a time of easily accessible quantum chemical results, there remains a need for empirically based estimates of either bond lengths or stretching frequencies. There are three primary reasons why such empirical rules and relationships are still valuable tools for modern research: (1) Established relationships between bond properties add to our understanding of the chemical bond, especially if they can be rationalized on a quantum mechanical basis because bonding between atoms is a quantum mechanical phenomenon. (2) There are experimental situations in which it is relatively easy to measure one bond property but difficult to obtain other bond properties. For example, it is easier to measure the vibrational spectra of a compound than to carry out a structural analysis. This is especially true for solid materials that do not crystallize, molecules on a surface, or molecules in some form of aggregation. If quantum chemical calculations are feasible only for model systems rather than the actual targets of chemical research, then vibrational spectroscopy may be the only tool for obtaining information that provides an insight into bond properties. (3) In the realm of computational chemistry, there is also a need for empirical relationships. They may be used to determine suitable bonding force fields for molecular mechanics utilizing force constant–bond length relationships. For quantum chemical geometry optimization, there is the need to set up a guess matrix of energy second derivatives (the Hessian matrix corresponding to the force constant matrix of a molecule), which is best done with the help of available geometry information and a suitable force constant–bond length relationship. For example, the standard procedure to calculate the geometry of a molecule is based on an initial guess of the energy Hessian derived with the help of the Badger rule [42, 55]. It is due to these three reasons that there is ongoing research exploring the relationships between bond length r, bond stretching

---

Computational Spectroscopy: Methods, Experiments and Applications. Edited by Jörg Grunenberg
Copyright © 2010 WILEY-VCH Verlag GmbH & Co. KGaA, Weinheim
ISBN: 978-3-527-32649-5

Table 4.1 Relationships between spectroscopic and geometrical constants of a bond in diatomic, quasidiatomic, and polyatomic molecules.

No	Year	Authors	Equation	Molecules	Comment	Ref.
1	1920	A. Kratzer	$\omega_e^2 I = $ const. $k_e r_e^2 = $ const.	D	I: moment of inertia, valid for hydrogen halides	[1]
2	1925	R. T. Birge	$\omega_e r_e^2 = $ const.	D	Different electronic states of a given D	[2]
3	1925	R. Mecke	$\omega_e r_e^2 = $ const.	D	Different electronic states of a given D	[3]
4	1929	P.M. Morse	$\omega_e r_e^3 = $ const.	D	Different D of similar kind	[4]
5	1934	C.H.D. Clark	$r_e^2 = C_{nm}(\mu/nk_e)^{1/3}$ $\omega_e r_e^3 n^{1/2} = $ const.	D	μ: reduced mass; m,n: group numbers for hydrides and nonhydrides	[5]
6	1934	R.M. Badger	$k_e(r_e - d_{ij})^3 = $ const.	D	d_{ij} is typical of AB with A in period i and B in period j; $r_e - d_{ij}$: effective bond length	[6]
7	1935	R.M. Badger	$r_e = (C_{nm}/k)^{1/3} + d_{nm}$	D	Constants typical of AB with A in group m and B in group n	[7]
8	1935	H.S. Allen, A.K. Longair	$r_e^2 = c_{ij}/k_e^{1/3}$ $k_e r_e^6 = $ const.	D	Constant typical of AB from a given period	[8]
9	1935	R.M. Badger	Same as 7	QD	Bonds in small symmetric polyatomic molecules	[9]
10	1935	M.L. Huggins	Equations for D_e, r_e, ω_e, k_e	D	Derived from modified Morse potential	[10]
11	1936	M.L. Huggins	$r_e = a - b\log(ck_e)$	D	Constants derived for periods, rule 6 explained	[11]
12	1938	G.B.B.M. Sutherland	$k_e r_e^6 n^{3/2} = $ const.	D	Related to 5	[12]
13	1940	G.B.B.M. Sutherland	$D_e = kr_e^2/nm$ $k_e = ma(n-m)r_e^{-(m+2)}$	D	$V = -ar^{-m} + br^{-n}$ (double-reciprocal V)	[13]

#	Year	Author	Formula	Type	Description	Ref.
14	1940	R.A. Newing	$r_e = r_e(\omega_e, x_e, D_e, \mu)$	D	Quantum mechanical arguments for existence of relationships	[14]
15	1940	J.W. Linnett	$k_e r_e^6 = $ const.	D	Testing of $V = -ar^{-m} - be^{-nr}$	[15]
16	1941	C.D. Clark, K.R. Webb	$k_e r_e^6 n = $ const. $k_e(r_e - d_{ij})^3 n^{1/2} = $ const.	D	Valid for molecules with similar μ modification of Badger rule; n as in 5	[16]
17	1942	J.W. Linnett	$k_e r_e^5 = $ const.	D	Same V as in 15	[17]
18	1944	C.K. Wu, C.T. Yang	$k_e = ar_e^{-(m+2)} + br_e^{-(m+1)}$	D	Constants typical of period	[18]
19	1945	J.W. Linnett	$k_e r_e^{7.6} = $ const.	P	For molecules with NH bond; for CH bonds 15 is better	[19]
20	1946	W. Gordy	$k_e = aN(x_A x_B/r_e^2)^{3/4}$	D	N: bond order; x_A, x_B: electronegativity of A, B	[20]
21	1946	K.M. Guggenheimer	$k = aN(z_A z_B)^{1/2} r_e^{-b}$	P, c-mode	N: bond order; z_A, z_B: valence electrons	[21]
22	1950	K.M. Guggenheimer		P (AX_n) c-mode	$b = 2.46, 1.84, 2.06$, and so on typical of bond polarity	[22]
23	1947	C.K. Wu, S.C. Chao	$\log k_e = ar_e + b$	D	Simplified version of 11	[23]
24	1950	G. Herzberg	Various formulas	D	Discussion of empirical rules, r_e and k_e changes over periods are shown for HX and OX bonds	[24]
25	1951	G. Lovera	$k_e = a(r_e + 1)^{-3.75}$	P, c-mode	Molecules of groups 4b–6b	[25]
26	1953	H. Siebert	$k_{CX} = aZ_X/n_X^3$ $r_X = a(n_X^3/Z_X)$ $N = a(k_N/k_1) + b$	P(XMe$_n$)	Z_X: atomic number; n_X principal quantum number of valence electrons for X; formula for k_{XY} r_X: covalent radius of X; N: bond order	[26]
27	1955	E.R. Lippincott, R. Schroeder	General formulas	D	Use of a modified Morse potential	[27]
28	1957	H.O. Jenkins	$k_e = ar_e^{-2} - b$	D	Homonuclear D	[28]

(*Continued*)

Table 4.1 (Continued)

No	Year	Authors	Equation	Molecules	Comment	Ref.
29	1958	Y.P. Varshmi	Equations for k_e, D_e, X_e	D	Derived from $V = -a(r-d)^{-m} + b(r-d)^{-n}$	[29]
30	1958	Y.P. Varshmi	$k_e(r_e-d)^p = \text{const}$ $r_e = ak_e^{-1/2} + b$	D	Review; $p = 2$ turns out to be best	[30]
31	1961	D.R. Hershbach, V.W. Laurie	$k_e = 10^{-(r_e-a_{ij})/b_{ij}}$ $r_e = d_{ij} + (a_{ij}-d_{ij})k_e^{-1/3}$	D	Constants derived for seven formulas for cubic and quartic force constants	[31]
32	1964	H.S. Johnston	$r_e = a_{ij} - b_{ij}\log(k)$	P, c-mode	Taken from 31	[32]
33	1964	J.A. Ladd, W.J. Orville-Thomas, B.C. Cox	$k = ar^{-b}$ or $\log(k) = -b\log(r) + c$	P, c-mode	$k = 35.5r^{-5.79}$ for CO bonds	[33]
34	1966	J.A. Ladd, W.J. Orville-Thomas	$k = ar^{-b}$	P, c-mode	$k = 34.4r^{-5.97}$ for NO bonds	[34]
35	1966	M.J.S. Dewar, G.J. Gleicher	$k = ar_e^{-2} + br_e^{-4} + cr_e^{-6}$	P, c-mode	Assumed; used in PPP	[35]
36	1966	J.C. Decius	$k = ar^{-6}$	P, c-mode	$k = 38.5r^{-6}$ for NN bonds (GS and ES of P)	[36]
37	1966	J. Goubeau	Review of formulas	P, c-mode	Comparison of Badger, Gordy, and Siebert formulas	[37]
38	1968	R.F. Borkman, R.G. Parr	$r_e^2 \omega = \text{const.}$	D	Use of virial theorem; quantum mechanical analysis k related to overlap population, r_e, D_e, and AO	[38]
39	1968	R.S. Roy	$r_e^6 = a\, m_1 m_2\, n^{-2}\, k^{-1}$	D	n: group number; m_1, m_2: atomic masses; a is considered a universal constant	[39]
40	1970	J. Stals	k–r formulas reviewed	P, c-mode	Material on CC, CN, CO, NN, NO, OO bonds	[40]
41	1978	D.C. McKean	$r(\text{CH}) = a\omega(\text{CH}) + b$	P isolated ω	Isolated stretching frequencies as bond descriptors	[41]

42	1984	H.B. Schlegel	Badger rule 6	P, **c**-mode	*Ab initio* calculations; use of **c**-vectors prediction of Hessian for geometry optimization	[42]
43	1987	D.M. Byler, H. Susi, W.C. Damert	33 tested	P, **c**-mode	$k = 37.3r^{-5.35}$ for CN bonds	[43]
44	1987	H.B. Bürgi, J.D. Dunitz	$r = a - b\ln(k)$	P, **c**-mode	Rationalized with modified Morse potential used in connection with reaction rates	[44]
45	1987	D.J. Swanton, B.R. Henry	$k_e(r_e - d_{ij})^n = c_{ij}$	D, P	Justification of Badger-type rules with modified Morse potential; local mode approximation for overtones	[45]
46	1987	A.A. Zavitsas	$k = a\mu(D-b)$	P, **c**-mode	For CC and other bonds tested	[46]
47	1987	V.M. Miskowski *et al.*	$r = a + be^{-k/c}$	P, **c**-mode	For M—M bonds in transition metal compounds	[47]
48	1989	J.C. Weisshaar	Badger rule 6	D	Third row diatomics	[48]
49	1990	P.K. Mallick *et al.*	Test of 31	P	Extended to frequency shifts and bond length changes	[49]
50	1992	T.H. Fischer, J. Almlöf	$k_{AB} = ae^{-b(r_{AB}-r_{ref})}$	P	Assumed; estimates of Hessian for geometry optimization	[50]
51	1992	A. Rutkowski *et al.*	$-\Delta r_e(\text{rel}) \propto \Delta k(\text{rel})$	D	Relativistic changes in r_e and k	[51]
52	1993	R.G. Pearson	$k_e r_e = aN(x_a x_b) + b$		N: bond order; x: electronegativity	[52]
53	1995	R. Lindh *et al.*	$k_{AB} = ae^{-b(r_{AB}-r_{ref})^2}$	P	Assumed; estimates of Hessian for geometry optimization	[53]
54	1996	P.D. Harvey	$r = a\ln(k) + b$	P, **c**-mode	Reparameterization of 31 for M—M bonds in transition metals	[54]
55	1997	J.M. Wittbrodt, H. B. Schlegel	Badger rule 6 $k_{AB} = a/(r_{AB}-b)^3$	P, **c**-mode	Estimates of Hessian for geometry optimization, relationships for b established	[55]
56	1999	M. Stichler, D. Menzel	$\omega = a'(r-b)^{-3/2}$	P, **c**-mode	Badger rule tested for NO on Ru-surface	[56]

(*Continued*)

Table 4.1 (Continued)

No	Year	Authors	Equation	Molecules	Comment	Ref.
57	1999	J.A. Larsson, D. Cremer	$r = a\omega + b$ $r = a\omega^2 + b\omega + c$	P, a-mode	41 tested for 66 CH bonds, extension to 40 CC bonds, requires quadratic relationship	[57]
58	2000	J. Cioslowski et al.	$k = A(r-b)^{-c}$	D	Badger-type rules are not universally shown for 108 diatomics	[58]
59	2001	P. Schwerdtfeger and coworkers	Test of 31	P, c-mode	Applied to Cu–Cu bonds	[59]
60	2003	J. Jules, J.R. Lombardi	Test of 6, 31	D	N, r, and k related for M–M bond in metal dimers	[60]
61	2004	K. Ohno and coworkers	$k = aR^{-3}$	P, c-mode	For 74 molecules with CX bonds R: effective bond length	[61]
62	2006	T. Green and coworkers	$r = c_{ij}\omega^{-2/3} + d_{ij}$	P, c-mode	Badger rule applied to FeO bonds	[62, 63]
64	2008	Y.W. Lin et al.	$r = c_{ij}\omega^{-2/3} + d_{ij}$	P, c-mode	Badger rule tested for CO interacting with amino acids	[64]
65	2008	U. Das, K. Raghavachari	$r = a\omega + b$	P, isolated ω	Isolated stretching frequencies used to predict r(PH) for molecules on surface	[65]
66	2008	T.H. Morton and coworkers	$\omega = a(r-b)^{-3/2}$	P, a-mode	CF^+ bonds	[66]
67	2009	E. Kraka, D. Cremer	r, ω, k, n, BDE related generalized Badger	P, a-mode	Relationships for CO and CF^+ bonds	[67]

A bond AB of a D (diatomic), QD (quasidiatomic), P (polyatomic molecule) is considered where A is located in period i or group m and B in period j or group n of the periodic table. The derivation of force constants for polyatomic molecules has been based on c-modes, adiabatic modes, isolated stretching modes, local modes from overtone spectroscopy, or simply assumed. D_e: bond dissociation energy; r_e: equilibrium bond length; ω_e: equilibrium bond stretching frequency; k_e: equilibrium bond stretching force constant; N: bond order; x_e: equilibrium anharmonicity constant; μ: reduced mass.

frequency ω or force constant k, bond order N, and bond dissociation energy D as is documented by a significant number of research papers on this topic (for recent work, see Refs [56–67]).

Investigations focusing on relationships between bond properties such as r, k, ω, N, and D are summarized in Table 4.1. Originally, such relationships were established for diatomic molecules and later extended to polyatomic molecules. Attempts have been made to verify and rationalize these relationships via model potentials for diatomics as, for example, Morse potentials, modified Morse potentials, double-reciprocal potentials (see entry 13, Table 4.1), single-reciprocal exponential potentials (entry 15, Table 4.1), or more complex forms of the potential (entry 29, Table 4.1). These relationships eventually led to the formulation of universal diatomic potentials [68, 69] that attempt to define energy and spectroscopic properties of a universal bond, which can be considered the equivalent of "the hydrogen atom in atom spectroscopy" [70]. Clearly, the derivation and rationalization of fundamental relationships between various bond properties led to a better understanding of the chemical bond. Therefore, it is appropriate to sketch the major steps in this development stretching now over almost 90 years.

After preliminary work by Kratzer [1], Birge [2], and Mecke [3], Morse [4] was the first to derive an empirical relationship between bond length and bond stretching frequency in 1929 (entry 4, Table 4.1) for diatomic molecules. Badger criticized the Morse relationship as being too limited in its practical application [6]. In 1934, he proposed a new relationship (entry 6) for diatomic molecules that relates the stretching force constant k to an effective bond length R obtained as the difference between equilibrium bond length r_e and an empirical parameter d_{ij} characteristic of the distance of closest contact between the bonded atoms. The experimental data available to Badger suggested that d_{ij} is the same for all atoms of period i bonded to atoms of period j. In 1935, Badger [9] generalized the relationship between k_e and r_e to polyatomic molecules by introducing an additional parameter (entry 7, c_{ij} or c_{nm}), which also depends on the location of the bonded atoms in the periodic table. Despite numerous alternative relationships suggested by various authors (entries 8–30; interesting extensions by Huggins (11), Linnett (15), Gordy (20), Guggenheimer (22)), the Badger rule was widely used until the early 1960s. In 1961, Herschbach and Laurie [31] pointed out that the Badger rule was not providing reliable predictions for heavier elements. Therefore, they suggested two major extensions of the Badger rule (see entry 31), one of which expresses the bond length r_e as a logarithmic function of the stretching force constant again using parameters that depend just on the periods i and j of the bonded atoms rather than the properties of these atoms themselves. The other extends Badger's rule also to cubic and quartic force constants thus confirming that both harmonic and fundamental stretching frequencies can be related to effective bond lengths. A cautious extension to polyatomic molecules was also discussed for bonds of similar type, that is, which do not suffer "an abrupt change" [31] in their properties when compared in a series of molecules.

The extension of the Badger–Herschbach–Laurie equations to metal–metal bonds required extensive reparameterization as carried out by Harvey in 1996 (entry 54) or the use of more elaborate exponential functions (Miskowski et al., entry 47). In 2000,

Cioslowski and coworkers [58] investigated the applicability of Badger-type equations to a test set of 108 diatomic molecules. They found that in a large number of cases the Badger rule does not lead to satisfactory predictions of stretching force constants and therefore cannot be considered a reliable tool for setting up the initial guess Hessian matrix in quantum chemical geometry optimizations. This result was contrasted by an investigation of Ohno and coworkers (entry 61), who could derive a simplified Badger-type equation by using effective bond lengths in the study of 74 CX bonds (X=C, Si, Ge, N, P, As, O, S, Se, F, Cl, Br) contained in polyatomic molecules. Kraka and Cremer [67] made a similar observation when investigating some 46 isoelectronic CO and CF^+ bonds. For the purpose of resolving these contradictory claims on the applicability of the Badger–Herschbach–Laurie equations between bond length and bond stretching force constant, there is the need to reinvestigate the physical basis of Badger-type relationships and to obtain a reliable assessment of their predictive value. We will approach this problem in two steps by first considering diatomic and then polyatomic molecules. We will identify those physical effects that influence the length of a chemical bond and, by doing so, clarify whether a relationship between bond length and stretching force constant exists. Then, we will determine those vibrational properties that lead to a description of chemical bonds in polyatomic molecules. Clearly, these cannot be the normal modes measured in infrared or Raman spectroscopy because they are in most cases delocalized, that is, they reflect the movement of larger structural units of the molecule (if not to say the whole molecule) rather than that of a specific bond within the molecule. In view of the limited usefulness of measured vibrational data, it is questionable whether an extension of the Badger rule to polyatomic molecules, as it was attempted in the past, can be successful on a larger scale. To solve this problem, we will discuss the difference between localized and delocalized vibrational modes, how the former can be derived from the latter, and how they lead to an extension of the Badger rule applicable to molecules.

4.2
Applicability of Badger-Type Relationships in the Case of Diatomic Molecules

A more general type of the Badger rule is given by Equation 4.1 [41, 71, 87]

$$k_e(r_e-d)^p = c \qquad (4.1)$$

where the quantity (r_e-d) is the effective bond length R, d and c are constants depending on the nature of atoms A and B, and the exponent p can take values between 2 and 8 thus embedding the original Badger rule (Table 4.1: entries 6, 7, 9) with $p = 3$. Alternative relationships with $p = 2$ (Table 4.1: entries 1, 28–30), $p = 4$ (2, 3, 38), $p = 5$ (17), $p = 6$ (4, 5, 8, 12, 15, 16, 36, 39), or noninteger values of p between 2 and 6 (e.g., 21, 22, 25, 33, 34, 43) are summarized in Table 4.1. Apart from Equation 4.1, relationships in the form of a power series (35), a logarithmic (11, 23, 31, 33, 44) or exponential dependence (50, 53) of r_e on k_e or vice versa were also used as already mentioned in the introduction. Previous research could not clarify

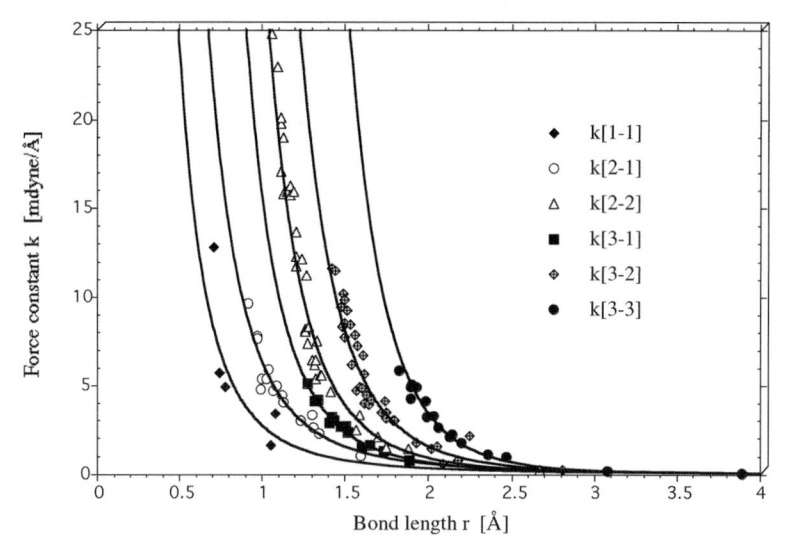

Figure 4.1 Experimental bond stretching force constants k in mdyne/Å are given in dependence of the bond length r in Å for diatomic molecules AB where atoms A and B belong to the first three periods. Symbol k[i–j] denotes that an atom A from period i is bonded to atom B from period j.

which of the relationships (4.1) or their extensions in the form of logarithmic or exponential functions is the most reliable and useful one.

In Figure 4.1, measured stretching force constants of 120 diatomic molecules in their ground state composed of atoms out of the first three rows of the periodic table and taken from the compilation of Huber and Herzberg [71] are plotted against the corresponding experimental bond lengths. The essence of the Badger rule becomes obvious from the diagram since it reveals that the data points cluster into six groups, each of which can be connected by a function according to Equation 4.1. The six groups correspond to the six possible ij combinations of periods (1–1, 2–1, 2–2, 3–1, 3–2, and 3–3). The corresponding bond lengths and stretching force constants are listed in Table 4.2.

In all cases, exponent p is a fractional quantity, which increases from 3.18 (1–1) to 7.44 (3–3), thus revealing a strong dependence on the number of electron shells of A and B in molecule AB, that is, on indices i, j and $i + j$. Clearly, by choosing appropriate effective bond lengths with the help of close-contact parameters, it will be possible to merge the six curves of Figure 4.1 into one. This, however, could become problematic because of a large variation in the prefactor increasing from 2.8 to 602. Testing various sets of Badger parameters given in the literature confirms that it is not possible to obtain one generally applicable form of the Badger rule, the Herschbach–Laurie variation, or any of the other forms suggested in the literature (Table 4.1). There is the general trend that with the number of data points in a group scattering increases and the reliability of any Badger-type relationship to predict either bond length or force constant, once the other quantity has been determined, decreases. Closer inspection reveals that especially cations and anions deviate from

Table 4.2 Experimental bond lengths r and bond stretching force constants k of diatomic molecules AB[a)].

No.	Molecule	r (Å)	k(mdyne Å^{-1})	No.	Molecule	r (Å)	k(mdyne Å^{-1})
1	$H_2(^1\Sigma_g^+)$	0.741	5.75	61	$SiH(^2\Pi_i)$	1.520	2.39
2	$H_2^+(^2\Sigma_g^+)$	1.052	1.60	62	$SiH^-(^3\Sigma^-)$	1.474	2.71
3	$He_2^+(^2\Sigma_u^+)$	1.081	3.40	63	$SiH^+(^1\Sigma^+)$	1.504	2.67
4	$He_2 2+(^1\Sigma_g^+)$	0.704	12.80	64	$PH(^3\Sigma^-)$	1.422	3.22
5	$HeH+(^1\Sigma^+)$	0.774	4.94	65	$PH^-(^2\Pi_i)$	1.407	2.86
6	$LiH(^1\Sigma^+)$	1.596	1.03	66	$PH^+(^2\Pi_r)$	1.435	3.04
7	$BeH(^2\Sigma^+)$	1.343	2.27	67	$SH(^2\Pi_i)$	1.341	2.43
8	$BeH^+(^1\Sigma^+)$	1.312	2.64	68	$HCl(^1\Sigma^+)$	1.275	5.16
9	$BH(^1\Sigma^+)$	1.232	3.05	69	$HCl^+(^2\Pi_i)$	1.315	4.13
10	$CH(^2\Pi_r)$	1.120	4.48	70	$LiNa(^1\Sigma)$	2.810	0.21
11	$CH^-(^3\Sigma^-)$	1.089	4.48	71	$Nao(^2\Pi)$	2.050	1.54
12	$CH^+(^1\Sigma^+)$	1.131	4.11	72	$NaF(^1\Sigma^+)$	1.926	1.76
13	$NH(^3\Sigma^-)$	1.036	5.97	73	$MgO(^1\Sigma^+)$	1.749	3.48
14	$NH^+(^2\Pi_g)$	1.070	4.73	74	$MgF(^2\Sigma^+)$	1.750	3.16
15	$OH(^2\Pi_i)$	0.970	7.80	75	$AlN(^3\Pi_i)$	1.768	3.03
16	$OH^-(^1\Sigma^+)$	0.970	7.65	76	$AlO(^2\Sigma^+)$	1.618	5.67
17	$OH^+(^3\Sigma^-)$	1.029	5.41	77	$AlF(^1\Sigma^+)$	1.654	4.23
18	$HF(^1\Sigma^+)$	0.917	9.66	78	$Si_2(^3\Sigma^-)$	2.246	2.15
19	$HF^+(^2\Pi_i)$	1.001	5.38	79	$SiN(^2\Sigma^+)$	1.571	7.29
20	$NeH^+(^1\Sigma^+)$	0.989	4.81	80	$SiO(^1\Sigma^+)$	1.510	9.24
21	$HeNe^+(^1\Sigma^+)$	1.300	3.36	81	$SiF(^2\Pi_i)$	1.601	4.90
22	$Li_2(^1\Sigma_g^+)$	2.67	0.26	82	$CP(^2\Sigma^+)$	1.562	7.83
23	$LiO(^2\Pi_i)$	1.695	2.08	83	$PN(^1\Sigma^+)$	1.491	10.16
24	$LiF(^1\Sigma^+)$	1.564	2.50	84	$PO(^2\Pi_r)$	1.476	9.45
25	$BeO(^2\Sigma^+)$	1.331	7.51	85	$PO^-(^3\Sigma^-)$	1.540	6.21

Table 4.2 (Continued)

No.	Molecule	r (Å)	k(mdyne Å^{-1})	No.	Molecule	r (Å)	k(mdyne Å^{-1})
26	BeF($^2\Sigma^+$)	1.361	5.60	86	PF($^3\Sigma^-$)	1.590	4.97
27	B$_2$($^3\Sigma_g^-$)	1.590	3.38	87	PF$^+$($^2\Pi_r$)	1.500	7.70
28	BN($^3\Pi$)	1.281	8.33	88	NS($^2\Pi_r$)	1.494	8,52
29	BO($^2\Sigma^+$)	1.205	13.66	89	NS$^+$($^1\Sigma^+$)	1.440	11.49
30	BO$^+$($^1\Sigma$)	1.205	12.27	90	BeS($^1\Sigma^+$)	1.742	4.13
31	BF($^1\Sigma^+$)	1.263	8.07	91	BS($^2\Sigma^+$)	1.609	6.72
32	C$_2$($^1\Sigma_g^+$)	1.243	12.16	92	CS($^2\Sigma^+$)	1.535	8.49
33	C$_2^-$($^2\Sigma_g^+$)	1.268	11.21	93	CS$^+$($^2\Sigma^+$)	1.495	9.85
34	C$_2^+$($^2\Pi_u$)	1.301	6.44	94	SO($^3\Sigma^-$)	1.481	8.30
35	CN($^2\Sigma^+$)	1.172	16.29	95	SO$^+$($^2\Pi_r$)	1.424	11.62
36	CN$^+$($^1\Sigma$)	1.173	15.74	96	LiCl($^1\Sigma^+$)	2.021	1.42
37	CO($^1\Sigma^+$)	1.115	19.80	97	LiCl$^-$($^2\Sigma^+$)	2.180	0.79
38	CO$^+$($^2\Sigma^+$)	1.115	19.80	98	BeCl($^2\Sigma^+$)	1.797	3.03
39	CF($^1\Sigma^+$)	1.272	7.42	99	BCl($^1\Sigma^+$)	1.716	3.47
40	N$_2$($^1\Sigma_g^+$)	1.098	22.95	100	CCl($^2\Pi_{1/2}$)	1.645	3.95
41	N$_2^-$($^2\Pi_g$)	1.193	15.98	101	NCl($^3\Sigma^-$)	1.614	4.03
42	N$_2^+$($^2\Sigma_g^+$)	1.116	20.09	102	ClO($^2\Pi_i$)	1.570	4.71
43	N$_2^{2+}$($^1\Sigma_g^+$)	1.132	15.85	103	ClF($^2\Pi_r$)	1.638	4.48
44	NO($^2\Pi_r$)	1.151	15.95	104	BeAr$^+$($^2\Sigma^+$)	2.085	0.59
45	NO$^-$($^3\Sigma^-$)	1.063	8.17	105	Na$_2$($^1\Sigma^+$)	3.079	0.17
46	NO$^+$($^1\Sigma^+$)	1.116	24.84	106	NaCl($^1\Sigma^+$)	2.361	1.09
47	NF($^3\Sigma^-$)	1.317	6.19	107	Mg$_2$($^1\Sigma_g^+$)	3.890	0.02
48	O$_2$($^3\Sigma^-$)	1.208	11.76	108	MgS($^1\Sigma^+$)	2.142	2.26
49	O$_2^-$($^2\Pi_{g,i}$)	1.350	5.60	109	MgCl($^2\Sigma^+$)	2.119	1.79
50	O$_2^+$($^2\Pi_g$)	1.116	17.09	110	Al$_2$($^3\Sigma_g^-$)	2.466	0.97

(Continued)

Table 4.2 (Continued)

No.	Molecule	r (Å)	k (mdyne Å^{-1})	No.	Molecule	r (Å)	k (mdyne Å^{-1})
51	$F_2(^1\Sigma_g^+)$	1.412	4.70	111	$AlS(^2\Sigma^+)$	2.029	3.28
52	$F_2^-(^2\Sigma_u^+)$	1.880	1.46	112	$AlCl(^1\Sigma^+)$	2.130	2.08
53	$F_2^+(^2\Pi_{g,i})$	1.322	6.45	113	$SiCl(^2\Pi_r)$	2.058	2.63
54	$FO(^2\Pi)$	1.326	5.41	114	$SiS(^1\Sigma^+)$	1.929	4.94
55	$Ne_2^+(^2\Sigma_u^+)$	1.750	1.53	115	$P_2^+(^2\Pi_{u3/2})$	1.986	4.12
56	$NaH(^1\Sigma^+)$	1.887	0.78	116	$PS(^2\Pi_r)$	1.900	5.06
57	$MgH(^2\Sigma^+)$	1.730	1.27	117	$S_2(^3\Sigma_g^-)$	1.889	4.96
58	$MgH^+(^1\Sigma^+)$	1.652	1.65	118	$S_2^+(^2\Pi_{g,r})$	1.825	5.88
59	$AlH(^1\Sigma^+)$	1.648	1.62	119	$Cl_2(^1\Sigma_g^+)$	1.988	3.23
60	$AlH^+(^2\Sigma^+)$	1.602	1.50	120	$Cl_2^+(^2\Pi_{3/2g})$	1.891	429

a) Experimental values from Ref. [71].
Molecules AB are listed according to A(period i)–B(period j) combinations in the order 1–1, 2–1, 2–2, 3–1, 3–2, 3–3.

the Badger-type relationships obtained in the least squares sense. This is most obvious for the 1–1 group that consists of just five data points, four of which belong to cations (Figure 4.1, Table 4.2). Scattering is in this case so strong that the k–r function given is no longer meaningful although it largely parallels those obtained for the five other groups.

Henry and Swanton [45, 72] provided some evidence suggesting the existence of a relationship between the bond length $r_e(AB)$ of bond AB and its associated stretching force constant $k_e(AB)$. They used a modified Morse potential that fulfilled in the case of diatomic molecules the following conditions:

1) The potential energy V must approach infinity for $r \to 0$, which is not the case for the general form of the Morse potential. Therefore, a hard sphere distance r_a is introduced, which leads to $V(r < r_a) = \infty$.
2) V measured relative to the separated atoms A and B must approach zero for $r \to \infty$.
3) V must approach the value of D_e for $r \to r_e$.

The potential (4.2) fulfills conditions (1), (2), and (3).

$$V(r > r_a) = D_e(1 - e^{-a_e(r-r_e)})^2 - D_e \tag{4.2a}$$

$$V(r_a) = V_a = D_e(1 - e^{-a_e(r_a-r_e)})^2 - D_e \tag{4.2b}$$

Henry and Swanton used Equation 4.2 to derive relationship (4.3) between the harmonic frequency ω_e of a diatomic molecule and its bond length r_e [72]:

$$(r_e - r_a)\omega_e = 2\hbar(D_e/2\mu)^{1/2}\ln[1 + (V_a/D_e + 1)^{1/2}] \qquad (4.3)$$

Taking the derivative with regard to k_e leads to

$$\frac{\partial(r_e - r_a)}{\partial k_e} = \frac{\partial(r_e - r_a)}{\partial \omega_e} \frac{\partial \omega_e}{\partial k_e} = \frac{\omega_e}{2k_e} \frac{\partial(r_e - r_a)}{\partial \omega_e} \qquad (4.4a)$$

$$\frac{\partial(r_e - r_a)}{\partial k_e} = \frac{(-V_a/\hbar)(2\mu x_e)^{1/2}}{k_e^2\{(4V_a x_e \mu/\hbar^2 k_e) + 1\}^{1/2}\{1 + [(4V_a x_e \mu/\hbar^2 k_e) + 1]\}^{1/2}} \qquad (4.4b)$$

where μ is the reduced mass and x_e the anharmonicity constant. For the case where k_e is large, the derivative (4.4) varies with k_e^{-2}, whereas for a small value of the force constant variation takes place with k_e^{-1} and otherwise with k_e^{-p} for $1 < p < 2$. By this, all possibilities of Equation 4.1 are accounted for, as becomes obvious when calculating the derivative of Equation 4.1 leading to

$$\frac{\partial(r_e - d)}{\partial k_e} = c\, k_e^{-(1 + 1/p)} \qquad (4.5)$$

where the exponent is between -1 and -2 depending on the value of p. The Badger rule for diatomic molecules will be obtained if the hard sphere distance r_a and the associated potential V_a do not change within a period, which would also require that the bond dissociation energy D_e varies only slightly within a period of the periodic table. This, however, is generally not the case and therefore Badger-type rules will hold only for closely related bonding situations in the case of diatomic molecules.

Next, we consider the physical effects determining the length of a chemical bond and its associated bond stretching force constant. The latter reflects the strength of the chemical bond, which in the general case is the result of a covalent contribution (depending on the overlap of the atomic orbitals forming the bonding and anti-bonding diatomic orbitals, their electron occupation, and the energy splitting between them) and an ionic (polar) contribution (depending on the electronegativity difference between A and B and the charge transfer resulting therefrom). Covalent and ionic contributions also impact bond length. However, contrary to the bond strength and the stretching force constant, the bond length depends on a third quantity that can be related to the size of the atomic core or, alternatively, its hard sphere size, which is related to the core size and also includes the effects of the valence electrons. It is this third quantity that determines the magnitude of the Badger parameter d_{ij}. Badger's assumption that d_{ij} is constant for all bonds formed from period i atoms and period j atoms is not justified. The hard sphere size of an atom depends on its charge and, accordingly, will be smaller for the cation and larger for the anion compared to the size of the neutral atom. This fact is reflected in Figure 4.1 where some of the strongly scattered data points correspond to charged molecules. Smaller variations will also result if open and closed shell systems of the same bond type are compared. In general, positively or negatively charged AB bonds and excited states of a given molecule AB should involve atoms A and B with different hard sphere sizes than those of the neutral ground state of AB.

In the set of diatomic molecules investigated in this work (Table 4.2), there are 20 molecules for which, besides the neutral state, there is also a charged state. Some of them do not lead to a significant change in the hard sphere size because an electron is added to a lone pair (or π) orbital not involved in bonding (see, for example, entries 15 and 16 in Table 4.2: HO and HO$^-$; also 10 and 11). Therefore, the variations found for a bond AB in polyatomic molecules cannot be reflected by the limited number of diatomic molecules investigated in this and previous studies. This would be given only if, besides the ground-state molecules, a large body of data would also be available for charged and excited states. Therefore, it is necessary to extend the investigation of the Badger rule to polyatomic molecules and verify the following two predictions based on the investigation of the diatomic molecules.

1) Badger-type relationships depending simply on period characteristic parameters such as c_{ij} and d_{ij}, as is the case with diatomic molecules, are no longer applicable to polyatomic molecules. They split up in AB bond-specific relationships that are fulfilled for AB bonds with closely related electronic structure. AB bonds with different hard sphere sizes will have to be described with other relationships.
2) In special cases, bond-specific Badger rules can collapse to a single rule. This is likely to occur for bond types that share a common atom, for example, AB and AC, provided the electronic structures of these bonds are related in such a way that the hard sphere sizes of atoms B and C can be described with just one parameter. In the following chapters, we will investigate these predictions in detail. For this purpose, we have to clarify first how to determine bond-stretching vibrations for polyatomic molecules that are localized in a bond and are not contaminated due to the coupling with other vibrational modes.

4.3
Dissection of a Polyatomic Molecule into a Collection of Quasi-Diatomic Molecules: Local Vibrational Modes

Vibrational modes are in most cases delocalized within a molecule. The properties of these modes (frequencies or force constants) are not suitable for investigating Badger-type relationships. Instead, there is the need for local mode information that provides bond stretching frequencies or force constants, which are no longer contaminated by contributions from other vibrational modes. For the purpose of clarifying the relationship between delocalized normal and local internal coordinate modes, we present here the theory of the adiabatic internal coordinate modes (AICoMs), recently used to set up bond length–bond stretching force constant relationships by Kraka and Cremer [67].

The standard method for calculating the vibrational spectra of polyatomic molecules with K atoms is based on two major approximations [73, 74]. First, the Born–Oppenheimer approximation is used, which leads to the separation of the nuclear motion from the electronic motion and by this to the concept of the potential energy surface (PES). The assumption is made that the nuclei of the molecule move as classical particles on the PES. Second, the vicinity of the minimum occupied by the vibrating molecule in question is described by a Taylor expansion of $V(\mathbf{x})$

$$V(\mathbf{x}) = V(0) + \sum_i^{3K} \left(\frac{\partial V}{\partial x_i}\right)_0 x_i + \frac{1}{2!} \sum_{i,j}^{3K} \left(\frac{\partial^2 V}{\partial x_i \partial x_j}\right)_0 x_i x_j$$
$$+ \frac{1}{3!} \sum_{i,j,k}^{3K} \left(\frac{\partial^3 V}{\partial x_i \partial x_j \partial x_k}\right)_0 x_i x_j x_k + \cdots \quad (4.6)$$

In Equation 4.6, $\mathbf{x}$ describes the displacements of the nuclei from the equilibrium positions at the minimum of the PES in the form of Cartesian displacement coordinates (i.e., $x = r - r_e$ in case of a diatomic molecule)

$$\mathbf{x} = (x_1, y_1, z_1, \ldots, x_{3K}, y_{3K}, z_{3K})^\dagger \quad (4.7)$$

The Taylor series is truncated after the quadratic term and since the first-order term is zero at the minimum,

$$\left(\frac{\partial V}{\partial x_i}\right)_0 = 0 \quad \text{for} \quad i = 1, \ldots, 3K \quad (4.8)$$

one obtains Equation 4.9

$$V(\mathbf{x}) = \frac{1}{2} \sum_{i,j}^{3K} \left(\frac{\partial^2 V}{\partial x_i \partial x_j}\right)_0 x_i x_j = \frac{1}{2} \sum_{i,j}^{3K} (f_{ij})_0 x_i x_j \quad (4.9)$$

which forms the basis of the harmonic (mechanical) approximation for describing vibrational modes. In Equation 4.9, the constants f_{ij} represent the force constants, which are collected in the Cartesian force constant matrix $\mathbf{f}$.

If the molecule behaves as a classical particle on the PES, Newton's second law applies:

$$K_{x,\text{I}} = m_\text{I} \frac{d^2 x_\text{I}}{dt^2} = \frac{d}{dt}(m_\text{I} \dot{x}_\text{I}) = \frac{d}{dt} p_{x,\text{I}} \quad (4.10)$$

where $K_{x,\text{I}}$ is the x-component of the force exerted on a nucleus I with mass m_I, t the time, $\dot{x}_\text{I}$ the velocity, and $p_{x,\text{I}}$ the corresponding momentum. Newton's second law can be expressed in terms of kinetic energy T and potential energy V

$$-\frac{\partial V}{\partial x_\text{I}} = \frac{d}{dt}\left(\frac{\partial T}{\partial \dot{x}_\text{I}}\right) \quad (4.11)$$

Using mass-weighted coordinates,

$$\xi_i = m_\text{I}^{1/2} \quad (4.12)$$

Equation 4.11 can be simplified via (4.13)–(4.15)

$$\frac{d}{dt}\left(\frac{\partial T}{\partial \dot{\xi}_i}\right) + \frac{\partial V}{\partial \xi_i} = 0 \quad \text{for} \quad i = 1, \ldots, 3K \quad (4.13)$$

$$\frac{d}{dt}\dot{\xi}_i + \frac{1}{2}\frac{\partial}{\partial \xi_i}\sum_{j,k}^{3K} f_{jk}\xi_j\xi_k = 0 \tag{4.14}$$

$$\frac{d^2\xi_i}{dt^2} + \sum_{j}^{3K} f_{ij}\xi_j = 0 \tag{4.15}$$

Equation 4.15 represents the vibrational equations, which can be solved by using standard mathematical procedures.

It is advantageous to revert to Equation 4.11 and to consider two simplifications. First, the representation of the harmonic potential is changed by a coordinate transformation, which leads from the bilinear form (4.9) to a linear form depending on a new set of coordinates, the so-called normal coordinates. Second, the vibrational equations resulting from Equation 4.11 are rewritten in matrix notation using the vector of Cartesian displacements $\mathbf{x}$, the mass matrix $\mathbf{M}$, and the force constant matrix $\mathbf{f}$. For example, the kinetic energy T of the vibrating molecule and the harmonic potential V are expressed in this notation as

$$T(\dot{\mathbf{x}}) = \frac{1}{2}\dot{\mathbf{x}}^\dagger \mathbf{M}\dot{\mathbf{x}} \tag{4.16}$$

$$V(\mathbf{x}) = \frac{1}{2}\mathbf{x}^\dagger \mathbf{f}\mathbf{x} \tag{4.17}$$

Using the matrix notation and the normal coordinates, the vibrational problem can be written in the form of the pseudoeigenvalue problem

$$\mathbf{fL} = \mathbf{ML\Lambda} \tag{4.18}$$

in which $\mathbf{\Lambda}$ is the eigenvalue matrix with the $N_{\text{vib}} = 3K-L$ vibrational eigenvalues λ_μ on the diagonal

$$\lambda_\mu = 4\pi^2 c^2 (\omega_\mu)^2 \quad \text{for} \quad \mu = 1, \ldots, 3K-L = N_{\text{vib}} \tag{4.19}$$

where ω_μ is the harmonic vibrational frequency. The eigenvector matrix $\mathbf{L}$ contains N_{vib} normal mode eigenvectors $\mathbf{l}_\mu$ as column vectors. In $\mathbf{L}$ and $\mathbf{\Lambda}$, L eigenvectors and eigenvalues correspond to overall translation and rotation of the molecule ($L = 5$ for linear and $L = 6$ for nonlinear molecules), respectively. These eigenvalues are equal to zero provided translational and rotational motions are completely independent of the vibrational modes. This is true for the translational motions, but not for the rotational motions, which couple with the vibrational motions because of cubic terms in the potential energy function (4.6). Consequently, one finds eigenvalues close to zero, which correspond to the overall rotation of the molecule. The columns that correspond to translational and rotational modes are omitted from matrix $\mathbf{L}$.

Equation 4.18 reveals that the mass matrix represents a metric, which has to be eliminated to convert Equation 4.18 to an eigenvalue problem. This leads to using

4.3 Dissection of a Polyatomic Molecule into a Collection of Quasi-Diatomic Molecules

mass-weighted Cartesian displacement coordinates as shown in Equations 4.12–4.15. Normal coordinates **Q** are related to Cartesian coordinates according to

$$\mathbf{x} = \mathbf{LQ} \tag{4.20}$$

The vibrational equations can also be formulated in an analogous way using internal displacement coordinates **q**, which describe changes of internal coordinates (bond length, bond angle, dihedral angle, etc.) instead of changes in atomic positions as expressed by **x**.

$$\mathbf{q} = (q_1, \ldots, q_{N_{\text{vib}}})^\dagger \tag{4.21}$$

For the transformation from internal to Cartesian coordinates, **L** additional coordinates corresponding to external motions (rotations and translations) are derived, which possess eigenvalues λ_i close to or equal to zero. The transformation from Cartesian to internal coordinates is done with the matrix **C**

$$\mathbf{C} = \mathbf{M}^{-1}\mathbf{B}^\dagger \mathbf{G}^{-1} \tag{4.22}$$

where **B** is defined by Equation 4.23

$$B_{ni} = \left(\frac{\partial q_n(\mathbf{x})}{\partial x}\right)_{\mathbf{x}=\mathbf{x}_e} \tag{4.23}$$

and **G** is the Wilson matrix [75].

$$\mathbf{G} = \mathbf{B}\mathbf{M}^{-1}\mathbf{B}^\dagger \tag{4.24}$$

The dynamics of the nuclear motions can be made independent of translations and rotations and the vibrational problem is solved in the internal coordinates only. The internal kinetic energy is given by

$$T(\dot{\mathbf{q}}) = \frac{1}{2}\dot{\mathbf{q}}^\dagger \mathbf{G}^{-1}\dot{\mathbf{q}} \tag{4.25}$$

and the potential energy is approximated in accordance with Equation 4.9 by

$$V(\mathbf{q}) = \frac{1}{2}\mathbf{q}^\dagger \mathbf{F}\mathbf{q} \tag{4.26}$$

where **F** is the internal force constant matrix, given by

$$\mathbf{F} = \mathbf{C}^\dagger \mathbf{f}\mathbf{C} \tag{4.27}$$

The vibrational equation in internal coordinates is given in Equation 4.28.

$$\mathbf{F}\mathbf{D} = \mathbf{G}^{-1}\mathbf{D}\mathbf{\Lambda} \tag{4.28}$$

where **D** contains the normal mode vectors $\mathbf{d}_\mu$ ($\mu = 1, \ldots, N_{\text{vib}}$) given as column vectors and expressed in internal coordinates. Equation 4.27 no longer contains the translational and rotational solutions and, consequently, **D** directly gives the transformation from normal coordinates to internal coordinates.

$$\mathbf{q} = \mathbf{DQ} \tag{4.29}$$

The relationship between eigenvectors $\mathbf{l}_\mu$ and eigenvectors $\mathbf{d}_\mu$ is provided by matrix **C** according to

$$\mathbf{l}_\mu = \mathbf{Cd}_\mu \tag{4.30}$$

The vibrational equations presented above show that the normal modes associated with the normal mode frequencies ω_μ are delocalized modes since each normal coordinate is a linear combination of internal coordinate displacements. In the following section, it has to be discussed under which circumstances one can expect normal modes to be localized within a given molecular fragment associated with a specific internal coordinate.

4.3.1
Localized Vibrational Modes

The degree of delocalization of a normal mode is primarily determined by the amount of coupling between the internal modes contained in the normal mode. In this way, the off-diagonal elements of the force constant matrix represent the coupling force constants. This becomes clear when realizing that the "**c**-vectors" of the transformation matrix **C**, each of which are associated with a given internal coordinate, can be used as internal localized modes [76]. Hence, a normal mode would be strictly localized if

$$(\mathbf{d}_\mu)_n = \delta_{n\mu} \tag{4.31}$$

with $\delta_{n\mu}$ being the Kronecker delta. Equation 4.31 leads to

$$\mathbf{l}_\mu = \mathbf{c}_n \tag{4.32}$$

where it is assumed that $\mu = n$. Equation 4.32 will be fulfilled only if all displacements along vectors $\mathbf{c}_n$ and $\mathbf{c}_m (m \neq n)$ do not couple and a diagonal force constant matrix **F** is obtained with all coupling force constants $F_{nm} = 0$. This implies that electronic coupling between the internal localized modes is zero. Second, there is always mass coupling (due to the kinetic energy) between the **c** vectors because the **G** matrix of Equation 4.28 is nondiagonal. Mass coupling can be suppressed to some extent if, for example, the reduced mass of a diatomic fragment is dominated by the mass of one of the atoms as in the case of a CH bond. However, if the two masses are comparable neither Equation 4.31 nor Equation 4.32 is true. Often, vibrational spectroscopists assume diagonal character of the **G** matrix provided there is a large mass difference between the atoms participating in the molecular motions since this assumption provides the only basis to discuss measured frequencies in terms of local mode frequencies.

Apart from mass coupling (coupling due to the kinetic energy), which is always present, there is electronic coupling (coupling due to the potential energy) as indicated by finite off-diagonal elements of the force constant matrix expressed in internal coordinates. Coupling constants are particularly large in the case of bond–bond interactions as they occur in delocalized π systems or in strained cyclic or polycyclic ring compounds. One observes that stretching force constants are the largest constants in a molecular force field and that these force constants also show the largest variation. Bending force constants are smaller than stretching force constants and torsional force constants are in turn smaller than bending force constants, at least as long as a torsional mode at a single bond is concerned. This qualitative ordering of the magnitude of the diagonal force constants provides an estimate of the coupling between stretching, bending, and torsional modes only if the stretching and torsional modes couple only weakly [75, 77, 78].

The various forms of stretch–stretch couplings can be described in the following way: (a) a coupling between symmetry-equivalent stretching modes, (b) coupling between stretching modes involving the same atom combinations, and (c) coupling between stretching modes involving different atom combinations. Only case (c) causes coupling when the internal force constant-reduced mass ratio of the different bond types are compatible, whereas cases (a) and (b) will always be present to some extent if there exist several bonds of the same type in a molecule. Case (b) coupling will be small if the internal force constants of two bonds are very different as in the case of AB bonds of different bond order (e.g., C–C versus C≡C). One internal stretching mode can be decoupled from other stretching modes of the same type by a change in mass as a consequence of isotope substitution so that the force constant–mass ratio considerably varies.

Localization of vibrational normal modes occurs in favorable cases with small electronic and mass coupling effects between the internal motions, for example, for a triatomic molecule such as HOCl where one internal stretching (OH stretching) is largely decoupled from the other stretching mode (OCl stretching) and the bending mode (HOCl bending). It is also reasonable to say that the bending vibration in HOCl is decoupled from both stretching modes, that is, in HOCl there are three normal modes, each of which is a largely localized vibrational mode associated with one of the three internal parameters.

For a general polyatomic molecule, localization of a normal mode within a particular molecular fragment is uncommon. For example, in aldehydes or ketones the normal mode that is dominated by the C=O stretching vibration is measured as a strong band in the area 1600–1800 cm^{-1} where the exact position of the band depends not only on the bond strength but also on the fact that the corresponding normal mode is not localized in the C=O group. In formaldehyde, acetaldehyde, and acetone, the internal C=O stretching mode contributes 89, 85, and 84%, respectively, to the normal mode considered to represent C=O stretching.

The discussion shows the dilemma of using normal mode properties to unravel geometric or electronic details of a molecule without being able to separate effects associated with different molecular fragments so that reliable information is gained. Because of these reasons, AICoMs [79] were introduced to obtain local modes that are associated with a specific structural unit of a molecule without being contaminated by coupling with other vibrational modes.

4.3.2
The Adiabatic Internal Coordinate Modes

Each AICoM of a molecule is associated with just one internal coordinate q_n, that is, it is independent of all other internal coordinates $q_m (m \neq n)$. The construction of an AICoM is based on how an internal coordinate mode v_n would vibrate if the associated internal coordinate were to be displaced by an amount q_n^* in such a way that the increase in the potential energy becomes minimal. To accomplish this objective, mode v_n, led by q_n^* (leading parameter principle [79]), must be constrained to the molecular fragment associated with q_n, that is, the rest of the molecule is allowed to relax upon applying a perturbation q_n^*. This is equivalent to minimizing the potential energy given in normal coordinates $\mathbf{Q}$ under the constraint that the internal coordinate displacement q_n is kept constant (Equation 4.33a):

$$V(\mathbf{Q}) = \min. \tag{4.33a}$$

$$q_n = \text{const.} = q_n^* \tag{4.33b}$$

The potential energy V and the internal coordinate q_n depend on the normal coordinates according to Equations 4.34 and 4.35.

$$V(\mathbf{Q}) = \frac{1}{2}\sum_{\mu=1}^{N_{vib}} k_\mu Q_\mu^2 \tag{4.34}$$

$$q_n(\mathbf{Q}) = \sum_{\mu=1}^{N_{vib}} D_{n\mu} Q_\mu \tag{4.35}$$

(see Equation 3.29) where k_μ is the force constant for normal mode $\mathbf{d}_\mu$ and $D_{n\mu}$ is an element of matrix $\mathbf{D}$ of Equation 4.28. Equation 4.33a is solved with the help of the method of Lagrange multipliers,

$$\frac{\partial}{\partial Q_\mu}\left[V(\mathbf{Q}) - \lambda(q_n(\mathbf{Q}) - q_n^*)\right] = 0 \tag{4.36}$$

where λ is the Lagrange multiplier. Equation 4.36 leads to (4.37) and (4.38):

$$\frac{\partial V(\mathbf{Q})}{\partial Q_\mu} - \frac{\partial \lambda}{\partial Q_\mu}(q_n(\mathbf{Q}) - q_n^*) - \lambda \frac{\partial(q_n(\mathbf{Q}) - q_n^*)}{\partial Q_\mu} = 0 \tag{4.37}$$

$$\frac{\partial V(\mathbf{Q})}{\partial Q_\mu} = \lambda \left(\frac{\partial q_n(\mathbf{Q})}{\partial Q_\mu} - \frac{\partial q_n^*}{\partial Q_\mu}\right) \tag{4.38}$$

4.3 Dissection of a Polyatomic Molecule into a Collection of Quasi-Diatomic Molecules

where in Equation 4.38 it is considered that $q_n(\mathbf{Q}) = q_n^*$ is a constant. When the expression (4.34) for $V(\mathbf{Q})$ and (4.35) for $q_n(\mathbf{Q})$ is inserted into (4.38), the result is

$$\frac{\partial}{\partial Q_\mu}\frac{1}{2}\sum_{\nu=1}^{N_{vib}} k_\nu Q_\nu^2 = \lambda \frac{\partial}{\partial Q_\mu}\sum_{\varrho=1}^{N_{vib}} D_{n\varrho} Q_\varrho \tag{4.39}$$

which leads to

$$k_\mu Q_\mu = \lambda D_{n\mu} \tag{4.40}$$

The solution of Equation 4.36 (which concerns internal parameter q_n) for the μth normal coordinate is

$$Q_\mu^{(n)} = \frac{D_{n\mu}}{k_\mu \lambda} \tag{4.41}$$

where the superscript (n) of Q denotes a solution obtained under constraint (4.33b) for q_n. There is one such solution for each normal coordinate. When these solutions are used to express the displaced internal parameter q_n^*, one gets

$$q_n^* = \sum_{\mu=1}^{N_{vib}} D_{n\mu} Q_\mu^{(n)} = \sum_{\mu=1}^{N_{vib}} \frac{D_{n\mu}^2}{k_\mu} \lambda \tag{4.42}$$

which leads to expression (4.43) for the Lagrange multiplier

$$\lambda = \frac{1}{\sum_{\mu=1}^{N_{vib}} \frac{D_{n\mu}^2}{k_\mu}} q_n^* \tag{4.43}$$

Equation 4.41 can be rewritten as

$$Q_\mu^{(n)} = \frac{\frac{D_{n\mu}}{k_\mu}}{\sum_{\nu=1}^{N_{vib}} \frac{D_{n\nu}^2}{k_\nu}} q_n^* \tag{4.44}$$

which means that the constraint for internal coordinate q_n leads to a change in the normal coordinates. The adiabatic internal mode $\mathbf{a}_n^Q$ for internal coordinate q_n expressed in terms of normal coordinates follows from (4.45):

$$Q_\mu^{(n)} = (a_n^Q)_\mu q_n^* \tag{4.45}$$

The AICoM $\mathbf{a}_n^Q$ can be transformed into an AICoM expressed in Cartesian coordinates, $\mathbf{a}_n$, with the help of the $\mathbf{L}$ matrix.

$$\mathbf{a}_n = \mathbf{L}\mathbf{a}_n^Q \tag{4.46}$$

Hence, Equations 4.45 and 4.46 completely specify the form of an AICoM.

4.3.3
Properties of Adiabatic Internal Coordinate Modes

Once an AICoM vector is known, one can define a force constant that corresponds to the AICoM motion.

$$k_n^a = \mathbf{a}_n^\dagger \mathbf{f} \mathbf{a}_n \tag{4.47}$$

For deriving an AICoM frequency with the help of k_n^a, one has to define the mass m_n^a that is associated with the AICoM. The latter has to fulfill two criteria. First, the AICoM mass m_n^a has to be extractable from the functional form of the internal coordinate q_n. Second, m_n^a has to be directly connected to the vibrational motion $\mathbf{a}_n$ caused by a change in q_n. While the potential energy has already been used to derive the AICoM vectors, so far nothing has been said with regard to the kinetic energy T. It has been shown that upon perturbation of the equilibrium geometry caused by a change in the leading parameter, q_n^*, the atoms of the molecule move in such a way that the kinetic energy adopts a minimum and the generalized velocity $\dot{q}_n$ becomes identical to $\dot{q}_n^*$. Again, this leads to a constrained minimization problem, the solution of which is found with the help of another Lagrange multiplier [79]. The results of the derivation are

$$T(\dot{q}_n^*) = \frac{1}{2} m_n^a (\dot{q}_n^*)^2 \tag{4.48}$$

and

$$m_n^a = \frac{(\mathbf{b}_n^\dagger \mathbf{a}_n)^2}{\mathbf{b}_n^\dagger \mathbf{M}^{-1} \mathbf{b}_n} \tag{4.49}$$

where vector $\mathbf{b}_n$ corresponds to the nth column of the $\mathbf{B}$ matrix and where

$$\mathbf{b}_n^\dagger \mathbf{a}_n = 1 \tag{4.50}$$

since the AICoMs are properly normalized. Hence, the AICoM mass can be recognized to be identical to a diagonal element G_{nn} of the $\mathbf{G}$ matrix, which is a generalization of the reduced mass to internal parameters connecting more than two atoms. This is an indirect proof that the constraints put on V and T to get the AICoMs are well chosen. With the AICoM force constant and the AICoM mass, it is straightforward to obtain the AICoM frequency

$$\omega_n^a = (\mathbf{a}_n^+ \mathbf{f} \mathbf{a}_n G_{nn})^{1/2} = \left(\frac{k_n^a}{m_n^a}\right)^{1/2} \tag{4.51}$$

The force constant, frequency, and mass associated with a given AICoM for internal coordinate q_n provide the most important properties for its characterization. This information can be used to investigate normal modes by considering them as being composed of AICoMs. If one knows the decomposition of a normal mode in terms of AICoMs, then one can clarify whether the normal modes are more or less delocalized and what electronic or geometric information they contain.

4.3.4
Characterization of Normal Modes in Terms of AICoMs

A chemist investigates and understands the molecular geometry and conformation in terms of internal coordinates rather than in terms of Cartesian or normal coordinates. All molecular structure information is detailed by listing the corresponding internal coordinates. Therefore, it is justified to add to the static representation of a molecule provided by the internal coordinates, a dynamic representation provided by the AICoMs. Accordingly, the AICoMs can be used as the dynamic counterparts of the internal coordinates to describe the normal modes and by this the dynamic behavior of a molecule. The problem is that there are no rules to define an amplitude that specifies to what extent a particular adiabatic mode a_n is active in normal mode l_μ. Therefore, criteria were set up that should be fulfilled by a given definition of an amplitude, $A_{n\mu}$, to guarantee a physically meaningful *characterization of normal modes (CNM)* in terms of AICoMs. These criteria are (1) the symmetry criterion, (2) the stability criterion, and (3) the dynamic criterion.

1) The symmetry criterion expresses the necessity that symmetry-equivalent adiabatic modes have to have the same amplitude in a normal mode provided the normal mode retains this symmetry.
2) The stability of results concerns the independence of the AICoM amplitudes from the choice of the internal coordinate set. The amplitudes should not change significantly for a normal mode if they are calculated with different redundant internal coordinate sets and the differences in the parameter sets only concern coordinates irrelevant to the normal mode.
3) There must be a relationship between the amplitude $A_{n\mu}$ of an AICoM contained in a normal mode to the difference $\Delta\omega_{n\mu} = \omega_n - \omega_\mu$ in the way that a small difference implies large amplitudes while large differences lead to very small amplitudes. In other words, the scattering of points $A_{n\mu}$ versus $\Delta\omega_{n\mu} = \omega_n - \omega_\mu$ should be enveloped by a Lorentzian curve. If this is the case, one can say that the dynamical origin of the normal mode principle (*dynamical origin of normal mode concept*) is fulfilled.

The amplitude that fulfills the above three criteria and performs best is defined in Equation 4.52

$$A_{n\mu} = \frac{\langle l_\mu | f | a_n \rangle^2}{\langle l_\mu | f | l_\mu \rangle \langle a_n | f | a_n \rangle} \tag{4.52}$$

Amplitude $A_{n\mu}$ of Equation 4.52 can be considered an absolute amplitude, but it is common practice to renormalize $A_{n\mu}$ and to express it in percentage:

$$A_{n\mu}^\% = \frac{A_{n\mu}}{\sum_m A_{m\mu}} 100 \tag{4.53}$$

It should be noted that the renormalized amplitudes of Equations 4.52 and 4.53 lead to a description of the normal modes as it is often performed with the help of the

C⋯C, adiabatic mode   **C⋯C, c-vector mode**

CC$_{bridge}$, adiabatic mode   **CC$_{bridge}$, c-vector mode**

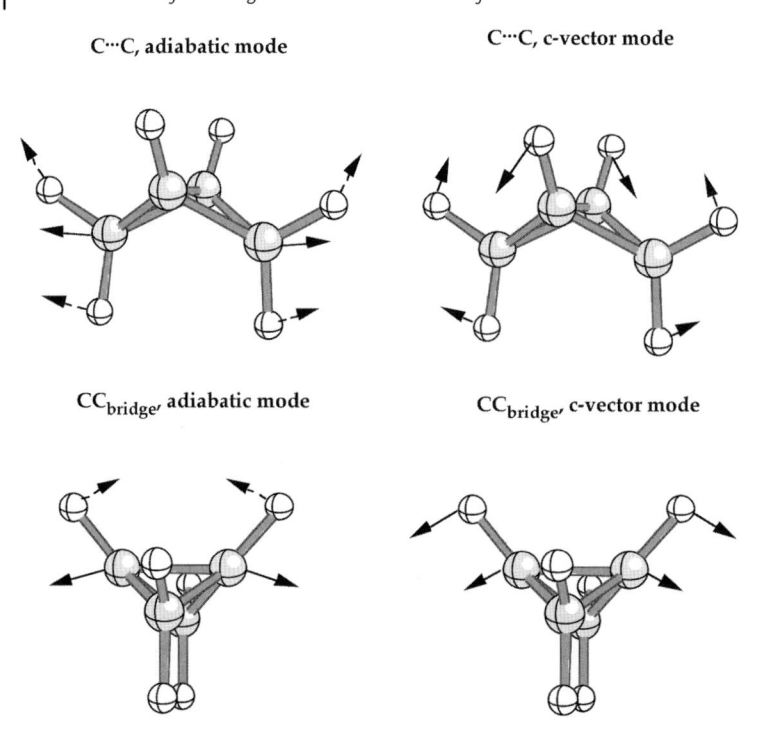

Figure 4.2 Two internal vibrational modes of bicyclobutane as described by adiabatic (left) and **c**-vector modes (right). *Top*: Folding motion of the ring (C⋯C). *Bottom*: Stretching of the bridge bond CC$_{bridge}$. Atom movements are indicated by solid arrows (strong movements) or dashed arrows (weak movements). (All CH stretches on the left) Very small displacements are not shown for the sake of clarity (B3LYP/6-31G(d,p) calculations).

potential energy distribution (PED) analysis [80–83]. However, the PED analysis suffers from several deficiencies that can lead to nonphysical results as is demonstrated by the following example.

PED and the CNM analyses with adiabatic amplitudes were carried out for bicyclobutane (Figure 4.2), for which the normal modes had been calculated at the B3LYP/6-31G(d,p) level of theory. For this purpose, a parameter set containing all CC and CH stretches, the nonbonded "C⋯C stretching" interaction (see Figure 4.2) for the description of the ring bending (puckering), and two HCC bends for each hydrogen (in total 24 parameters) was constructed. If one compares the results of the PED and the CNM analyses, a major difference in the description of the ring folding and the bridge stretching motion is observed.

Figure 4.2 shows that the adiabatic modes for the ring folding (C⋯C stretching) and the CC$_{bridge}$ stretching are localized in the corresponding molecular fragments. The movements of the hydrogen atoms follow that of the C atoms in the energetically optimal way without carrying out a coupled CH stretching motion. In Figure 4.2, this is indicated by dashed arrows giving the direction of the H atom movements. The

CC$_{bridge}$ stretching motion is similarly described by the **c**-vector motions even though the movement of the H atoms is now stronger since it has to comply with fixed CH bond distances, that is, all relaxations in the geometry because of CC$_{bridge}$ stretching are suppressed for those internal coordinates defined in the parameter set for bicyclobutane. However, **c**-vector motion for ring folding (C$\cdots$C stretching) differs considerably from the corresponding AICoM in the way that the CH$_2$ carbons hardly move. Instead, the hydrogens at the bridging carbons strongly move keeping the CC$_{bridge}$ distance and those CCH bending angles defined in the parameter set constant, while changing the angle HCC$_{bridge}$ (not contained in the parameter set), which is of course a consequence of the construction of **c**-vector modes. Clearly, the folding motion is not correctly described and this has serious consequences for the PED analysis.

Both methods predict that for normal mode 1, the dominant contribution is the folding motion of the ring (C$\cdots$C stretching): It amounts to 34.8% according to the CNM analysis and to 48.3% according to the PED analysis. However, for the latter description ring folding is also dominant for normal mode 5. Actually, normal mode 5 consists of a vibration of the CC$_{bridge}$ and C$\cdots$C fragments in such a way that when the CC$_{bridge}$ bond becomes longer, the C$\cdots$C distance becomes shorter, and the hydrogens follow the vibration of the carbon atoms to which they are attached. The CNM analysis based on adiabatic amplitudes describes mode 5 as being composed of 40.1% stretching of the CC bridge bond and 28.5% of ring folding as reflected by a vibration of the C$\cdots$C unit. Hence, mode 1 possesses more ring folding character whereas mode 5 is dominated by a vibration of the CC bridge. However, the PED amplitudes suggest a contribution of 60% of ring folding C$\cdots$C and 17.8% of CC$_{bridge}$ stretching simply because the **c**-vectors provide a misleading description of the folding motion as shown in Figure 4.2. Hence, the PED analysis suggests that there is more than one ring folding motion in bicyclobutane, which makes little sense and indicates that PED can lead to physically unreasonable descriptions due to the mechanical behavior of **c**-vectors. This is confirmed by the CNM analysis based on the AICoM amplitudes.

The example given (many more examples are found for mono- and polycyclic molecules) reveals the deficiencies of the PED analysis, which primarily result from the use of **c**-vectors. The disadvantages of the latter have explicitly been discussed in the literature [76]. The CNM analysis based on the amplitude A$_{n\mu}$ of Equation 4.52 has been applied with success in various investigations [84].

4.3.5
Advantages of AICoMs

AICoMs have the advantage that they are derived from a clear dynamic principle, namely, the leading parameter principle, which points out that a single internal coordinate q_n (in general, a single internal parameter) defines the displacements of the nuclei from their equilibrium positions and, by this, leads the internal mode **a**$_n$. The *leading parameter principle* [79] implies a new set of Euler–Lagrange equations because the generalized momenta for all other internal coordinates $q_m (m \neq n)$

become zero [79], which can be pictured in the way that all atomic masses outside the molecular fragment are considered as massless points. Hence, the derivation of the AICoMs follows the same procedure as the derivation of normal modes and it has been shown that the solutions of the Euler–Lagrange equations for the AICoMs are obtained by requiring that the potential energy V is minimized for a geometric perturbation under the constraint that the perturbation is defined by q_n^* [79].

The second advantage of the AICoMs is that their properties, namely, adiabatic force constant, adiabatic mass, and adiabatic frequency, are clearly defined and easy to calculate. The adiabatic mass is a generalization of the reduced mass for diatomic molecules and corresponds to G_{nn}^{-1}, which adds credibility to the physical basis of the AICoMs.

The third advantage of the AICoMs is that they lead to the CNM analysis of normal modes in a more sound and physically meaningful way than, for example, provided by the PED analysis. This is due to a clear definition of the amplitudes $A_{n\mu}$ [85]. The CNM analysis provides an easy way of analyzing vibrational spectra and quantitatively specifying the degree of delocalization of each vibrational mode.

As the fourth advantage, it has to be mentioned that the CNM analysis simplifies the correlation of the vibrational spectra of different molecules.

The fifth advantage is that an AICoM intensity can be derived that can be used to investigate the charge distribution within a molecule.

AICoMs are discussed in this chapter for the equilibrium geometry of a molecule. However, they can also be defined and applied to a reacting molecule. In this case, the AICoMs are based on generalized modes and are separately discussed in a one-dimensional subspace, the reaction path, and a $3K\text{-}(L+1)$-dimensional subspace orthogonal to the reaction path. As has been shown by Konkoli, Kraka, and Cremer [86, 87], the AICoMs lead in this case to a wealth of information and help describe the reaction mechanism in great detail.

An important advantage of AICoMs is that they can also be derived from experimental vibrational spectra and establish in this way a solid connection between theory and experiment. This is pointed out in the following.

Calculated AICoM frequencies and force constants suffer in the same way as the frequencies and force constants of normal vibrational modes from the deficiencies of the quantum chemical method used and the harmonic approximation employed in standard calculations of vibrational spectra. Even when applying efficient scaling procedures, there is no guarantee that *ab initio* frequencies accurately reproduce the fundamental frequencies of the experiment. In view of this, it seems to be much more useful to calculate the adiabatic frequencies in such a way that the experimental frequencies of the fundamental vibrations are exactly reproduced. In this way, each adiabatic internal frequency is the exact local mode counterpart of the measured vibrational frequency.

Since an experimental vibrational spectrum can only provide the frequencies of the fundamentals, it raises the question how the vibrational modes are to be obtained. In principle, this is achieved by setting up a force field from available experimental information and then using the theory described above. Alternatively, the force field can be calculated by correlation-corrected *ab initio* or DFT methods. Combining the two sources of information, namely, experimental frequencies and calculated normal modes, it is possible to determine that force constant matrix that in the harmonic approximation would reproduce fundamental frequencies exactly. Clearly, the force field obtained in this way is made up by *effective* force constants rather than pure quadratic force constants since the elements of the force field not only have absorbed all deficiencies of the quantum chemical calculation (correlation errors, basis set errors) but also cover all anharmonicity effects, normally described by cubic and quartic force constants. The use of this force field in the AICoM calculation leads to adiabatic force constants and adiabatic frequencies, which directly correspond to the measured vibrational spectrum and, therefore, can be used for analysis of the vibrational spectrum and for the description of bond properties.

The theory needed to obtain experimentally based AICoM properties is described in standard books on vibrational spectroscopy and can be summarized in the following way [88]. If one considers the difference between experimental fundamental frequencies ω_μ^{exp} and calculated harmonic frequencies ω_μ as a relatively small error caused by a similarly small error in the force constant matrix, then one can assume the changes in the normal mode vectors to be negligible and use first-order perturbation theory to set up the corrected vibrational secular Equation 4.54

$$\mathbf{D}(\mathbf{F}_0 + \Delta \mathbf{F})\mathbf{D}^\dagger = \Lambda + \Delta\Lambda \tag{4.54}$$

where $\mathbf{F}_0$, $\mathbf{D}$, and Λ correspond to the force constant matrix, the eigenvector matrix, and the eigenvalue matrix, respectively, of the *ab initio* or DFT calculation. It holds that

$$\mathbf{D}^T \mathbf{F}_0 \mathbf{D} = \Lambda \tag{4.55}$$

because the eigenvectors $\mathbf{D}$ are normalized with regard to $\mathbf{G}$, that is, $\mathbf{DD}^T = \mathbf{G}$. Accordingly, one can write the equation for the first-order correction as

$$\mathbf{D}^T \Delta \mathbf{F} \mathbf{D} = \Delta\Lambda \tag{4.56}$$

from which the correction for the force constant matrix results as

$$\Delta \mathbf{F} = (\mathbf{D}^{-1})^\dagger \Delta\Lambda (\mathbf{D})^{-1} = \mathbf{G}^{-1} \mathbf{D} \Delta\Lambda \mathbf{D}^T \mathbf{G}^{-1} \tag{4.57}$$

Hence, diagonalization of the experimentally determined correction matrix $\Delta\Lambda$ leads to $\Delta\mathbf{F}$ and the force constant matrix $\mathbf{F}_0 + \Delta\mathbf{F}$, which correctly reproduces experimental frequencies. Once the force constant matrix $\mathbf{F}_0 + \Delta\mathbf{F}$ is determined, one can apply the adiabatic mode analysis in the same way as it is applied to calculated vibrational spectra.

Equation 4.54 can also be used if only part of a vibrational spectrum of a given molecule has been measured. Those frequencies, which have not been experimentally observed, can be taken from calculated spectra after appropriate scaling. In this way, experimental adiabatic frequencies can be determined for any molecule, for which sufficient infrared and/or Raman information is available.

4.4
Local Mode Properties Obtained from Experiment

As was shown in the previous section, both the mass and the electronic coupling are responsible for the delocalized nature of the normal vibrational modes. Apart from this, there is the problem of Fermi resonances. The transformation that leads to a separation of the quadratic terms in Equation 4.9 does not separate the cubic and quartic terms of expansion (4.6). These anharmonic terms are responsible for the fact that an overtone or a combination band can mix with the fundamental of a vibrational mode. This phenomenon is called Fermi resonance and plays an important role in vibrational spectroscopy [77, 78]. For example, the CH stretching modes undergo Fermi resonances with the first overtone of the CH_3 and CH_2 bending modes (both CCH and HCH bending). This leads to a shift in the CH stretching frequency that makes the determination of a localized CH stretching frequency rather uncertain [41, 89–94].

Actually, the CH stretching motion and other XH stretching motions might be considered as being ideally suited to represent localized modes. For example, the stretching of a terminal bond couples always less than the motion of a bond in a central position of the molecule. Second, the mass ratio of a heavy atom X and H is optimal to reduce mass coupling. Finally, there are little electronic coupling effects between XH bonds with other bonds (with the exception of hyperconjugation and anomeric effects). If one compares the coupling between symmetry-equivalent stretching modes with the coupling between stretching modes involving the same atom combinations, and the coupling between stretching modes involving different atom combinations, then the first will always be present in symmetric molecules and, therefore, it will be the most important coupling effect for CH bonds whereas the second and third effects have smaller importance. Considering also Fermi resonances, the chance of observing a localized CH or XH stretching motion in a molecule is much smaller than might be expected. However, in certain situations one can obtain local mode information nevertheless, which will be discussed in the following section.

4.4.1
Isolated Stretching Modes

McKean considered the problem of deriving isolated CH stretching motions and provided a simple solution by replacing in a given molecule all H atoms but the target H by their D isotope thus yielding CD_2H and CDH groups [41, 89–94]. The change in mass decouples the remaining CH stretching mode from all CD stretching modes and particularly those that previously (as CH stretching modes) coupled strongly because of symmetry. In addition, one can make three other assumptions:

1) Due to isotope substitution, the CH stretching mode is largely isolated, which means that it is decoupled not only from the CD stretching vibrations but also from other stretching, bending, or torsional modes.

2) For transition from asymmetrical/symmetrical CH_n vibrations to an isolated CH stretching mode, all anharmonicity effects stay the same.
3) After D isotope substitution, all Fermi resonances for the CH stretching mode are suppressed.

As a result, the CH stretching mode is largely localized and the corresponding mode frequency, that is, the isolated stretching frequency $\omega_{iso(CH)}$, can be considered to accurately reflect the value of a local mode frequency.

McKean prepared a large number of isotopomers to measure isolated CH stretching frequencies and to investigate their dependence on geometric and electronic features of a given molecule [41, 89–94]. He showed that in this way CH bonds can be used as sensitive antennae or probes testing the properties of molecules. While his first work just focused on CH bonds, he and his coworkers studied later also other XH bonds (X: Si, Ge). In addition, other authors used McKean's approach to describe local XH stretching modes [95–97].

Investigations involving other than CH bonds revealed the large difficulties an experiment faces when a generalization of McKean's approach is attempted. For the purpose of decoupling one internal stretching mode from other stretching modes of the same type, the change in mass by isotope substitution must be so large that it significantly modifies the mass ratio. Replacement of H by deuterium results in a doubling of the mass and a satisfactory suppression of coupling and Fermi resonances so that any residual coupling for the isolated CH stretching modes is estimated to be less than 5 cm^{-1}. For a CC bond, one would obtain a very small effect if ^{12}C is replaced by ^{13}C or even ^{14}C since the change in the mass ratio is too small in these cases to play any significant role in the localization of the CC stretching motion. Hence, this example demonstrates that the isolated stretching frequencies are very useful quantities for the description of XH bonding in terms of local modes. However, a generalization of this approach faces too many difficulties to play any important role in the description of general AB bonds or when trying to verify the applicability and limitations of the Badger rule. In this situation, theory has made an important contribution.

Isolated stretching modes can be calculated for a given molecule containing a CH_n group by simply replacing for the calculation of the **G** matrix the masses of the H atoms by those of the D isotopes but keeping for the isotopomer the force field of the parent molecule. In this way, harmonic isolated CH stretching modes ω_n^{iso} are calculated, which can easily be compared with both McKean's experimental ω_n^{iso} values for CH stretchings and AICoM CH stretchings. B3LYP/6-31G(d,p) calculations produce isolated CH stretching frequencies, which correlate well ($R^2 = 0.985$ [57]) with McKean's experimental ones. Calculating also the corresponding AICoM CH frequencies reveals that AICoMs are the theoretical equivalences of McKean's isolated stretching modes:

a) There is a linear relationship between isolated CH stretching frequencies and AICoM CH stretching frequencies with a correlation coefficient of 0.997.
b) Since the isolated CH stretching modes can also be calculated, the equivalence of AICoMs and isolated stretching modes can be quantified by the CNM analysis.

With just a few exceptions discussed in the following, the overlap between AICoM and isolated modes is above 99% (in two cases 98%).

c) The isolated CH stretching modes are local modes that imply a relaxation of the electron density distribution upon perturbation of the CH equilibrium bond length. We find very similar relaxation effects for calculated stretching modes and AICoM stretching modes.

Exceptions are found for alkines for which a residual coupling between CH stretching and triple bond stretching is observed. Acetylenic hydrogens are not completely isolated from the C≡C stretching as reflected by an adiabatic overlap amplitude $A_{n\mu}$, which is smaller than 96% when the CH stretching AICoM of H–CX is compared with the normal mode representing an isolated CH stretching vibration. This explains why isolated CH frequencies ω_{CH}^{iso} at a triply bonded C deviate from the linear relation between CH(AICoM) and isolated CH stretching frequencies [41]. Coupling leads to an error of 45 cm^{-1} in the isolated CH stretching frequency of acetylene. Similar, but much smaller, residual mass and electronic couplings probably exist for other CH couplings and explain the small scattering of isolated stretching frequencies (errors about 5 cm^{-1}) relative to the corresponding AICoM frequencies. Hence, one has to consider the residual coupling when one uses isolated stretching frequencies as a tool for structural analysis.

The fact that the isolated stretching frequencies can be replaced by AICoM stretching frequencies as their theoretical counterparts leads to a number of advantages: (1) The linear relationship between isolated CH stretching frequencies and CH bond lengths found by McKean and used to predict unknown CH bond lengths with an accuracy of ±0.001 Å, once the isolated CH stretching frequency is measured, is confirmed for AICoM CH stretching frequencies [57] where both experimental or calculated values can be used. (2) Contrary to the determination of isolating stretching frequencies, which so far could be carried out only for XH bonds [41, 89–94], AICoM stretching frequencies can be easily determined for each bond of a molecule and McKean relationships can be established for all types of bonds. For example, it could be demonstrated that a quadratic relationship between the AICoM CC stretching frequencies and the CC bond length exists [57].

4.4.2
Local Mode Frequencies from Overtone Spectroscopy

Another way of obtaining information on localized XH stretching modes is to record overtone spectra of these vibrational modes [98]. Henry has pioneered this work by showing that the higher overtones of an XH mode can be reasonably well described with an anharmonic potential of a quasidiatomic molecule [98, 99]. Higher overtones ($\Delta v \geq 3$) of XH stretching modes reveal considerable local mode character. For overtones with $\Delta v = 5, 6$, one observes mostly one band for each unique XH bond, even if there are several symmetry-equivalent XH bonds in the molecule. In fundamental and lower overtone modes, there is always a splitting of the frequency into, for example, a symmetric and an antisymmetric mode frequency of two symmetry-equivalent XH stretching modes, but this splitting virtually disappears for overtones with $\Delta v \geq 5$. In general, the different linear combinations of symmetry-equivalent XH stretchings become effectively degenerate for the higher overtones.

Since the overtone intensity decreases for each higher overtone level, conventional spectroscopy cannot be used for overtones with Δv larger than 4. In gas-phase investigations, the higher overtones are recorded by intracavity dye laser photoacoustic spectroscopy, which uses sophisticated techniques to enhance the signal to noise ratio in the overtone spectra [98]. The local XH stretching modes are highly anharmonic. The very presence of overtones indicates that XH modes are anharmonic where the more the anharmonicity increases the higher the overtone is.

The local mode behavior of the fifth overtone ($v = 6$) of CH stretching modes can be verified by comparison with the corresponding AICoM frequency. In Table 4.3, frequencies for the fifth overtone of CH stretchings of alkane, alkenes, and aromatic molecules are listed [100–102]. The values for thiophene are taken from liquid-phase spectra whereas all other spectra were measured for the gas phase at room temperature. For isoxazole, there are only two overtone frequencies for the fifth overtone, which suggests that the difference between the overtones of the stretching motions of the C(4)H and C(5)H bonds are too small to be detected in the spectra. Similarly for toluene, the overtone spectra cannot resolve any difference between the overtones of the *meta* and *para* CH stretchings.

In Figure 4.3, AICoM frequencies $\omega_n^a(CH)$ are correlated with frequencies taken from overtone spectroscopy (see Table 4.3). There is a linear relationship between the two quantities (correlation coefficient $R^2 = 0.990$), which again confirms that AICoMs are suitable local vibrational modes that are related to the local modes of overtone spectroscopy.

The use of overtone spectroscopy as a means of obtaining information on local vibrational modes and their properties is limited to terminal bonds, of which so far only XH (X=C, N, O, S, and so on [103, 104]) bonds were investigated. Although coupling due to the potential energy (electronic coupling) is significant only for delocalized bonds, there is always coupling between local modes due to the kinetic energy (mass coupling). This coupling will be weak in the overtone spectra if the mass ratio between the two atoms of the bond considered is small as in the case of XH bonds. The investigation of isotopomers improves the situation, but these improvements are limited to XH bonds where the D,H mass ratio is favorable. Hence, a generalization of the local mode description by overtone spectroscopy is not possible.

4.4.3
Local Mode Information via an Averaging of Frequencies: Intrinsic Frequencies

Spectroscopists have often tried to assess the properties of local modes by simple averaging methods. For example, if two CH stretching modes in a CH_2 group interact to give a symmetric and an antisymmetric stretching mode, one can estimate the frequency of the corresponding local CH mode by taking the arithmetic mean of the frequencies of the symmetric and antisymmetric CH stretching vibration. A theoretical approach based on this idea was suggested by Boatz and Gordon [105], who

Table 4.3 Comparison of measured overtone frequencies for CH stretching ($\Delta v = 6$) with the corresponding B3LYP/6-31G(d,p) AICoM frequencies for various organic molecules.[a)]

Molecule	Bond	ω_n^a	$\Delta v = 6$
Methane		3129	16150
Fluoromethane		3074	15972
Chloromethane		3152	16216
Ethane		3085	15824
Ethene		3188	16550
Ethyne		3437	18430
Propane	CH_3, ip	3085	15845
	CH_3, op	3074	15746
	CH_2	3047	15562
Cyclopropane		3180	16504
Benzene		3186	16550
Furan	C(O)–H	3285	17223
	C(C)–H	3261	17121
Isoxazole	C(O)–H	3273	17143[b)]
	C(C)–H	3284	17143[b)]
	C(N)–H	3248	16911[b)]
Thiophene	C(S)–H	3260	16890[c)]
	C(C)–H	3218	16700[c)]
Propene	=C–H(trans)	3195	16569
	=C–H(cis)	3175	16395
	H–(Me)C=	3138	16236
	CH_3, ip	3099	15895
	CH_3, op	3055	15681
n-Butane	CH_3, ip	3086	15829
	CH_3, op	3074	15751
	CH_2	3036	15473
Isobutene	CH_2	3186	16474
	CH_3, ip	3107	15978
	CH_3, op	3050	15628
Isobutane	CH	3015	15305
	CH_3, ip	3065	15683
	CH_3, op	3079	15804
Toluene	C(ortho)–H	3170	16430
	C(meta)–H	3184	16543
	C(para)–H	3188	16543
	CH_3, ip	3055	
	CH_3, op	3062	

a)b)c) Measured frequencies from Refs [100–102]. ip and op denote in-plane and out-of-plane hydrogen atoms. All frequencies in cm^{-1}.

derived the intrinsic frequencies ω_n^{BG} as representatives of local mode frequencies associated with an internal coordinate q_n.

$$(\omega_n^{BG})^2 = \sum_{\mu=1}^{N_{vib}} \sum_{m=1}^{N_{Parm}} P_{nm}^\mu \omega_\mu^2 \qquad (4.58)$$

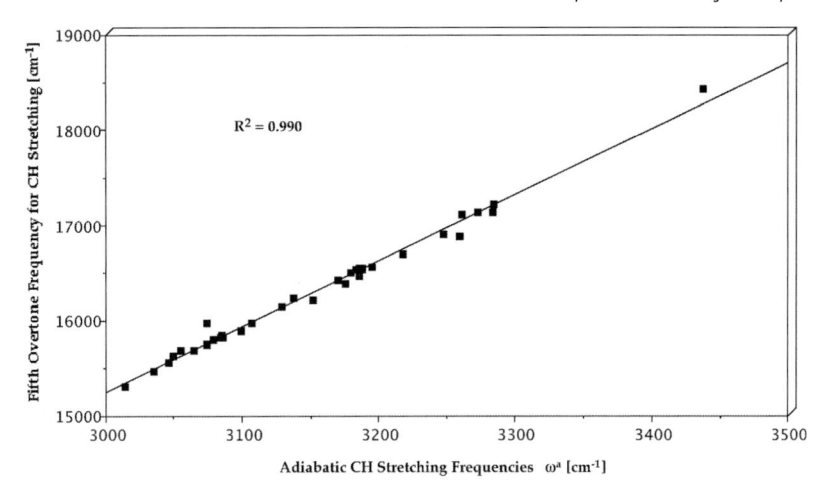

Figure 4.3 Correlation of measured overtone frequencies $\Delta v = 6$ for CH stretching modes with the corresponding calculated adiabatic CH stretching frequencies. For details, see Table 4.3.

where P^μ_{nm} leads to the PED amplitudes [105] and N_Parm defines the number of internal parameters used in the set of internal coordinates. N_Parm will be equal to N_vib if a nonredundant parameter set is used; however, in general, N_Parm can be larger than N_vib for the calculation of the intrinsic frequencies.

Equation 4.58 reveals that the intrinsic frequencies ω_n^BG are constructed as an average of those normal mode frequencies that have nonzero PED amplitudes for the internal parameters q_n. This averaging approach leads to problems when trying to obtain reliable local mode information, which becomes obvious when comparing intrinsic modes with AICoMs.

1) The intrinsic frequency is a frequency without a vibrational mode. This has to do with the fact that the derivation of ω_n^BG does not explicitly revert to a dynamic principle. AICoM frequencies correspond to AICoM vectors, which in turn are based on the leading parameter principle (the dynamic principle [79]) and the modified Euler–Lagrange equations for the vibrational problem expressed in terms of local modes.
2) The intrinsic frequencies are parameter set dependent whereas the AICoM frequencies are completely independent of the size and the composition of the set of internal coordinates used for the description of the molecular geometry.
3) Intrinsic frequencies can become negative for a true equilibrium geometry, which is not the case for AICoMs.
4) Intrinsic frequencies reflect the molecular symmetry only when redundant coordinate sets are used. AICoM frequencies comply with the molecular symmetry, no matter whether redundant or nonredundant parameter sets are used to describe the internal vibrations.
5) For the intrinsic frequencies, electronic and mass effects are not correctly separated, which is a problem when discussing electronic effects in terms of

4 Generalization of the Badger Rule Based on the Use of Adiabatic Vibrational Modes

Table 4.4 Intrinsic frequencies ω_n^{BG} and AICoM frequencies ω_n^a in cm^{-1} given for 10 different parameter sets of CH$_4$ as calculated at the HF/6-31G(d,p) level of theory.[a]

	1	2	3	4	5	6	7	8	9	10
ω_n^{BG}										
CH$_1$	3461	3452	3421	3322	3306	3286	3261	3261	3261	3261
CH$_2$		3452	3421	3322	3306	3293	3276	3272	3261	3261
CH$_3$			3421	3322	3306	3293	3276	3272	3261	3261
CH$_4$				3322	3306	3286	3276	3261	3261	3261
H$_2$CH$_1$					1570	1563	1551	1529	1577	1419
H$_3$CH$_1$						1563	1551	1529	1460	1419
H$_4$CH$_1$							1551	1577	1577	1419
H$_2$CH$_3$								1577	1577	1419
H$_3$CH$_4$									1577	1419
H$_4$CH$_2$										1419
ω_n^a										
All CH	3255	3255	3255	3255	3255	3255	3255	3255	3255	3255
All HCH					1560	1560	1560	1560	1560	1560

a) The composition of parameter sets 1–10 (top line) is obtained by adding internal coordinates in the first column, that is, parameter set 1 contains just the CH$_1$ bond length, parameter set 2 the bond lengths CH$_1$, CH$_2$, and so on.

intrinsic frequencies. However, for the AICoM frequencies electronic and mass effects are clearly separated.

6) As a consequence of (5), intrinsic frequencies do not only lack an *intrinsic mode* but also an *intrinsic force constant*, which are both defined for AICoMs.

For the purpose of showing some of the deficiencies of the intrinsic frequencies, in Table 4.4 intrinsic and AICoM frequencies for methane are listed employing HF/6-31G(d,p) theory and using eight incomplete (nonredundant) parameter sets (1–8 internal coordinates), one complete, nonredundant parameter set ($3K - L = 9$ internal coordinates), and one overcomplete, redundant parameter set with 10 internal coordinates. From Table 4.4, it can be seen that the intrinsic frequencies adopt different values for different numbers of coordinates in the parameter set. Even worse, for a given parameter set the intrinsic frequencies can take different values for symmetry-equivalent stretching and bending modes. Even for the complete, nonredundant parameter set with nine parameters, the intrinsic frequencies for the bending motions are different. When this is remedied in the way suggested by Boatz and Gordon [105], the bending frequencies ω_n^{BG} become identical. Upon increasing the parameter set, the intrinsic frequencies decrease toward the values of the redundant set.

For the AICoM frequencies of methane, just two values are obtained, namely, 3255 cm^{-1} for CH stretchings and 1560 cm^{-1} for HCH bendings, no matter how many coordinates are used in the parameter set (see Table 4.4). It is obvious that the intrinsic frequencies of the redundant parameter set are the only one that should be compared with the AICoM frequencies and used for the discussion of electronic structure. However, it is by no means clear how intrinsic frequencies of similar reliability are obtained for a larger molecule with no symmetry at all. We have found in this work that for normal acyclic molecules using complete, nonredundant basis sets the intrinsic frequencies of stretching modes may agree well with the adiabatic frequencies. Problems arise with bending frequencies and even more with torsional frequencies, which can become negative.

4.4.4
Compliance Force Constants

A way of obtaining local mode information, although it does not appear so on first sight, is to use compliance force constants. The latter are obtained when expressing the potential energy of a molecule in terms of generalized displacement forces rather than internal displacement coordinates (see Equation 4.26) [106, 107]:

$$V(\mathbf{g}) = \frac{1}{2}\mathbf{g}_q \mathbf{C}\mathbf{g} \tag{4.59}$$

where the elements of the compliance matrix $\mathbf{C}$ are given as the partial second derivatives of the potential energy V with regard to forces $f_i = -g_i$ and $f_j = -g_j$:

$$C_{ij} = \frac{\partial^2 V}{\partial f_i \partial f_j} \tag{4.60}$$

The gradient vector $\mathbf{g}_q$ of Equation 4.59 can be obtained by differentiation of Equation 4.26:

$$\mathbf{g}_q = \mathbf{F}\mathbf{q} \tag{4.61}$$

thus yielding

$$V(\mathbf{g}) = \frac{1}{2}\mathbf{q}^\dagger \mathbf{F}^\dagger \mathbf{C}\mathbf{F}\mathbf{q} \tag{4.62}$$

Comparing Equation 4.62 with Equation 4.26 clarifies that the compliance matrix $\mathbf{C}$ is identical with the inverse of the force constant matrix:

$$\mathbf{C} = \mathbf{F}^{-1} \tag{4.63}$$

From Equation 4.61, one sees that

$$\mathbf{q} = \mathbf{F}^{-1}\mathbf{g} \tag{4.64}$$

Hence, the diagonal compliance force constant C_{ii} gives the displacement of internal coordinate q_i under the impact of a unit force while all other forces are allowed to relax [107]. This leads to the fact that off-diagonal elements of $\mathbf{C}$ are largely reduced. Although the compliance force constants are force constants without a vibrational mode, there seems to be some relationship with adiabatic force constants and as a result also with the local modes associated with the adiabatic force constants.

The compliance force constants are largely independent of the internal coordinate set used. In a similar way as the adiabatic force constants describe the strength of a bond, the compliance force constants measure its weakness (the larger $C(AB)$ the weaker is bond AB, the smaller $C(AB)$ the stronger is bond AB). Compliance force constants have been used to describe the gallium, gallium triple bond [108], the strength of the NN and CO bond in NNH^+ and COH^+, respectively [109] or H-bonding in Watson–Crick base pairs [110].

It will be interesting to derive the relationship between adiabatic and compliance force constants and to investigate whether both force constants can be used in a confirmative or even complementary way when describing chemical bonding.

4.5
Badger-type Relationships for Polyatomic Molecules

The vibrational spectra of 51 polyatomic molecules with a total of 170 different bonds were analyzed for the purpose of determining adiabatic and c-vector vibrational modes. Exclusively, those molecules were considered for which measured vibrational data are available so that either directly or with the approach described in Section 4.3.5 experimental AICoMs and c-vector modes could be determined and compared with the corresponding modes calculated at the B3LYP/6-31G(d,p) level of theory. This objective of the current analysis limited the molecules investigated exclusively to neutral closed shell systems with normal bonding situations. This has to be considered when discussing the generalization of the Badger rule to polyatomic molecules.

The analysis of the stretching force constants and frequencies provided new insights into the usefulness of AICoMs. Calculated and experimental adiabatic frequencies correlate with a correlation coefficient R^2 of 0.997. The harmonic AICoM frequencies can be scaled down to experimental AICoM frequencies using a factor of 0.963. Similarly, calculated AICoM stretching force constants, if multiplied by 0.928, satisfactorily agree with experimental AICoM stretching force constants. Hence, it will be possible to base future studies on calculated adiabatic stretching modes. The correlation of AICoM stretching frequencies and force constants with the corresponding c-modes values led to a somewhat lower correlation coefficient R^2 of 0.988. Analysis of the data revealed that c-mode stretching force constants are always somewhat larger than AICoM force constants where the difference $k^c - k^a$ can be considered as a measure for the degree of mode coupling of the bond stretching vibration. For CH stretching force constants, deviations are between 0.05 and 0.10 mdyne/Å, whereas for CC stretching force constants deviations increase to 0.2–0.3 mdyne/Å. If carbon is bonded to a hetero atom, a further increase in the deviation from AICoM force constants is found. However, deviations decrease when comparing CX single bonds with double and triple bonds. Adiabatic and c-vector force constants for triple bonds do hardly differ.

Not surprisingly, deviations as large as 1.9 mdyne/Å are found for CC and CX bonds in conjugated five- and six-membered rings. In general, strained cyclic and polycyclic systems lead to a relatively large coupling effect of bond stretching motions with other stretching and bending motions. These observations clearly show that c-mode vibrations mostly used in vibrational analysis are not suited to study bond properties. It remains to be clarified why studies based on these modes could lead to Badger-type relationships. A typical example is the study of Ohno and coworkers who could derive a Badger-type relationship for 74 different CX bonds using c-vector vibrational modes [61]. The CX double and triple bonds investigated outnumbered the CX single bonds by a factor 2, that is, only a relatively small number of single bonds were considered. Also, the number of cases with X belonging to the third or the fourth period was large. Finally, all molecules with divalent Si or Ge were excluded from the test set as was also the case for triply bonded Si and Ge. Hence, the set of investigated molecules did not contain any "problem" cases and c-vector stretching force constants, although contaminated by coupling contributions, seemed to verify the Badger rule.

In Figure 4.4, the AICoM stretching force constants of 51 polyatomic molecules are given in dependence of the corresponding bond lengths. Hence, Figure 4.4 is based on the description of each polyatomic molecule as a collection of N quasi-diatomics where N is the number of bonds in the molecule. In so far it is not surprising that the diagrams in Figure 4.4 are closely related to those obtained for diatomic molecules (Figure 4.1). However, one essential difference between the Badger-type diagrams for diatomics and those for polyatomics becomes obvious: In the case of the latter molecules, there is one k_e-r_e curve for each type of bond, that is, the curves for OH, NH, CH, and BH bonds are all different, even though they seem to be closely related. This also holds for the relationships describing CC, CN, or CO bonds. We note that this observation is in agreement with the prediction made in Chapter 4.2 and suggests that for each bond type AB specified by atoms A and B and

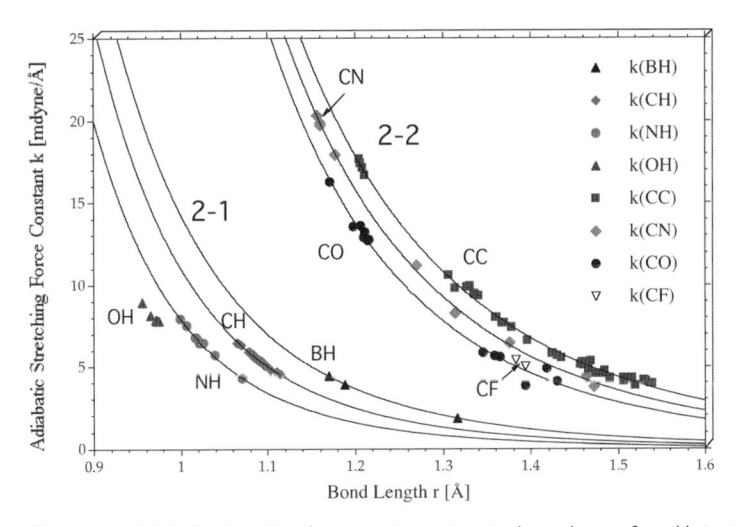

Figure 4.4 Adiabatic stretching force constants given in dependence of equilibrium bond lengths both calculated at the B3LYP/6-31G(d,p) level of theory for polyatomic molecules.

the electronic state of the polyatomic molecule an individual curve can be expected. The curves grouped according to A(period i)–B(period j) combinations, however, do not coincide.

We have used various ways of fitting the data in Figure 4.4: The original Badger formula was tested, that is, $(k^a_{AB})^{-1/3}$ was plotted against r_e, but other possible relationships were also tested, which led to $(k^a_{AB})^{-1/p}$ with $p = 2, 4, 5,$ or 6 in the general form of (4.65):

$$k^{-1/p} = ar_e + b \quad p = 2\text{--}6 \tag{4.65}$$

The results of these tests can be summarized as follows:

1) Badger's rule is fulfilled for individual bonds of polyatomic molecules provided that they are described by the AICoM concept and all bonds considered possess similar electronic features. The correlation coefficients are between 0.98 and 0.99 or even higher. However, if cations, anions, or open shell cases are included, the scattering of data points will increase as seen in the case of the diatomics.
2) According to the calculated correlation coefficients, there is hardly any difference whether $p = 3, 4, 5,$ or 6 is used in relationship (4.65). One can avoid relationships of the form (4.65) by using the exponential form (4.66) previously suggested by other authors (entries 31, 33, 47, 50, and 53 in Table 4.1):

$$k^a_n = ae^{-br_e} \tag{4.66}$$

The exponential dependence of the adiabatic force constant on the calculated bond length in Equation 4.66 accounts for all possibilities provided by Equation 4.65.

By defining an effective bond length, the curves of Figure 4.4 can be merged in one XH and one CX curve (see, for example, Figure 4.5). This provides evidence that

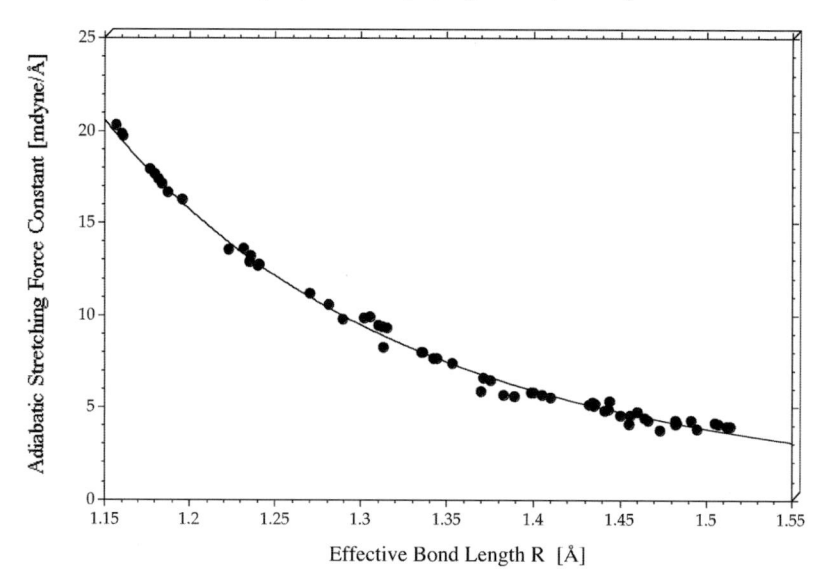

Figure 4.5 Merging of CX Badger-type relationships from Figure 4.4 by introducing an effective bond length R.

a universal Badger-type relationship can be derived on the basis of hard sphere adjustment parameters d_{AB} characteristic of A and B, their charge, and spin situation rather than just the location of A and B in periods i and j. Work is in progress to determine these parameters.

4.6 Conclusions

A universal relationship between bond length $r(AB)$ and bond stretching force constant $k(AB)$ valid for both diatomic and polyatomic molecules can be derived only if two major prerequisites are fulfilled.

1) The bond stretching force constant must correspond to a local stretching mode that is characteristic of the bond AB only. A generally applicable way of deriving local modes is provided by the adiabatic internal coordinate mode concept. As described in this work, AICoMs can be determined for all bonds of a molecule using either calculated or experimental vibrational mode frequencies. AICoMs have been verified in this work as suitable local modes by comparing them with McKean's isolated XH stretching modes and the local CH modes from overtone spectroscopy. They differ from c-vector modes because the latter are contaminated by the coupling with other vibrational modes. The averaging of vibrational frequencies to obtain local mode information is also not suitable because it can lead to physically not meaningful local mode frequencies.

2) In view of the increasing size of an atom A or B with the number of its electrons, a universal bond length–force constant relationship must be based on effective bond lengths, which are corrected for the different hard sphere sizes of A and B. This work has shown that correction parameters cannot be uniformly defined for all atoms of a period i and all atoms of a second period j. Instead, they have to consider the charge and spin multiplicity of bond AB (or the molecule containing bond AB). In addition, one has to consider that in strained molecules such as cyclopropane the actual bond path is significantly longer than the internuclear distance because of the (concave or convex) bond bending. Because of this, the stretching force constant turns out to be smaller (indicating a weaker bond) than the length of the bond might suggest.

The results obtained in this work indicate that different bonding situations (single, double, triple bonded AB) of closed shell molecules can be described with just one hard sphere parameter. Test calculations for cations and anions (not described in this chapter) suggest that the number of parameters for a given bond AB will not be larger than 3 or 4 because in this way cationic, anionic, closed, and open shell situations can be described for a large body of molecules containing bond AB. Work is in progress to determine hard sphere parameters for bonds formed by atoms of the first three periods in a systematic way. This work will lead to a universal Badger-type relationship that will facilitate the derivation of suitable initial guesses of the Hessian matrix in quantum chemical geometry optimizations, the construction of force fields, and the prediction of bond lengths from measured vibrational data in surface studies, for molecular aggregates, and catalysis. Conversely, calculated bond lengths can be used to predict via the stretching force constant the strength of the bond.

Acknowledgment

This work was financially supported by the National Science Foundation, Grant CHE 071893. We thank SMU for providing computational resources. Proofreading and useful comments by Robert Kalescky are acknowledged.

References

1 Kratzer, A. (1920) Die ultraroten rotationsspektren der halogenwasserstoffe. *Z. Physik*, **3**, 289–307.
2 Birge, R.T. (1925) The quantum structure of the OH bands, and notes on the quantum theory of band spectra. *Phys. Rev.*, **25**, 240–254.
3 Mecke, R. (1925) Zum aufbau der bandenspektra. *Z. Physik*, **32** (1925), 823–834.
4 Morse, P.M. (1929) Diatomic molecules according to the wave mechanics. II. Vibrational levels. *Phys. Rev.*, **34**, 57–64.
5 Clark, C.H.D. (1934) The relation between vibration frequency and nuclear separation for some simple non-hydride diatomic molecules. *Phil. Mag.*, **18**, 459–470.
6 Badger, R.M. (1934) A relation between internuclear distances and bond force constants. *J. Chem. Phys.*, **2**, 128–132.
7 Badger, R.M. (1935) The relation between internuclear distances and the force constants of diatomic molecules. *Phys. Rev.*, **48**, 284–285.
8 Allen, H.S. and Longair, A.K. (1935) Internuclear distance and vibration frequency for diatomic molecules. *Nature*, **135**, 764–764.
9 Badger, R.M. (1935) The relation between the internuclear distances and force constants of molecules and its application to polyatomic molecules. *J. Chem. Phys.*, **3**, 710–715.
10 Huggins, M.L. (1935) molecular constants and potential energy curves for diatomic molecules. *J. Chem. Phys.*, **3**, 473–479.
11 Huggins, M.L. (1936) Molecular constants and potential energy curves for diatomic molecules. II. *J. Chem. Phys.*, **4**, 308–312.
12 Sutherland, G.B.B.M. (1938) *Proc. Indian Natl. Sci. Acad.*, **8**, 341.
13 Sutherland, G.B.B.M. (1940) The determination of internuclear distances and of dissociation energies from force constants. *J. Chem. Phys.*, **8**, 161–165.
14 Newing, R.A. (1940) On the interrelation of molecular constants for diatomic molecules – II. *Phil. Mag.*, **29**, 298–301.
15 Linnett, J.W. (1940) The relation between potential energy and interatomic distance in some diatomic molecules. *Trans. Faraday Soc.*, **36**, 1123–1135.
16 Clark, C.H.D. and Webb, K.R. (1941) Systematics of band-spectral constants. Part VI. Interrelation of equilibrium bond constant and internuclear distance. *Trans. Faraday Soc.*, **37**, 293–298.
17 Linnett, J.W. (1942) The relation between potential energy and interatomic distance in some di-atomic molecules II. *Trans. Faraday Soc.*, **38**, 1–9.
18 Wu, C.K. and Yang, C.T. (1944) The relation between the force constant and the interatomic distance of a diatomic linkage. *J. Phys. Chem.*, **48**, 295–303.

19 Linnett, J.W. (1945) The force constants of some CH, NH and related bonds. *Trans. Faraday Soc.*, **41**, 223–232.
20 Gordy, W. (1946) A relation between bond force constants, bond orders, bond lengths, and the electronegativities of the bonded atoms. *J. Chem. Phys.*, **14**, 305–321.
21 Guggenheimer, K.M. (1946) New regularities in vibrational spectra. *Proc. Phys. Soc.*, **58**, 456–468.
22 Guggenheimer, K.M. (1950) General discussion. *Discuss. Faraday Soc.*, **9**, 207–222.
23 Wu, C.K. and Chao, S.C. (1947) The relation between the force constant and the interatomic distance of a diatomic linkage. II. A modified Huggins relation. *Phys. Rev.*, **71**, 118–121.
24 Herzberg, G. (1950) *Spectra of Diatomic Molecules*, 2nd edn, D. Van Nostand Co. Inc., Princeton, NJ, p. 453.
25 Lovera, G. (1951) Regolarita nei dati spettroscopici di molecole biatomiche formate con atomi del 4 e 6 gruppo. *Nuovo Cimento*, **8**, 1014–1015.
26 Siebert, H. (1953) Kraftkonstante und strukturchemie. I. Über die verwendung der molekularen kraftkonstanten zu strukturchemischen aussagen. *Z. Anorg. Allg. Chem.*, **273**, 170–182.
27 Lippincott, E.R. and Schroeder, R. (1955) General relation between potential energy and internuclear distance for diatomic and polyatomic molecules. I. *J. Chem. Phys.*, **23**, 1131–1142.
28 Jenkins, H.O. (1955) Bond energies, internuclear distances, and force constants in series of related molecules. *Trans. Faraday Soc.*, **51**, 1042–1051.
29 Varshni, Y.P. (1958) Correlation of molecular constants. I. Deduction of relations. *J. Chem. Phys.*, **28**, 1078–1081.
30 Varshni, Y.P. (1958) Correlation of molecular constants. II. Relation between force constant and equilibrium internuclear distance. *J. Chem. Phys.*, **28**, 1081–1089.
31 Hershbach, D.R. and Laurie, V.W. (1961) Anharmonic potential constants and their dependence upon bond length. *J. Chem. Phys.*, **35**, 458–463.
32 Johnston, H.S. (1964) Continuity of bond force constants between normal molecules and Lennard-Jones pairs. *J. Am. Chem. Soc.*, **86**, 1643–1645.
33 Ladd, J.A., Orville-Thomas, W.J., and Cox, B.C. (1964) Molecular parameters and bond structure III. Carbon–oxygen bonds. *Spectrochim. Acta*, **20**, 1771–1780.
34 Ladd, J.A. and Orville-Thomas, W.J. (1966) Molecular parameters and bond structure V. Nitrogen–oxygen bonds. *Spectrochim. Acta*, **22**, 919–925.
35 Dewar, M.J.S. and Gleicher, G.J. (1966) Ground states of conjugated molecules. VII. Compounds containing nitrogen and oxygen. *J. Chem. Phys.*, **44**, 759–774.
36 Decius, J.C. (1966) Relation between force constant and bond length for the nitrogen–nitrogen bond. *J. Chem. Phys.*, **45**, 1069–1071.
37 Goubeau, J. and Sawodny, W. (1966) Kraftkonstanten und bindungsgrade von stickstoffverbindungen. *Angew. Chem.*, **78**, 565–576.
38 Borkman, R.F. and Parr, R.G. (1968) Toward an understanding of potential–energy functions for diatomic molecules. *J. Chem. Phys.*, **48**, 1116–1127.
39 Roy, R.S. (1968) New relationship between bond length and force constant. *Proc. Phys. Soc. Ser.*, **2** (1), 445–448.
40 Stals, J. (1970) Empirical correlations of molecular constants for carbon–carbon, carbon–nitrogen, carbon–oxygen, nitrogen–nitrogen, nitrogen–oxygen and oxygen–oxygen bonds. *Rev. Pure Appl. Chem.*, **20**, 2–5.
41 McKean, D.C. (1978) Individual CH bond strengths in simple organic compounds: effects of conformation and substitution. *Chem. Soc. Rev.*, **7**, 399–422.
42 Schlegel, H.B. (1984) Estimating the Hessian for gradient-type geometry optimizations. *Theor. Chim. Acta*, **66**, 333–340.
43 Byler, D.M., Susi, H., and Damert, W.C. (1987) Relation between force constant and bond length for carbon nitrogen bonds. *Spectrochim. Acta A*, **43**, 861–863.
44 Bürgi, H.B. and Dunitz, J.D. (1987) Fractional bonds: relations among their lengths, strengths, and stretching force

constants. *J. Am. Chem. Soc.*, **109**, 2924–2926.

45 Swanton, D.J. and Henry, B.R. (1987) A theoretical basis for the correlation between bond length and local mode frequency. *J. Chem. Phys.*, **86**, 4801–4808.

46 Zavitsas, A.A. (1987) Quantitative relationship between bond dissociation energies, infrared stretching frequencies, and force constants in polyatomic molecules. *J. Phys. Chem.*, **91**, 5573–5577.

47 Miskowski, V.M., Dallinger, R.F., Christoph, G.G., Morris, D.E., Spies, G.H., and Woodruff, W.H. (1987) Assignment of the rhodium–rhodium stretching frequency in $Rh_2(O_2CCH_3)_4L_2$ complexes and the crystal and molecular structure of $[C(NH_2)_3]_2[Rh_2(O_2CCH_3)4Cl_2]$. Relationship between vibrational spectra and structure. *Inorg. Chem.*, **26**, 2127–2139.

48 Weisshaar, J.C. (1989) Application of badger rule to third row metal diatomics. *J. Chem. Phys.*, **90**, 1429–1434.

49 Mallick, P.K., Strommen, D.P., and Kincaid, J.R. (1990) Molecular structure determination of transient species from vibrational frequency data. Application to the 3MLCT state of tris(bipyridine) ruthenium(II). *J. Am. Chem. Soc.*, **112**, 1686–1690.

50 Fischer, T.H. and Almlöf, J. (1992) General methods for geometry and wave function optimization. *J. Phys. Chem.*, **96**, 9768–9774.

51 Rutkowski, A., Rutkowska, D., and Schwarz, W.H.E. (1992) Relativistic perturbation theory of molecular structure. *Theor. Chem. Acc.. Theor. Comput. Model.*, **84**, 105–114.

52 Pearson, R.G. (1993) Bond-energies, force-constants and electronegativities. *J. Mol. Struct.*, **300**, 519–525.

53 Lindh, R., Bernhardsson, A., Karlström, G., and Malmqvist, P.-Å. (1995) On the use of a Hessian model function in molecular-geometry optimizations. *Chem. Phys. Lett.*, **241**, 423–428.

54 Harvey, P.D. (1996) Reparameterized Herschbach–Laurie empirical relationships between metal–metal distances and force constants applied to homonuclear bi- and polynuclear complexes (M = Cr, Mo, Rh, Pd, Ag, W, Re, Ir, P, A, Hg). *Coord. Chem. Rev.*, **153**, 175–198.

55 Wittbrodt, J.M. and Schlegel, H.B. (1997) Estimating stretching force constants for geometry optimization. *J. Mol. Struct. (Theochem)*, **398–399**, 55–61.

56 Stichler, M. and Menzel, D. (1999) A systematic investigation of the geometric structures of four oxygen/nitric oxide coadsorbate layers on Ru(001). *Surf. Sci.*, **419**, 272–290.

57 Larsson, J.A. and Cremer, D. (1999) Theoretical verification and extension of the McKean relationship between bond lengths and stretching frequencies. *J. Mol. Struct.*, **485**, 385–408.

58 Cioslowski, J., Liu, G., and Castro, A.M. (2000) Badger rule revised. *Chem. Phys. Lett.*, **331**, 497–501.

59 Hermann, H.L., Boche, G., and Schwerdtfeger, P. (2001) Metallophilic interactions in closed-shell copper(I) compounds: a theoretical study. *Chem. Eur. J.*, **24**, 5333–5342.

60 Jules, J.L. and Lombardi, J.R. (2003) Transition metal dimer internuclear distances from measured force constants. *J. Phys. Chem. A*, **107**, 1268–1273.

61 Kurita, E., Matsuura, H., and Ohno, K. (2004) Relationship between force constants and bond lengths for CX (X = C, Si, Ge, N, P, As, O, S, Se, F, Cl, and Br) single and multiple bonds: formulation of Badger's rule for universal use. *Spectrochim. Acta A*, **60**, 3013–3023.

62 Stone, K.L., Behan, R.K., and Green, M.T. (2006) Resonance Raman spectroscopy of chloroperoxidase compound II provides direct evidence for the existence of an iron(IV)-hydroxide. *Proc. Natl. Acad. Sci. USA*, **103**, 12307–12310.

63 Behan, R.K. and Green, M.T. (2006) On the status of ferryl protonation, *J. Inorg. Biochem*, **100**, pp. 448–459.

64 Lin, Y.W., Nie, C.M., and Liao, L.F. (2007) Probing the weak interactions between amino acids and carbon monoxide. *Chin. Chem. Lett.*, **19**, 119–122.

65 Das, U. and Raghavachari, R. (2008) Predicting PH vibrations of gas phase molecules and surface-adsorbed species using bond length-frequency correlations. *J. Comp. Chem.*, **30**, 1872–1881.

66 Oomens, J., Kraka, E., Nguyen, M.K., and Morton, T.H. (2008) Structure, vibrational spectra, and unimolecular dissociation of gaseous 1-fluoro-1-phenethyl cations. *J. Phys. Chem. A*, **112**, 10774–10783.

67 Kraka, E. and Cremer, D. (2009) Characterization of CF bond with multiple-bond character: bond lengths, stretching force constants, and bond dissociation energies. *Chem. Phys. Chem.*, **10**, 689–698.

68 Xie, R.-H. and Gong, J. (2005) Simple three-parameter model potential for diatomic systems: from weakly to strongly bound molecules to metastable molecular ions. *Phys. Rev. Lett.*, **95**, 263202-1–263202-4.

69 Xie, R.-H. and Hsu, P.S. (2006) Universal reduced potential function for diatomic systems. *Phys. Rev. Lett.*, **96**, 243201-1–243201-4.

70 Hooydonk, G.V. (1999) A universal two-parameter Kratzer-potential and its superiority over Morse's for calculating and scaling first-order spectroscopic constants of 300 diatomic bonds. *Eur. J. Inorg. Chem.*, 1617–1642.

71 Huber, K.P. and Herzberg, G. (1979) *Molecular Spectra and Molecular Structure IV. Constants of Diatomic Molecules*, Van Nostrand Rheinhold, London.

72 Henry, B.R. (1989) The frequency bond length correlation in local-mode overtone spectra. *J. Mol. Struct. (Theochem)*, **202**, 193–201.

73 Hess, B.A., Jr., Schaad, L., Čársky, R., and Zahradník, J.P. (1986) Ab initio calculations of vibrational-spectra and their use in the identification of unusual molecules. *Chem. Rev.*, **86**, 709–730.

74 Wilson, S. ed. (1992) *Methods in Computational Chemistry: Molecular Vibrations*, vol. 4, Plenum Press, New York.

75 Wilson, E.B., Jr., Decius, J.C., and Cross, P.C. (1955) *Molecular Vibrations*, McGraw-Hill, New York.

76 Konkoli, Z., Larsson, J.A., and Cremer, D. (1997) A new way of analyzing vibrational spectra. II. Comparison of internal mode frequencies. *Int. J. Quantum Chem.*, **67**, 11–27.

77 Herzberg, G. (1946) *Infrared and Raman Spectra of Polyatomic Molecules*, Van Nostrand Rheinhold Co., New York.

78 Califano, S. (1976) *Vibrational States*, John Wiley & Sons, Inc., New York.

79 Konkoli, Z. and Cremer, D. (1998) A new way of analyzing vibrational spectra. 1. Derivation of adiabatic internal modes. *Int. J. Quantum Chem.*, **67**, 1–9.

80 Torkington, P. (1949) The general solution of the secular equation of second degree, with application to the class-A1 vibrations of the symmetrical triatomic molecule. *J. Chem. Phys.*, **17**, 357–369.

81 Morino, Y. and Kuchitsu, K. (1952) A note on the classification of normal vibrations of molecules. *J. Chem. Phys.*, **20**, 1809–1810.

82 Pulay, P. and Török, F. (1966) On parameter form of matrix F. 2. Investigation of assignment with aid of parameter form. *Acta Chim. Hung.*, **47**, 273–279.

83 Keresztury, G. and Jalsovszky, G. (1971) Alternative calculation of vibrational potential energy distribution. *J. Mol. Struct.*, **10**, 304.

84 Cremer, D., Larsson, J.A., and Kraka, E. (1998) New developments in the analysis of vibrational spectra: on the use of adiabatic internal vibrational modes, in *Theoretical Organic Chemistry Theoretical and Computational Chemistry*, vol. 5 (ed. C. Parkany), Elsevier Science, New York, pp. 259–327.

85 Konkoli, Z. and Cremer, D. (1998) A new way of analyzing vibrational spectra. III. Characterization of normal vibrational modes in terms of internal vibrational modes. *Int. J. Quantum Chem.*, **67**, 29–40.

86 Kraka, E. (1998) Reaction path Hamiltonian and its use for investigating reaction mechanism, in *Encyclopedia of Computational Chemistry*, vol. 4 (eds E.V.R. Schleyer, N.L. Allinger, T. Clark, J. Gasteiger, E.A. Kollman, H.F. Schaefer, III, and E.R. Schreiner), John Wiley & Sons Ltd., Chichester, UK, p. 2437.

87 Konkoli, Z., Kraka, E., and Cremer, D. (1997) Unified reaction valley approach mechanism of the reaction $CH_3 + H_2 \rightarrow CH_4 + H$. *J. Phys. Chem. A*, **101**, 1742–1757.

88 Vijay, A. and Sathyanarayana, D.N. (1994) Use of L-matrix to obtain reliable *ab-initio* force-constants of polyatomic molecules: ethylene as a test. *J. Mol. Struct.*, **328**, 269–276.

89 McKean, D.C. (1975) CH stretching frequencies, bond lengths and strengths in halogenated ethylenes. *Spectrochim. Acta A*, **31**, 1167–1186.

90 McKean, D.C. (1989) CH bond dissociation energies, isolated stretching frequencies, and radical stabilization energy. *Int. J. Chem. Kinet.*, **21**, 445–464.

91 McKean, D.C. and Torto, I. (1982) CH stretching frequencies and bond strengths, and methyl-group geometry in CH_3CXO compounds (X = H, Me, F, Cl, Br, CN, OMe) and CH_3CH_2CN. *J. Mol. Struct.*, **81**, 51–60.

92 Duncan, J.L., Harvie, J.L., McKean, D.C., and Cradock, S. (1986) The ground-state structures of disilane, methyl silane and the silyl halides, and an SiH bond length correlation with stretching frequency. *J. Mol. Struct.*, **145**, 225–242.

93 McKean, D.C. (1989) CH bond-dissociation energies, isolated stretching frequencies, and radical stabilization energy. *Int. J. Chem. Kinet.*, **21**, 445–464.

94 Murphy, W.F., Zerbetto, F., Duncan, J.L., and McKean, D.C. (1993) Vibrational-spectrum and harmonic force-field of trimethylamine. *J. Phys. Chem.*, **97**, 581–595.

95 Caillod, J., Saur, O., and Lavalley, J.-C. (1980) Etude par spectroscopie infrarouge des vibrations (CH) de composes cycliques: dioxanne-1,4, dithianne-1,4, oxathianne-1,4 et cyclohexane. *Spectrochim. Acta A*, **36**, 185–191.

96 Snyder, R.G., Aljibury, A.L., Strauss, H.L., Casal, H.L., Gough, K.M., and Murphy, W.J. (1984) Isolated C-H stretching vibrations of N-alkanes: assignments and relation to structure. *J. Chem. Phys.*, **81**, 5352–5361.

97 Aljibury, A.L., Snyder, R.G., Strauss, H.L., and Raghavachari, K. (1986) The structure of N-alkanes: high-precision *ab initio* calculations and relation to vibrational-spectra. *J. Chem. Phys.*, **84**, 6872–6878.

98 Henry, B.R. (1987) The local mode model and overtone spectra: a probe of molecular-structure and conformation. *Acc. Chem. Res.*, **20**, 429–435 and references therein.

99 Pople, J.A., Schlegel, H.B., Krishnan, R., DeFrees, D.J., Binkley, J.S., Frisch, M.J., Whiteside, R.A., Hout, R.F., and Hehre, W.J. (1981) Molecular-orbital studies of vibrational frequencies. *Int. J. Quantum Chem. Quant. Chem. Symp.*, **15**, 269–278.

100 Mizugai, Y. and Katayama, M. (1980) The 5th overtone of the C-H stretching vibrations and the bond lengths in some heterocyclic-compounds. *Chem. Phys. Lett.*, **73**, 240–243.

101 Wong, J.S. and Moore, C.B. (1982) Inequivalent C-H oscillators of gaseous alkanes and alkenes in laser photo-acoustic overtone spectroscopy. *J. Chem. Phys.*, **77**, 603–615.

102 Sbrana, G. and Muniz-Miranda, M. (1998) High overtones of C-H stretching vibrations in isoxazole, thiazole, and related methyl and dimethyl derivatives. *J. Phys. Chem. A*, **102**, 7603–7608.

103 Hippler, M. and Quack, M. (1997) Intramolecular energy transfer from isotope selective overtone spectroscopy by vibrationally assisted dissociation and photofragment ionization. *Ber. Bunsenges. Phys. Chem.*, **101**, 356–362.

104 Hollensteun, H., Luckhaus, D., and Quack, M. (1993) Dynamics of the CH chromophore in CHX_3: a combined treatment for a set of isotopic-species. *J. Mol. Struct.*, **294**, 65–70.

105 Boatz, J.A. and Gordon, M.S. (1989) Decomposition of normal-coordinate vibrational frequencies. *J. Phys. Chem.*, **93**, 1819–1826.

106 Decius, J.C. (1963) Compliance matrix and molecular vibrations. *J. Chem. Phys.*, **38**, 241–248.

107 Brandhorst, K. and Grunenberg, J. (2008) How strong is a bond? The interpretation of force and compliance constants as

bond strength descriptors. *Chem. Soc. Rev.*, **37**, 1558–1567.
108 Grunenberg, J. and Goldberg, N. (2000) How strong is the gallium gallium triple bond? Theoretical compliance matrices as a probe for intrinsic bond strengths. *J. Am. Chem. Soc.*, **122**, 6046–6047.
109 Grunenberg, J., Streubel, R., Frantzius, G.V., and Marten, W. (2003) The strongest bond in the universe? Accurate calculation of compliance matrices for the ions N_2H^+, HCO^+, and HOC^+. *J. Chem. Phys.*, **119**, 165–169.
110 Grunenberg, J. (2004) Direct assessment of interresidue force in Watson–Crick base pairs using theoretical compliance constants. *J. Am. Chem. Soc.*, **126**, 16310–16311.

5
The Simulation of UV-Vis Spectroscopy with Computational Methods
Benedetta Mennucci

5.1
Introduction

In the past 10 years, quantum chemistry has developed toward the modeling of systems of increasing complexity. Among the most active research lines, there is one aiming at an accurate description of molecular (and supramolecular) systems in their electronically excited states [1]. The enormous progress in time-dependent spectroscopies, in fact, has allowed a detailed study of the process of formation and relaxation of excited states and has pushed theoretical chemists to develop new methods and efficient computational strategies able to simulate such a process. In parallel, the development of new branches of technological research focusing on the process of capture and conversion of light has shown that a reliable quantum description of the phenomenon of photoexcitation in molecular systems and materials is of fundamental importance to designing and optimizing efficient devices.

Although quantum chemistry has since long developed theoretical methods and computational approaches to describe excited states [2], however the older quantum methods were either very expensive or extremely approximated and thus they could not be applied to systems of real interest for spectroscopic and/or technological applications. In addition, to successfully apply these formulations it was often necessary to have a large experience in both quantum theories and computational techniques. For years, all these aspects limited the modeling of excited states and related phenomena to a few very specialized research groups. Things have started to really change after the enormous success of methods based on the density functional theory (DFT) [3] for ground-state systems. This success, in fact, encouraged a search for a time-dependent theory of a similar flavor (TDDFT) [4] that could be applied to excited states. It is exactly the development of TDDFT approaches that, by combining computational efficiency with ease if use, has pushed many researchers (not only with quantum or computational background) to shift their studies toward processes and phenomena involving electronic excitations.

Computational Spectroscopy: Methods, Experiments and Applications. Edited by Jörg Grunenberg
Copyright © 2010 WILEY-VCH Verlag GmbH & Co. KGaA, Weinheim
ISBN: 978-3-527-32649-5

In this new (or enhanced) interest in the modeling of electronically excited states, an important role is played by UV-Vis spectroscopy. This, in fact, is the first fundamental test of any theoretical method aiming at describing photoinduced processes and at the same time at becoming a useful tool to calibrate computational methodologies so as to obtain not only accuracy and reliability but also efficiency. In this calibration process, a further modelistic aspect that has to be carefully considered is the environment and its effects on the structure and the electronic properties of the molecular systems undergoing excitation. The coupling between quantum mechanical methods and solvation models represents a very active research line for many years, and now an important part of this research has been shifted to approaches that can describe formation and relaxation of excited states of solvated (or more generally embedded) systems [5].

From this brief excursus comes out clearly the large complexity involved in characterizing any description of excited states. Indeed, a reliable and accurate approach for realistic molecules in real-life environments is still an open problem. Some effective strategies based on cleverly chosen approximations are, however, available. These are exactly the focus of this chapter in which a review of the main aspects of the computational simulation of UV-Vis spectra of molecular systems is presented. The first section of this chapter will focus on the QM methods used to describe electronic excitations, the second will present models to include effects that environment has on such excitations, the third will focus on the computational strategies developed to simulate the entire spectra, and the fourth will summarize some studies appeared in the literature and selected here as illustrative examples of applications. Finally, in the fifth section some conclusive remarks will be made together with some comments on the main aspects that still require further development both in the theoretical and in the computational tools.

5.2
Quantum Mechanical Methods

The quantum chemical approaches for the calculation of excited states and their properties are generally divided into three groups: (1) wavefunction-based *ab initio* methods, (2) semiempirical methods, and (3) density functional theory-based approaches.

Wavefunction-based *ab initio* methods for the calculation of excited states of molecular systems can be divided into single reference and multireference methods, on the one hand, and into configuration interaction and coupled cluster methods, on the other hand.

In configuration interaction-type calculations, the electronic many-body wavefunction is constructed as a linear combination of the ground-state Slater determinant and "excited" determinants, which are obtained by replacing occupied orbitals with virtual ones. In this framework, the exact numerical solution of the Schrödinger equation within the chosen atomic basis set would correspond to include all possible "excited" determinants (full CI). Such calculations are extremely expensive and at present feasible only for very small systems. As a consequence, one has to truncate the CI expansion; in particular, if one stops after the "singly" excited determinants, the widespread method called CIS (configuration interaction singles) is obtained [6]. The obvious limits of such a description may be partially overcome if a linear combination of all possible single excited determinants is used to build the excited state wavefunction. The main problem of the CIS method is the basic lack of correlation energy. As a result, excitation energies computed with the CIS method are usually overestimated up to 0.5–2 eV compared to their experimental values. Despite these limitations, the CIS method possesses some useful properties; for example, it is variational and size-consistent. In addition, the excited state energies are analytically differentiable with respect to external parameters, for example, nuclear displacements and external fields, which makes possible the application of analytic gradient techniques for the calculation of excited state properties such as equilibrium geometries and vibrational frequencies.

To make CI calculations computationally feasible, a different approximation with respect to the CI truncation is made defining an active space of occupied and virtual orbitals in which all possible "excited" determinants are constructed. Since such an approximation can significantly reduce the flexibility of the CI wavefunction, in general one has to reoptimize the molecular orbitals during the minimization procedure to improve the accuracy. The resulting approach is known as the complete active space self-consistent field (CASSCF) [7]. In an analogy to CIS, CASSCF analytical derivatives are available allowing efficient optimization of excited state geometries and localization of conical intersections. However, from a computational point of view, CASSCF calculations become quickly very expensive since the computational effort increases exponentially with the size of the active space. In addition, three general problems exist with CASSCF calculations.

First, the choice of the active space is not unique, but a careful analysis of both the chemical nature of the system and the physical nature of the orbitals is required. Second, the relative energies of the calculated excited states and their energetic ordering strongly depend on the choice of the state for which the orbitals are optimized. One way to partially circumvent this problem is to perform a state-averaged CASSCF calculation, in which the orbitals are optimized with respect to a weighted mean of all states of interest. Third, too small active spaces and concomitant neglect of large parts of dynamic electron correlation can lead to significant errors and an unbalanced treatment of electronic states of different nature. A very successful, but also expensive, approach to include dynamic electron correlation is a second-order perturbation theory on a wavefunction obtained from a state-averaged CASSCF calculation: this approach is known as CASPT2 [8]. Analytical CASPT2 gradients have been also developed [9].

Correlation can also be included in single-reference wavefunction methods, for instance, through coupled cluster (CC) theory [10]. CC is exact when all the possible excitations are included, but in practice the expansion is truncated at a given order. The most widely used truncation order implies single and double excitations (CCSD). Because of the nonlinearity of the exponential wave operator, higher order excitations are included at this level of truncation, so that CCSD is a dramatic improvement over configuration interaction singles and doubles (CISD). The success and limitation of CCSD are also related to its computational cost, as it scales iteratively at $O(N^6)$, where N is the number of basis functions, so this method is feasible only for small- and medium-sized molecules. CC theory has been extended to excited state calculation through the equation of motion (EOM) formalism [11] or alternatively through the linear response (LR) formalism [12] (these two approaches are equivalent for transition energies). EOM-CC scales as the ground state at a certain level of truncation; thus, the same systems that are accessible at CCSD level can be in general also treated at EOM-CCSD level for the excited states.

Closely related to these coupled cluster theories is the symmetry-adapted cluster configuration interaction (SAC-CI) approach [13] and approximate coupled cluster schemes of second or third order (CC2, CC3) [14]. At present, analytical gradients for excited states in single configuration methods are available, at a high computational cost, at the SAC-CI [15], EOM-CCSD [16], and CC2 [17] levels.

Meanwhile, semiempirical approaches were also successfully applied to the calculation of photoinitiated processes in molecular systems; the most common approach is the intermediate neglect of differential overlap (INDO/S) method in combination with a CIS formalism developed by Zerner (the method is also known as ZINDO) [18]. The simple structure of the one- and two-electron integrals allows an efficient and fast computation of excited states of very large molecular systems. The accuracy of the method, however, is unpredictable, and careful comparison with experiment and/or higher level computations must be made. However, the quantitative aspect of the results obtained was found to be highly system dependent and thus practically unpredictable. More recently, calculations carried out for organic dyes have indicated that PM5 could be a promising approach [19], but such a claim remains to be tested on a broader set of transitions and molecules.

As anticipated in the introduction, in the last few years there has been an explosion of interest in DFT driven largely by its applications in quantum chemistry. This is because of the recent progress made in the accuracy of available approximations and the wealth of chemical problems that can be tackled with such a computationally inexpensive tool. The success of DFT methods for ground-state systems has pushed researchers to extend the theory to excited states. This has led to the development of the linear response regime within a time-dependent DFT (TDDFT) approach. The linear response of any system can, in fact, be used to determine its excitation energies. In fact, by simply applying an oscillating potential and varying its frequency, a resonance occurs whenever the frequency equals the difference of two energy eigenvalues of the system. The basic idea is to apply time-dependent perturbation theory to first order and to describe the time-dependent linear response of the one-particle density to a time-dependent oscillating electric field. Before the time-dependent electric field is applied, the molecular system is assumed to be in its electronic ground state, that is, to obey the ground-state Kohn–Sham equation. Now a time-dependent oscillating electric field is applied, and as a consequence, the Kohn–Sham orbitals, the electron density, and the Kohn–Sham operator will change, since the latter depends on the orbitals. If the field is a small perturbation, the new density can in first order be written as

$$\varrho(r,t) = \varrho^0(r) + \delta\varrho(r,t) \qquad (5.1)$$

Equation 5.1 when inserted into the time-dependent Kohn–Sham equation yields, after Fourier transformation into the energy space, the TDDFT equation for the excitation energies and transition vectors. In compact matrix notation, the TDDFT equations read [20]

$$\begin{pmatrix} A & B \\ B^* & A^* \end{pmatrix} \begin{pmatrix} X \\ Y \end{pmatrix} = \omega \begin{pmatrix} 1 & 0 \\ 0 & -1 \end{pmatrix} \begin{pmatrix} X \\ Y \end{pmatrix} \qquad (5.2)$$

where the matrices A and B are the Hessians of the electronic energy; for example, for a hybrid exchange correlation functional, their elements become

$$\begin{aligned} A_{ia,jb} &= \delta_{ab}\delta_{ij}(\varepsilon_a - \varepsilon_i) + \langle ja|ib\rangle - c_{HF}\langle ji|ab\rangle + (1 - c_{HF})\langle ja|f_{xc}|ib\rangle \\ B_{ia,jb} &= \langle ja|bi\rangle - c_{HF}\langle jb|ai\rangle + (1 - c_{HF})\langle ja|f_{xc}|bi\rangle \end{aligned} \qquad (5.3)$$

The leading term on the diagonal of the A matrix is the difference of the energies of the orbitals i and a, which are the ones from which and to which the electron is excited, respectively (indices $i, j, \ldots$, label occupied, $a, b, \ldots$, virtual orbitals). The second term of the A matrix and the elements of the B matrix contain the antisymmetrized two-electron integrals and they follow from the linear response of the Coulomb and exchange correlation (xc) operators to the first-order changes in the single-particle orbitals. In Equation 5.3, we have used the hybrid mixing parameter c_{HF} that allows us to interpolate between the limits of "pure" density functionals ($c_{HF} = 0$, no "exact" exchange) and HF theory ($c_{HF} = 1$, full exchange, no correlation).

Equation 5.2 is a non-Hermitian eigenvalue equation, the solution of which yields excitation energies ω and transition vectors $|XY\rangle$ determining the first-order change in the density and thereby the excited state wavefunction. Although the exchange correlation kernel f_{xc} of Equations 5.3 formally depends on energy, in practical calculations standard ground-state xc functionals are employed to evaluate those terms. This is a consequence of the so-called adiabatic local density approximation (ALDA) [21].

It is worth noting that Equations 5.2–5.3 contain different but closely related schemes for the calculation of excited states. If the coefficient c_{HF} is equal to 1, the scheme reduces to the time-dependent Hartree–Fock (TDHF). In addition, if the Tamm–Dancoff approximation (TDA) is introduced, which means that the B matrix is neglected in Equation 5.2, the TDHF scheme reduces to CIS. This shows that the CIS scheme can be obtained either via the CI formalism as presented previously or via linear response theory.

So far, simulation and assignment of vertical electronic absorption spectra have been the main task of TDDFT calculations in chemistry. Most benchmark studies agree that low-lying valence excitations are predicted with errors of about 0.4 eV by local density approximation (LDA) and generalized gradient approximation (GGA) functionals. Hybrid functionals can be more accurate but display a less systematic error pattern [22] (see Section 5.5 for more details).

The reason for the accuracy of TDDFT excitation energies is that the difference of the Kohn–Sham orbital energies, which are the leading term of the diagonal elements of the **A** matrix in Equation 5.3, is usually an excellent approximation for excitation energies. This is because the virtual KS orbital energies are evaluated for the N-electron system and, thus, correspond more to the single-particle energy of an excited electron than to the energy of an additional electron as in Hartree–Fock theory, where the virtual orbital energies are evaluated for the $N + 1$ electron system. Consequently, orbital energy differences are a much better estimate for valence-excited states in KS-DFT than in HF theory.

Although TDDFT performs usually very well for valence-excited states, it is now well known that it has severe problems with the correct description of Rydberg states, valence states of molecules exhibiting extended π systems, doubly excited states, and charge transfer excited states [23]. For such states, the errors in the excitation energies can be as large as a few electron volts, and the potential energy surfaces (PES) can exhibit incorrect curvature.

The problems with Rydberg states and extended π-systems can be attributed to the wrong long-range behavior of current standard xc functionals. In the case of excited charge transfer (CT) states, the excitation energies are quite too low (by up to 1 eV) and the potential energy curves do not exhibit the correct $1/R$ asymptote when R corresponds to a distance coordinate between the positive and the negative charges of the CT state. The $1/R$ failure of TDDFT employing pure standard xc functionals has, in fact, been understood as an electron transfer self-interaction error [24]. This electron transfer self-interaction effect is canceled in TDHF by the response of the HF exchange term. At present, several different pathways are starting to emerge to address this substantial failure of TDDFT for CT states and to correct for it. The most obvious way is to improve the xc functional by including exact exchange in the unperturbed Hamiltonian, either in the form of nonlocal Hartree–Fock exchange or in the form of the exact local Kohn–Sham exchange potential. Inclusion of nonlocal Hartree–Fock exchange has been realized in a few schemes so far. In all these schemes, the Coulomb operator of the Hamiltonian is split into two parts: a short-range (SR) and a long-range (LR) part [25]:

$$\frac{1}{u} = \left\{\frac{1-\mathrm{erf}(\omega u)}{u}\right\}_{SR} + \left\{\frac{\mathrm{erf}(\omega u)}{u}\right\}_{LR} \quad (5.4)$$

where the screening parameter ω defines the range separation and u represents the interelectronic distance. Short-range exchange is then treated primarily using a local functional; long-range exchange is treated primarily using exact orbital exchange. For example, the Coulomb-attenuated CAM-B3LYP [26] functional contains just 20% exact exchange at short range, like a conventional hybrid, but 65% at long range.

Recently, an extensive assessment of the performance of the CAM-B3LYP for the calculation of local, Rydberg, and intramolecular CT excitation energies was presented [27]. By comparing CAM-B3LYP with standard GGA and hybrid functionals, a quantification of the extent to which excitation energy errors correlate with the spatial overlap between the occupied and the virtual orbitals is found. The results suggest that it is inappropriate to say that standard GGA functionals fail to describe CT excitations. Indeed, the degree of CT – in the sense of how much the occupied and virtual orbitals overlap – must be quantified before judgment can be made.

Simulation and assignment of vertical electronic absorption spectra surely constitute the main application of TDDFT calculations in chemistry; however, by solving the same TDDFT equations, other spectroscopies can be simulated. In particular, electronic circular dichroism (ECD) is easily obtained. The CD effect is, in fact, expressed as the difference between the molar extinction coefficients for left and right circularly polarized light that is related to the rotatory strength of the transition between ground and excited states. In turn, the rotatory strength can be derived from quantum mechanical theory as the product of the electric dipole and magnetic dipole transition moments. Both of them are obtained in a straightforward manner from TDDFT, as it will be shown in Chapter 9; by combining rotatory strengths with excitation energies, the full ECD spectra of many molecular systems have been calculated.

In recent years, TDDFT analytical gradients have been presented for the excited states, making possible the efficient calculation of excited state properties such as equilibrium geometries and dipole moments [28]. Owing to the more limited number of available studies so far, however, it is still not clear if TDDFT behaves with respect to properties/geometries of the excited states exactly as for excitation energies (and transition properties). It could, in fact, be possible that cancellations of errors that often make UV absorption spectra well reproduced by the standard hybrid functionals disappear with subsequent failure of the same functional in the evaluation of the state properties.

5.3
Modeling Solvent Effects

Solvents strongly influence the electronic spectral bands of individual species measured by various spectrometric techniques (UV-Vis, fluorescence spectroscopy, etc.) [29]. Significant shifts in absorption and emission bands can be induced by a change in solvent nature or composition; these shifts, called solvatochromic shifts, are experimental evidence of changes in solvation energy. In fact, when a solute is surrounded by solvent molecules, its ground state and its excited state are differently stabilized by solute–solvent interactions, depending on the chemical nature of both solute and solvent molecules.

In addition to shifts, the solvent generally induces broadening of the absorption and fluorescence bands as a result of the fluctuations in the structure of the solvation shell around the solute. This effect, called inhomogeneous broadening, superimposes homogeneous broadening because of the existence of a continuous set of vibrational sublevels.

Numerous methods of describing solvatochromism in terms of solvent characteristics have been proposed. It is worth mentioning here that the use of solvatochromism of betaine dyes was proposed by Reichardt as a probe of solvent polarity [30]. The exceptionally strong solvatochromism shown by these compounds can be explained by considering that in their ground state they are zwitterions while, upon excitation, electron transfer occurs exactly in the direction of canceling this charge separation. As a result, the dipole moment that is rather large in the ground state becomes almost zero in the excited state and thus solvent interactions change markedly leading to the observed negative solvatochromism.

An alternative approach to quantifying polarity effects was proposed by Kamlet et al. [31]. According to this approach, the positions of the bands in UV-Vis absorption and fluorescence spectra can be determined as

$$\bar{\nu} = \bar{\nu}^0 + s\pi^* + a\alpha + b\beta \tag{5.5}$$

where $\bar{\nu}$ and $\bar{\nu}^0$ are the wave numbers of the band maxima in the solvent considered and in the reference solvent (generally cyclohexane), respectively, π^* is a measure of the polarity/polarizability effects of the solvent, α is an index of solvent hydrogen bond donor acidity, and β is an index of solvent hydrogen bond acceptor basicity. The coefficients s, a, and b describe the sensitivity of a process to each of the individual contributions.

The π^* scale of Kamlet et al. deserves special recognition not only because it has been successfully applied in many studies (not limited to UV or fluorescence spectra, and including many other physical or chemical parameters such as reaction rate, equilibrium constant, etc.) but also because it gives a very clear introduction of the problem. Namely, Equation 5.5 indicates that the two main aspects to consider when modeling solvent effects on excitation energies are polarity/polarizability effects and hydrogen bonding.

In addition to these "time-independent" effects, the processes of formation and relaxation of the electronic excited states will be strictly coupled with the dynamics of the solvent molecules. A well-known *extreme* example of such a coupling is the distinction between "nonequilibrium" and "equilibrium" solvation regimes following an electronic transition in the solute. The differences in the characteristic response time of the various degrees of freedom of the solvent, in fact, may lead to a solvation regime in which the slow components (i.e., those arising from molecular translations and rotations) are not equilibrated with the excited state electronic redistribution upon vertical excitation. The resulting nonequilibrium regime will then relax into a new equilibrium in which the solvent is allowed to completely equilibrate, that is, to reorganize all its degrees of freedom including the slow ones. Especially for highly polar solvents, these two different regimes can influence the properties of the solute excited states in very different ways.

This brief excursus on the various specificities introduced by the presence of the solvent is sufficient to underline the complexity of the modeling of excited states in solution. From such a large complexity, it follows that strong approximations are necessary to be able to treat molecular systems of real interest and not only simple model cases.

An effective and well-diffused strategy to overcome most of the difficulties is to introduce a focused approach, that is, a more accurate description of the molecular system of interest (the chromophore, possibly including small portions of the environment) and a less accurate description of the remainder. There are different formulations of the focused approach; the most common ones are the hybrid QM/molecular mechanics (QM/MM) [32] and the continuum solvation models [33]. Both of them use a classical description for the environment, but while in the former the microscopic nature of the solvent molecules is maintained, in the latter a macroscopic dielectric is used. In both cases, we can introduce a similar QM picture in terms of an effective Hamiltonian giving rise to an effective Schrödinger equation for the solvated solute. Introducing the standard Born–Oppenheimer approximation, the solute electronic wavefunction will in fact satisfy the following equation:

$$H_{\text{eff}}|\Psi\rangle = (H_0 + H_{\text{env}})|\Psi\rangle = E|\Psi\rangle \tag{5.6}$$

where H_0 is the Hamiltonian of the isolated solute system and the operator H_{env} introduces the coupling between the solute and the solvent. The form of the operator H_{env} depends on the method used; in particular, for the two alternative schemes analyzed here, we have

$$H_{\text{env}} = \begin{cases} H_{\text{QM/MM}} + H_{\text{MM}} & \text{QM/MM} \\ H_{\text{QM/cont}} & \text{QM/continuum} \end{cases} \tag{5.7}$$

In more common formulations of the QM/MM approaches, the MM system is represented through atomic point charges; as a result, the first term in Equation 5.7 is the electrostatic interaction between the QM system and the point charges in the MM part of the system:

$$H_{\text{QM/MM}} \rightarrow H^{\text{el}} = \sum_m q_m(\mathbf{r}_m) V(\mathbf{r}_m) \tag{5.8}$$

where $V(\mathbf{r}_m)$ is the electrostatic potential operator due to solute electrons and nuclei at the MM charges q_m. This term is directly included in the one-electron part of the vacuum Hamiltonian. The H_{MM} introduced in Equation 5.7 is the classical MM energy; this term, however, is a contribution only to the energy and does not affect the wavefunction.

Moving now to QM/continuum approaches, we shall limit our exposition to the so-called apparent surface charges (ASC) version of such approaches, and in particular to the family known with the acronym, PCM (polarizable continuum model) [34]. In this family of methods, the reaction potential $H_{\text{QM/cont}}$ introduced in Equation 5.7 has a form completely equivalent to the electrostatic part of the $H_{\text{QM/MM}}$ operator defined in Equation 5.8, namely,

$$H_{\text{QM/cont}} \rightarrow H^{\text{PCM}} = \sum_s q_s^{\text{PCM}}(\mathbf{r}_s) V(\mathbf{r}_s) \tag{5.9}$$

Now, the point charges q^{PCM} are no longer centered on the solvent nuclei as in the MM description, but they are placed on selected points $(\mathbf{r}_s)$ on the surface of the molecular cavity containing the solute. In addition, such charges are not fixed but are "apparent" in the sense that they exist only when the solute exists. These charges are determined by the electrostatic potential acting at the selected points on the surface (i.e., the potential due to the solute and the charges themselves), but now they also depend on the dielectric properties of the solvent, on the geometry of the cavity, and on the number and position of the points chosen to map the cavity surface.

As for the QM/MM description also for PCM, nonelectrostatic (or van der Waals) terms can be added to the $H_{QM/cont}$ operator; in this case, besides the dispersion and repulsion terms, a new term has to be considered, namely, the energy required to build a cavity of the proper shape and dimension in the continuum dielectric. This continuum-specific term is generally indicated as cavitation. In general, all the nonelectrostatic terms are expressed using empirical expressions and thus their effect is only on the energy and not on the solute wavefunction.

The different philosophy beyond the two classes of solvation methods leads to important differences both in the physical and the computational aspects of their applications and in their range of applicability. The methods based on explicit representations of the environment yield information on specific configurations of the environment around the chromophore, whereas the continuum models give only an averaged picture of it. On the other hand, QM/MM requires many more calculations than continuum models to obtain a correct statistical description. This much larger computational cost of QM/MM is particularly disadvantageous in the study of excited states, as the QM level required is generally quite expensive even for a single calculation on an isolated system; thus, the necessity to repeat the calculation many times makes the approach very expensive (or even not feasible). For this reason, most of the QM/MM calculations on excited states are done using semiempirical QM methods [35]. On the contrary, the level of the QM description can be any when continuum models are used, as the additional cost with respect to gas-phase calculations remains very limited. In addition, continuum solvation models include effects of mutual polarization between the solute and the environment (also those due to a possible nonequilibrium solvation), whereas standard QM/MM methods are based on nonpolarizable force fields. As a matter of fact, QM/MM approaches including environment polarization have been proposed and also applied to the study of excited states of solvated systems [36]. However, among the available approaches, the most popular for this kind of study is still represented by continuum solvation models [37].

Within the solvation framework, as in the case of isolated molecules, excitation energies can be obtained with two different approaches: the state-specific method and the linear response method. The former has a longer tradition, being related to the classical theory of solvatochromic effects, whereas the latter has been introduced more recently in connection with the development of the linear response theory for excited states. In both approaches, it is generally possible to introduce the nonequilibrium effects described before appearing in any vertical electronic transition (absorption and emission). For example, in the PCM method this is easily obtained by partitioning the apparent charges into an "inertial" and a "dynamic" component corresponding to the slow and the fast part of the solvent polarization, respectively. If we assume a Franck–Condon-like response of the solvent, exactly as for the solute molecule, the nuclear motions inside and among the solvent molecules will not be able to follow immediately the fast changes in the solute electronic charge distribution. The corresponding part of the solvent response (the inertial charges) will remain frozen in the state immediately prior to the transition while the fast part (the dynamic charges) will immediately change according to the new state.

Also, QM/MM methods account for nonequilibrium solvation even if not always in a complete way: in fact, a standard nonpolarizable QM/MM includes the inertial part of the solvent response (the MM charges that remain constant in the excitation), but it lacks the dynamic part of the response. The latter can be recovered if polarizable dipole moments are also used to describe the solvent molecules such as in polarizable QM/MM approaches [38].

5.4
Toward the Simulation of UV-Vis Spectra

In the previous sections, we have sketched the key methodological steps for the calculation of transition energies of molecular systems eventually including solvent effects. In such a description, however, we have not said anything about the simulation of the intensities of signals corresponding to the electronic excitations.

As a first approximation, to simulate the intensities of UV-Vis signals, the square of the transition dipole moment integral between the initial and the final electronic state has to be evaluated. Typically, semiempirical approaches or CIS methods have been used for estimating absorption intensities, often along with empirical scaling procedures to correct for the overestimation of transition dipoles predicted by such methods. Of course, it would be desirable to avoid such scaling, and recently more accurate QM methods including electron correlation effects have begun to be used to obtain transition dipoles. As a matter of fact, a more systematic study has shown that transition dipole moments are much less sensitive to the particular level of QM theory (and even less with respect to the basis set) than excitation energies [39].

The approximation of using transition dipoles to simulate bands, however, cannot provide information on the spectrum shape, and the computed (vertical) transition energies give only a rough estimate of the spectrum band maximum when it is characterized by sensible vibrational progressions.

In general, the distribution of intensity within the spectra sensitively depends on the details of the ground- and excited state potential energy surfaces, transition dipole moment coordinate dependence, temperature, spin-orbit coupling, and nonadiabatic effects. Though the nonadiabatic effects have been found to be rather important in some cases, the Born–Oppenheimer (BO) adiabatic approximation appears to satisfactorily work for the description of various electronic spectra of molecules. The Franck–Condon (FC) approximation is usually employed within the BO *ansatz* to interpret strongly dipole-allowed electronic bands for which it has been found to be sufficiently accurate. In the FC approximation, the electronic transition dipole moment coordinate dependence is neglected and the simulation of absorption spectra is reduced to evaluation of the well-known FC factors that represent the overlap between the ground- and the excited state vibrational wavefunctions. In the previous sections, we underlined that reliable energy minima and (numerical) vibrational modes can be obtained for excited states thanks to the extension of linear response QM methods (mainly CIS and TDDFT) to analytical gradients. It is exactly the development of these theoretical tools that has made the simulation of spectral shapes possible [40]. In general, this is done in the FC approximation using three main computational steps, namely, (1) electronic structure calculations of both ground and excited electronic states providing reliable potential energy surfaces, (2) computation of vibrational wave functions using these PES, and (3) computation of FC factors. The resulting vibronic lines are convoluted with Gaussian (or Lorentzian) functions of given full width at half maximum to generate the final spectrum.

In this procedure, however, it is implicitly assumed that bandwidths are determined entirely by homogeneous broadening. This is valid for the idealized case of a gas at low pressure. In more realistic situations, inhomogeneous broadening, which arises from fluctuations in the molecular environment, plays an important role. Each solute molecule experiences a slightly different solvent environment and therefore has a slightly different absorption spectrum. The observed absorption spectrum is made up of all the different spectra for the different molecular environments; it is said to be inhomogeneously broadened [41]. The resulting broadening of the spectral band is generally associated with the solvent reorganization energy corresponding to the optical excitation and the solvent reorganization energy is the thermodynamic quantity associated with change in the solvation free energy passing from the nonequilibrium to the equilibrium solvation regimes described before. These quantities are easily obtained with solvation continuum methods [42].

A different approach to obtain solvent-induced broadening in UV-Vis signals is represented by molecular dynamics coupled to QM calculations. If a classical MD simulation is used, one has to extract snapshots from trajectories, introduce a cutoff to define the solvation shells, and extract the resulting solute–solvent clusters. On each of these clusters, the excitation energies are finally calculated at a QM/MM level. Alternatively, the Car–Parrinello molecular dynamics (CPMD) scheme [43] can be used; in such a case, a fully quantum mechanical treatment of the solvent is, in principle, possible, with a subsequent TDDFT calculation of the excitation energies on selected snapshots. In both cases, the fluctuations of the environment and the resulting shifts in the transition energies are automatically included in the analysis: a convolution of the excitation energies calculated for the different local environments will mimic the inhomogeneous broadening.

5.5
Some Numerical Examples

In this section, we summarize three studies appeared in the literature in the past few years about the QM simulation of UV-Vis spectra of molecular systems. These summaries should help the reader to directly evaluate the real performances and limitations of the QM methodologies described in the previous sections. As nowadays, TDDFT approaches represent the large majority of the QM applications in the simulation of UV-Vis spectra, in this brief overview of examples we shall focus only on TDDFT studies. In particular, we have selected two among the many studies recently appeared in the literature about benchmarking analyses of TDDFT for excitation energies of isolated molecules and an example of the study of the solvent effect on the same properties. As the summary reported here is necessarily limited, the reader may refer to the original papers in order to get a more complete picture of each of the three studies.

The first study we summarize was recently published by Jacquemin et al. [22b]. The study consists in an extensive analysis of the merits of a large number of DFT functionals including LDA, GGA (and meta-GGA), global hybrids, and range-separated hybrids in predicting vertical absorption energies of a broad range of organic molecules and dyes.

In Figure 5.1, we report a part of the results of the study, namely, the one focused on an internal comparison between TDDFT and CASPT2 and CC2 benchmarks reported by Thiel and coworkers [22a] for 28 medium-sized organic molecules (104 singlet vertical excitations). It has to be remarked that here the use of an internal comparison is to be preferred to a comparison with experiments for various reasons. Experimental data, in fact, are generally in solution and this should be taken into account in the calculation by introducing a solvation model; this, however, prevents us from an evaluation of the real performances of the QM method as they always couple with those of the solvation model used. In addition, a proper comparison with experimental spectra requires the inclusion of vibronic effects;

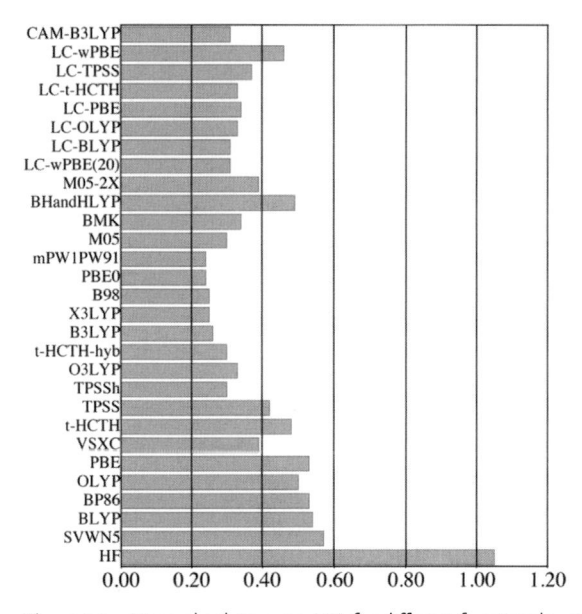

Figure 5.1 Mean absolute error (eV) for different functionals with respect to QM benchmarks. Reprinted with permission from Ref. [22b].

also in this case, as for the solvent effects, we have to introduce a model that necessarily makes the analysis of the merits of the QM method difficult.

From the figure, it turns out that, as expected, TDHF provides very large errors (MAE > 1.00 eV) and overestimates the transition energies in nearly 90% of the cases. Using any TDDFT scheme does reduce the errors by a factor ranging from two to four. As expected, the pure functionals tend to provide too small transition energies, with MAE > 0.2 eV. Global hybrids are more accurate than the GGA, and adding more and more exact exchange tends to shift the transition energies to larger values. The MAE of all long-range corrected hybrids is close to 0.3 eV, due to an overestimation of the transition energies. In particular, CAM-B3LYP remains slightly less efficient than B3LYP. A possible explanation of this behavior is the small size of the set of molecules used in which charge transfer states are not significantly represented.

A better appraisal of the potentialities of CAM-B3LYP is represented by a study published in 2008 by Peach et al. [27]. As already reported in Section 5.2, the authors of the study defined a parameter (indicated as Λ) that measures the degree of spatial overlap between the occupied and the virtual orbitals involved in an excitation and examined how it correlates with the accuracy of the excitation energy. Such a parameter Λ is defined as

$$\Lambda = \frac{\sum_{i,a} \varkappa_{ia}^2 O_{ia}}{\sum_{i,a} \varkappa_{ia}^2} \quad \text{with} \quad 0 \leq \Lambda \leq 1 \tag{5.10}$$

where the spatial overlap factors O_{ia} are defined as the inner product of the moduli of the occupied (i) and virtual (a) orbital involved in the excitation whereas $\varkappa_{ia} = X_{ia} + Y_{ia}$, with X_{ia} and Y_{ia} being elements of the solution of Equation 5.2.

The molecules considered in the study were chosen to include a wide range of excitations, many of which have been shown to be a challenge for TDDFT. An extract of the results of the study is summarized in Figure 5.2 in which the error in the excitation energies is plotted against the associated λ values for B3LYP and CAM-B3LYP. Each point corresponds to a single excitation, with different symbols and colors, for the three categories of excitation.

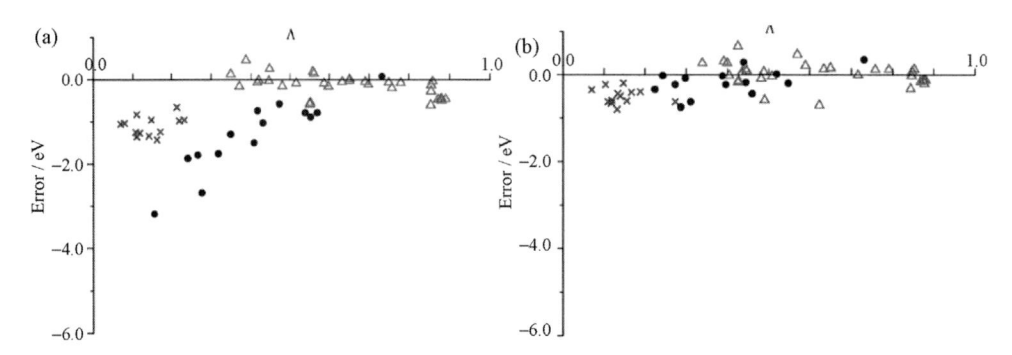

Figure 5.2 Excitation energy errors plotted against λ using B3LYP (a) and CAM-B3LYP (b) for local excitations ($\triangle$), Rydberg excitations ($\times$), and CT excitations ($\bullet$). Reprinted with permission from [27].

These graphs clearly show that for standard hybrid functionals such as B3LYP, the accuracy significantly decreases when the excitation presents a Rydberg (very low Λ) or CT (low to medium Λ) character. In particular, when the overlap dropped too low then excitations became significantly underestimated. By contrast, CAM-B3LYP provides by far a more homogeneous behavior for all the excitations, with essentially no correlation between errors and spatial orbital overlap. From this analysis, it comes out that low overlap could be used in a diagnostic manner to identify the problematic excitations. Obviously, the quantity Λ is not unique, and its diagnostic value is qualitative rather than quantitative. However, it captures the essential physics of the problem and may prove useful in practical calculations. In addition, it can be trivially computed from quantities available in a regular TDDFT calculation.

To complete this short overview on applications of TDDFT, we present an extract from a study conducted by our group and published in 2005 [44]. The original study focuses on the TDDFT simulation of various spectroscopic properties of solvated N-methyl-acetamide (NMA), namely, IR, NMR, and UV, but here we report an extract limited only to UV absorption.

The electronic absorption spectrum of NMA has been measured in gas-phase and in a variety of solvents [45]. The gas-phase spectrum presents two well-separated bands that have been assigned to transitions in the π system of the amide group. In nonpolar solvents such as cyclohexane or dioxane, a $n\pi^*$ also appears as a shoulder. The lowest $\pi\pi^*$ transition (the one of interest here) comprises an electron transfer from the nitrogen to the carbonyl carbon atom. This transition is redshifted from gas-phase to polar solution and to water. By contrast, for the $n\pi^*$ transition a blueshift is observed passing from cyclohexane to water. In fact, the $n\pi^*$ transition reduces the permanent dipole moment of the corresponding excited state that is thus less strongly stabilized in polar solvents with respect to the ground state.

Absorption energies are calculated at TDDFT: B3LYP/d95v + (d) level and solvent effects are introduced using PCM solvation model.

For both $\pi\pi^*$ and $n\pi^*$ transitions, PCM results reproduce the observed trend in solvent shift, namely, a blueshift on $n\pi^*$ and a redshift on $\pi\pi^*$; in both cases, however, there is an underestimation of the shift, especially for $n\pi^*$ transition. The possible discrepancies between computed and observed shift can be due to different reasons. The main one is clearly due to the fact that PCM results do not include explicit hydrogen bonding effects and these can differentially affect the two transitions.

The most straightforward way to include these effects is to compute solute–solvent clusters including one, two, and three water molecules. For NMA, three water molecules are enough to saturate the three H-bonding donor and acceptor sites, namely, two on (C)O and one on (N)H.

An alternative strategy to build NMA–water clusters is MD simulation: water molecules having any H-atom closer to (C)O-atom than the corresponding first minimum of the O-HW radial distribution function will be included in the cluster. An analogous criterion was applied for the (N)H–OW pair. The selected structures were sampled at equal time intervals long enough to avoid correlation.

The comparison of the two alternative sets of clusters (either QM or MD) to represent the effects of the first solvation shell(s) should give further hints on the

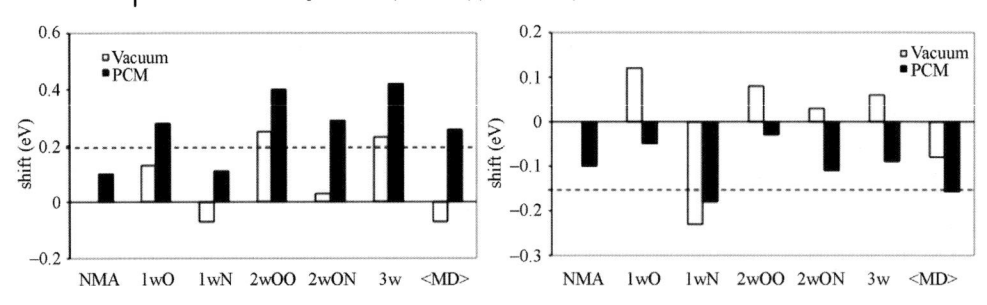

Figure 5.3 Analysis of absorption energies for the $n\pi^*$ (left) and $\pi\pi^*$ (right) transitions in the NMA-water QM and MD clusters in terms of the corresponding shifts with respect to the isolated NMA. Dotted line shows the experimental value. Revised with permission from Ref. [44].

nature of the solvation around NMA, and in particular it should help in analyzing if these effects are better represented in terms of rigid structures obtained as minima of the potential energy surfaces or if, by contrast, dynamics, and therefore a variety of different and representative structures, are needed to properly describe the hydration effects.

In order to have a more direct picture of solvent effects in Figure 5.3, we present the analysis of absorption energies for the $n\pi^*$ and $\pi\pi^*$ transitions in terms of the corresponding shifts with respect to the isolated NMA: as reference excitation energies of isolated NMA, we have considered those obtained for the most stable conformer while the dotted lines indicate the experimental data (namely, $+ 0.2\,eV$ for $n\pi^*$ and $-0.14\,eV$ for $\pi\pi^*$). For MD-derived clusters the value reported in the figure ($\langle MD \rangle$) is obtained as an average on a set of 40 different clusters.

The analysis of Figure 5.3 involves different aspects.

First, let us consider the effects of H-bonds alone by analyzing isolated NMA-nw clusters. H-bonding effects on the two transitions are opposite if the H-bond acts on (C)O or on (N)H. In fact, including one or two water molecules close to carbonyl oxygen induces a blueshift on both transitions, while when only the (N)H—Ow bond is included (1wN), the total effect is a redshift instead of a blueshift. When the two different H-bonds are combined (2wON and 3w clusters) the global effect is again a blueshift but smaller than in the 1(2)wO(O) cases. In the case of $\pi\pi^*$ transition also, the largest 3w cluster presents a very small shift due to the two opposite effects of the two H-bonds that almost cancel out. The inclusion of an external continuum (PCM) improves the description leading to the correct sign in the shift for all the possible clusters: for both transitions, the long-range electrostatic effects of the bulk solvent appear to be more important than the local specific H-bond. The set of PCM results, however, give additional information. First, $n\pi^*$ results show that summing the effects of two H-bonds on carbonyl oxygen together with the long-range interactions leads to an overestimation of the observed solvent effect (around twice the experimental value); much better descriptions are given by the 1wO clusters and by the mixed 2wON clusters (even if slightly overestimated in this latter case). The 1wN result remains too low and in practice it does not add anything more to the continuum-only result.

Concerning the $\pi\pi^*$ transition, the best results are represented by the 2wON clusters, but in any case we do not observe a significant improvement with respect to the continuum-only model (NMA column). Once again by assuming two (C)O...HOH interactions (2wOO), we obtain a noncorrect description even if the inclusion of the continuum partially improves the wrong picture obtained in the corresponding isolated clusters. Finally, the 1wN cluster leads to a slightly too large effect.

If we pass from the static picture of QM-optimized clusters to that of statistical type obtained in terms of MD clusters and we also include bulk effects with the external continuum, a very good result is obtained for $\pi\pi^*$ transition while, for $n\pi^*$, we find a slightly too large shift. This result, however, has to be reappraised if we consider that the experimental shift refers to cyclohexane and not to gas phase and thus it has to be read as a lower limit. We point out that the same clusters without the continuum give a negligible shift for $n\pi^*$ and a too small one for $\pi\pi^*$. It is also interesting to note that when the same clusters are described at a simpler level, namely, by representing the water molecules in terms of classical charges, the description we obtain for the two transitions is different: while the $n\pi^*$ transition is reasonably described, the $\pi\pi^*$ one is not reproduced at all (small and wrong sign shift). Therefore, pure electrostatic effects are not enough to describe the processes and a more complex picture of the interactions involving a quantum mechanics electronic solvent description is needed.

These results strongly indicate that short-range and long-range solvation effects are required in order to properly describe the observed shifts. For short-range effects, a better picture is obtained by including explicit solvent molecules (namely, those representing the first solvation shells) and mediating on different configurations (i.e., introducing a statistical description), while long-range effects are properly and effectively described by adding a reaction field through a continuum model.

5.6
Conclusions and Perspectives

In the past few years, there has been a rapid development of methods to describe electronic excitations in systems of increasing complexity. In this chapter, we have tried to give a short but still exhaustive summary of the QM methodologies that are dominating the computational research in this field.

Obviously, this review reflects the situation at the present moment and the fast evolution still in act will probably significantly change it in few years. However, some of the aspects considered in this chapter will remain as hot topics in the near future. In particular, the complexity issue (i.e., considering chromophores embedded in environment of increasing complexity) is one of the most important challenges in the QM modeling. Many novel experimental techniques have, in fact, clearly shown that the signal of each single molecule is affected by its local environment, but, in many cases, the signal itself exists (or it becomes detectable) only in the presence of a specific environment capable of activating or enhancing some specific molecular responses. It is sufficient here to recall all the surface-enhanced (SE) spectroscopies in which a spectroscopic signal (either Raman, IR, absorption, or fluorescence) is amplified on orders of magnitude due to the presence of a metal nanoparticle close to the molecular probe [46]. Another aspect that will be of enormous interest in the near future is the explicit consideration of the dynamics. In all the methods I have presented and discussed, in fact, time is not explicitly taken into account and the picture one can get is always static. Indeed, time-dependent spectroscopies have shown that the signal reflects the complex coupling of different relaxation modes involving many electronic and nuclear degrees of freedom of the molecular probe. As a matter of fact, the intrinsic dynamics of the probe can further be complicated by additional couplings with relaxation modes of the environments. These, in turn, can give rise to coherence effects that strongly affect the final spectroscopic features. Such coherence effects, for example, seem to play a fundamental role in determining the efficiency of excitation energy transfers in light-harvesting photosynthetic proteins [47]. Unfortunately, the quantum mechanical description of dynamical processes such as those involving excited states is very difficult, especially if one also has to include relaxation processes within the environment. It is, thus, of fundamental importance to combine different computational descriptions for the different components of the whole system using a multiscale approach in which, however, careful attention has to be paid to possible couplings between the different parts and their evolution with time.

References

1 (a) Serrano-Andrés, L. and Merchan, M. (2005) *J. Mol. Struct.: THEOCHEM*, **729**, 99; (b) Dreuw, A. (2006) *Chem. Phys. Chem.*, **7**, 2259.

2 Klessinger, M. and Michl, J. (1994) *Excited States and Photochemistry of Organic Molecules*, John Wiley & Sons, Inc., New York.

3 (a) Parr, R.G. and Yang, W. (1989) *Density-Functional Theory of Atoms and Molecules*, Oxford Science Publication, New York; (b) Dreizler, R.M. and Gross, E.K.U. (1995) *Density Functional Theory*, Springer, Heidelberg; (c) Koch, W. and Holthausen, M.C. (2000) *A Chemist's Guide to Density Functional Theory*, Wiley-VCH Verlag GmbH, Weinheim.

4 (a) Casida, M.E. (1995) *Recent Advances in Density Functional Methods*, Part I (ed. D.P. Chong), World Scientific, Singapore, p. 155; (b) Gross, E.K.U., Dobson, J.F., and Petersilka, M. (1996) *Top. Curr. Chem.*, **181**, 81;

(c) Bauernschmitt, R. and Ahlrichs, R. (1996) *Chem. Phys. Lett.*, **256**, 454.

5 (a) Mennucci, B. and Cammi, R. (eds) (2007) *Continuum Solvation Models in Chemical Physics: from Theory to Applications*, John Wiley & Sons Ltd., Chichester, UK; (b) Canuto F S. (ed.) (2008) *Solvation Effects on Molecules and Biomolecules: Computational Methods and Applications*, vol. 6, Series: Challenges and Advances in Computational Chemistry and Physics, Springer.

6 Foresman, J.B., Head-Gordon, M., Pople, J.A., and Frisch, M.J. (1992) *J. Phys. Chem.*, **96**, 135.

7 (a) Roos, B.O. (1980) *Int. J. Quantum Chem.*, **14**, 175; (b) Siegbahn, P.E.M., Almlof, J., Heiberg, A., and Roos, B.O. (1981) *J. Chem. Phys.*, **74**, 2384; (c) Roos, B.O. and Andersson, K. (1995) *Modern Electronic Structure Theory* (ed. D.R. Yarkony), World Scientific, Singapore, p. 55.

8 (a) Andersson, K., Malmqvist, P.-Å., and Roos, B.O. (1992) *J. Chem. Phys.*, **96**, 1218; (b) McDouall, J.J.W., Peasley, K., and Robb, M.A. (1988) *Chem. Phys. Lett.*, **148**, 183.

9 Celani, P. and Werner, H.-J. (2003) *J. Chem. Phys.*, **119**, 5044.

10 (a) Paldus, J. (1992) *Methods in Computational Physics* (eds S. Wilson and G.H.F. Diercksen), Plenum, New York; (b) Bartlett F R.J. (ed.) (1997) *Modern Ideas in Coupled-Cluster Methods*, World Scientific, Singapore; (c) Gauss, J. (1998) *Encyclopedia of Computational Chemistry* (ed. P. von Rague Schleyer), John Wiley & Sons, Inc., New York.

11 Stanton, J.F. and Bartlett, R.J. (1993) *J. Chem. Phys.*, **98**, 7029.

12 Christiansen, O., Koch, H., and Jørgensen, P. (1995) *Chem. Phys. Lett.*, **243**, 409.

13 (a) Nakatsuji, H. (1979) *Chem. Phys. Lett.*, **67**, 344; (b) Das, A., Hasegawa, J., Miyahara, T., Ehara, M., and Nakatsuji, H. (2003) *J. Comput. Chem.*, **24**, 1421; (c) Hasegawa, J., Fujimoto, K., Swerts, B., Miyahara, T., and Nakatsuji, H. (2007) *J. Comput. Chem.*, **28**, 2443.

14 Christiansen, O., Koch, H., and Jørgensen, P. (1995) *J. Chem. Phys.*, **103**, 7429.

15 Nakajima, T. and Nakatsuji, H. (1997) *Chem. Phys. Lett.*, **280**, 79.

16 Stanton, J.F. and Gauss, J. (1994) *J. Chem. Phys.*, **100**, 4695.

17 Kohn, A. and Hattig, C. (2003) *J. Chem. Phys.*, **119**, 5021.

18 (a) Ridley, J.E. and Zerner, M.C. (1973) *Theor. Chim. Acta*, **32**, 111; (b) Kotzian, M., Rosch, N., and Zerner, M.C. (1992) *Theor. Chim. Acta*, **81**, 201.

19 Matsuura, M., Sato, H., Sotoyama, W., Takahashi, A., and Sakurai, M. (2008) *J. Mol. Struct. THEOCHEM*, **860**, 119.

20 (a) Stratmann, R.E., Scuseria, G.E., and Frisch, M.J. (1998) *J. Chem. Phys.*, **109**, 8218; (c) Furche, F. and Ahlrichs, R. (2002) *J. Chem. Phys.*, **117**, 7433.

21 (a) Marques, A.L. and Gross, E.K.U. (2004) *Annu. Rev. Phys. Chem.*, **55**, 427; (b) Appel, H., Gross, E.K.U., and Burke, K. (2003) *Phys. Rev. Lett.*, **90**, 043005; (c) Gonze, X. and Scheffler, M. (1999) *Phys. Rev. Lett.*, **82**, 4416.

22 (a) Silva-Junior, M.R., Schreiber, M., Sauer, S.P.A., and Thiel, W. (2008) *J. Chem. Phys.*, **129**, 104103; (b) Jacquemin, D., Wathelet, V., Perpete, E.A., and Adamo, C. (2009) *J. Chem. Theory Comput.*, **5**, 2420; (c) Elliott, P., Burke, K., and Furche, F. (2009) *Recent Advances in Density Functional Methods*, vol. **26** (eds K.B. Lipkowitz and T.R. Cundari), John Wiley & Sons, Inc., Hoboken, NJ, p. 91.

23 (a) Dreuw, A., Weisman, J.L., and Head-Gordon, M. (2003) *J. Chem. Phys.*, **119**, 2943; (b) Tozer, D.J. (2003) *J. Chem. Phys.*, **119**, 12697; (c) Gritsenko, O. and Baerends, E.J. (2004) *J. Chem. Phys.*, **121**, 655; (d) Maitra, N.T. (2005) *J. Chem. Phys.*, **122**, 234104; (e) Neugebauer, J., Gritsenko, O., and Baerends, E.J. (2006) *J. Chem. Phys.*, **124**, 214102.

24 Dreuw, A. and Head-Gordon, M. (2005) *Chem. Rev.*, **105**, 4009.

25 (a) Savin, A. (1996) *Recent Developments and Applications of Modern Density Functional Theory* (ed. J.M. Seminario), Elsevier, Amsterdam; (b) Ikura, H., Tsuneda, T., Yanai, T., and Hirao, K. (2001) *J. Chem. Phys.*, **115**, 3540; (c) Toulouse, J., Colonna, F., and Savin, A. (2004) *Phys. Rev. A*, **70**, 062505; (d) Baer, R. and Neuhauser, D. (2005) *Phys. Rev. Lett.*, **94**, 043002; (e) Heyd, R.J., Scuseria, G.E., and Ernzerhof, M. (2003) *J. Chem. Phys.*, **118**, 8207.

26 (a) Yanai, T., Tew, D.P., and Handy, N.C. (2004) *Chem. Phys. Lett.*, **393**, 51; (b) Tawada, Y., Tsuneda, T., Yanagisawa, S., Yanai, T., and Hirao, K. (2004) *J. Chem. Phys.*, **120**, 8425.

27 Peach, M.J.G., Benfield, P., Helgaker, T., and Tozer, D.J. (2008) *J. Chem. Phys.*, **128**, 044118.

28 (a) Caillie, C.V. and Amos, R.D. (1999) *Chem. Phys. Lett.*, **308**, 249; (b) Caillie, C.V. and Amos, R.D. (2000) *Chem. Phys. Lett.*, **317**, 159; (c) Furche, F. and Ahlrichs, R. (2002) *J. Chem. Phys.*, **117**, 7433 (erratum: 121 (2004) 12772); (d) Scalmani, G., Frisch, M.J., Mennucci, B., Tomasi, J., Cammi, R., and Barone, V. (2006) *J. Chem. Phys.*, **124**, 094107.

29 (a) Reichardt, C. (2003) *Solvents and Solvent Effects in Organic Chemistry*, Wiley-VCH Verlag GmbH, Weinheim; (b) Wang, C.H. (1985) *Spectroscopy of Condensed Media*, Academic Press, New York; (c) Suppan, P. and Ghoneim, N. (1997) *Solvatochromism*, The Royal Society of Chemistry, Cambridge, UK; (d) Douhal, A. and Santamaria, J. (eds) (2002) *Femtochemistry and Femtobiology: Ultrafast Dynamics in Molecular Science*, World Scientific, Singapore; (e) Valeur, B. (2001) *Molecular Fluorescence: Principles and Applications*, Wiley-VCH Verlag GmbH, Weinheim.
30 Reichardt, C. (1979) *Angew. Chem. Int. Ed. Engl.*, **18**, 98.
31 Kamlet, M.J., Abboud, J.L., and Taft, R.W. (1977) *J. Am. Chem. Soc.*, **99**, 6027.
32 (a) Warshel, A. and Levitt, M. (1976) *J. Mol. Biol.*, **103**, 227; (b) Singh, U.C. and Kollman, P.A. (1986) *J. Comput. Chem.*, **7**, 718; (c) Field, M.J., Bash, P.A., and Karplus, M. (1990) *J. Comput. Chem.*, **11**, 700; (d) Gao, J. (1995) *Reviews in Computational Chemistry* (eds E.K.B. Lipkowitz and D.B. Boyd), John Wiley & Sons, Inc., New York, pp. 119–185; (e) Monard, G. and Merz, K.M. (1999) *Acc. Chem. Res.*, **32**, 904.
33 (a) Tomasi, J. and Persico, M. (1994) *Chem. Rev.*, **94**, 2027; (b) Cramer, C.J. and Truhlar, D.G. (1999) *Chem. Rev.*, **99**, 2161; (c) Orozco, M. and Luque, F.J. (2000) *Chem. Rev.*, **100**, 4187; (d) Tomasi, J., Mennucci, B., and Cammi, R. (2005) *Chem. Rev.*, **105**, 2999.
34 (a) Miertus, S., Scrocco, E., and Tomasi, J. (1981) *Chem. Phys.*, **55**, 117; (b) Cances, E., Mennucci, B., and Tomasi, J. (1997) *J. Chem. Phys.*, **107**, 3032.
35 (a) Gao, J. (1994) *J. Am. Chem. Soc.*, **116**, 9324; (b) Thompson, M.A. (1996) *J. Phys. Chem.*, **100**, 14492; (c) Rajamani, R. and Gao, J. (2002) *J. Comput. Chem.*, **23**, 96; (d) Coutinho, K. and Canuto, S. (2000) *J. Chem. Phys.*, **113**, 9132.
36 (a) Thompson, M.A. and Schenter, G.K. (1995) *J. Phys. Chem.*, **99**, 6374; (b) Osted, A., Kongsted, J., Mikkelsen, K.V., Astrand, P.-O., and Christiansen, O. (2006) *J. Chem. Phys.*, **124**, 124503; (c) Nielsen, C.B., Christiansen, O., Mikkelsen, K.V., and Kongsted, J. (2007) *J. Chem. Phys.*, **126**, 154112; (e) Ohrn, A. and Karlstrom, G. (2006) *Mol. Phys.*, **104**, 3087; (f) Muñoz-Losa, A., Galván, I.F., Aguilar, M.A., and Martín, E. (2007) *J. Phys. Chem. B*, **111**, 9864.
37 (a) Karelson, M.M. and Zerner, M.C. (1990) *J. Am. Chem. Soc.*, **112**, 9405; (b) Kim, H.J. and Hynes, J.T. (1992) *J. Chem. Phys.*, **96**, 5088; (c) Pappalardo, R.R., Reguero, M., Robb, M., and Frisch, M.J. (1993) *Chem. Phys. Lett.*, **212**, 12; (d) Klamt, A. (1996) *J. Phys. Chem.*, **100**, 3349; (e) Christiansen, O. and Mikkelsen, K.V. (1999) *J. Chem. Phys.*, **110**, 8348; (f) Li, J., Cramer, C.J. and Truhlar, D.G. (2000) *Int. J. Quantum Chem.*, **77**, 264; (g) Cossi, M. and Barone, V. (2000) *J. Chem. Phys.*, **112**, 2427; (h) Ferrighi, L., Frediani, L., Fossgaard, E., and Ruud, K. (2007) *J. Chem. Phys.*, **127**, 244103; (i) Mennucci, B., Cappelli, C., Guido, C.A., Cammi, R., and Tomasi, J. (2009) *J. Phys. Chem. A*, **113**, 3009.
38 (a) Jensen, L., van Duijnen, P.Th., and Snijders, J.G. (2003) *J. Chem. Phys.*, **119**, 3800; (b) Curutchet, C., Munoz-Losa, A., Monti, S., Kongsted, J., Scholes, G.D., and Mennucci, B. (2009) *J. Chem. Theory Comput.*, **5**, 1838.
39 Munoz-Losa, A., Curutchet, C., Galvan, I.F., and Mennucci, B. (2008) *J. Chem. Phys.*, **129**, 034104.
40 (a) Santoro, F., Improta, R., Lami, A., Bloino, J., and Barone, V. (2007) *J. Chem. Phys.*, **126**, 084509; (b) Dierksen, M. and Grimme, S. (2005) *J. Chem. Phys.*, **122**, 244101. (c) Jankowiak, H.-C., Stuber, J.L., and Berger, R. (2007) *J. Chem. Phys.*, **127**, 234101; (d) Petrenko, T. and Neese, F. (2007) *J. Chem. Phys.*, **127**, 164319.
41 Myers, A.B. (1996) *Chem. Rev.*, **96**, 911.
42 Matyushov, D.V. and Newton, M.D. (2001) *J. Phys. Chem. A*, **105**, 8516.
43 Car, R. and Parrinello, M. (1985) *Phys. Rev. Lett.*, **55**, 2471.
44 (a) Mennucci, B. and Martinez, J.M. (2005) *J. Phys. Chem. B*, **109**, 9818; (b) Mennucci, B. and Martinez, J.M. (2005) *J. Phys. Chem. B*, **109**, 9830.
45 (a) Kaya, K. and Nakagura, S. (1967) *Theor. Chim. Acta*, **7**, 117; (b) Nielsen, E. and

Schellman, J. (1967) *J. Chem. Phys.*, **71**, 2297.

46 (a) Kneipp, K., Kneipp, H., Itzkan, I., Dasari, R.R., and Feld, M.S. (1999) *Chem. Rev.*, **99**, 2957; (b) Schatz, G.C. and Van Duyne, R.P. (2002) *Handbook of Vibrational Spectroscopy* (eds J. Chalmers and P.R. Griffiths), John Wiley & Sons, Inc., New York; (c) Kneipp, K., Moskovits, M., and Kneipp, H. (eds) (2006) *Surface-Enhanced Raman Scattering: Physics and Applications*, vol. **103**, Topics in Applied Physics Series, Springer, Berlin.

47 (a) Engel, G.S., Calhoun, T.R., Read, E.L., Ahn, T.-K., Mancal, T., Cheng, Y.-C., Blankenship, R.E., and Fleming, G.R. (2007) *Nature*, **446**, 782; (b) Lee, H., Cheng, Y.-C., and Fleming, G.R. (2007) *Science*, **316**, 1462.

6
Nonadiabatic Calculation of Dipole Moments
Francisco M. Fernández and Julián Echave

6.1
Introduction

Many properties of molecular aggregates are attributed to the distribution of charges in the constituent individual molecules or what is commonly called the molecular dipole moment [1]. This molecular property is responsible not only for the behavior of the molar polarization with temperature [1] but also for the most salient features of the absorption and emission spectra of molecules [2–4]. Molecular dipole moments can be derived, for example, from the molar polarization at varied temperatures [1], from microwave spectra [2–4], or from first principles [5].

The purpose of this chapter is to outline the calculation of molecular dipole moments by means of quantum mechanical approaches and to compare such theoretical results with corresponding experimental measurements. It is not intended to be exhaustive because of the enormous number of worthy works on the subject. However, we hope to succeed in giving an idea of the difficulties encountered in such an endeavor.

In Section 6.2, we consider the nonrelativistic molecular Hamiltonian and discuss the separation of the center of mass in detail. We derive general expressions that may be useful for most of the various nonadiabatic approaches that appear in the literature. In Section 6.3, we outline the dynamical symmetry of the molecular Hamiltonian and some of the properties of the molecular stationary states. In Section 6.4, we discuss the well-known Hellmann–Feynman theorem for variational wavefunctions because it is relevant to the calculation of dipole moments. Although we are interested mainly in nonadiabatic calculations of dipole moments, in Section 6.5 we consider the separation of the electronic and nuclear motions that leads to the Born–Oppenheimer (BO) and adiabatic approaches to the calculation of molecular properties. For completeness, in Section 6.6 we outline the interaction between a system of point charges and an external electric field with the purpose of defining dipole moment, polarizabilities, and other molecular properties. In Section 6.7, we briefly describe the models used to derive the dipole moment from the Stark effect in rotational spectra. They are useful for comparison between

Computational Spectroscopy: Methods, Experiments and Applications. Edited by Jörg Grunenberg
Copyright © 2010 WILEY-VCH Verlag GmbH & Co. KGaA, Weinheim
ISBN: 978-3-527-32649-5

experimental measurements and theoretical calculations. In Section 6.8, we briefly review the theoretical calculation of molecular dipole moments under the Born–Oppenheimer approximation. In Section 6.9, we discuss some of the existing nonadiabatic calculations of dipole moments that are the main topic of this chapter. In Section 6.10, we briefly describe the more rigorous nonadiabatic calculations of dipole moments in the molecule-fixed reference frame. For simplicity, we restrict ourselves to diatomic molecules because they appear to be the only ones accessible to current nonadiabatic approaches. In Section 6.11, we outline the application of perturbation theory to the Schrödinger equation for a molecule in an external electric field with the purpose of discussing a more rigorous connection between the measured Stark shift and the molecular dipole moment. Finally, in Section 6.12 we summarize the main conclusions of this chapter.

6.2
The Molecular Hamiltonian

In this section, we consider the nonrelativistic molecular Hamiltonian as a system of N charged point particles with only Coulomb interactions

$$\hat{H} = \hat{T} + \hat{V}$$

$$\hat{T} = \sum_{i=1}^{N} \frac{\hat{p}_i^2}{2m_i} \tag{6.1}$$

$$V = \frac{1}{4\pi\varepsilon_0} \sum_{i=1}^{N-1} \sum_{j=i+1}^{N} \frac{q_i q_j}{r_{ij}}$$

In this expression, m_i is the mass of particle i, $q_i = -e$ or $q_i = Z_i e$ denotes the charges of either an electron or a nucleus, and $r_{ij} = |\mathbf{r}_i - \mathbf{r}_j|$ is the distance between particles i and j located at the points $\mathbf{r}_i$ and $\mathbf{r}_j$, respectively, from the origin of the laboratory coordinate system. In the coordinate representation, $\hat{\mathbf{p}}_i = -i\hbar\nabla_i$.

Since the Coulomb potential V is invariant under translations $\hat{U}(\mathbf{a})\mathbf{r}_i \hat{U}(\mathbf{a})^\dagger = \mathbf{r}_i + \mathbf{a}$ ($\hat{U}(\mathbf{a})V\hat{U}(\mathbf{a})^\dagger = V$), the eigenfunctions of the translation invariant Hamiltonian operator (6.1) are not square integrable. For that reason, we first separate the motion of the center of mass by means of a linear coordinate transformation

$$\mathbf{r}'_j = \sum_i t_{ji} \mathbf{r}_i \tag{6.2}$$

that leads to

$$\nabla_j = \sum_i t_{ij} \nabla'_i \tag{6.3}$$

and

$$\sum_i \frac{1}{m_i}\nabla_i^2 = \sum_i \frac{t_{1i}^2}{m_i}\nabla_1'^2 + 2\sum_i\sum_{j>1} \frac{t_{ji}t_{1i}}{m_i}\nabla_1'\nabla_j' + \sum_i\sum_{j>1}\sum_{k>1} \frac{t_{ji}t_{ki}}{m_i}\nabla_j'\nabla_k' \quad (6.4)$$

It is our purpose to keep the transformation (6.2) as general as possible so that it applies to all the nonadiabatic approaches discussed in this chapter. To uncouple one of the new coordinates $\mathbf{r}_1'$ from the remaining ones, we require that the coefficients of the transformation (6.2) satisfy

$$\sum_i \frac{t_{ji}t_{1i}}{m_i} = 0, \quad j > 1 \quad (6.5)$$

The new coordinates transform under translations as follows:

$$\hat{U}(\mathbf{a})\mathbf{r}_j'\hat{U}(\mathbf{a})^\dagger = \mathbf{r}_j' + \mathbf{a}\sum_i t_{ji} \quad (6.6)$$

If we require that $\mathbf{r}_1'$ transforms exactly as the original variables and that the remaining $\mathbf{r}_j'$ are translationally invariant, we have

$$\sum_i t_{ji} = \delta_{j1} \quad (6.7)$$

If we choose $t_{1i} = \xi m_i$, where ξ is an arbitrary real number, then Equation 6.5 becomes Equation 6.7 for $j > 1$. If we then substitute $t_{1i} = \xi m_i$ into Equation 6.7 with $j = 1$, we conclude that

$$t_{1i} = \frac{m_i}{M}, \quad M = \sum_i m_i \quad (6.8)$$

and $\mathbf{r}_1'$ results to be the well-known coordinate of the center of mass of the molecule [6]. The choice of the coefficients of the transformation (6.2) for the remaining variables $\mathbf{r}_j'$, $j > 1$ is arbitrary as long as they satisfy Equation 6.7.

Finally, the total Hamiltonian operator reads

$$\hat{H} = -\frac{\hbar^2}{2M}\nabla_1'^2 + \hat{H}_M$$

$$\hat{H}_M = -\frac{\hbar^2}{2}\sum_{j>1}\sum_{k>1}\left(\sum_i \frac{t_{ji}t_{ki}}{m_i}\right)\nabla_j'\nabla_k' + \frac{1}{4\pi\varepsilon_0}\sum_{i=1}^{N-1}\sum_{j=i+1}^{N}\frac{q_i q_j}{r_{ij}} \quad (6.9)$$

where $\hat{H}_M$ is the internal or molecular Hamiltonian operator. The explicit form of the interparticle distances r_{ij} in terms of the new coordinates $\mathbf{r}_k'$ may be rather cumbersome in the general case. We consider it in the particular applications discussed in subsequent sections.

For future reference, it is convenient to define the center of mass and relative kinetic energy operators

$$\hat{T}_{CM} = -\frac{\hbar^2}{2M}\nabla_1'^2 \tag{6.10}$$

and

$$\hat{T}_{rel} = -\frac{\hbar^2}{2}\sum_{j>1}\sum_{k>1}\left(\sum_i \frac{t_{ji}t_{ki}}{m_i}\right)\nabla_j'\nabla_k' \tag{6.11}$$

respectively, so that $\hat{T} = \hat{T}_{CM} + \hat{T}_{rel}$, $\hat{H}_M = \hat{T}_{rel} + V$, and $\hat{H} = \hat{T}_{CM} + \hat{H}_M$.

The inverse transformation $\mathbf{t}^{-1}$ exists and gives us the old coordinates in terms of the new ones:

$$\mathbf{r}_i = \sum_j (\mathbf{t}^{-1})_{ij}\mathbf{r}_j' \tag{6.12}$$

According to Equations 6.6 and 6.7, we have $\hat{U}(\mathbf{a})\mathbf{r}_i\hat{U}(\mathbf{a})^\dagger = (\mathbf{t}^{-1})_{i1}\mathbf{a} + \mathbf{r}_i$ from which we conclude that

$$(\mathbf{t}^{-1})_{i1} = 1, \; i = 1, 2, \ldots, N \tag{6.13}$$

To understand the meaning of this result, notice that the momentum conjugate to $\mathbf{r}_i'$ is given by the transformation

$$\hat{\mathbf{p}}_i' = \sum_j (\mathbf{t}^{-1})_{ji}\hat{\mathbf{p}}_j \tag{6.14}$$

so that the linear momentum of the center of mass

$$\hat{\mathbf{p}}_1' = \sum_j \hat{\mathbf{p}}_j \tag{6.15}$$

is precisely the total linear momentum of the molecule. We also appreciate that $T_{CM} = \hat{\mathbf{p}}_1'^2/(2M)$ and that the inverse transformation of the momenta is (see Equation 6.3)

$$\mathbf{p}_j = \sum_i t_{ij}\mathbf{p}_i' \tag{6.16}$$

Equation 6.13 is also relevant to the behavior of the dipole moment in internal coordinates. We have

$$\mu = \sum_i q_i \mathbf{r}_i \tag{6.17}$$

in the laboratory-fixed frame, and

$$\hat{U}(\mathbf{a})\mu\hat{U}(\mathbf{a})^\dagger = \mu + q\mathbf{a}, \; q = \sum_i q_i \tag{6.18}$$

clearly shows that μ is invariant under translation of the origin for a neutral molecule $q = 0$. In internal coordinates, we have

$$\mu = \sum_j q'_j \mathbf{r}'_j, \quad q'_j = \sum_i q_i(\mathbf{t}^{-1})_{ij} \tag{6.19}$$

and $\hat{U}(\mathbf{a})\mu\hat{U}(\mathbf{a})^\dagger = \mu + q'_1 \mathbf{a}$ consistent with Equation 6.18 because $q'_1 = q$ by virtue of Equation 6.13.

We have kept the transformation (6.2) as general as possible to have a suitable expression of the molecular Hamiltonian that applies to all the nonadiabatic approaches discussed in this chapter. In what follows, we illustrate some particular ways of determining the remaining transformation coefficients. More precisely, we choose a point in the molecule as the origin of the new coordinate axis and refer the positions of $N-1$ particles to it (remember that three coordinates have been reserved for the location of the center of mass).

For example, consider an arbitrary set of particle labels I and choose

$$\mathbf{r}'_j = \mathbf{r}_j - \frac{1}{M_I}\sum_{i \in I} m_i \mathbf{r}_i, \quad j > 1, \quad M_I = \sum_{i \in I} m_i \tag{6.20}$$

That is to say

$$t_{ji} = \delta_{ji} - \frac{m_i}{M_I}\delta_{iI}, \quad j > 1, \quad \delta_{iI} = \begin{cases} 1 & \text{if } i \in I \\ 0 & \text{otherwise} \end{cases} \tag{6.21}$$

These coefficients t_{ji} already satisfy Equation 6.7. If I is the set of nuclear labels, then we refer the positions $\mathbf{r}'_j, j > 1$ to the nuclear center of mass. If the set I contains only one nuclear label we refer the positions of the remaining particles to that particular nucleus. Various authors have already chosen one or the other coordinate origin as shown in Sections 6.9 and 6.10.

If we take into account that

$$\tau_{jk} = \sum_i \frac{t_{ji}t_{ki}}{m_i} = \frac{\delta_{jk}}{m_j} - \frac{1}{M_I}(\delta_{kI} + \delta_{jI}) + \frac{1}{M_I} \tag{6.22}$$

then we realize that the transformation (6.20) uncouples the coordinates of the particles that belong to I from the remaining ones in the kinetic operator:

$$\tau_{jk} = \begin{cases} \dfrac{\delta_{jk}}{m_j} - \dfrac{1}{M_I} & \text{if } j, k \in I \\ 0 & \text{if } j \in I \text{ and } k \notin I \text{ or } j \notin I \text{ and } k \in I \\ \dfrac{\delta_{jk}}{m_j} + \dfrac{1}{M_I} & \text{if } j, k \notin I \end{cases} \tag{6.23}$$

The equations derived in this section are sufficiently general to cover all the strategies commonly followed in the separation of the center of mass prior to the solution of the Schrödinger equation for atoms and molecules. Other general expressions for the translation-free internal coordinates have already been discussed by Sutcliffe and Woolley [7–10]. It is a most important topic in the nonadiabatic quantum mechanical approach to atoms and molecules.

6.3
Symmetry

In this section, we discuss the symmetry of a molecular system that is determined by all the operators that commute with its Hamiltonian. For example, the total Hamiltonian operator commutes with the total linear momentum

$$[\hat{H}, \hat{\mathbf{p}}] = 0, \quad \hat{\mathbf{p}} = \sum_{j=1}^{N} \hat{\mathbf{p}}_j = \hat{\mathbf{p}}'_1 \tag{6.24}$$

and also with the total angular momentum

$$[\hat{H}, \hat{\mathbf{J}}] = 0, \quad \hat{\mathbf{J}} = \sum_{j=1}^{N} \hat{\mathbf{J}}_j \tag{6.25}$$

It follows from Equations 6.12 and 6.16 that

$$\hat{\mathbf{J}} = \sum_i \hat{\mathbf{r}}_i \times \hat{\mathbf{p}}_i = \sum_i \hat{\mathbf{r}}'_i \times \hat{\mathbf{p}}'_i = \hat{\mathbf{r}}'_1 \times \hat{\mathbf{p}}'_1 + \hat{\mathbf{J}}' \tag{6.26}$$

where $\hat{\mathbf{J}}'$ is the internal or molecular angular momentum. It is clear that

$$[\hat{H}_M, \hat{\mathbf{J}}'] = 0, \quad \hat{\mathbf{J}}' = \sum_{i=2}^{N} \hat{\mathbf{r}}'_i \times \hat{\mathbf{p}}'_i \tag{6.27}$$

The Hamiltonian operator is invariant under permutation of identical particles. We may formally write this symmetry as

$$\hat{P}\hat{H}\hat{P}^{-1} = \hat{H} \Rightarrow [\hat{H}, \hat{P}] = 0 \tag{6.28}$$

where $\hat{P}$ stands for any permutation of electrons or identical nuclei. Since $\hat{T}_{CM}$ is invariant under permutation of identical particles, we conclude that

$$\hat{P}\hat{H}_M\hat{P}^{-1} = \hat{H}_M \Rightarrow [\hat{H}_M, \hat{P}] = 0 \tag{6.29}$$

Besides, the molecular Hamiltonian operator is also invariant under inversion

$$\hat{\Im}\hat{H}_M\hat{\Im}^{-1} \Rightarrow [\hat{H}_M, \hat{\Im}] = 0 \tag{6.30}$$

where the inversion operator $\hat{\Im}$ produces the transformations $\hat{\Im}\hat{\mathbf{r}}_j\hat{\Im}^{-1} = -\hat{\mathbf{r}}_j$ and $\hat{\Im}\hat{\mathbf{p}}_j\hat{\Im}^{-1} = -\hat{\mathbf{p}}_j$. Also, notice that $[\hat{\mathbf{J}}', \hat{\Im}] = 0$, $[\hat{\mathbf{J}}', \hat{P}] = 0$, and $[\hat{\Im}, \hat{P}] = 0$ so that we can choose a stationary state Ψ to be a simultaneous eigenfunction of such set of commuting operators:

$$\begin{aligned}
\hat{H}_M \Psi &= E\Psi, \\
\hat{J}'^2 \Psi &= \hbar^2 J'(J'+1)\Psi, \quad J' = 0, 1, \ldots \\
\hat{J}'_z \Psi &= \hbar M'_J \Psi, \quad M'_J = 0, \pm 1, \ldots, \pm J' \\
\hat{\mathcal{P}} \Psi &= \sigma \Psi \\
\hat{\mathcal{S}} \Psi &= \pm \Psi
\end{aligned} \qquad (6.31)$$

where $\sigma = 1$ or $\sigma = -1$ for identical bosons or fermions, respectively. A somewhat more rigorous discussion of the kinematics and dynamics of molecules was provided, for example, by Woolley [11]. The simplified discussion in this section is sufficient for the present purposes.

6.4
The Hellmann–Feynman Theorem

In this section, we derive the Hellmann–Feynman theorem for optimized variational wavefunctions because it is most important for the accurate calculation of dipole moments [5, 12].

Consider a trial function Φ and the variational energy W given by

$$W\langle\Phi|\Phi\rangle = \langle\Phi|\hat{H}|\Phi\rangle \qquad (6.32)$$

An arbitrary variation $\delta\Phi$ leads to

$$\delta W\langle\Phi|\Phi\rangle = \langle\delta\Phi|\hat{H}-W|\Phi\rangle + \langle\delta\Phi|\hat{H}-W|\Phi\rangle^* \qquad (6.33)$$

where the asterisk denotes complex conjugation. The right-hand side of this equation vanishes when $\delta W = 0$; that is to say, when W is stationary with respect to the infinitesimal change $\delta\Phi$ in the trial function. The variational or optimal trial function satisfies this condition.

If the Hamiltonian operator depends on a parameter λ, which may be, for example, a mass, charge, force constant, strength of a external field, and so on, then the variational function also depends on it. Therefore, differentiation of Equation 6.32 with respect to λ leads to

$$\frac{\partial W}{\partial \lambda}\langle\Phi|\Phi\rangle = \left\langle\Phi\left|\frac{\partial \hat{H}}{\partial \lambda}\right|\Phi\right\rangle + \left\langle\frac{\partial \Phi}{\partial \lambda}\bigg|\hat{H}-W\bigg|\Phi\right\rangle + \left\langle\frac{\partial \Phi}{\partial \lambda}\bigg|\hat{H}-W\bigg|\Phi\right\rangle^* \qquad (6.34)$$

It is clear from this equation that if the set of variations $\delta\Phi$ includes $(\partial\Phi/\partial\lambda)\delta\lambda$, then the optimized trial function Φ satisfies the well-known Hellmann–Feynman theorem [13, 14]:

$$\frac{\partial W}{\partial \lambda} = \left\langle\frac{\partial \hat{H}}{\partial \lambda}\right\rangle \qquad (6.35)$$

where $\langle\hat{A}\rangle = \langle\Phi|\hat{A}|\Phi\rangle/\langle\Phi|\Phi\rangle$.

6.5
The Born–Oppenheimer Approximation

Since the vast majority of quantum mechanical studies of molecular properties are based on the Born–Oppenheimer approximation [15], we discuss it here in some detail, in spite of the fact that we are more interested in non-BO approaches. Before proceeding, we mention the curious fact that the modern and most useful form of that approach appears to have been first developed by Slater [16] in his study of the helium atom as early as 1927. However, we adopt the common practice to attribute it to Born and Oppenheimer. In this section, we begin with the traditional procedure proposed by Slater [16] and Born and Huang [15] and then mention other authors' approaches and criticisms.

For simplicity and clarity, we write the Schrödinger equation for a molecule as

$$\hat{H}\Psi_\alpha(\mathbf{r}^e, \mathbf{r}^n) = E_\alpha \Psi_\alpha(\mathbf{r}^e, \mathbf{r}^n) \tag{6.36}$$

where $\mathbf{r}^e$ and $\mathbf{r}^n$ denote all the electronic and nuclear coordinates, respectively, and the subscript α stands for the collection of all the quantum numbers necessary for the description of the molecular state. We use the symbols $\hat{T}_e$, $\hat{T}_n$, V_{ee}, V_{ne}, and V_{nn} to indicate the kinetic energy operators for the electrons and nuclei, and the electron–electron, nuclei–electron, and nuclei–nuclei Coulomb interactions, respectively:

$$\hat{H} = \hat{T}_n + \hat{H}_e + V_{nn}, \quad \hat{H}_e = \hat{T}_e + V_{ee} + V_{ne} \tag{6.37}$$

The BO approximation is based on the ansatz

$$\Psi_\alpha(\mathbf{r}^e, \mathbf{r}^n) = \sum_j \chi_{\alpha j}(\mathbf{r}^n) \varphi_j(\mathbf{r}^e, \mathbf{r}^n) \tag{6.38}$$

and we assume that $\langle \varphi_j | \varphi_k \rangle_e = \delta_{jk}$, where the subscript e indicates integration over electronic coordinates. It follows from Equations 6.36, 6.37, and 6.38 that

$$\langle \varphi_k | \hat{H} | \Psi_\alpha \rangle_e = \hat{T}_n \chi_{\alpha k} + \sum_j (U_{kj} + W_{kj}) \chi_{\alpha j} = E_\alpha \chi_{\alpha k} \tag{6.39}$$

where

$$U_{kj} = \langle \varphi_k | \hat{H}_e | \varphi_j \rangle_e + V_{nn} \delta_{kj}, \quad W_{kj} = \langle \varphi_k | [\hat{T}_n, \varphi_j] \rangle_e \tag{6.40}$$

We have introduced the formal commutator $[\hat{T}_n, \varphi_j]$ for the sole purpose of making the resulting equation more compact.

In the standard BO approximation, we choose the functions φ_j to be eigenfunctions of the electronic Hamiltonian $\hat{H}_e$

$$\hat{H}_e \varphi_j(\mathbf{r}^e, \mathbf{r}^n) = E_{ej}(\mathbf{r}^n) \varphi_j(\mathbf{r}^e, \mathbf{r}^n) \tag{6.41}$$

so that

$$U_{kj}(\mathbf{r}^n) = U_j(\mathbf{r}^n) \delta_{kj}, \quad U_j(\mathbf{r}^n) = E_{ej}(\mathbf{r}^n) + V_{nn}(\mathbf{r}^n) \tag{6.42}$$

In the first approximation, we assume the couplings W_{kj} between "electronic states" to be small and keep just one term for the ground state

$$\Psi_\alpha(\mathbf{r}^e, \mathbf{r}^n) \approx \chi_{\alpha 0}(\mathbf{r}^n)\varphi_0(\mathbf{r}^e, \mathbf{r}^n) \tag{6.43}$$

and

$$[\hat{T}_n + U_0(\mathbf{r}^n)]\chi_{\alpha 0}(\mathbf{r}^n) = E_\alpha^{BO}\chi_{\alpha 0}(\mathbf{r}^n) \tag{6.44}$$

Addition of the diagonal term W_{00} to the nuclear operator in this equation gives rise to the adiabatic approximation.

Sutcliffe and Woolley have criticized the main assumptions of this approach in several papers [8–11, 17]. For example, they point out that the functions in the expansion on the right-hand side of Equation 6.38 are assumed to be square integrable, whereas the left-hand side is not because we have not removed the motion of the center of mass. Such criticism, which also applies to the approximate and widely used ansatz (6.43), does not appear to be justified because the nuclear functions $\chi_{\alpha j}$ are not square integrable. According to these authors, another drawback of the BO approximation is that it does not take into account the permutational symmetry of identical nuclei that are treated as distinguishable particles clamped in space to define a framework geometry unambiguously. However, in principle we can introduce such permutational symmetry in the approximate wavefunction (6.43) by means of appropriate projection operators such as those mentioned in Section 6.9. In our opinion, the Born–Oppenheimer approximation is a consistent way of deriving a suitable approximate solution to the Schrödinger equation.

It is interesting how some popular books on quantum chemistry introduce this subject. For example, Szabo and Ostlung [5] state that "Our discussion of this approximation is qualitative. The quantitative aspects of this approximation are clearly discussed by Sutcliffe," and they give a reference and simply write the single product (6.43). On the other hand, Pilar [18], referring to a Hamiltonian operator such as (6.1), says: "...where it is assumed that all nuclear and electronic coordinates have been referred to the center of mass of the system." However, the Hamiltonian operator exhibits all the electronic and nuclear coordinates and no coupling terms coming from the separation of the center of mass. Therefore, it cannot be the internal Hamiltonian operator as we have already seen above. These are just two examples of how most authors blindly accept the clamped nuclei approximation without further analysis.

In some cases, mostly for diatomic molecules, the separation of the motion of the center of mass has been carried out rigorously prior to the application of the BO approximation [13]. Sutcliffe and Woolley have already discussed this issue in detail in several papers [8–11, 17].

Although it is not customary to remove the motion of the center of mass of the whole molecule before the application of BO approximation, it appears to be common practice to separate the nuclear center of mass at the second stage (6.44). Since the nuclei are much heavier than the electrons, we may assume that the error is small. A rigorous discussion of this issue and a simple exactly solvable example are given elsewhere [19, 20].

A most important by-product of the clamped nuclei approximation is that it enables one to introduce the familiar (classical) chemical concept of molecular structure into the quantum mechanical calculations. The molecular geometry is given by the equilibrium nuclear configuration $\mathbf{r}^n_{eq}$ at the minimum of $U(\mathbf{r}^n)$:

$$\nabla U(\mathbf{r}^n)|_{\mathbf{r}^n=\mathbf{r}^n_{eq}} = 0 \tag{6.45}$$

At this point, it is worth noticing that $U(\mathbf{r}^n)$ would not appear in a straightforward rigorous solution of the Schrödinger Equation 6.36. Sutcliffe and Woolley [7–11, 17, 21] have also pointed out the difficulty in discussing molecular structure without the BO approximation (or any of its variants).

6.6
Interaction between a Molecule and an External Field

To make the present chapter sufficiently self-contained, we briefly develop the main equations for the interaction between a molecule and an external potential $\Phi(\mathbf{r})$. The energy of that interaction is given by [22]

$$W = \int \varrho(\mathbf{r})\Phi(\mathbf{r})\,d\mathbf{r} \tag{6.46}$$

where $\varrho(\mathbf{r})$ is the molecular charge density. If the potential varies slowly in the region where the charge density is nonzero we can expand it in a Taylor series about the coordinate origin located somewhere in the molecule:

$$\Phi(\mathbf{r}) = \Phi(0) + \sum_u u \frac{\partial \Phi}{\partial u}(0) + \frac{1}{2}\sum_u \sum_v uv \frac{\partial^2 \Phi}{\partial u \partial v}(0) + \cdots \tag{6.47}$$

where $u, v = x, y, z$. Since $\nabla^2 \Phi(\mathbf{r}) = 0$, we subtract $r^2 \nabla^2 \Phi(0)/6$ from this equation and rewrite the result as

$$\Phi(\mathbf{r}) = \Phi(0) + \sum_u u \frac{\partial \Phi}{\partial u}(0) + \frac{1}{6}\sum_u \sum_v (3uv - r^2 \delta_{uv}) \frac{\partial^2 \Phi}{\partial u \partial v}(0) + \cdots \tag{6.48}$$

If we now take into account that the external field is given by $\mathbf{F}(\mathbf{r}) = -\nabla \Phi(\mathbf{r})$, we have

$$\Phi(\mathbf{r}) = \Phi(0) - \mathbf{r}\cdot\mathbf{F}(0) - \frac{1}{6}\sum_u \sum_v (3uv - r^2 \delta_{uv}) \frac{\partial F_u}{\partial v}(0) + \cdots \tag{6.49}$$

For a set of point charges q_i located at $\mathbf{r}_i$, $i = 1, 2, \ldots, N$, we have

$$\varrho(\mathbf{r}) = \sum_{i=1}^{N} q_i \delta(\mathbf{r}-\mathbf{r}_i) \tag{6.50}$$

and the interaction energy becomes

$$W = q\Phi(0) - \mu \cdot F(0) - \frac{1}{6}\sum_u \sum_v Q_{uv} \frac{\partial F_u}{\partial v}(0) + \cdots \quad (6.51)$$

where the net charge q, the dipole moment μ, and the quadrupole moment $\mathbf{Q}$ are given by

$$\begin{aligned} q &= \sum_{i=1}^N q_i \\ \mu &= \sum_{i=1}^N q_i \mathbf{r}_i \\ Q_{uv} &= \sum_{i=1}^N q_i(3u_i v_i - r_i^2 \delta_{uv}) \end{aligned} \quad (6.52)$$

If the applied field is uniform in the region of interest, then all the terms beyond the first two on the right-hand side of Equation 6.51 vanish.

In one of his well-known discussions of long-range intermolecular forces, Buckingham [23] considers a Hamiltonian operator of the form

$$\hat{H} = \hat{H}^0 - \mu \cdot \mathbf{F} - \frac{1}{3}\sum_u \sum_v \Theta_{uv} F_{uv} + \cdots \quad (6.53)$$

where $\Theta = \mathbf{Q}/2$ and $F_{uv} = (\partial F_u/\partial v)(0)$. Thus, the perturbation expansion for the energy of the molecule in the external field leads to [23]

$$\begin{aligned} W = \langle \Psi|\hat{H}|\Psi\rangle = W^{(0)} &- \sum_u \mu_u^{(0)} F_u - \frac{1}{2}\sum_u \sum_v \alpha_{uv} F_u F_v - \cdots \\ &- \frac{1}{3}\sum_u \sum_v \Theta_{uv}^{(0)} F_{uv} - \cdots \end{aligned} \quad (6.54)$$

where

$$\begin{aligned} \mu_u^{(0)} &= \langle \Psi^{(0)}|\hat{\mu}_u|\Psi^{(0)}\rangle \\ \Theta_{uv}^{(0)} &= \langle \Psi^{(0)}|\hat{\Theta}_{uv}|\Psi^{(0)}\rangle \end{aligned} \quad (6.55)$$

are the permanent dipole and quadrupole moments, respectively. Buckingham [23] does not write the Hamiltonian operator $\hat{H}^{(0)}$ explicitly, but he refers to the energy of "separate molecules for fixed molecular positions and orientations." Therefore, one may assume that he probably means the BO Hamiltonian operator. On the other hand, Bishop [24] considers that Equation 6.54 can be used irrespective of whether one is considering the electronic or the total molecular energy. He explicitly indicates that q_i is an element of charge at the point $\mathbf{r}_i$ relative to an origin fixed at some point in the molecule. As shown in Sections 6.9 and 6.10, this choice of reference frame is not convenient for the nonadiabatic calculation of the dipole moment.

In a most interesting and comprehensive review of the electric moments of molecules, Buckingham [25] argues, by means of simple and rigorous symmetry arguments, that a diatomic molecule in some stationary states does not possess a dipole moment. We come back to this point in Sections 6.9 and 6.10.

6.7
Experimental Measurements of Dipole Moments

Since long ago scientists have estimated the permanent dipole moment of a given molecule from the Clausius–Mosotti equation

$$P = \frac{\varepsilon-1}{\varepsilon+2} \frac{M}{\varrho} \tag{6.56}$$

where ε is the dielectric constant of the medium (mainly a gas or vapor), ϱ its density, and M the molar mass. The Debye equation gives us the behavior of the polarization with temperature

$$P = \frac{4\pi}{3} N_A \left(\alpha_0 + \frac{\mu^2}{3kT} \right) \tag{6.57}$$

where N_A is the Avogadro constant, α_0 takes account of distortion effects, and μ is the permanent dipole moment [1]. According to this equation, the plot of P as a function of $1/T$ should give a straight line, the slope of which is proportional to the square of the dipole moment [1]. The term $\mu^2/(3kT)$ is the contribution to the polarizability of the orientation of the permanent dipole moment due to an external electric field [1].

In this chapter, we are mainly interested in spectroscopic measurements of dipole moments that are commonly based on microwave spectra, and almost invariably on the Stark effect and the model of a rigid (or almost rigid) rotating dipole. If $\hat{H}_{rot}$ is the Hamiltonian of a rigid rotator and $-\mu \cdot \mathbf{F}$ is the interaction between the molecular dipole moment and the uniform electric field (as shown in Section 6.6), then the Stark rotational energies are given by

$$\hat{H}_{rot}(F)\psi^{rot} = E^{rot}(F)\psi^{rot}, \quad \hat{H}_{rot}(F) = \hat{H}_{rot}(0) - \mu F \cos(\theta) \tag{6.58}$$

where θ is the angle between the external field $\mathbf{F}$ and the dipole moment μ.

The external electric field produces both shift and split of the molecular rotational energies. The magnitude of the shift of the spectral lines $\tilde{v}(F) = \Delta E^{rot}(F)/(hc)$ changes with the field. Rayleigh–Schrödinger perturbation theory provides analytical expressions for the Stark shifts in the form of a power series of the field intensity:

$$\Delta\tilde{v}(F) = \tilde{v}(F) - \tilde{v}(0) = \tilde{v}^{(1)}\mu F + \tilde{v}^{(2)}(\mu F)^2 + \cdots \tag{6.59}$$

where the coefficients $\tilde{v}^{(j)}$ are known functions of the rotational quantum numbers [2–4]. If we measure the Stark shifts for known values of the field intensity and then fit selected experimental data to a polynomial function of the field, we can obtain the dipole moment from the polynomial coefficients.

To account for the hyperfine structure of the spectra, one should add the terms arising from the spin–rotational interaction, the quadrupole interaction of the nuclei, and the spin–spin magnetic interactions [2–4].

Molecular beam electric resonance experiments are also based on the same model of rotating dipole [4].

There have been many experimental studies of a wide variety of molecules. We restrict to the simplest ones that are accessible to existing nonadiabatic approaches, such as, obviously, diatomics [26–29]. In this case, $\tilde{v}^{(1)} = 0$ and the Stark effect is quadratic. It is worth paying attention to the discrepancy in notation and presentation of the empirical models. For example, the model Hamiltonian may be either an actual operator [28] or a scalar function of the quantum numbers [29]. The interaction between the dipole and the field may also be written in somewhat varied ways [28, 29] or the effective Hamiltonian may omit the rotational kinetic energy [27].

Although symmetric tops appear to be beyond present nonadiabatic treatments, we quote them here as another example of the use of the model of a rotating dipole outlined above [30, 31]. In this case, $\tilde{v}^{(1)} = 0$ when $K = 0$ and $\tilde{v}^{(1)} \neq 0$ when $K \neq 0$, where K is the quantum number for the projection of the angular momentum along the symmetry axis.

It is clear from the above discussion that the accuracy and reliability of the experimental determination of the molecular dipole moment not only depend on the accuracy of the measured Stark line shifts but also on the validity of the theoretical, semiempirical model of the molecule as a rotating electric dipole. (Notice that the same model is assumed in the calculation of dipole moments from measurements of the dielectric constant.) We will come back to this important point in Section 6.11.

6.8
The Born–Oppenheimer Calculations of Dipole Moments

Although we are mainly interested in the nonadiabatic calculation of dipole moments, it is worth comparing it with the more popular BO approach to the problem. For this reason, in this section we outline the latter. If we are able to solve Equation 6.41 for an appropriate set of nuclear configurations and determine the equilibrium geometry of the molecule given by Equation 6.45, we can then calculate the dipole moment as follows:

$$\mu^{BO} = -e \int \varphi(\mathbf{r}^e, \mathbf{r}^n_{eq})^* \left(\sum_{j=1}^{N_e} \mathbf{r}^e_j \right) \varphi(\mathbf{r}^e, \mathbf{r}^n_{eq}) \, d\mathbf{r}^e + e \sum_{j=1}^{N_n} Z_j \mathbf{r}^n_{j\,eq} \qquad (6.60)$$

for a molecule with N_e electrons and N_n nuclei [5]. In this equation, $d\mathbf{r}^e$ denotes the volume element for all the electronic coordinates.

An alternative approach consists in solving the electronic BO equation for the molecule in an electric field

$$\hat{H}_e(\mathbf{F}) = \hat{H}_e(0) + e\mathbf{F} \cdot \sum_{j=1}^{N_e} \mathbf{r}_j^e \quad (6.61)$$

and then calculate the electronic part of the dipole moment as

$$\mu_u^e = -\left(\frac{\partial E_e(\mathbf{r}_{eq}^n, \mathbf{F})}{\partial F_u}\right)_{F=0} \quad (6.62)$$

where $E_e(\mathbf{r}_{eq}^n, \mathbf{F})$ is the lowest eigenvalue of the electronic operator (6.61) for the equilibrium geometry.

Szabo and Ostlund [5] discuss the reasons of the noticeable discrepancy between both approaches. It is known to be due to the fact that approximate wavefunctions that are not well optimized fail to satisfy the Hellmann–Feynman theorem [12] already discussed in Section 6.4. Since the formally correct definition of properties such as the dipole moment is as a response function to an external field (see Section 6.6), Swanton et al. [12] proposed a calculation based on the following expression:

$$\mu_u^e = \langle \Phi(0)|\hat{\mu}_u^e|\Phi(0)\rangle - 2\left\langle \frac{\partial \Phi}{\partial F_u}(0) \middle| \left[\hat{H}_e(0) - E_e(\mathbf{r}_{eq}^n, 0)\right] \middle| \Phi(0)\right\rangle \quad (6.63)$$

in such cases where the variational function is not fully optimized and one does not have appropriate analytical expressions for the derivatives $(\partial E_e/\partial F_u)_{F=0}$. This equation is a particular case of (6.34) when the electronic approximate wavefunction is real and normalized to unity $\langle \Phi(0)|\Phi(0)\rangle = 1$.

The nuclear configuration given by the set of equilibrium coordinates $\mathbf{r}_{eq}^n$ determines what we usually call the geometry of the molecule and thereby the orientation of the dipole moment. For example, we know that the dipole moment is directed along the molecular axis in a linear molecule without an inversion center or along the symmetry axis in a symmetric top. The analysis is not so simple in the case of the nonadiabatic calculations that we discuss in Section 6.9.

According to Cade and Huo [32], the dipole moment calculated by means of this quantum mechanical approach for just one internuclear distance (say, the theoretical or experimental R_e) is not strictly comparable to the experimental one that commonly corresponds to a particular vibrational state or an average over a set of vibrational states. In spite of this apparent deficiency of the BO approach, it is worth noticing that when Wharton et al. [26] determined the dipole moment of LiH experimentally to be $\mu = 5.9D$, there were as many as eight previous reliable quantum mechanical calculations that agreed with it to $\pm 0.3D$ [33].

6.9
Nonadiabatic Calculations of Dipole Moments

Most theoretical calculations of dipole moments are based on the BO approximation for several reasons: first, and most important, non-BO calculations require far more computer time; second, most theoretical chemists are unwilling to go beyond the BO approximation that they assume to know well; and third, non-BO calculations give rise to some additional theoretical difficulties as we show below. It is therefore not surprising that non-BO calculations of dipole moments are scarce and few. We describe some of them in this section and in Section 6.10.

Tachikawa and Osamura [34] proposed a dynamic extended molecular orbital (DEMO) approach based on SCF wavefunctions of the form

$$\Psi^{SCF} = \prod_I \Phi^I \tag{6.64}$$

where each Φ^I is a function of the coordinates of a set of identical particles with the appropriate permutation symmetry. These functions are expressed in terms of generalized molecular orbitals ϕ_j^I that are linear combinations of floating Gaussians χ_r^I:

$$\phi_j^I = \sum_r c_{rj}^I \chi_r^I \tag{6.65}$$

It is worth noticing that Tachikawa and Osamura [34] did not separate the motion of the center of mass, and this omission gives rise to considerable errors if the SCF orbitals depend on laboratory-fixed coordinates [35]. Another limitation of this approach is that the SCF wavefunction does not take into account particle correlation that may be quite strong between nuclei [36, 37].

If the variational approach were based on a trial function $\varphi(\mathbf{r}_2', \mathbf{r}_3', \ldots, \mathbf{r}_N')$ of just internal, translation-free coordinates $\mathbf{r}_j'$ (see Section 6.2), it would not be necessary to separate the motion of the center of mass explicitly because $\langle \varphi | \hat{H} | \varphi \rangle = \langle \varphi | \hat{H}_M | \varphi \rangle$. But it is not the case of the SCF wavefunction (6.64) so that we have $E^{SCF} = \langle \Psi^{SCF} | \hat{H} | \Psi^{SCF} \rangle = \langle \Psi^{SCF} | \hat{T}_{CM} | \Psi^{SCF} \rangle + \langle \Psi^{SCF} | \hat{H}_M | \Psi^{SCF} \rangle > \langle \Psi^{SCF} | \hat{H}_M | \Psi^{SCF} \rangle$ Therefore, the estimated energy is always worse than when the trial function depends only on internal coordinates, even though the SCF wavefunction may satisfy the virial theorem $2 \langle \Psi^{SCF} | \hat{T} | \Psi^{SCF} \rangle = - \langle \Psi^{SCF} | V | \Psi^{SCF} \rangle$ [14, 34]. The reader may find a more detailed discussion of this issue elsewhere [35].

Tachikawa and Osamura [34] calculated the dipole moments of the ^{m}H^{n}H and ^{m}Li^{n}H isotopomer series but, unfortunately, they did not show the expression they used. This is not a minor issue as discussed in what follows.

Before proceeding further, it is convenient to discuss the failure of the naive approach to the nonadiabatic calculation of dipole moments. In quantum mechanics, one obtains the average of an observable O as the expectation value $\langle \Psi | \hat{O} | \Psi \rangle$ of the corresponding operator $\hat{O}$. In Section 6.3, we showed that $\hat{\Im} \Psi = \pm \Psi$ because the molecular Hamiltonian is invariant under inversion. Since $\hat{\Im} \hat{\mu} \hat{\Im}^{-1} = -\hat{\mu}$, we conclude that

$$\mu = \langle \Psi | \hat{\mu} | \Psi \rangle = 0 \tag{6.66}$$

for any nondegenerate molecular state Ψ. This result known since long ago [25] applies to *any* molecule in its ground state and renders fruitless the calculation of its dipole moment as a straightforward expectation value in a set of axes parallel to those of the laboratory (like the one discussed in Section 6.2).

6 Nonadiabatic Calculation of Dipole Moments

One of the problems arising from the separation of the center of mass discussed in Section 6.2 is that the transformation (6.12) can make the Coulomb potential rather messy. One way of keeping a simple form of the potential–energy function is to choose one of the particles as coordinate origin. From a practical point of view, it appears to be convenient to choose the heaviest nucleus for that purpose [36]. Thus, the transformation

$$\mathbf{r}'_1 = \sum_{i=1}^{N} \frac{m_i}{M} \mathbf{r}_i$$

$$\mathbf{r}'_j = \mathbf{r}_j - \mathbf{r}_1, \, j = 2, 3, \ldots, N \tag{6.67}$$

leads to

$$\hat{H}_M = -\frac{\hbar^2}{2} \sum_i \frac{1}{m_i} \nabla_i'^2 - \frac{\hbar^2}{2m_i} \sum_{j>1} \sum_{k>1} \nabla_j' \nabla_k' + \frac{1}{4\pi\varepsilon_0} \sum_{i=1}^{N-1} \sum_{j=i+1}^{N} \frac{q_i q_j}{r_{ij}} \tag{6.68}$$

where m_1 is the mass of the heaviest nucleus and $\mathbf{r}_1$ its location in the laboratory reference frame. The transformed Coulomb potential does not make the calculation of matrix elements unnecessarily complicated as follows from the fact that $|\mathbf{r}_i - \mathbf{r}_1| = |\mathbf{r}'_i|$, $i = 2, 3, \ldots, N$, and $|\mathbf{r}_i - \mathbf{r}_j| = |\mathbf{r}'_i - \mathbf{r}'_j|$, $i, j = 2, 3, \ldots, N$.

For example, we consider a four-particle molecule. The transformation (6.67) is given by the matrix

$$\mathbf{t} = \begin{pmatrix} \frac{m_1}{M} & \frac{m_2}{M} & \frac{m_3}{M} & \frac{m_4}{M} \\ -1 & 1 & 0 & 0 \\ -1 & 0 & 1 & 0 \\ -1 & 0 & 0 & 1 \end{pmatrix} \tag{6.69}$$

with inverse

$$\mathbf{t}^{-1} = \begin{pmatrix} 1 & -\frac{m_2}{M} & -\frac{m_3}{M} & -\frac{m_4}{M} \\ 1 & 1-\frac{m_2}{M} & -\frac{m_3}{M} & -\frac{m_4}{M} \\ 1 & -\frac{m_2}{M} & 1-\frac{m_3}{M} & -\frac{m_4}{M} \\ 1 & -\frac{m_2}{M} & -\frac{m_3}{M} & 1-\frac{m_4}{M} \end{pmatrix} \tag{6.70}$$

This transformation applies, for example, to the H_2 isotopomer series [38]. The only members that exhibit dipole moments are HD, HT, and DT. In this case, we choose the labels 1, 2, 3, and 4 to denote the heaviest nucleus, the lightest one, and the two electrons, respectively.

The fact is that $m_1 > m_2$ gives rise to a charge asymmetry and a small dipole moment. Obviously, the straightforward BO approximation cannot account for it because the electronic Hamiltonian $\hat{H}_e$ does not depend on nuclear masses (see

Section 6.5) and the resulting electronic charge density is exactly the same for all isotopomers. Therefore, to obtain the dipole moment of diatomic molecules of the form ^{m}A^{n}A, one has to take into account nonadiabatic corrections [39–41]. An alternative approach is based on the fact that there is no unique way of implementing the BO approximation. In fact, appropriate canonical transformations of the coordinates prior to the application of the BO approximation may force the required asymmetry and provide a suitable way of calculating the dipole moment of ^{m}A^{n}A molecules [42, 43].

Cafiero and Adamowicz [36, 38] calculated nonadiabatic dipole moments for some small diatomic molecules; in what follows, we outline the variational method proposed by these authors for a diatomic molecule with $N-2$ electrons. The core of the approach is a basis set of floating s-type explicitly correlated Gaussian functions of the form

$$g_k(\mathbf{r}) = \exp[-(\mathbf{X}-\mathbf{s}_{kx}) \cdot \mathbf{A}_k \cdot (\mathbf{X}-\mathbf{s}_{kx})^t - (\mathbf{Y}-\mathbf{s}_{ky}) \cdot \mathbf{A}_k \cdot (\mathbf{Y}-\mathbf{s}_{ky})^t \\ -(\mathbf{Z}-\mathbf{s}_{kz}) \cdot \mathbf{A}_k \cdot (\mathbf{Z}-\mathbf{s}_{kz})^t] \quad (6.71)$$

where $\mathbf{A}_k$ is a $N' \times N' (N' = N-1)$ symmetric matrix, $\mathbf{X}, \mathbf{Y}$, and $\mathbf{Z}$ are $1 \times N'$ matrices of the form $\mathbf{X} = (x'_2, x'_3, \ldots, x'_N)$, $\mathbf{s}_k$ is a $1 \times N'$ matrix that determines the location of the center of the Gaussian in space, and t stands for transpose. To assure that the Gaussians are square integrable, they chose $\mathbf{A}_k$ to be of Cholesky-factored form $\mathbf{A}_k = \mathbf{L}_k \cdot \mathbf{L}_k^t$, where $\mathbf{L}_k$ is a $N' \times N'$ lower triangular matrix.

To have the correct permutation symmetry of identical particles, they resorted to appropriate projection operators of the form [36]

$$\hat{E} = \prod_i \hat{E}_i \quad (6.72)$$

and constructed the variational function

$$\Psi = \sum_{k=1}^m c_k \hat{E} g_k(\mathbf{r}) \quad (6.73)$$

so that the permutation of any pair of identical particles leads to either Ψ or $-\Psi$ if they are bosons or fermions, respectively. Then, they minimized the variational energy

$$E = \min \frac{\langle \Psi | \hat{H} | \Psi \rangle}{\langle \Psi | \Psi \rangle} \quad (6.74)$$

Notice that the matrix $\mathbf{L}_k$ has $N'(N'+1)/2$ independent variational parameters and each $\mathbf{s}_k$ contributes with N'; therefore, there are $N'(N'+1)/2 + 3N' + 1$ adjustable variational parameters for every basis function. In a most comprehensive review, Bubin et al. [44] discussed this type of calculation.

Since the straightforward expectation value of the dipole moment operator $\hat{\mu}$ does not produce any physically meaningful result, as discussed above in this section,

Cafiero and Adamowicz [36–38] resorted to an alternative approach based on the energy of the molecule in an external electric field **F**, given by the Hamiltonian operator:

$$\hat{H}(\mathbf{F}) = \hat{H}_M - \mathbf{F} \cdot \hat{\boldsymbol{\mu}} \tag{6.75}$$

They fitted energy values to a polynomial function of the field [36–38]:

$$E(F_z) = E(0) - \mu_z F_z - \frac{1}{2}\alpha_{zz}F_z^2 - \cdots \tag{6.76}$$

and obtained the dipole moment from the linear term. The dipole moments of HD, HT, LiH, and LiD calculated in this way proved to be very accurate and in remarkable agreement with available experimental values [36, 38]. In particular, the rate of convergence of the theoretical results for the LiH [36] toward the corresponding experimental dipole moment [27] is astonishing. Table 6.1 shows some dipole moments calculated by Cafiero and Adamowicz [36, 38] and other authors [34, 45, 46] as well as the corresponding experimental values [27, 47].

However, if the approximate variational function (6.73) were fully optimized, then it should satisfy the Hellmann–Feynman theorem discussed in Section 6.4 and in that case the only possible result would be

$$\mu_z = \left.\frac{\partial E(F_z)}{\partial F_z}\right|_{F_z=0} = \langle \Psi | \hat{\mu}_z | \Psi \rangle = 0 \tag{6.77}$$

Table 6.1 Dipole moments for some diatomic molecules.

Ref.	μ(D)	Method
LiH		
[36]	5.8816	Non-BO
[46]	5.879	BO
[34]	6.072	Non-BO
[47]	5.8820 (4)	Exp.
[27]	5.882 ± 0.003	Exp.
LiD		
[36]	5.8684	Non-BO
[34]	6.080	Non-BO
[47]	5.8677 (5)	Exp.
[27]	5.868 ± 0.003	Exp.
HD		
[45]	1.54×10^{-3}	Non-BO
[38]	0.831×10^{-3}	Non-BO
HT		
[38]	1.111×10^{-3}	Non-BO
DT		
[38]	2.77×10^{-4}	Non-BO

The only way that Cafiero and Adamowicz [36, 38] could obtain a nonzero dipole moment by means of this approach is that their variational wavefunction did not become spherically symmetric when $F_z \to 0$. In other words, their variational ansatz was not sufficiently accurate at small values of the field strength where it would reveal the permanent molecular dipole moment. In that case, it is not clear how those authors [36, 38] obtained such remarkable agreement between their theoretical results and the experimental values of the dipole moment [27]. A more detailed discussion of this baffling agreement has been published elsewhere [48] (Fernández, F.M., On nonadiabatic calculation of dipole moments, arXiv:0808.3714v4 [math-ph].) A most speculative explanation is that a biased placement of the centers of the floating Gaussians, inspired by the classical view of a polar diatomic molecule [36, 38], somehow mimicked the use of a body-fixed set of axes (see Section 6.10). However, not only did the authors never mention this possibility but, curiously, in a later paper they stated that "This spherical symmetry for the ground-state wavefunction implies several things about molecules that may go against common chemical intuition. First of all, no molecule in the ground state will have a dipole moment, just as atoms do not. Similarly, the molecule will have only one unique polarizability, an isotropic polarizability. The current authors have presented several papers which discuss these phenomena" [49]. However, they did not explain how they had obtained the dipole moments in two of those earlier papers [36, 38]. Besides, in Section 6.10 we show that the statement "no molecule in the ground state will have a dipole moment, just as atoms do not" is false.

It would be interesting to investigate to what extent the variational ansatz (6.73) optimized by Cafiero and Adamowicz [36, 38] satisfies the Hellmann–Feynman theorem that in the present case gives us an exact theoretical relation between the molecule's response to the field and the dipole moment. Exact relationships like those given by the hypervirial and Hellmann–Feynman theorems are useful to determine the accuracy of approximate wavefunctions [13, 14]. The questionable success of the method devised by these authors is obviously based on the floating nature of the Gaussian functions that one can place conveniently to get the desired result [36, 38]. Notice that Bubin et al. [50] resorted to one-center Gaussian functions of the form

$$\phi_k(\mathbf{r}) = r_1^{m_k} \exp[-\mathbf{X} \cdot \mathbf{A}_k \cdot \mathbf{X}^t - \mathbf{Y} \cdot \mathbf{A}_k \cdot \mathbf{Y}^t - \mathbf{Z} \cdot \mathbf{A}_k \cdot \mathbf{Z}^t] \tag{6.78}$$

to determine the charge asymmetry of the rotationless states of the HD molecule. Clearly, this kind of basis functions cannot be placed at will to force an axial symmetry and a nonzero dipole moment. It is for this reason that we mentioned above in this section that it is unfortunate that Tachikawa and Osamura [34] did not show the expression that they used for the calculation of the dipole moment of the ^{m}H^{n}H and ^{m}Li^{n}H molecules.

The nonadiabatic calculations of molecular dipole moments described in this section are bound to fail because they are based on a set of axes parallel to the laboratory one. Therefore, the expectation value of the dipole moment operator vanishes for any molecule in a nondegenerate state. In Section 6.10, we discuss more judicious calculations based on molecule-fixed coordinate systems.

6.10
Molecule-Fixed Coordinate System

It is convenient to discuss the motion of a system of particles in space by means of three sets of axes. One set of axes is fixed somewhere in the laboratory. A second set of axes with its origin fixed on the system center of mass and parallel to the laboratory one. It was discussed in Section 6.2 with the purpose of separating the motion of the entire system as a point particle with mass equal to the total mass of the system. A third set of axes with origin on the center of mass and somehow completely fastened to the system enables us to describe its rotational motion. It is not difficult to define such set of axes for a rigid body, but it is not so obvious when the distance between the system particles change rather arbitrarily. It is clear that we should define any intrinsic molecular property, such as the electric dipole moment, moment of inertia, and so on, with respect to this body-fixed reference frame. If we use the second set of axes, we expect that the expectation value of the dipole moment operator for the molecule in a nondegenerate state vanishes as discussed in Section 6.9. This result is an obvious consequence of the average over the angular degrees of freedom of the whole system. The proper use of the body-fixed set of axes was clearly addressed by Blinder [39, 40] in his studies on the HD molecule, and Sutcliffe [7–9, 51] discussed the more general case of polyatomic molecules. For simplicity, here we restrict ourselves to Blinder's proposal for diatomic molecules [39, 40].

For generality, we first consider a diatomic molecule with the nuclei located at r_1 and r_2 and the electrons at $r_2, r_3, \ldots, r_N$. Following Blinder [39], we use relative coordinates for the nuclei and refer the electron coordinates to the midpoint between the nuclei according to

$$\begin{aligned} \mathbf{r}'_1 &= \frac{1}{M} \sum_i m_i \mathbf{r}_i \\ \mathbf{r}'_2 &= \mathbf{r}_2 - \mathbf{r}_1 \\ \mathbf{r}'_i &= \mathbf{r}_i - \frac{1}{2}(\mathbf{r}_1 + \mathbf{r}_2), \quad i > 2 \end{aligned} \tag{6.79}$$

that is a particular case of Equation 6.2 with

$$\begin{aligned} t_{1j} &= \frac{m_j}{M} \\ t_{2j} &= \delta_{j2} - \delta_{j1} \\ t_{ij} &= \delta_{ij} - \frac{1}{2}(\delta_{j1} + \delta_{j2}), \quad j > 2 \end{aligned} \tag{6.80}$$

The resulting molecular Hamiltonian operator in the coordinate representation is

$$\hat{H}_M = -\frac{\hbar^2}{2m_s} \nabla'^2_2 - \frac{\hbar^2}{2m_e} \sum_{j=3}^N \nabla'^2_j - \frac{\hbar^2}{8m_s} \left(\sum_{j=3}^N \nabla'_j \right)^2 + \frac{\hbar^2}{2m_a} \nabla'_2 \cdot \sum_{j=3}^N \nabla'_j + V \tag{6.81}$$

where m_e is the electronic mass, $m_s = m_1 m_2/(m_1 + m_2)$, and $m_a = m_1 m_2/(m_1 - m_2)$. These coordinates are most convenient to calculate the dipole moment of diatomic molecules of the form $^mA^nA$ because the fourth term on the right-hand side of the internal Hamiltonian (6.81) vanishes when $m_1 = m_2$ and can therefore be treated as a perturbation that converts the symmetric case to an asymmetric one.

To place the molecule-fixed set of axes, we take into account the unit vectors generated by

$$\mathbf{R} = \mathbf{r}_2 - \mathbf{r}_1 = R(\cos\phi \sin\theta\, \mathbf{e}_x + \sin\phi \sin\theta\, \mathbf{e}_y + \cos\theta\, \mathbf{e}_z) \quad (6.82)$$

where $\mathbf{e}_x$, $\mathbf{e}_y$, and $\mathbf{e}_z$ are the space-fixed orthonormal Cartesian vectors (second set of axes). Notice that we renamed the vector $\mathbf{r}_2'$ to match Blinder's notation [39].

We next define the body-fixed orthonormal vectors

$$\mathbf{e}_x' = \frac{\mathbf{e}_\theta}{|\mathbf{e}_\theta|},\ \mathbf{e}_\theta = \frac{\partial \mathbf{R}}{\partial \theta}$$

$$\mathbf{e}_y' = \frac{\mathbf{e}_\phi}{|\mathbf{e}_\phi|},\ \mathbf{e}_\phi = \frac{\partial \mathbf{R}}{\partial \phi} \quad (6.83)$$

$$\mathbf{e}_z' = \mathbf{e}_R,\ \mathbf{e}_R = \frac{\partial \mathbf{R}}{\partial R}$$

and the new particle coordinates with respect to them

$$\mathbf{r}_i' = x_i' \mathbf{e}_x + y_i' \mathbf{e}_y + z_i' \mathbf{e}_z = x_i'' \mathbf{e}_x' + y_i'' \mathbf{e}_y' + z_i'' \mathbf{e}_z' \quad (6.84)$$

The transformation between these two sets of coordinates

$$\begin{pmatrix} x_i'' \\ y_i'' \\ z_i'' \end{pmatrix} = \mathbf{C} \begin{pmatrix} x_i' \\ y_i' \\ z_i' \end{pmatrix} \quad (6.85)$$

is obviously given by the orthogonal matrix

$$\mathbf{C} = \begin{pmatrix} \mathbf{e}_x \cdot \mathbf{e}_x' & \mathbf{e}_y \cdot \mathbf{e}_x' & \mathbf{e}_z \cdot \mathbf{e}_x' \\ \mathbf{e}_x \cdot \mathbf{e}_y' & \mathbf{e}_y \cdot \mathbf{e}_y' & \mathbf{e}_z \cdot \mathbf{e}_y' \\ \mathbf{e}_x \cdot \mathbf{e}_z' & \mathbf{e}_y \cdot \mathbf{e}_z' & \mathbf{e}_z \cdot \mathbf{e}_z' \end{pmatrix} = \begin{pmatrix} \cos\phi\cos\theta & \sin\phi\cos\theta & -\sin\theta \\ -\sin\phi & \cos\phi & 0 \\ \cos\phi\sin\theta & \sin\phi\sin\theta & \cos\theta \end{pmatrix} \quad (6.86)$$

We do not show the explicit form of the Hamiltonian operator in this molecule-fixed reference frame because it is not necessary for the present discussion. Blinder [39] derived it for the HD molecule.

The new nuclear coordinates are

$$\mathbf{R}_1 = \mathbf{r}_1 - \frac{1}{2}(\mathbf{r}_1 + \mathbf{r}_2) = -\frac{\mathbf{R}}{2} = \left(0, 0, -\frac{R}{2}\right)$$

$$\mathbf{R}_2 = \mathbf{r}_2 - \frac{1}{2}(\mathbf{r}_1 + \mathbf{r}_2) = \frac{\mathbf{R}}{2} = \left(0, 0, \frac{R}{2}\right) \quad (6.87)$$

and the electronic ones are simply given by $\mathbf{r}''_i$, $i > 2$. Assuming that both nuclei have identical charges, the classical dipole moment in the body-fixed set of axes is given by a purely electronic contribution

$$\boldsymbol{\mu}'' = -e \sum_{i=3}^{N} \mathbf{r}''_i \qquad (6.88)$$

For the particular case of HD ($N = 4$), the above expressions agree with those developed by Blinder [39], except for the different notation and labeling of nuclear and electronic coordinates. They allow one to calculate the dipole moment as the expectation value of the corresponding operator by means of an eigenstate of the Hamiltonian operator in the molecule-fixed set of axes.

The effect of the inversion operator on the rotation angles is given by $\theta \to \pi - \theta$ and $\phi \to \phi + \pi$. Therefore, the first row of the matrix $\mathbf{C}$ remains unchanged and the other two change sign. For that reason, the inversion operation $(x'_i, y'_i, z'_i) \to (-x'_i, -y'_i, -z'_i)$ in the laboratory-fixed frame results in $(x''_i, y''_i, z''_i) \to (-x''_i, y''_i, z''_i)$ in the body-fixed one, and, consequently, we do not expect that $\langle \hat{\mu}''_z \rangle$ vanishes because of inversion symmetry. However, in the case of identical nuclei $m_1 = m_2$, we cannot have a net dipole moment along the internuclear axis because the additional permutational symmetry leads to $\langle \hat{\mu}''_z \rangle = 0$. To explain the occurrence of a dipole moment in a diatomic molecule of the form $^m A^n A$, we resort to perturbation theory and write $\hat{H}_M = \hat{H}_0 + \lambda \hat{H}'$, where

$$\hat{H}_0 = -\frac{\hbar^2}{2m_s}\nabla'^2_2 - \frac{\hbar^2}{2m_e}(\nabla'^2_3 + \nabla'^2_4) - \frac{\hbar^2}{8m_s}(\nabla'_3 + \nabla'_4)^2 + V \qquad (6.89)$$

$\lambda = m_s/m_a$, and

$$\hat{H}' = \frac{\hbar^2}{2m_s}\nabla'_2 \cdot (\nabla'_3 + \nabla'_4) \qquad (6.90)$$

Notice that $\hat{H}_0$ is the molecular Hamiltonian for identical nuclei, as follows from the fact that $\lambda = 0$ when $m_1 = m_2$.

We can thus expand the eigenfunction in a λ-power series $\Psi = \Psi^{(0)} + \Psi^{(1)}\lambda \ldots$ that leads to a similar expansion for the dipole moment $\mu_z = \mu_z^{(1)}\lambda + \cdots$, where

$$\mu_z^{(1)} = 2\langle \Psi^{(0)} | \hat{\mu}''_z | \Psi^{(1)} \rangle \qquad (6.91)$$

provided that the eigenfunction is chosen to be real. This simple argument shows how the different nuclear masses produce a charge asymmetry and a net dipole moment.

Blinder [39] estimated the dipole moment of HD by means of a rather more complicated perturbation approach and later Kolos and Wolniewicz [45, 52] and Wolniewicz [53, 54] improved Blinder's calculation by means of a variational perturbation method based on Equation 6.91.

It is clear that any rigorous nonadiabatic calculation of the molecular dipole moment should be carried out in the molecule-fixed set of axes. Consequently, the approaches outlined in Section 6.9 are unconvincing (to say the least). Cafiero and Adamowicz [36, 38] must have placed the floating Gaussians in a convenient way to obtain nonzero dipole moments in good agreement with the experimental ones and Tachikawa and Osamura [34] did not explain how they obtained their results. Besides, the latter authors even forgot to remove the motion of the center of mass.

6.11
Perturbation Theory for the Stark Shift

As we outline in Section 6.7, the experimental determination of the molecular dipole moment relies on the validity of the model of a rotating quasi-rigid polar body. The procedure consists of fitting a polynomial function of the field strength to the observed Stark shift lines. In principle, we should derive more rigorous theoretical expressions for those line shifts by means of perturbation theory and the actual quantum mechanical Hamiltonian operator for a molecule in an electric field.

Instead of the rigid rotator model, we should choose the Hamiltonian operator, $\hat{H}_0 = \hat{H}_M$, for the isolated molecule and the perturbation $\hat{H}' = -\mathbf{F} \cdot \hat{\mu}$. If we set $\mathbf{F}$ along the z-axis in the laboratory frame ($\mathbf{F} = F\mathbf{e}_z$), then $\hat{H}' = -F\hat{\mu}_z$. In this way, perturbation theory gives us the well-known quantum mechanical expressions for the Stark shifts

$$E_n = E_n^{(0)} + E_n^{(1)} + E_n^{(2)} + \cdots \tag{6.92}$$

where n is a collection of quantum numbers that completely specifies a given (nonadiabatic) molecular state. For the ground state, we have

$$E_0^{(1)} = -\langle \Psi_0^{(0)} | \hat{\mu}_z | \Psi_0^{(0)} \rangle F = 0$$

$$E_0^{(2)} = F^2 \sum_{m>0} \frac{|\langle \Psi_0^{(0)} | \hat{\mu}_z | \Psi_m^{(0)} \rangle|^2}{E_0^{(0)} - E_m^{(0)}} \tag{6.93}$$

where we clearly appreciate that the field-reduced splitting $\Delta E/F^2$ does not give us what we may call the square of the dipole moment but a kind of energy-weighted average over those molecular states with nonzero matrix element $\langle \Psi_0^{(0)} | \hat{\mu}_z | \Psi_m^{(0)} \rangle$, including the continuum part of the spectrum. This important issue was already discussed by Brieger et al. [55] and Brieger [56] more than 20 years ago in their BO study of the Stark effect of heteronuclear diatomic molecules in $^1\Sigma$ states.

In the molecule-fixed reference frame, we have

$$\mu = \mu'_x \mathbf{e}'_x + \mu'_y \mathbf{e}'_y + \mu'_z \mathbf{e}'_z \tag{6.94}$$

and a similar expression for **F**. Brieger [56] argued that the spherical vector components

$$u'_0 = u'_z$$
$$u'_{\pm 1} = \mp \frac{1}{\sqrt{2}}(u'_x \pm u'_y) \tag{6.95}$$

are more convenient for the calculation of the matrix elements in Equation 6.93. It is not difficult to prove that

$$\mu \cdot \mathbf{F} = \sum_p (-1)^p \mu'_p F'_p \tag{6.96}$$

However, since the laboratory-fixed and molecule-fixed sets are the natural frames for the external field and the molecule, respectively, Brieger [56] chose $\mathbf{F} = F\mathbf{e}_z$ in the former and μ in the latter as in Equation 6.94. Thus, in the simple notation of Section 6.10, we have

$$\mu_z = \mu'_x \mathbf{e}_z \cdot \mathbf{e}'_x + \mu'_y \mathbf{e}_z \cdot \mathbf{e}'_y + \mu'_z \mathbf{e}_z \cdot \mathbf{e}'_z$$
$$= -\sin\theta\, \mu'_x + \cos\theta\, \mu'_z = \frac{\sin\theta}{\sqrt{2}}(\mu'_{+1} - \mu'_{-1}) + \cos\theta\, \mu'_0 \tag{6.97}$$

In this way, Brieger [56] showed that the perturbation $\hat{H}'$ connects the ground electronic state $^1\Sigma$ to excited $^1\Sigma$ and $^1\Pi$ states (in general, those with electronic angular momentum quantum numbers Λ and $\Lambda \pm 1$). Under the BO approximation, Brieger et al. [55] and Brieger [56] separated the Stark shift (6.93) into four contributions: (1) coupling of rotational states within the same $^1\Sigma$ vibronic one, (2) coupling of vibrational–rotational states within the same $^1\Sigma$ electronic state, (3) coupling of vibrational–rotational states between two $^1\Sigma$ electronic ones, and (4) coupling of vibrational–rotational states between the given $^1\Sigma$ and $^1\Pi$ ones. The first three contributions are due to μ_0 and the fourth one due to $\mu_{\pm 1}$ [55, 56].

As far as we know, there is no nonadiabatic calculation of the Stark shift by means of Equation 6.93. A straightforward comparison of BO and non-BO calculations may lead to some difficulties as mentioned by Wolniewicz [54]: "Since the familiar classification of electronic states of diatomic molecules is based on the Born–Oppenheimer approximation, some difficulties arise if one tries to use the standard nomenclature to describe nonadiabatic functions."

6.12
Conclusions

Quantum mechanics is a theory for observables; it gives us a recipe for the calculation and prediction of what can be experimentally measured. When we apply this point of view to molecular properties such as the dipole moment, we face some problems.

First, we should define precisely what we understand for the dipole moment of a molecule and calculate it accordingly. As we said above, the nonadiabatic calculations of dipole moments are scarce and few. Those of Tachikawa and Osamura [34] and Cafiero and Adamowicz [36, 38] are suspicious because they are not based on the molecule-fixed Hamiltonian operator. In our opinion, the only serious attempts at nonadiabatic calculations of dipole moments are those of Blinder [39, 40], Kolos and Wolniewicz [45, 52], and Wolniewicz [53, 54]. All other calculations of dipole moments are based on the BO approximation.

Once we calculate the dipole moment according to such an agreed theoretical definition, we realize that it may not be directly comparable to available experimental data. Brieger [56] judiciously argued that the field-reduced line splittings do not give us the square of the molecular dipole moment, as predicted by the oversimplified rigid rotor model, but a kind of energy-weighted average. Thus, the well-known expression for the Stark shift in $^1\Sigma$ levels dominated by the square of the dipole moment is reasonably accurate only for sufficiently large dipole moments [56].

Cafiero and Adamowicz [36, 38] stated that "Our calculations simulate experiment more closely than any previous calculations." However, they calculated the Stark shift for the ground-state energy whereas spectroscopic experiments provide the Stark shifts for the frequencies. In principle, the method developed by these authors could be adapted to the calculation of the shift of the spectral lines that one can compare directly with experiment. At this point, it is worth mentioning that the Hamiltonian operator with the Coulomb potential plus the interaction with the field [36, 38] is unbounded from below and therefore it exhibits only a continuous spectrum [48]. However, if the field strength is not too large then it is possible to apply the variational method for bound states [36, 38]. Since there has not yet been any sound nonadiabatic calculation of excited states, it is not clear whether it is feasible to obtain reliable nonadiabatic Stark shifts. Even in the most favorable scenario, one would at best be able to compare field-reduced splittings because there does not appear to be a rigorous procedure for extracting dipole moments from them [55, 56].

We appreciate that there is much to be done in this field and we hope that this discussion will contribute to motivate such work.

References

1 Debye, P. (1929) *Polar Molecules*, Dover Publications, Inc., New York.
2 Gordy, W., Smith, W.V., and Trambarulo, R.F. (1953) *Microwave Spectroscopy*, John Wiley & Sons, Inc., New York.
3 Townes, C.H. and Schawlow, A.L. (1955) *Microwave Spectroscopy*, McGraw-Hill, New York.
4 Ramsey, N.F. (1956) *Molecular Beams*, Oxford University Press, London.
5 Szabo, A. and Ostlund, N.S. (1996) *Modern Quantum Chemistry*, Dover Publications, Inc., Mineola, NY.
6 Margenau, H. and Murphy, G.M. (1946) *The Mathematics of Physics and Chemistry*, Van Nostrand, New York.
7 Sutcliffe, B.T. (1992) The concept of molecular structure, in *Theoretical Models of Chemical Bonding. Part 1: Atomic Hypothesis and the Concept of Molecular Structure* (ed. Z.B. Maksic), Springer-Verlag, Berlin, pp. 1–28.
8 Sutcliffe, B.T. (1993) *J. Chem. Soc. Faraday Trans.*, **89** (4), 2321–2335.
9 Sutcliffe, B.T. (1994) The decoupling of nuclear from electronic motions in

molecules, in *Conceptual Trends in Quantum Chemistry* (eds E.S. Kryachkoand J.L. Calais), Kluwer Academic Publishers, Dordrecht, pp. 53–85.

10. Sutcliffe, B.T. and Woolley, R.G. (2005) *Phys. Chem. Chem. Phys.*, **7** (21), 3664–3676.

11. Woolley, R.G. (1991) *J. Mol. Struct. (Theochem)*, **230**, 17–46.

12. Swanton, D.J., Bacskay, G.B., and Hush, N.S. (1986 *J. Chem. Phys.*, **84** (10), 5715–5727.

13. Hirschfelder, J.O. and Meath, W.J. (1967) The nature of intermolecular forces, in *Intermolecular Forces* (ed. J.O. Hirschfelder), John Wiley & Sons, Inc., New York, pp. 107–142.

14. Fernández, F.M. and Castro, E.A. (1987) *Hypervirial Theorems*, Springer, Berlin.

15. Born, M. and Huang, K. (1954) *Dynamical Theory of Crystal Lattices*, Oxford University Press, New York.

16. Slater, J.C. (1927) *Proc. Natl. Acad. Sci.*, **13** (6), 423–430.

17. Woolley, R.G. and Sutcliffe, B.T. (1977) *Chem. Phys. Lett.*, **45** (2), 393–398.

18. Pilar, F.L. (1968) *Elementary Quantum Chemistry*, McGraw-Hill, New York.

19. Kutzelnigg, W. (1997) *Molec. Phys.*, **90** (6), 909–916.

20. Fernández, F.M., Born–Oppenheimer Approximation for a Harmonic Molecule, arXiv: 0810.2210v1 [math-ph].

21. Sutcliffe, B.T. and Woolley, R.G. (2005) *Chem. Phys. Lett.*, **408** (4–6), 445–447.

22. Jackson, J.D. (1990) *Classical Electrodynamics*, John Wiley & Sons, Inc., Singapore.

23. Buckingham, A.D. (1967) Permanent and induced molecular moments and long-range intermolecular forces, in *Intermolecular Forces* (ed. J.O. Hirschfelder) John Wiley & Sons, Inc., New York, pp. 107–142.

24. Bishop, D.M. (1990) *Rev. Mod. Phys.*, **62** (2), 343–374.

25. Buckingham, A.D. (1970) Electric moments of molecules, in *Physical Chemistry: An Advanced Treatise* (ed. D. Henderson), Academic Press, New York, pp. 249–386.

26. Wharton, L., Gold, L.P., and Klemperer, W. (1960) *J. Chem. Phys.*, **33** (4), 1255.

27. Wharton, L., Gold, L.P., and Klemperer, W. (1962) *J. Chem. Phys.*, **37** (9), 2149.

28. Rothstein, E. (1969) *J. Chem. Phys.*, **50** (4), 1899–1900.

29. Muenter, J.S. and Klemperer, W. (1970) *J. Chem. Phys.*, **52** (12), 6033–6037.

30. Larkin, D.M. and Gordy, W. (1963) *J. Chem. Phys.*, **38** (10), 2329–2333.

31. Steiner, P.A. and Gordy, W. (1966) *J. Mol. Spectrosc.*, **21** (1–4), 291–301.

32. Cade, P.E. and Huo, W.M. (1966) *J. Chem. Phys.*, **45** (3), 1063–1065.

33. Matsen, F.A. (1960) *J. Chem. Phys.*, **34** (1), 337–338.

34. Tachikawa, M. and Osamura, Y. (2000) *Theor. Chem. Acc.*, **104** (1), 29–39.

35. Fernández, F.M. *J. Phys. B*, **43** (02), 025101.

36. Cafiero, M. and Adamowicz, L. (2002) *Phys. Rev. Lett.*, **88** (3), 033002.

37. Cafiero, M., Adamowicz, L., Duran, M., and Luis, J.M. (2003) *J. Mol. Struct.*, **633** (2–3), 113–122.

38. Cafiero, M. and Adamowicz, L. (2002) *Phys. Rev. Lett.*, **89** (7), 073001.

39. Blinder, S.M. (1960) *J. Chem. Phys.*, **32** (1), 105–110.

40. Blinder, S.M. (1961) *J. Chem. Phys.*, **35** (3), 974–981.

41. Ford, A.L. and Browne, J.C. (1977) *Phys. Rev. A*, **16** (5), 1992–2001.

42. Thorson, W.R., Choi, J.H., and Knudson, S.K. (1985) *Phys. Rev. A*, **31** (1), 22–33.

43. Thorson, W.R., Choi, J.H., and Knudson, S.K. (1985) *Phys. Rev. A*, **31** (1), 34–42.

44. Bubin, S., Cafiero, M., and Adamowicz, L. (2005) *Adv. Chem. Phys.*, **131**, 377–475.

45. Kolos, W. and Wolniewicz, L. (1966) *J. Chem. Phys.*, **45** (3), 944–946.

46. Papadopoulos, M.G., Willets, A., Handy, N.C., and Underhill, A.E. (1996) *Molec. Phys.*, **88** (4), 1063–1075.

47. Lovas, F.J. and Tiemann, E. (1974) *J. Phys. Chem.*, **3** (3), 609–769.

48. Fernández, F.M. (2009) *J. Chem. Phys.*, **130** (16), 166101-1–166101-3.

49 Cafiero, M. and Adamowicz, L. (2004) *Chem. Phys. Lett.*, **387** (1–3), 136–141.
50 Bubin, S., Leonarski, F., Stanke, M., and Adamowicz, L. (2009) *J. Chem. Phys.*, **130** (12), 124120 (6 pp.).
51 Sutcliffe, B.T. (1999) *Int. J. Quantum Chem.*, **74** (2), 109–121.
52 Kolos, W. and Wolniewicz, L. (1963) *Rev. Mod. Phys.*, **55** (3), 473–483.
53 Wolniewicz, L. (1975) *Can. J. Phys.*, **53** (13), 1207–1214.
54 Wolniewicz, L. (1976) *Can. J. Phys.*, **54** (6), 672–679.
55 Brieger, M., Renn, A., Sodiek, A., and Hese, A. (1983) *Chem. Phys.*, **75** (1), 1–9.
56 Brieger, M. (1984) *Chem. Phys.*, **89** (2), 275–295.

7
The Search for Parity Violation in Chiral Molecules
Peter Schwerdtfeger

7.1
Introduction

It is well known that symmetry through group and representation theory is an indispensable tool for the interpretation of atomic, molecular, and solid-state spectra [1, 2]. Thus, symmetry plays an important role in chemistry and physics [3, 4]. One of the most fundamental laws concerning symmetry in physics is Noether's first theorem [5], which states that any differentiable symmetry of the action of a physical system has a corresponding conservation law. The action of a physical system is obtained from its known Lagrangian function (currently obtained from the so-called *standard model* (SM)) [6], which by the principle of least action should completely describe the system's physical behavior. Beside the kinematic symmetries like space–time transformations, permutation symmetry (leading to the famous Pauli principle for fermions), and gauge symmetry describing the dynamical symmetries of particle fields, we also have charge conjugation (C), time reversal (T), and space inversion (P, parity) symmetry in nature. The conservation of all physical laws under simultaneous charge, space, and time inversion is known as the CPT theorem, or the Schwinger–Lüders–Pauli theorem [7–9]. More exactly, the CPT theorem implies that any Lorentz invariant local quantum field theory with a self-adjoint Hamiltonian must have CPT symmetry. Violation of the CPT theorem has far reaching consequences as it implies violation of Lorentz invariance (sometimes called relativity violation), but not vice versa (!) [10]. So far no violation of CPT symmetry has been found in nature, although experiments are currently in progress to rigorously test this important conservation law [11]. CPT violation could only be explained beyond the SM and, consequently, any positive outcome of such experiments implies new physics.

In order to preserve CPT symmetry, every violation of the combined symmetry of two of its components (such as CP) must have a corresponding violation in the third component (such as T); mathematically, these are completely equivalent. Thus, violation of T symmetry is more often referred to as CP violation. CP violation is most likely responsible for the matter/antimatter asymmetry in our

Computational Spectroscopy: Methods, Experiments and Applications. Edited by Jörg Grunenberg
Copyright © 2010 WILEY-VCH Verlag GmbH & Co. KGaA, Weinheim
ISBN: 978-3-527-32649-5

universe [12], although the SM does not predict enough CP violation to explain the antimatter/matter upper limit ratio of about 10^{-6} in the universe [13]. Despite the search for more than 40 years of CP violating effects, starting from the discovery of CP violation in the K_0 meson decay into 2π (1 in 500 decays) [14] and much later the B_0 meson, no further examples of CP violation have been found [15]. P and T violation results in an electric dipole moment for the neutron, proton, and electron. These effects might be larger than the SM predicts and could, therefore, result in new physics. The current experimental limit for the electron–electric dipole moment (EEDM) from an experiment on the thallium atom is $|d_e| \leq 1.6 \times 10^{-27}$ e cm [16, 17] (the SM through the Kobayashi–Maskawa mechanism predicts $\sim 10^{-41}$ e cm) [16]. For the EEDM in diatomic molecules such as YbF, HI^+, TlF, and PbO, see Refs [18–21].

In 1956, Lee and Yang suggested that parity might not be conserved in nuclear β-decay [22]. The experimental proof of parity nonconservation (PNC, or PV for P violation) came 1 year later from Wu et al. [23] by measuring the electron distribution during the β-decay of the ^{60}Co isotope. The electrons were predominantly emitted with negative helicity with their spin being opposite to the nuclear spin axis, that is, they were left-handed. Further support came from the anti-β-decay of ^{58}Co [24], where the angular distribution of the emitted positrons showed exactly the opposite behavior, that is, they were right-handed. Parity violation is described by the SM (or Weinberg–Glashow–Salam theory) of elementary particle physics utilizing $SU(3) \times SU(2) \times U(1)$ symmetry transformations [25–27]. This model considers three of the four fundamental forces (strong, electromagnetic, and weak) between quarks and leptons through the carriers of the forces, the photon, the $W^{\pm}$ and Z bosons, and the gluons. However, the SM fails to provide a deeper insight into the nature of these interactions. Fundamental questions remain open, such as why fermions and quarks come in groups of three, what determines the mass of an elementary particle (the Higgs boson responsible for the masses of elementary particles in SM theory has not been detected yet), and what determines the values of the fundamental constants in nature. For example, the SM is not capable of correctly explaining the Big-Bang baryogenesis from CP violation, and it is therefore widely assumed that the SM is only a low-energy manifestation of a more complete theory, for example, the so-called theory of everything (TOE) [28].

A stringent test of the SM in the low-energy regime is by measuring PV effects in heavy atoms. It originates mainly from the weak neutral current, that is, the Z_0 boson exchange between electrons and nucleons. The weak interaction causes a mixing between opposite parity states in atoms, for example, it mixes $s_{1/2}$ and $p_{1/2}$ states, so for the hydrogen $2p_{1/2}$ state, for example, we get

$$\Psi = \Psi_{2p_{1/2}} + i\varepsilon \Psi_{2s_{1/2}} \quad (7.1)$$

with ε being of the order of 10^{-11},

$$\varepsilon = \frac{i\langle \Psi_{2s_{1/2}} | H_{PV} | \Psi_{2p_{1/2}} \rangle}{E_{2s_{1/2}} - E_{2p_{1/2}}} \quad (7.2)$$

and H_{PV} being the weak interaction Hamiltonian between the electrons and nucleons. For this Hamiltonian, the main P-odd contributions come from the nuclear spin-independent (SI) and spin-dependent (SD) part of the weak current (P-even contributions are neglected here since they lead only to small corrections in the isotope shift and hyperfine structure) given by

$$H_{PV} = H_{PV}^{SI} + H_{PV}^{SD} = \frac{G_F}{2\sqrt{2}} \sum_n Q_w(n) \sum_i \gamma_i^5 \varrho_n(r_{in})$$
$$+ \frac{G_F}{\sqrt{2}} \sum_n (\varkappa_{NC}(n) + \varkappa_{HF}(n) + \varkappa_A(n)) \sum_i (\boldsymbol{\alpha}_i \cdot \mathbf{I}_n) \varrho_n(r_{in})$$
(7.3)

with the Fermi coupling constant $G_F/(\hbar c)^3 = 1.16637(1) \times 10^{-5}$ GeV^{-2}, which gives a value of $G_F = 2.22250(2) \times 10^{-14}$ in atomic units (a.u.). The summation is taken over all electrons i and nuclei $n \cdot \varrho$ is the normalized nuclear charge density, $\mathbf{I}_n$ is the spin of nucleus n, and the γ^5 Dirac pseudoscalar (chirality operator) is defined as

$$\gamma^5 = \begin{pmatrix} 0_2 & 1_2 \\ 1_2 & 0_2 \end{pmatrix} \quad \boldsymbol{\alpha} = \begin{pmatrix} 0_2 & \boldsymbol{\sigma} \\ \boldsymbol{\sigma} & 0_2 \end{pmatrix} \quad \beta = \begin{pmatrix} 1_2 & 0_2 \\ 0_2 & -1_2 \end{pmatrix}$$
(7.4)

with 1_2 and 0_2 being the 2×2 identity and zero matrix, respectively. Here the Dirac matrices $\boldsymbol{\alpha}$ and β are in the standard (Dirac–Pauli) representation, and $\boldsymbol{\alpha}$ contains the 2×2 Pauli spin matrices $\boldsymbol{\sigma}$. The weak charge Q_W for a specific isotope is defined as

$$Q_W = -N + Z(1 - 4\sin^2\theta_W) \approx -N \tag{7.5}$$

where N and Z are the numbers of neutrons and protons in the nucleus, respectively. The Weinberg mixing angle θ_W is given as $\sin^2\theta_W = 0.2397(13)$ [29]. Radiative corrections change the weak charge and one approximately obtains [36] (Johnson, W. (2007), personal communication)

$$Q_W = -0.9857 N + 0.0675 Z \tag{7.6}$$

$\varkappa_{NC}$, $\varkappa_{HF}$, and $\varkappa_A$ are the nuclear-dependent factors (nuclear spin-dependent (NC), hyperfine (HF), and anapole moment (A) components), where the nuclear anapole moment [30] is the largest among all three contributions [31]. The anapole arises due to the charged weak current inside the nucleus and interacts with the electrons by the usual magnetic interaction (PV hyperfine interaction). Typically, one has $|Q_W| \sim 100|\varkappa|$ with $\varkappa = \varkappa_{NC} + \varkappa_{HF} + \varkappa_A$. The $\varkappa_{NC}$ parameters are determined from nuclear structure calculations [32] and we approximately have $\varkappa_A \sim 5\varkappa_{NC}$ and $\varkappa_{HF} \sim \varkappa_{NC}/2$. For example, theoretical estimates for ^{133}Cs gave $\varkappa_A = 0.063 - 0.084$ [33] in good agreement with experiment ($(\varkappa_A = 0.09(2))$ [34, 35]. More details can be found in Ref. [28].

It is obvious that H_{PV} does not commute with the parity operator, that is,

$$[H_{PV}, P] \neq 0 \tag{7.7}$$

G_F and consequently ε are rather small, and early predictions of weak neutral currents in atoms showed rather small effects [30]. However, in 1974, Bouchiat and

Bouchiat predicted a Z^3 scaling with the nuclear charge Z for parity violation effects in atoms and proposed optical rotation experiments for highly forbidden transitions in atomic spectra of heavy elements [37–41]. The first observation of atomic PV came in 1978 from Barkov and Zolotorevin at Novosibirsk from measurements on the ^{209}Bi isotope [42–44]. The most accurate optical rotation experiments are currently from groups at Seattle (^{205}Tl, 1283 nm transition and ^{207}Pb, 1279 nm transition) [45, 46] and Oxford (^{209}Bi, 876 nm transition) [47]. Here the laser frequency is tuned to the vicinity of a magnetic dipole resonance. The interference between the highly forbidden PV-induced E1 amplitude and, in this case, the dominant M1 amplitude, leads to optical activity. Further successful experiments came from Berkeley using a Stark interference technique for ^{205}Tl [48]. The most accurate experiments to date come from the Colorado group in 1997 using a Stark interference technique for ^{133}Cs [49, 50]. These results are in unprecedented agreement with the SM to an uncertainty of 0.35% [51–53]. Even more interestingly, these experiments are so precise that nuclear spin-dependent effects arising mainly from the nuclear anapole moment ($\varkappa_A$ in Equation 7.3) have been confirmed (first and only), thus providing new insight into neutral current weak interactions in the hadron sector, which otherwise is difficult to obtain. From Equation 7.6 we obtain for the weak charge of ^{133}Cs ($Z=55$, $N=78$), $Q_W = -73.172$. This agrees with the experiment and theory ($Q_W = -73.16(29)_{exp}(20)_{theor}$ [54, 55]) to 1σ accuracy. More recently, Tsigutkin et al. found atomic PV in the $6s^2$ ($^1S_0$) → $5d^16s^1$ ($^3D_1$) 408 nm forbidden transition in ytterbium to be two orders of magnitude larger than in cesium [56]. For reviews on atomic PV, see Refs [41, 57–59].

The SM also predicts that PV originating from the weak neutral current between electrons and nucleons introduces a tiny energy difference between left (S) and right (R) symmetric (mirror image) molecules (enantiomers) of a chiral compound [60–63]. Consider the positive $|+\rangle$ and negative $|-\rangle$ parity eigenstates of a parity-conserving molecular Hamiltonian H_{PC}, which are usually split by tunneling in a double-minimum potential between the two enantiomers,

$$P|+\rangle = +|+\rangle, \quad P|-\rangle = -|-\rangle, \quad H_{PC}|+\rangle = E_+|+\rangle, \quad H_{PC}|-\rangle = E_-|-\rangle \tag{7.8}$$

with the tunneling splitting $\Delta E_T = E_- - E_+ > 0$. For almost all chiral molecules, the tunneling splitting can be completely neglected compared to the parity violation energy shift, $\Delta E_T \ll |E_{PV}|$. In this case, we form the left–right superpositions of the parity states,

$$|S\rangle = (|+\rangle + |-\rangle)/\sqrt{2}, \quad |R\rangle = (|+\rangle - |-\rangle)/\sqrt{2}, \quad |S\rangle = P|R\rangle, \quad |R\rangle = P|S\rangle \tag{7.9}$$

In the standard Dirac picture, the parity operator is given as [64]

$$P = \gamma^0 = \beta \tag{7.10}$$

and we have for the Dirac matrices the anticommutation rules,

$$\{P, \gamma^\mu\} = 0 \quad (\mu = 1, 2, 3), \quad \{P, \gamma^5\} = 0 \tag{7.11}$$

with

$$\gamma^5 = i\gamma^0\gamma^1\gamma^2\gamma^3 \quad \text{and} \quad \gamma^\mu = (\beta, \beta\alpha) \quad (\mu = 0, 1, 2, 3) \tag{7.12}$$

From the parity conserving and nonconserving (Equation 7.3) Hamiltonians, we therefore get

$$[P, H_{PC}] = 0, \qquad \{P, H_{PV}\} = 0 \tag{7.13}$$

or

$$P^{-1}H_{PC}P = H_{PC}, \qquad P^{-1}H_{PV}P = -H_{PV} \tag{7.14}$$

and thus we obtain for the left–right expectation values

$$E^S_{PV} = \langle S|H_{PV}|S\rangle = \langle R|P^{-1}H_{PV}P|R\rangle = -\langle R|H_{PV}|R\rangle = -E^R_{PV} \tag{7.15}$$

and we have used $P^{-1} = P^\dagger = P$. E^S_{PV} is the PV energy shift for the S-enantiomer. The PV energy difference between both enantiomers is therefore

$$\Delta E^{SR}_{PV} = E^S_{PV} - E^R_{PV} = 2E^S_{PV} \tag{7.16}$$

Equation 7.16 of course includes the unlikely case that $\Delta E^{SR}_{PV} = 0$ for a chiral molecule in its equilibrium geometry. Therefore, strictly speaking, nonzero PV transforms enantiomers into diastereomers. PV in chiral molecules led to some speculation that it is the cause for biomolecular homochirality in nature, the fact that life on earth is completely dominated by left-handed amino acids and right-handed sugars [60, 61] (see Refs [65–70] for critical reviews on this highly debatable hypothesis). Despite many claims in the past, PV effects in chiral molecules have never been unambiguously observed [71]. The PV energy differences between enantiomers are currently estimated to be in the range of below 1 Hz for experimentally accessible chiral molecules [71]. Moreover, precise calculations for PV energy shifts are rather difficult at the molecular level and, therefore, offer no competition for testing the SM unlike the atomic case. We mention, however, a recent study by DeMille et al. who suggested to measure the nuclear spin-dependent PV through a Stark interference technique to determine the mixing between opposite–parity rotational/hyperfine levels of ground-state diatomic molecules such as BaF [72]. Nevertheless, the search for thermodynamically stable chiral compounds with large parity violation effects, together with the development of ultrahigh-resolution spectroscopic techniques, constitutes one of the most exciting and challenging areas in molecular physics, which will be discussed in the following sections. There are a number of excellent review articles already available on both experimental and theoretical aspects of parity violation in chiral molecules [71, 73–78].

7.2
Experimental Attempts

Letokov predicted in 1975 [62] that PV leads to energy differences in molecular spectra of chiral molecules. In 1976, Kompanets et al. suggested the use of narrow saturated absorption resonances in the vibrational–rotational spectrum of CHFClBr induced in a continuous-wave CO_2 laser field [63]. It was speculated that a resolution of $\Delta\nu_{PV}/\nu = 10^{-15}$ can be achieved ($\Delta E_{PV} = h\Delta\nu_{PV}$). There have been many more suggestions of how to measure P-odd effects in chiral molecules since then, but only few experiments have been carried out so far that stand on safe scientific ground. The first serious attempt to find differences in transition frequencies was carried out by Arimondo and coworkers in 1977 [79] using D- and L-camphor and a 9.22 μm R(28) CO_2 laser line, which conveniently lies in the C–C*–CO bending mode range of camphor (C* denotes a chiral carbon in the camphor molecule). They concluded that for this compound, PV effects must be smaller than 300 kHz, which gave the upper limit $\Delta\nu_{PV}/\nu < 10^{-8}$. Almost 30 years later, theoretical work carried out in Lazzeretti's group (Modena) and our group (Auckland) demonstrated that for camphor, the PV shift is more than 10 orders of magnitude smaller, that is, $\Delta\nu_{PV}/\nu \approx 10^{-19}$ [80, 81]. Because of the weak charge and in general the Z scaling behavior for PV proposed by Bouchiat and Bouchiat for atoms [37–41] and by Zel'dovich et al. or later Hegstrom and Wiesenfeld for chiral molecules [82–84], it is clear that one has to choose chiral molecules containing heavy elements. In fact, it was shown by us in 1999 that for H_2X_2 (X=O, S, Se, Te, and Po, torsion angle of 45°) at the Dirac-Coulomb–Hartree–Fock (DC-HF) level of theory, the PV energy shift E_{PV}^{SR} scales approximately like Z^6 [85]. Another important aspect to consider is the single-center theorem of Hegstrom et al. [84], which implies that it is desirable to involve two or more neighboring heavy elements in the chiral compound, preferably with PV contributions of the same sign such that they amplify. Next, chiral compounds should consist of as few atoms as possible leading to a more favorable partition function. Moreover, the compound must be reasonably stable, volatile, and synthesizable, and an enantioselective synthesis or enantiomeric enrichment is highly desirable for independent measurements. It is now evident that finding suitable chiral molecules for PV measurements is less than trivial. In the following sections, we review some attempts to measure P-odd frequency shifts in chiral molecules. We mention that one very promising future alternative experimental method is to trap molecules at ultracold temperatures (in the mK range or below) [86], for example, by pulsed electric fields [87, 88], and subsequently perform ultrahigh-resolution spectroscopic measurements of vibration–rotation or electronic transitions. Tunable lasers for high-resolution measurements are already available in the 1–20 μm range, which can reach resolutions of 1 Hz or below [76, 89–91]. Cold molecules are ideally suited for high-resolution spectroscopy [92].

7.2.1
Vibration–Rotation Spectroscopy

PV effects in vibrational transitions are usually an order of magnitude or more smaller than the direct PV shift in the total electronic energy. Perturbation theory shows that enhancement effects in the PV contribution are expected if the first and second derivatives of the PV energy shift with respect to the normal coordinate of the vibrational mode are of the same sign. Previous attempts by the Paris group focused on Letokov's suggestion of using CHFClBr as the chiral target molecule. It took until 1989 when finally CHFClBr was synthesized in high optical purity [93], and the absolute configuration was determined by Collet and coworkers to be S(+) and R (−) [94, 95]. In 1999, the first high-resolution PV experiment has been carried out by Chardonnet's group in Paris [74, 89]. In this experiment, a single CO_2 laser feeds an electro-optic modulator generating tunable frequency-stabilized side bands. The stabilized laser directly feeds two 3 m long Fabry–Perot cavities containing the separated enantiomeric compounds in high optical purity. The laser was locked on one hyperfine component of the ν_4 C–F fundamental frequency in $CHF^{37}Cl^{81}Br$ with a peak-to-peak line width of 90 kHz. Five hundred and eighty measurements were carried out over 10 days resulting in a negative outcome, that is, $\Delta\nu_{PV}^{R(-)/S(+)} = 9.4$ Hz with a statistical and systematic uncertainty of 5.1 and 12.7 Hz, respectively, and $\Delta\nu_{PV}/\nu < 1.6 \times 10^{-13}$ [74, 89]. These experiments were repeated in 2002 with samples of higher enantiomeric excess and a slightly modified setup with a line width of the signal of 60 kHz, which gave $\Delta\nu_{PV}^{R(-)/S(+)} = -4.2 \pm 0.6 \pm 1.6$ Hz [96] corresponding to a frequency ratio of $\Delta\nu_{PV}/\nu < 5 \times 10^{-14}$. Current theoretical estimates provide $\Delta\nu_{PV}/\nu = 8 \times 10^{-17}$ [97], hence, the resolution needs to improve by three orders of magnitude for this compound. The Paris group is therefore moving to a molecular beam experiment using two-photon Ramsey-fringe spectroscopy, which will narrow the line widths and improve the sensitivity by one or two orders of magnitude [96]. Moreover, larger PV shifts in chiral molecules containing heavier elements, either as the chirality center or as ligands, are required for this experiment, which will be discussed in the theoretical section. For example, CHFClI has been synthesized and characterized recently by Cuisset et al. [98, 99].

7.2.2
Mössbauer Spectroscopy

In atomic Mössbauer transitions, PV manifests itself in the emission of circular polarized γ-radiation. Inzhechik et al. claimed to have seen circular polarization of $(0.9 \pm 0.1) \times 10^{-3}$ for the 23.9 keV transition in ^{119}Sn and $(0.58 \pm 0.12) \times 10^{-3}$ for the 14.4 keV transition in ^{57}Fe [100–102]. This, however, was disputed later by Shuskov and Telitsin who obtained five orders of magnitude smaller values from the theory [103]. This clearly underlines the importance of an independent theoretical confirmation of experimental PV observations. No other atomic PV Mössbauer experiments have been carried out so far.

On the molecular side, Khriplovich suggested Mössbauer experiments on heavy nuclei such as ^{181}Ta [104]. In 1994, Mössbauer measurements on the L- and D-tris(1,2-ethanediamine) iridium(III) complex provided an upper limit for the PV energy difference of 4×10^{-9} eV [105]. More recently, Compton and coworkers at Tennessee reported Mössbauer spectra for the enantiomers of $Fe(phen)_3Sb_2(C_4H_2O_6)_2 \times 8 H_2O$ [106]. Four independent experiments showed a small (but reproducible) energy shift in the ^{57}Fe Mössbauer spectra of 1.9×10^{-10} eV. Besides the fact that ^{57}Fe is a relatively light element and rather small PV effects are therefore expected, a theoretical analysis concluded that for closed-shell systems, PV energy shifts in Mössbauer spectra should be very small due to Kramers symmetry [58, 106]. Here PV contributions between different Kramers pairs of orbitals cancel out. This is easily shown, that is, consider the Kramers operator K for the Kramers degenerate orbitals (ϕ^K, ϕ),

$$K = -\gamma^1\gamma^3 C^* = -i\begin{pmatrix} \sigma_2 & 0_2 \\ 0_2 & \sigma_2 \end{pmatrix} C^*, \quad \phi^K = K\phi \qquad (7.17)$$

where C^* denotes the complex conjugation. K commutes with the parity-conserving Dirac operator H_{PC},

$$KK^\dagger = 1, \quad [K, H_{PC}] = 0 \qquad (7.18)$$

and we have

$$\langle \phi^K | H_{PV}^{SI} | \phi^K \rangle = \langle \phi | H_{PV}^{SI} | \phi \rangle, \quad \langle \phi^K | H_{PV}^{SD} | \phi^K \rangle = -\langle \phi | H_{PV}^{SD} | \phi \rangle \qquad (7.19)$$

The first term comes from the nuclear spin-independent PV Hamiltonian in Equation 7.3. Here a change in the nuclear density is required to account for a PV shift, which would lead to extremely small PV level splitting between the enantiomers. More important, the nuclear spin-dependent contributions cancel out unless strong magnetic fields are applied to lift Kramers' degeneracy. Moreover, as these are solid-state measurements, imperfections in the crystal and impurities can lead to spurious energy shifts between the enantiomers. Hence, the observed level splitting is most likely not due to PV.

7.2.3
NMR Spectroscopy

As mentioned before, most of the effect in the nuclear spin-dependent PV Hamiltonian comes from the nuclear anapole moment. In 1986, Barra, Robert, and Wiesenfeld discussed the possibility of measuring PV in NMR observables such as the nuclear magnetic shielding, the nuclear spin–spin coupling, and the spin–rotation coupling [107–111]. From relativistic extended Hückel calculations, they estimated that PV shifts in the nuclear magnetic shielding are in the mHz range for chiral molecules containing heavy elements [107, 108]. Ultrahigh-resolution high-field NMR is indeed capable to achieve mHz resolution. However, there are certain restrictions. One has to avoid line broadening from nuclear quadrupole coupling, restricting the nuclei to nuclear spin $I = 1/2$. There are not too many promising heavy element isotopes available satisfying this condition, and in addition forming stable chiral compounds, and possible choices are ^{117}Sn, ^{119}Sn, ^{187}Os, and ^{183}W as chiral centers [112]. Another possibility is to attach a heavy spin 1/2 nucleus to a chiral center. The gas-phase NMR will be the best method, as the liquid-phase leads again to line broadening due to dynamic effects, and the solid-state may introduce spurious effects due to imperfections in the sample. Nevertheless, Mukhamedjanov et al.

suggested recently to use insulating garnets doped by rare earth ions to measure PV shifts in the NMR frequency, which arises from the lattice crystal field and the nuclear anapole moment of the rare earth nucleus such as ^{169}Tm or ^{141}Pr [113, 114]. Moreover, for the nuclear spin-dependent term in Equation 7.3 the Z scaling might not be as favorable as for the spin-independent case [115, 116]. A recent article by Ledbetter et al. [117] reported on optical detection of NMR J-spectra at zero magnetic field. The experimental setup was a microfabricated optical atomic magnetometer providing high sensitivity. They obtained 0.1 Hz line widths and measured scalar-coupling parameters of the form $(JI_1 I_2)$ between two nuclei with 4 mHz statistical uncertainty. This high-precision J-spectroscopy technique can provide a new technique for measuring PV in nuclear spin coupling. We mention that recently Fujiki reported on the observation of PV effects in the NMR spectra of the ^{29}Si signal (as well as in the UV and CD spectrum) of symmetrically substituted helical polysilylene, poly [bis(S)-3,7-dimethyloctylsilylene] and its enantiomer, in $CDCl_3$ solution [118]. However, such compounds are not suitable for high-precision measurements. As there are currently no other reports on measurements of PV effects in NMR properties, the high-resolution NMR remains a challenge to future experimental and theoretical investigations.

7.2.4
Electronic Spectroscopy

In principle, electronic spectroscopy would be the most promising way to observe PV effects in chiral molecules, as PV shifts in the total electronic energy are largest, and are predicted to be in the range of a few 100 Hz for heavy element containing chiral compounds [119]. However, narrow line widths are only obtained in symmetry- or spin-forbidden electronic transitions with lifetimes of the excited electronic state of 1 ms or more. Enhancements in PV energy differences are expected if the PV shifts of the ground and electronic excited states are of opposite sign. There has been little to no work published on PV energy shifts in electronic transitions, and to find suitable candidates remains a challenge. We mention that in 1986, Quack and coworkers proposed an experiment making use of an achiral excited intermediate state of well-defined parity [120, 121], based on an earlier study by Harris and Stodolsky on tunneling in the presence of weak interactions [122]. They suggested to investigate molecules like Cl_2S_2 [123]. This allows one to isolate the PV energy shift in the ground state as a spectroscopic combination difference. A similar scheme has been introduced by Berger, but suggesting an achiral electronic ground state and a chiral excited state, which, for example, is the case for the molecule HFCO [124].

7.2.5
Other Experiments

Very recently, Bargueño et al. proposed to measure PV energy shifts between enantiomers of chiral molecules by measuring changes induced by an external chiral field such as circularly polarized light in the time-averaged relative optical activity of a molecular sample prepared with chiral purity at the initial time [125]. This proposal follows earlier work by Harris and coworkers [122, 126–131]. Here the tunneling splitting must be of the same order as the PV energy shift, which substantially narrows down the choice for suitable chiral candidates. If such an experiment is feasible, remains to be seen. In the following, we briefly report on various experiments carried out in the past to detect PV effects in chiral molecules, some of them with claims of positive outcomes, which are highly questionable and to our opinion, difficult to justify on theoretical grounds.

In a series of papers, Wang et al. [133–136] claimed evidence of a reversible second-order Salam phase transition [132] at 247 K caused by PV in a D-alanine crystal using a variety of spectroscopic techniques. However, the Salam hypothesis of a PV-induced Bose–Einstein condensation in a crystal is to our opinion seriously flawed (to phrase it mildly), as it completely neglects the huge barrier of inversion between two enantiomers in an amino acid crystal [137, 138]. Moreover, some of the experiments have been carefully repeated in Compton's group in Tennessee with a (not surprisingly) negative outcome [138]. More recently, Wilson and coworkers carried out neutron diffraction experiments for L- and D-alanine at different temperatures, with no evidence at all for a Salam-type phase transition [139]. Bolik et al. presented a Raman spectra of D- and L-enantiomers of an RNA duplex in water solution and found differences in the intensity of the vibrational modes [140]. They concluded that their work is the first to provide experimental evidence at the macromolecular level that the D-enantiomer of RNA has an energy level scheme different from that of its counterpart in the L-configuration. Again, such experimental findings are highly debatable and to our opinion not linked to PV effects.

There have also been various claims of PV-induced preferred enantiomeric crystallization by the Keszthelyi group [105, 141, 142]. A statistical analysis from CD spectra of tris(1,2-ethanediamine)cobalt(III) and tris(1,2-ethanediamine)iridium (III) crystals revealed an asymmetric distribution of D- and L-crystals, which the authors see as a proof of PV forces acting in chiral molecules [141]. However, crystallization is a complex process and such experiments at the macroscopic scale will barely prove any PV effects in molecules, as a theoretical treatment is out of reach to current computational methods. It comes therefore at no surprise that such experiments are under constant criticisms [69].

In a more recent paper by Shinitzky et al., unexpected differences in the solubilities of D- and L-tyrosine in water were reported [143]. They speculated that minute energy differences between D- and L-tyrosine originating from PV, or other nonconservative chiral discriminatory rules, could account for their observations. Their results were again challenged by Lahav et al. pointing out the crucial role of impurities in the samples [144]. An interesting note was added to Shinitzky's paper in the journal *Chirality*. In a commentary, the journal's editor A. W. Schwartz explains why he accepted Shinitsky's manuscript despite one referee turning it down [145]. He asks the question, "Is this good science or wishful thinking? Will this paper turn out to be another example of 'polywater' or 'cold fusion' or is it a fundamental discovery of great importance to theories of the origin of life?"

7.3 Theoretical Predictions

The first crude PV calculations for chiral molecules were performed by Hegstrom et al. in 1980 on twisted ethylene and A-nor-2-thiacholestane [84]. The calculated PV shifts were extremely small, and 2×10^{-20} a.u. for twisted ethylene. However, *ab initio* calculations for PV in chiral molecules really started with Mason and Tranter's work 25 years ago, who investigated a variety of molecules, mainly amino acids and sugars, to address the problem of the origin of biomolecular homochirality [146–150]. They used a nonrelativistic approach, that is, a unitary transformation from the Dirac to the Schrödinger picture, leading to the so-called Bouchiat–Hamiltonian (in atomic units) [77],

$$\tilde{H}_{PV}^{NR} = \tilde{H}_{PV}^{SI} + \tilde{H}_{PV}^{SD} = \frac{G_F}{4c\sqrt{2}} \sum_n Q_W(n) \sum_i \{\sigma_i p_i, \varrho_n(r_{in})\}$$
$$+ \frac{G_F}{2c\sqrt{2}} \sum_n \varkappa(n) \sum_i (\{p_i I_n, \varrho_n(r_{in})\}$$
$$- i(\sigma_i \times I_n) \cdot [p_i, \varrho_n(r_{in})]) + \cdots \quad (7.20)$$

For most nonrelativistic calculations, it is sufficient to apply a point-charge nucleus $(\varrho_n(r_{in}) = \delta^{(3)}(r_{in}))$ [85]. Spin–orbit coupling must be included in these nonrelativistic PV calculations, otherwise the matrix elements over the Hamiltonian (7.20) would be exactly zero [84]. This can be done within a relativistic two-component framework, or by using linear response theory. For the latter method, we have [151]

$$E_{PV} = \langle\langle \tilde{H}_{PV}^{NR}; H_{SO} \rangle\rangle_0 \quad (7.21)$$

where H_{SO} usually contains both the spin–orbit and spin–other–orbit term. Berger pointed out that for the lighter elements, the Breit interaction giving rise, for example, to the spin–other–orbit term cannot be neglected anymore [152]. Laerdahl and Schwerdtfeger demonstrated that while the largest PV contributions come from the valence space where the chiral field is largest, the main amount comes from the inner core region as one expects from the form of the PV operator (7.20) [85]. This implies that most standard contracted Gaussian basis sets used in standard molecular quantum calculations cannot be applied anymore, as the wavefunction close to the nucleus needs to be more flexible. This is similar to calculations of other typical core properties like electric field gradients [153]. Indeed, it was first pointed out by Kikuchi and Wang in 1987 that popular contracted basis sets used at that time, like STO-3G or 6-31G, are not very useful in PV calculations [154]. A careful basis set study by Kikuchi and Wang for glycine and alanine showed that the PV energy shift is very sensitive to the various basis sets applied [155]. This was later reinvestigated by Quack and coworkers, who pointed out that in the early work by Hegstrom et al. [84] as well as Mason and coworkers [147–149], PV effects were underestimated by about one order of magnitude [156]. It is now clear that great care needs to be applied for the choice of the right basis set to accurately determine PV energy shifts. Further, test calculations on H_2O_2 and H_2S_2 indicated that electron correlation effects are rather small [157].

Table 7.1 PV energy shift E_{PV} for H_2X_2 (X=O, S, Se, Te, Po) at a dihedral angle of 45° (in atomic units).

Method	H_2O_2	H_2S_2	H_2Se_2	H_2Te_2	H_2Po_2
DC–HF	-7.06×10^{-19}	-2.24×10^{-17}	-2.45×10^{-15}	-3.67×10^{-14}	-1.55×10^{-12}
ZORA-HF	-7.93×10^{-19}	-2.35×10^{-17}	-2.50×10^{-15}	-3.71×10^{-14}	-1.55×10^{-12}
DC-MP2	-5.67×10^{-19}	-2.12×10^{-17}	-2.28×10^{-15}	-3.81×10^{-14}	—
DC-CCSD(T)	-6.12×10^{-19}	-2.11×10^{-17}	—	—	—
ZORA-LDA	-6.91×10^{-19}	-2.78×10^{-17}	-3.16×10^{-15}	-4.60×10^{-14}	-1.39×10^{-12}
ZORA-Xα	-6.84×10^{-19}	-2.74×10^{-17}	-3.14×10^{-15}	-4.59×10^{-14}	-1.45×10^{-12}
ZORA-BLYP	-6.54×10^{-19}	-2.75×10^{-17}	-3.06×10^{-15}	-4.45×10^{-14}	-1.39×10^{-12}
ZORA-B3LYP	-6.93×10^{-19}	-2.69×10^{-17}	-3.00×10^{-15}	-4.39×10^{-14}	-1.44×10^{-12}

The number of neutrons chosen for the weak charge is $N = 8$ (O), 16 (S), 46 (Se), 78 (Te), and 125 (Po). DC–HF, MP2, and coupled-cluster results from Refs [85, 157, 158] and ZORA results from Refs [159, 160].

Table 7.1 collects results from E_{PV} calculations on H_2X_2 (X=O to Po) from our research group (Dirac–Coulomb results) [85, 157], from DC-MP2 calculations of van Stralen et al. [158], and from zero-order regular approximation (ZORA) density functional calculations of Berger and van Wüllen [159, 160]. The H-X-X-H torsion angle is kept at 45°, where PV effects are largest (around 90° the PV energy shift goes through zero and changes sign) [85]. The data in this table show that ZORA-HF results are in good agreement with the earlier DC–HF calculations, and larger deviations are only seen for the lighter elements as the DC–HF calculations did not incorporate any Breit interactions. Nevertheless, both the HF and density functional theory (DFT) results are relatively close to the CCSD(T) results. However, this is generally not the case for other molecules as we shall see. Interestingly, a fit to a $f(Z) = aZ^n$ function gives basically the same Z scaling law of $n = 6.17$–6.18 for all levels of theory applied, and is therefore above the Z^5 scaling as suggested by Zel'dovich [82]. A number of research groups in the past studied PV on the "test" molecules H_2O_2 or H_2S_2 or similar systems such as H_2SO [151, 156, 161–166].

Concerning the nuclear spin-dependent operator in Equation 7.3 or 7.20, the work by Laubender and Berger [167, 168] and Soncini et al. [115] on PV contributions to NMR chemical shifts for H_2O_2, H_2S_2, and H_2Se_2 suggested a much lower $Z^{2.7}$ scaling, hence, at the time NMR did not seem to be the best experimental method to detect PV effects in chiral molecules. However, Bast et al. claimed a much larger relativistic enhancement factor for H_2Te_2 and H_2Po_2 at the DC–HF level of theory, bringing the scaling even up to $Z^{7.1}$ [112]. ZORA-DFT calculation by Nahrwold and Berger later showed only a $Z^{3.9}$ scaling indicating an instability in the Kramers restricted DC–HF wavefunction [116], or in other words, electron correlation effects cannot be neglected anymore in such calculations, and DFT seems to perform well. A comparison of four-component HF, MP2, and DFT results on R-CHFClBr [169] showed indeed the importance of electron correlation as large varying PV energy shifts for the total electronic energy (even with changing sign) were obtained between different approximations used, that is, one obtains for E_{PV} (in 10^{-18} a.u.) 5.53 (HF), 2.54 (MP2), 1.54 (B3LYP), -0.63 (BLYP), -0.66 (PW86), and -2.01 (LDA). Recent coupled-cluster calculations in our group gave 1.95×10^{-18} a.u., hence, a value somewhere between the MP2 and B3LYP result (Thierfelder, C., Rauhut, G., Schwerdtfeger, P. (2010) *Phys. Rev. A*, **81**, 032513.). As a consequence, for the determination of accurate PV energy shifts, one requires a higher level of electron correlation treatment. This will become especially important for the correct description of PV in electronic transitions. Concerning NMR shielding and spin–spin coupling constants, Manninen and coworkers performed HF and DFT calculations for CHFClBr and CHFBrI [172]. For the spin–spin coupling constants, PV effects were calculated in the nHz region or below, suggesting that heavier elements are required to detect such effects. Laubender and Berger found that PV in NMR shielding constants for the heavy elements in H_2O_2, H_2S_2, and H_2Se_2 computed at the CCSD level of theory deviate from their uncorrelated counterparts typically by approximately 20%, with varying more pronounced corrections at the equilibrium structures, while in 2-fluorooxirane, electron correlation changes the shielding constants by almost 100% [168]. For the same H_2X_2 molecules at their equilibrium structures, Weijo *et al.* found that coupled-cluster and DFT results for PV contributions differ significantly from the HF data, that is, for the ^{77}Se PV shift in the nuclear shielding constant for H_2Se_2, they obtain (in 10^{-10} ppm) 256.1 (HF), 139.9 (CCSD), 116.8 (B3LYP), 60.5 (BP86), 128.6 (PBE0), and 68.6 (PBE) [173]. Similarly, for the PV contribution to the Se–Se spin–spin coupling, they calculate (in nHz) -6.606 (HF), -3.181 (CCSD), -3.038 (B3LYP), -1.943 (BP86), -3.795 (PBE0), and -2.081 (PBE). They also showed that basis set effects are very pronounced, especially at the correlated level of theory, and that the choice of the nuclear charge distribution model (point charge or extended Gaussian) has a significant impact on the PV contribution in the spin–spin coupling constant [173]. Vibrational effects in parity-violating contributions to the isotropic nuclear magnetic resonance chemical shift were recently investigated for CHFClBr by Weijo *et al.* [174]. They found that zero-point vibrational corrections are less than 10% with respect to the PV contributions calculated at the equilibrium geometry, but become more important for vibrationally excited states as one expects. Weijo and Manninen also analyzed PV shifts in the g-tensor of electron spin resonance (ESR) spectra of CH_3XHO (X = N, P, As, and Sb) [175]. They concluded that it is unlikely that PV effects could be experimentally observed by ESR.

From the Z scaling behavior, it is clear that the best candidates for PV measurements in chiral molecules should include one or more heavy elements. For example, calculations for PV frequency shifts in fluorooxirane by Berger *et al.* show rather small values, that is, $\Delta\nu_{PV}/\nu = 1.2 \times 10^{-18}$ for the C–F stretching mode [176]. The first large PV energy shifts of up to 300 Hz were obtained for chiral organometallic compounds such as $[(\eta^5\text{-}C_5H_5)\text{Re(CO)(NO)I}]$ [119]. These were DC–HF calculations, and from the more recent results for CHFClBr, we expect that these HF results are probably not very reliable and could change significantly upon inclusion of electron correlation. Nevertheless, it was pointed out that the PV shifts to vibrational transitions are relatively unaffected, as the main contributions come from the first and second derivatives of the PV energy with respect to one particular normal coordinate q, and the total PV energy shift ($P_{PV}(q=0) \equiv E_{PV}(q=0)$) cancels out in vibrational transitions [169, 177],

$$P_{PV,n} = P_{PV}(q=0) + \frac{1}{2}\frac{\hbar}{\mu\omega_e}\left[\frac{\partial^2 P_{PV}(q)}{\partial q^2}\bigg|_{q=0} - \frac{1}{\mu\omega_e^2}\frac{\partial P_{PV}(q)}{\partial q}\bigg|_{q=0}\frac{\partial^3 V(q)}{\partial q^3}\bigg|_{q=0}\right] \times \left(n+\frac{1}{2}\right) + \cdots \qquad (7.22)$$

The derivatives are taken at the equilibrium geometry ($q=0$), and we formulated Equation 7.22 for the more general case of the vibrational dependence of any PV property shift P_{PV}. This expression can be easily extended to a multimode expression [170]. For the fundamental $0 \rightarrow 1$ C–F stretching mode in R-CHFClBr, one obtains (in 10^{-18} a.u.) −0.139 (HF), −0.186 (MP2), −0.198 (B3LYP), −0.184 (BLYP), −0.176 (PW86), and −0.192 (LDA) [169]. These calculations show that PV effects in vibrational transitions are about one order of magnitude smaller than the PV total energy shift. The importance to include anharmonicity effects in PV shifts in vibrational transitions was already pointed out by Laerdahl et al. [178], that is, a harmonic treatment can even lead to the wrong sign in $\Delta\nu_{PV}$ [179, 180]. Similar findings were obtained by Quack and Stohner who used a four-dimensional parity-violating potential energy hypersurface for CDBrClF and demonstrated that the multidimensional anharmonic couplings provide the dominant corrections [171]. Concerning the accurate treatment of the vibrational spectrum of CHFClBr and CDFClBr [181], Rauhut et al. [182] used coupled-cluster theory (CCSD(T)) for the total electronic energy and truncated vibrational configuration interaction to produce fundamental frequencies of a few wave number accuracy compared to experiment [183, 184]. Similarly, good results were achieved for CHFClI and CDFClI [185]. The vibrational wavefunctions obtained will be used for a complete vibrational PV treatment (Rauhut, G., Thierfelder, C., and Schwerdtfeger, P., to be published.). Two-component DFT was also used by Berger and Stuber to study PV-induced vibrational frequency shifts in these chiral polyhalomethanes [186]. Fokin et al. studied PV effects in pseudotetrahedral polyhalocubanes, thus shifting the ligands outward from the chirality center, resulting in PV energy shifts of two orders of magnitude smaller compared to the polyhalomethanes [187].

Faglioni and Lazzeretti studied PV effects in the vibrational spectrum of BiHFBr and BiHFI [188]. Barone and Viglione carried out anharmonic frequency analyses by perturbation theory for the PV contributions in PHFBr and AsHFBr [170]. They pointed out that even the coupling terms in the cubic force field between different normal modes are important, which are usually neglected. Isotope effects in chiral phosphorus compounds of the form PXYZ (XYZ = H, F, Cl, and Br with different isotopes) have been studied by Quack and coworkers [189]. These molecules are all good test cases for theoretical investigations, but have not yet been prepared, and as such may be thermodynamically unstable. Bast and Schwerdtfeger, therefore, suggested new target molecules, which can be synthesized and where important fundamental modes lie conveniently in the CO_2 laser frequency range [190, 191]. In fact, one compound investigated, $(\eta^5\text{-Cp}^*)(\text{Re}=\text{O})(\text{CH}_3)\text{Cl}$, shows PV effects for the

Re=O stretch in the Hz range [190]. Crassous and coworkers are currently trying to synthesize suitable chiral molecules with a rigid molecular framework containing a Re=O double bond [1]. More recently, Figgen and Schwerdtfeger investigated chiral selenium molecules of the form O=SeXY with (X,Y = F, Cl, Br, and I) [192, 193]. The PV energy shift for S-SeOClI ranges between −7.1 and −9.2 Hz, depending on the level of theory applied. For the Se=O fundamental stretching mode predicted at 968 cm^{-1}, the PV energy difference is 110 mHz between the two enantiomers, and therefore twice as large compared to the C−F stretching mode of CHFBrI [192].

We finally mention that besides molecules with point chirality (a chirality center) or a few systems with axial chirality (such as H_2O_2), little to no work has been done on systems with planar chirality. For larger helical systems, we mention the work by Kikuchi and Kiyonaga who investigated n-alkanes [194]. They concluded that the right-handed helix is more stable than the left-handed one, albeit by a very small amount of less than 10^{-20} a.u.

7.4 Conclusions

In the last decade, the methodology for treating PV in chiral molecules improved considerably. We have come a long way and learned that electron correlation is more important than originally anticipated, basis sets have to be more flexible in the core region to correctly account for the response to the PV Hamiltonian, Breit interactions need to be included for the lighter elements, the coupling between different vibrational modes need to be taken into account for the correct description of PV shifts, and relativistic effects need to be considered for the heavier elements. This is certainly all very challenging from both the theoretical and the computational point of view. On the experimental side, electronic spectroscopy would offer the largest PV shifts, but remains relatively unexplored to date. Perhaps the best chance for the first successful identification of PV effects in chiral molecules is by high-resolution vibrational spectroscopy as anticipated in Chardonnet's group in Paris. However, both high-resolution gas-phase NMR, and perhaps solid-state Mössbauer spectroscopy, should not be ruled out. On the practical side, it remains a challenge to find and synthesize suitable chiral molecules, including heavy elements (either at the chiral center or as ligands), with large PV energy shifts. As chirality is not a quantum mechanical observable, it remains to be seen if chirality measures can be linked to PV energy shifts [195], that is, we do not have a simple model in place to qualitatively predict which molecules show large PV effects (besides the Z^5 scaling law). Finally, we mention that one very promising new technique is to trap cold chiral molecules, which are ideally suited for high-resolution spectroscopy.

1) De Montigny, F., Bast, R., Gomes, A.S.P., Pilet, G., Vanthuyne, N., Roussel, C., Guy, L., Schwerdtfeger, P., Saue, T., and Crassous, J., Phys. Chem. Chem. Phys., DOI:10.1039/b925050f. In this paper the original PV shifts to the fundamental transitions published in Ref. [190] are revised.

Acknowledgments

The author was supported by the Marsden Fund Council (Contract No. MAU0606) from Government funding administered by the Royal Society of New Zealand. Critical comments by P. Bunker (Ottawa), R. Berger (Darmstadt), Sophie Nahrwold (Frankfurt), J. Brand (Auckland), and K. Beloy (Auckland) are much appreciated.

References

1 Bunker, P.R. and Jensen, P. (2006) *Molecular Symmetry and Spectroscopy*, 2nd edn, NRC Press, Ottawa.
2 Bunker, P.R. and Jensen, P. (2004) *Fundamentals of Molecular Symmetry*, Taylor & Francis, London.
3 Weyl, H. (1952) *Symmetry*, Princeton University Press, Princeton.
4 Ludwig, W. and Falter, C. (1988) *Symmetries in Physics*, Springer, Berlin.
5 Noether, E. (1918) *Nachr. D. König. Gesellsch. D. Wiss. Göttingen, Math-Phys. Klasse*, 235–257. Available at http://arxiv.org/abs/physics/0503066v1.
6 Peskin, M.E. and Schroeder, D.V. (1995) *An Introduction to Quantum Field Theory*, Westview Press, Boulder.
7 Schwinger, J. (1951) *Phys. Rev.*, **82**, 914.
8 Pauli, W. (1955) Exclusion principle, Lorentz group and reflexion of space-time and charge, in *Niels Bohr and the Development of Physics, Essays Dedicated to Niels Bohr on the Occasion of His Seventieth Birthday* (eds W. Pauli, L. Rosenfeld, and V. Weisskopf), Pergamon Press, New York, p. 30.
9 Lüders, G. (1957) *Ann. Phys. (New York)*, **2**, 1.
10 Greenberg, O.W. (2002) *Phys. Rev. Lett.*, **89**, 231602.
11 Eades, J. and Hartmann, F.J. (1999) *Rev. Mod. Phys.*, **71**, 373.
12 Buchmüller, W. and Plümacher, M. (1999) *Phys. Rep.*, **320**, 329.
13 Khlopov, M.Yu. (1998) *Grav. Cosmol.*, **4**, 69.
14 Christenson, J.H., Cronin, J.W., Fitch, V.L., and Turlay, R. (1964) *Phys. Rev. Lett.*, **13**, 138.
15 Carter, A.B. and Sanda, A.I. (1981) *Phys. Rev. D*, **23**, 1567.
16 Khriplovich, I.B. and Lamoreaux, S.K. (1997) *Cp Violation Without Strangeness*, Springer-Verlag, Berlin.
17 Commins, E.D., Ross, S.B., DeMille, D., and Regan, B.C. (1994) *Phys. Rev. A*, **50**, 2960.
18 Cho, D., Sangster, K., and Hinds, E.A. (1989) *Phys. Rev. Lett.*, **63**, 2559.
19 Hudson, J.J., Sauer, B.E., Tarbutt, M.R., and Hinds, E.A. (2002) *Phys. Rev. Lett.*, **89**, 023003.
20 Isaev, T.A., Petrov, A.N., Mosyagin, N.S., and Titov, A.V. (2005) *Phys. Rev. Lett.*, **95**, 163004.
21 Ravaine, B., Porsev, S.G., and Derevianko, A. (2005) *Phys. Rev. Lett.*, **94**, 013001.
22 Lee, T.D. and Yang, C.N. (1956) *Phys. Rev.*, **104**, 254.
23 Wu, C.S., Ambler, E., Hayward, R., Hoppes, D., and Hudson, R.P. (1957) *Phys. Rev.*, **105**, 1413.
24 Rodberg, L.S. and Weisskopf, V.F. (1957) *Science*, **125**, 627.
25 Glashow, S.L. (1961) *Nucl. Phys.*, **22**, 579.
26 Weinberg, S. (1967) *Phys. Rev. Lett.*, **19**, 1264.
27 Salam, A. (1968) *Elementary Particle Theory, Relativistic Groups, and Analyticity* (ed. N. Svartholm), Almqvist and Wiksells, Stockholm, p. 367.
28 Ginges, J.S.M. and Flambaum, V.V. (2004) *Phys. Rep.*, **397**, 63.
29 Anthony, P.L. et al. (SLAC E158 collaboration) (2005) *Phys. Rev. Lett.*, **95**, 081601.
30 Zel'dovich, B.Y. (1958) *Sov. Phys. JETP*, **33**, 1184.
31 Haxton, W.C., Liu, C.-P., and Ramsey-Musolf, M.J. (2002) *Phys. Rev. C*, **65**, 045502.

32 Haxton, W.C., Liu, C.-P., and Ramsey-Musolf, M.J. (2001) *Phys. Rev. Lett.*, **86**, 5247.

33 Flambaum, V.V., Khriplovich, I.B., and Sushkov, O.P. (1984) *Phys. Lett. B*, **146**, 367.

34 Flambaum, V.V. and Murray, D.W. (1997) *Phys. Rev. C*, **56**, 1641.

35 Haxton, W.C. and Wieman, C.E. (2001) *Ann. Rev. Nucl. Part. Sci.*, **51**, 261.

36 Milstein, A.I., Sushkov, O.P., and Terekhov, I.S. (2002) *Phys. Rev. Lett.*, **89**, 283003.

37 Bouchiat, M.-A. and Bouchiat, C. (1974) *Phys. Lett. B*, **48**, 111.

38 Bouchiat, M.-A. and Bouchiat, C. (1974) *J. Phys. France*, **35**, 899.

39 Bouchiat, M.A. and Bouchiat, C. (1975) *J. Phys. France*, **36**, 493.

40 Bouchiat, M.A. and Bouchiat, C. (1997) *Rep. Prog. Phys.*, **60**, 1351.

41 Bouchiat, M.-A. and Bouchiat, C. (1999) *Parity Violation in Atoms and Polarized Electron Scattering* (eds B. Froisand M-.A. Bouchiat), World Scientific, Singapore, p. 536.

42 Barkov, L.M. and Zolotorev, M. (1978) *JETP Lett.*, **27**, 357.

43 Barkov, L.M. and Zolotorev, M. (1978) *JETP Lett.*, **28**, 503.

44 Barkov, L.M. and Zolotorev, M. (1979) *Phys. Lett. B*, **85**, 308.

45 Vetter, P., Meekhov, D.M., Majumder, P.K., Lamoreaux, S.K., and Fortson, E.N. (1995) *Phys. Rev. Lett.*, **74**, 2658.

46 Meekhof, D.M., Vetter, P.A., Majumder, P.K., Lamoreaux, S.K., and Fortson, E.N. (1995) *Phys. Rev. A*, **52**, 1895.

47 McPherson, M.J.D., Zetie, K.P., Warrington, R.B., Stacey, D.N., and Hoare, J.P. (1991) *Phys. Rev. Lett.*, **67**, 2784.

48 Conti, R., Bucksbaum, P., Chu, S., Commins, E., and Hunter, L. (1979) *Phys. Rev. Lett.*, **42**, 343.

49 Wood, C.S., Bennett, S.C., Cho, D., Masterson, B.P., Roberts, J.L., Tanner, C.E., and Wieman, C.E. (1997) *Science*, **275**, 1759.

50 Wood, C.S., Bennett, S.C., Roberts, J.L., Cho, D., and Wieman, C.E. (1999) *Can. J. Phys.*, **77**, 7.

51 Dzuba, V.A., Flambaum, V.V., and Sushkov, O.P. (1989) *Phys. Lett. A*, **141**, 147.

52 Blundell, S.A., Johnson, W.R., and Sapirstein, J. (1990) *Phys. Rev. Lett.*, **65**, 1411.

53 Blundell, S.A., Johnson, W.R., and Sapirstein, J. (1992) *Phys. Rev. D*, **45**, 1602.

54 Shabaev, V.M., Pachucki, K., Tupitsyn, I.I., and Yerokhin, V.A. (2005) *Phys. Rev. Lett.*, **94**, 213002.

55 Porsev, S.G., Beloy, K., and Derevianko, A. (2009) *Phys. Rev. Lett.*, **102**, 181601.

56 Tsigutkin, K., Dounas-Frazer, D., Family, A., Stalnaker, J.E., Yashchuk, V.V., and Budker, D. (2009) *Phys. Rev. Lett.*, **103**, 071601.

57 Derevianko, A. and Porsev, S.G. (2007) *Eur. Phys. J. A*, **32**, 517.

58 Khriplovich, I.B. (1991) *Parity Non Conservation in Atomic Phenomena*, Gordon and Breach, Philadelphia.

59 Commins, E.D. (1993) *Phys. Scr.*, **T46**, 92.

60 Yamagata, Y. (1966) *J. Theor. Biol.*, **11**, 495.

61 Rein, D.W. (1974) *J. Mol. Evol.*, **4**, 15.

62 Letokhov, V.S. (1975) *Phys. Lett.*, **53**, 275.

63 Kompanets, O.N., Kukudzhanov, A.R., and Letokhov, V.S. (1976) *Opt. Commun.*, **19**, 414.

64 Halzen, F. and Martin, A.D. (1984) *Quarks and Leptons: An Introductory Course in Modern Particle Physics*, John Wiley & Sons, Inc., New York.

65 Keszthelyi, L. (1984) *Orig. Life Evol. Biosph.*, **14**, 375.

66 Buschmann, H., Thede, R., and Heller, D. (2000) *Angew. Chem. Int. Ed.*, **39**, 4033.

67 Compton, R.N. and Pagni, R.M. (2002) *Adv. At. Mol. Opt. Phys.*, **48**, 219.

68 Berger, R., Quack, M., and Tschumber, G. (2000) *Helv. Chim. Acta*, **83**, 1919.

69 Bonner, W.A. (2000) *Chirality*, **12**, 114.

70 Lente, G. (2006) *J. Phys. Chem. A*, **110**, 12711.

71 Crassous, J., Chardonnet, C., Saue, T., and Schwerdtfeger, P. (2005) *Org. Biomol. Chem.*, **3**, 2218.

72 DeMille, D., Cahn, S.B., Murphree, D., Rahmlow, D.A., and Kozlov, M.G. (2008) *Phys. Rev. Lett.*, **100**, 023003.

73 Quack, M. (1989) *Angew. Chem. Int. Ed. Engl.*, **28**, 571.

74 Chardonnet, C., Daussy, C., Marrel, T., Amy-Klein, A., Nguyen, C.T., and Bordé, C.J. (1999) *Parity Violation in Atomic Physics and Electron Scattering* (eds B. Froisand M.A. Bouchiat), World Scientific, New-York, p. 325.

75 Quack, M. (2002) *Angew. Chem. Int. Ed.*, **41**, 4619.

76 Crassous, J., Monier, F., Dutasta, J.-P., Ziskind, M., Daussy, C., Grain, C., and Chardonnet, C. (2003) *ChemPhysChem*, **4**, 541.

77 Berger, R. (2004) *Relativistic Electronic Structure Theory. Part 2: Applications* (ed. P. Schwerdtfeger), Elsevier, Amsterdam, p. 188.

78 Quack, M., Stohner, J., and Willeke, M. (2008) *Annu. Rev. Phys. Chem.*, **59**, 741.

79 Arimondo, E., Glorieux, P., and Oka, T. (1977) *Opt. Commun.*, **23**, 369.

80 Lazzeretti, P., Zanasi, R., and Faglioni, F. (1999) *Phys. Rev. E*, **60**, 871.

81 Schwerdtfeger, P., Kühn, A., Bast, R., Laerdahl, J.K., Faglioni, F., and Lazzeretti, P. (2004) *Chem. Phys. Lett.*, **383**, 496.

82 Ya Zel'dovich, B., Saakyan, D.B., and Sobel'man, I.I. (1977) *JETP Lett.*, **25**, 94.

83 Wiesenfeld, L. (1988) *Mol. Phys.*, **64**, 739.

84 Hegstrom, R.A., Rein, D.W., and Sandars, P.G.H. (1980) *J. Chem. Phys.*, **73**, 2329.

85 Laerdahl, J.K. and Schwerdtfeger, P. (1999) *Phys. Rev. A*, **60**, 4439.

86 Levi, B.G. (2000) *Physics Today*, **53** (September), 46.

87 Bethlem, H.L., Berden, G., van Roij, A.J.A., Crompvoets, F.M.H., and Meijer, G. (2000) *Phys. Rev. Lett.*, **84**, 5744.

88 Takekoshi, T., Patterson, B.M., and Knize, R.J. (1998) *Phys. Rev. Lett.*, **81**, 5105.

89 Daussy, C., Marrel, T., Amy-Klein, A., Nguyen, C.T., Bordé, C.J., and Chardonnet, C. (1999) *Phys. Rev. Lett.*, **83**, 1554.

90 Holzwarth, R., Udem, T., Hänsch, T.W., Knight, J.C., Wadsworth, W.J., and Russell, P.S.J. (2000) *Phys. Rev. Lett.*, **85**, 2264.

91 Shelkovnikov, A., Grain, C., Nguyen, C., Butcher, R.J., Amy-Klein, A., and Chardonnet, C. (2001) *Appl. Phys. B*, **73**, 93.

92 Schnell, M. and Meijer, G. (2009) *Angew. Chem., Int. Ed.*, **48**, 6010.

93 Doyle, T.R. and Vogl, O. (1989) *J. Am. Chem. Soc.*, **111**, 8511.

94 Constante, J., Hecht, L., Polaravapu, P.L., Collet, A., and Baron, L.D. (1997) *Angew. Chem., Int. Ed. Engl.*, **36**, 885.

95 Costante-Crassous, J., Marrone, T.J., Briggs, J.M., McCammon, J.A., and Collet, A. (1997) *J. Am. Chem. Soc.*, **119**, 3818.

96 Ziskind, M., Daussy, C., Marrel, T., and Chardonnet, C. (2002) *Eur. Phys. J. D*, **20**, 219.

97 Schwerdtfeger, P., Laerdahl, J.K., and Chardonnet, C. (2002) *Phys. Rev. A*, **65**, 042508.

98 Cuisset, A., Aviles Moreno, J.R., Huet, T.R., Petitprez, D., Demaison, J., and Crassous, J. (2005) *J. Phys. Chem. A*, **109**, 5708.

99 Soulard, P., Asselin, P., Cuisset, A., Moreno, J.R.A., Huet, T.R., Petitprez, D., Demaison, J., Freedman, T.B., Cao, X., Nafiec, L.A., and Crassous, J. (2006) *Phys. Chem. Chem. Phys.*, **8**, 79.

100 Inzhechik, L.V., Melnikov, E.V., Rogozev, B.I., Khlebnikov, A.S., and Tsinov, V.G. (1986) *Yad. Fiz.*, **44**, 1370; Inzhechik, L.V., Melnikov, E.V., Rogozev, B.I., Khlebnikov, A.S., and Tsinov, V.G. (1986) *Sov. J. Nucl. Phys.*, **44**, 890.

101 Inzhechik, L.V., Melnikov, E.V., Khlebnikov, A.S., Tsinov, V.G., and Rogozev, B.I. (1987) *Yad. Fiz.*, **93**, 800; Inzhechik, L.V., Melnikov, E.V., Khlebnikov, A.S., Tsinov, V.G., and Rogozev, B.I. (1987) *Sov. J. Nucl. Phys.*, **66**, 450.

102 Inzhechik, L.V., Khlebnikov, A.S., Tsinov, V.G., Rogozev, B.I., Silin, M.Yu., and Penkov, Yu.P. (1987) *Yad. Fiz.*, **93**, 1569; Inzhechik, L.V., Khlebnikov, A.S., Tsinov, V.G., Rogozev, B.I., Silin, M.Yu., and Penkov, Yu.P. (1987) *Sov. J. Nucl. Phys.*, **66**, 897.

103 Sushkov, O.P. and Telitsin, V.B. (1993) *Phys. Rev. C*, **48**, 1069.

104 Khriplovich, I.B. (1980) *Sov. Phys. JETP*, **52**, 177.

105 Keszthelyi, L. (1994) *J. Biol. Phys.*, **20**, 241.

106 Lahamer, A.S., Mahurin, S.M., Compton, R.N., House, D., Laerdahl, J.K., Lein, M.,

and Schwerdtfeger, P. (2000) *Phys. Rev. Lett.*, **85**, 4470.
107 Barra, A.L., Robert, J.B., and Wiesenfeld, L. (1986) *Phys. Lett.*, **115**, 443.
108 Barra, A.L. and Robert, J.B. (1996) *Mol. Phys.*, **4**, 875.
109 Barra, A.L., Robert, J.B., and Wiesenfeld, L. (1988) *Europhys. Lett.*, **5**, 217.
110 Barra, A.L., Robert, J.B., and Wiesenfeld, L. (1987) *Biosystems*, **20**, 57.
111 Robert, J.B. and Barra, A.L. (2001) *Chirality*, **13**, 699.
112 Bast, R., Schwerdtfeger, P., and Saue, T. (2006) *J. Chem. Phys.*, **125**, 064504.
113 Mukhamedjanov, T.N., Sushkov, O.P., and Cadogan, J.M. (2005) *Phys. Rev. A*, **71**, 012107.
114 Mukhamedjanov, T.N., Sushkov, O.P., and Cadogan, J.M. (2006) *J. Supercond. Nov. Magn.*, **20**, 1557.
115 Soncini, A., Faglioni, F., and Lazzeretti, P. (2003) *Phys. Rev. A*, **68**, 033402.
116 Nahrwold, S. and Berger, R. (2009) *J. Chem. Phys.*, **130**, 214101.
117 Ledbetter, M.P., Crawford, C.W., Pines, A., Wemmer, D.E., Knappe, S., Kitching, J., and Budker, D. (2009) *J. Magn. Reson.*, **199**, 25.
118 Fujiki, M. (2001) *Macromol. Rapid Commun.*, **22**, 669.
119 Schwerdtfeger, P., Gierlich, J., and Bollwein, T. (2003) *Angew. Chem. Int. Ed.*, **42**, 1293.
120 Gottselig, M. and Quack, M. (2005) *J. Chem. Phys.*, **123**, 084305.
121 Quack, M. (1986) *Chem. Phys. Lett.*, **132**, 147.
122 Harris, R.A. and Stodolsky, L. (1978) *Phys. Lett. B*, **78**, 313.
123 Berger, R., Gottselig, M., Quack, M., and Willeke, M. (2001) *Angew. Chem. Int. Ed.*, **40**, 4195.
124 Berger, R. (2003) *Phys. Chem. Chem. Phys.*, **5**, 12.
125 Bargueño, P., Gonzalo, I., and de Tudela, R.P. (2009) *Phys. Rev. A*, **80**, 012110.
126 Harris, R.A. and Stodolsky, L. (1981) *J. Chem. Phys.*, **74**, 2145.
127 Harris, R.A. and Silbey, R. (1983) *J. Chem. Phys.*, **78**, 7330.
128 Silbey, R. and Harris, R.A. (1989) *J. Chem. Phys.*, **93**, 7062.
129 Harris, R.A. (1994) *Chem. Phys. Lett.*, **223**, 250.
130 Harris, R.A., Shi, Y., and Cina, J.A. (1994) *J. Chem. Phys.*, **101**, 3459.
131 Harris, R.A. (2002) *Chem. Phys. Lett.*, **365**, 343.
132 Salam, A. (1991) *J. Mol. Evol.*, **33**, 105.
133 Wang, W.Q., Yi, F., Ni, Y.M., Zhao, Z.X., Jin, X.L., and Tang, Y.Q. (2000) *J. Biol. Phys.*, **26**, 51.
134 Wang, W.Q., Min, W., Bai, F., Sun, L., Yi, F., Wang, Z.M., Yan, C.H., Ni, Y.M., and Zhao, Z.X. (2002) *Tetrahedron Asymmetry*, **13**, 2427.
135 Wang, W.Q., Min, W., Liang, Z., Wang, L.Y., Chen, L., and Deng, F. (2003) *Biophys. Chem.*, **103**, 289.
136 Wang, W.Q., Min, W., Zhub, C., and Yi, F. (2003) *Phys. Chem. Chem. Phys.*, **5**, 4000.
137 Wesendrup, R., Laerdahl, J.K., Compton, R.N., and Schwerdtfeger, P. (2003) *J. Phys. Chem. A*, **107**, 6668.
138 Sullivan, R., Pyda, M., Pak, J., Wunderlich, B., Thompson, J., Pagni, R., Pan, H., Barnes, C., Schwerdtfeger, P., and Compton, R. (2003) *J. Phys. Chem. A*, **107**, 6674.
139 Wilson, C.C., Myles, D., Ghosh, M., Johnson, L.N., and Wang, W. (2005) *New. J. Chem.*, **29**, 1318.
140 Bolik, S., Rübhausen, M., Binder, S., Schulz, B., Perbandt, M., Genov, N., Erdmann, V., Klussmann, S., and Betzel, C. (2007) *RNA*, **13**, 1877.
141 Szabo-Nagy, A. and Keszthelyi, L. (1999) *Proc. Natl. Acad. Sci. USA*, **96**, 4252.
142 Keszthelyi, L. (2003) *Mendeleev Commun.*, **13**, 129.
143 Shinitzky, M., Nudelman, F., Barda, Y., Haimovitz, R., Chen, E., and Deamer, D.W. (2002) *Orig. Life Evol. Biosph.*, **32**, 285.
144 Lahav, M., Weissbuch, I., Shavit, E., Reiner, C., Nicholson, G.J., and Schurig, V. (2006) *Orig. Life Evol. Biosph.*, **36**, 151.
145 Schwartz, A.W. (2002) *Orig. Life Evol. Biosph.*, **32**, 283.
146 Mason, S.F. and Tranter, G.E. (1983) *Chem. Phys. Lett.*, **94**, 34.
147 Mason, S.F. and Tranter, G.E. (1984) *Mol. Phys.*, **53**, 1091.
148 Tranter, G.E. (1985) *Mol. Phys.*, **56**, 825.

149 MacDermott, A.J., Tranter, G.E., and Indoe, S.B. (1987) *Chem. Phys. Lett.*, **135**, 159.

150 Tranter, G.E. (1985) *Chem. Phys. Lett.*, **121**, 339.

151 Berger, R. and Quack, M. (2000) *J. Chem. Phys.*, **112**, 3148.

152 Berger, R. (2008) *J. Chem. Phys.*, **129**, 154105.

153 Pernpointner, M., Seth, M., and Schwerdtfeger, P. (1998) *J. Chem. Phys.*, **108**, 6722.

154 Kikuchi, O. and Wang, H. (1987) Report on Special Research Project on Evolution of Matter, Vol. 1, University of Tsukuba, p. 219, and Vol. 2, p. 215.

155 Kikuchi, O. and Wang, H. (1990) *Bull. Chem. Soc. Jpn*, **63**, 2751.

156 Bakasov, A., Ha, T.K., and Quack, M. (1998) *J. Chem. Phys.*, **109**, 7263.

157 Thyssen, J., Laerdahl, J.K., and Schwerdtfeger, P. (2000) *Phys. Rev. Lett.*, **85**, 3105.

158 van Stralen, J.N.P., Visscher, L., Larsen, C.V., and Jensen, H.J.Aa. (2005) *Chem. Phys.*, **311**, 81.

159 Berger, R. and van Wüllen, C. (2005) *J. Chem. Phys.*, **122**, 134316.

160 Berger, R., Langermann, N., and van Wüllen, C. (2005) *Phys. Rev. A*, **71**, 042105.

161 Quack, M. and Willeke, M. (2003) *Helv. Chim. Acta*, **86**, 1641.

162 MacDermott, A.J., Hyde, G.O., and Cohen, A.J. (2009) *Orig. Life Evol. Biosph.*, **39**, 439.

163 Lazzeretti, P. and Zanasi, R. (1997) *Chem. Phys. Lett.*, **279**, 349.

164 Hennum, A.C., Helgaker, T., and Klopper, W. (2002) *Chem. Phys. Lett.*, **354**, 274.

165 Bakasov, A., Berger, R., Ha, T.K., and Quack, M. (2004) *Int. J. Quantum Chem.*, **99**, 393.

166 Gottselig, M., Luckhaus, D., Quack, M., Stohner, J., and Willeke, M. (2001) *Helv. Chim. Acta*, **84**, 1846.

167 Laubender, G. and Berger, R. (2003) *ChemPhysChem*, **4**, 395.

168 Laubender, G. and Berger, R. (2006) *Phys. Rev. A*, **74**, 032105.

169 Schwerdtfeger, P., Saue, T., van Stralen, J.N.P., and Visscher, L. (2005) *Phys. Rev. A*, **71**, 012103.

170 Barone, V. and Viglione, R.G. (2005) *J. Chem. Phys.*, **123**, 234304.

171 Quack, M. and Stohner, J. (2003) *J. Chem. Phys.*, **119**, 11228.

172 Weijo, V., Manninen, P., and Vaara, J. (2005) *J. Chem. Phys.*, **123**, 054501.

173 Weijo, V., Bast, R., Manninen, P., Saue, T., and Vaara, J. (2007) *J. Chem. Phys.*, **126**, 074107.

174 Weijo, V., Hansen, M.B., Christiansen, O., and Manninen, P. (2009) *Chem. Phys. Lett.*, **470**, 166.

175 Weijo, V. and Manninen, P. (2006) *Chem. Phys. Lett.*, **433**, 37.

176 Berger, R., Quack, M., and Stohner, J. (2001) *Angew. Chem. Int. Ed.*, **40**, 1667.

177 Buckingham, A.D. (1962) *J. Chem. Phys.*, **36**, 3096.

178 Laerdahl, J.K., Schwerdtfeger, P., and Quiney, H.M. (2000) *Phys. Rev. Lett.*, **84**, 3811.

179 Quack, M. and Stohner, J. (2000) *Phys. Rev. Lett.*, **84**, 3807.

180 Viglione, R.G., Zanasi, R., Lazzeretti, P., and Ligabue, A. (2000) *Phys. Rev. A*, **62**, 052516.

181 Quack, M. and Stohner, J. (2001) *Chirality*, **13**, 745.

182 Rauhut, G., Barone, V., and Schwerdtfeger, P. (2006) *J. Chem. Phys.*, **125**, 054308.

183 Beil, A., Luckhaus, D., and Quack, M. (1996) *Ber. Bunsenges. Phys. Chem.*, **100**, 1853.

184 Beil, A., Hollenstein, H., Monti, O.L.A., Quack, M., and Stohner, J. (2000) *J. Chem. Phys.*, **113**, 2701.

185 Heislbetz, S., Schwerdtfeger, P., and Rauhut, G. (2007) *Mol. Phys.*, **105**, 1385.

186 Berger, R. and Stuber, J.L. (2007) *Mol. Phys.*, **105**, 41.

187 Fokin, A.A., Schreiner, P.R., Berger, R., Robinson, G.H., Wei, P., and Campana, C.F. (2006) *J. Am. Chem. Soc.*, **128**, 5332.

188 Faglioni, F. and Lazzeretti, P. (2003) *Phys. Rev. A*, **67**, 032101.

189 Berger, R., Laubender, G., Quack, M., Sieben, A., Stohner, J., and Willeke, M. (2005) *Angew. Chem. Int. Ed.*, **44**, 3623.

190 Schwerdtfeger, P. and Bast, R. (2004) *J. Am. Chem. Soc.*, **126**, 1652.

191 Bast, R. and Schwerdtfeger, P. (2003) *Phys. Rev. Lett.*, **91**, 023001.

192 Figgen D. and Schwerdtfeger P. (2008) *Phys. Rev. A*, **78**, 012511.
193 Figgen D. and Schwerdtfeger P. (2009) *J. Chem. Phys.*, **130**, 054306.
194 Kikuchi, O. and Kiyonaga, H. (1994) *J. Mol. Struct. (Theochem)*, **312**, 271.
195 Faglioni F. and Lazzeretti P. (2001) *Phys. Rev. E*, **65**, 011904.

8
Vibrational Circular Dichroism: Time-Domain Approaches

Hanju Rhee, Seongeun Yang, and Minhaeng Cho

8.1
Introduction

When light propagates through a medium, not a vacuum, its velocity and intensity are modulated depending on the medium's refractive index $n(\omega)$ and absorption coefficient $\varkappa(\omega)$, respectively, and they are intrinsic optical properties of the medium. As a result, frequency-dependent phase retardation and attenuation processes occur simultaneously [1]. If the medium is isotropic and nonchiral, these quantities are constant at a given frequency irrespective of the optical polarization, so the polarization state of the incident light remains unchanged after passing through the medium. In chiral media, however, this is not the case for circular polarization (left: LCP, right: RCP) because the parameters $n(\omega)$ and $\varkappa(\omega)$ leading the dispersion and absorption processes vary with its handedness, that is, $n_L(\omega) \neq n_R(\omega)$ and $\varkappa_L(\omega) \neq \varkappa_R(\omega)$ [2].

Circular dichroism (CD) and circular birefringence (CB), generally referred to as *optical activity*, are given by the frequency-dependent differential absorption coefficient $\Delta\varkappa(\omega) = \varkappa_L(\omega) - \varkappa_R(\omega)$ and the differential refractive index $\Delta n(\omega) = n_L(\omega) - n_R(\omega)$, respectively. Note that we will use the terms *optical rotatory dispersion* (ORD) and *circular birefringence* together when denoting Δn because the optical rotation of linearly polarized light is the observable effect caused by the CB. Since these chiroptical properties are manifested by almost all natural products, artificial drugs, and so on and are highly sensitive to their structural conformations and absolute configurations, the CD or ORD spectroscopy has been used to elucidate secondary structures of biomolecules such as polypeptides and proteins and to determine absolute configurations of chiral molecules and drugs in condensed phases [2]. However, as with the mutual dependence between $\varkappa(\omega)$ and $n(\omega)$, known as the *Kramers–Kronig* relations [3–5], the two observables CD ($\Delta\varkappa$) and ORD (Δn) are also linked to each other via such relations. Thus, if one of them is measured for the entire frequency range, the other can be in principle obtained by performing the Kramers–Kronig transformation. This means that the CD and ORD ultimately provide the same chirality-specific information of a given molecule with a specific handedness.

Computational Spectroscopy: Methods, Experiments and Applications. Edited by Jörg Grunenberg
Copyright © 2010 WILEY-VCH Verlag GmbH & Co. KGaA, Weinheim
ISBN: 978-3-527-32649-5

However, in practice, this is not the case because it is not possible to measure one of the two spectra in the frequency range from zero to infinity, which is the integration range for the Kramers–Kronig transformations.

Of the two, the CD spectroscopy has been more widely used than the ORD measurement technique in studying structural details of chiral molecules, which is not only because the measurement is less troublesome but also because a direct comparison between experimental and quantum mechanically computed results is possible. The conventional CD spectroscopy relies on a differential intensity measurement technique, where the absorbance difference of LCP and RCP lights by the chiral sample (CS) is selectively measured. In the case of the vibrational CD (VCD) [6, 7], which is just the vibrational analogue of electronic CD, the chiral susceptibilities for nuclear vibrations are far much smaller than the corresponding signals in the UV-Vis range. Consequently, it is by no means an easy measurement because one has to differentiate such a weak effect by using the differential measurement scheme with relatively largely fluctuating incident IR beams. This is why it takes a long time (typically a few hours) to acquire a statistically meaningful VCD spectrum with conventional VCD spectrometers.

In this chapter, we will introduce a novel time-domain approach characterizing the vibrational optical activity (VOA) of chiral molecule. Instead of measuring incoherent transmitted beam intensities, we have developed a chiroptical spectroscopic technique that can sensitively detect the transmitted beam at an amplitude level so that both magnitude and phase of the signal field are measured [8–11]. This additional phase information is used to simultaneously characterize the circular dichroic and birefringence effects induced by the field–chiral molecule interactions. Also, we have developed a computational method for calculating vibrational CD spectra of peptides and chiral molecules in solution, where quantum mechanical/molecular mechanical (QM/MM) molecular dynamics (MD) simulation method is used [12]. A notable difference between our method and the previous quantum chemistry calculation method is that our approach is based on the direct time-domain calculation of the electric dipole and magnetic dipole cross-correlation function. The imaginary part of the Fourier transform of the cross-correlation function corresponds to the vibrational CD. In this chapter, a brief account of time-correlation function formalism for both time-domain computational and experimental methods will be presented (Section 8.2). Next, QM/MM MD simulation method for directly calculating the corresponding time-correlation function will be described and some simulation results will be shown and discussed (Section 8.3). Finally, we will present a detailed description of femtosecond VOA measurement experiment for directly measuring the phase and amplitude of the time-domain chiral response function (Section 8.4).

8.2
Time-Correlation Function Theory

The radiation–matter interaction Hamiltonian in the minimal coupling scheme is given by the inner product of the vector potential of the electromagnetic field and the momentum operator of the charged particle. Then, the multipolar expansion form of the interaction Hamiltonian, which is valid up to the first order in the wave vector **k**, is [8, 13, 14]

$$H_I = -\boldsymbol{\mu} \cdot \mathbf{E}(\mathbf{r}, t) - \mathbf{M} \cdot \mathbf{B}(\mathbf{r}, t) - (1/2)\mathbf{Q} : \nabla \mathbf{E}(\mathbf{r}, t) \tag{8.1}$$

Here, **E** and **B** are the electric and magnetic fields, respectively. **μ**, **M**, and **Q** are the electric dipole, magnetic dipole, and electric quadrupole operators, respectively. Using the linear response function theory, one can find that the linear polarization is given as [8]

$$\mathbf{P}^{(1)}(\mathbf{r}, t) = \langle \boldsymbol{\mu} \varrho^{(1)}(\mathbf{r}, t) \rangle + \langle (\mathbf{M} \times \hat{\mathbf{k}}) \varrho^{(1)}(\mathbf{r}, t) \rangle - (i/2) \langle (\mathbf{k} \cdot \mathbf{Q}) \varrho^{(1)}(\mathbf{r}, t) \rangle \tag{8.2}$$

where $\hat{\mathbf{k}} \equiv \mathbf{k}/|\mathbf{k}|$ and $\varrho^{(1)}(\mathbf{r}, t)$ is the first-order perturbation-expanded density operator with respect to the above radiation–matter interaction Hamiltonian H_I in Equation 8.1. Then, the linear polarization in Equation 8.2 can be rewritten as, in terms of the corresponding linear response functions [14],

$$\mathbf{P}^{(1)}(\mathbf{r}, t) = \varrho_0 \int_0^\infty d\tau \{\phi_{\mu\mu}(\tau) + \phi_{\mu M}(\tau) + (i/2)\phi_{\mu Q}(\tau) + \phi_{M\mu}(\tau) - (i/2)\phi_{Q\mu}(\tau)\} \cdot \mathbf{e} E(\mathbf{r}, t-\tau) \tag{8.3}$$

where **e** is the unit vector in the polarization direction of electric field and ϱ_0 is the number density N/V, and

$$\phi_{\mu\mu}(\tau) \equiv \frac{i}{\hbar}\theta(\tau)\langle[\boldsymbol{\mu}(\tau), \boldsymbol{\mu}(0)]\varrho_{eq}\rangle$$

$$\phi_{\mu M}(\tau) \equiv \frac{i}{\hbar}\theta(\tau)\langle[\boldsymbol{\mu}(\tau), \mathbf{M}(0) \times \hat{\mathbf{k}}]\varrho_{eq}\rangle$$

$$\phi_{\mu Q}(\tau) \equiv \frac{i}{\hbar}\theta(\tau)\langle[\boldsymbol{\mu}(\tau), \mathbf{k} \cdot \mathbf{Q}(0)]\varrho_{eq}\rangle \tag{8.4}$$

$$\phi_{M\mu}(\tau) \equiv \frac{i}{\hbar}\theta(\tau)\langle[\mathbf{M}(\tau) \times \hat{\mathbf{k}}, \boldsymbol{\mu}(0)]\varrho_{eq}\rangle$$

$$\phi_{Q\mu}(\tau) \equiv \frac{i}{\hbar}\theta(\tau)\langle[\mathbf{k} \cdot \mathbf{Q}(\tau), \boldsymbol{\mu}(0)]\varrho_{eq}\rangle$$

Here, $\langle \ldots \rangle$ denotes the trace over the bath degrees of freedom and ϱ_{eq} is the thermal equilibrium density operator. Note that the first term on the right-hand side of Equation 8.3 is typically two to three orders of magnitude larger than the other terms for electronic transition and four to six orders for vibrational transition.

For an absorption process, without loss of generality, it is assumed that the incident field propagates along the z-axis in a space-fixed frame, that is, $\mathbf{k} = (\omega_0/c)\hat{\mathbf{z}}$, where ω_0 is the center frequency of the electric field, and that $\mathbf{e} = \hat{\mathbf{y}}$. Then, the rotationally averaged y-component of $\mathbf{P}^{(1)}(t)$, which is the temporal amplitude of $\mathbf{P}^{(1)}(\mathbf{r}, t)$, is

$$P_y^{(1)}(t) = \int_0^\infty d\tau \chi_{\mu\mu}(\tau) E(t-\tau) \tag{8.5}$$

where

$$\chi_{\mu\mu}(t) \equiv \varrho_0 \left(\frac{i}{\hbar}\right) \langle [\mu_y(t), \mu_y(0)] \varrho_{eq} \rangle \tag{8.6}$$

In this case, the magnetic dipole and electric quadrupole contributions to the linear polarization in Equation 8.3 vanish. The linear susceptibility is defined as

$$\chi(\omega) = \int_0^\infty d t \chi_{\mu\mu}(t) e^{i\omega t} = \chi'(\omega) + i\chi''(\omega) \tag{8.7}$$

The imaginary part of $\chi(\omega)$, denoted as $\chi''(\omega)$, can be rewritten as

$$\begin{aligned}\chi''(\omega) &= \frac{\pi \varrho_0}{\hbar} \sum_{a,b} P(a)|\mu_{ab}|^2 [\delta(\omega-\omega_{ba}) - \delta(\omega+\omega_{ba})] \\ &= \frac{\pi \varrho_0}{\hbar}(1-e^{-\beta\hbar\omega}) \sum_{a,b} P(a)|\mu_{ab}|^2 \delta(\omega-\omega_{ba})\end{aligned} \tag{8.8}$$

where $P(a)$ is the population of the state $|a\rangle$ and $\mu_{ab} = \langle a|\mu_y(0)|b\rangle$. Now, the absorption line shape function in an isotropic medium is given as

$$I(\omega) = \frac{3\hbar\varepsilon''(\omega)}{4\pi^2(1-e^{-\beta\hbar\omega})} = \frac{\varrho_0}{2\pi}\int_{-\infty}^\infty dt\, e^{i\omega t}\langle \boldsymbol{\mu}(t) \cdot \boldsymbol{\mu}(0)\rangle \tag{8.9}$$

where the imaginary part of the dielectric constant is $\varepsilon''(\omega) = 4\pi\chi''(\omega)$.

We next consider the optical activity, that is, circular dichroism and circular birefringence. When the incident radiation is circularly polarized, the unit vector **e** should be replaced with $\mathbf{e}_L = (\hat{x}+i\hat{y})/\sqrt{2}$ and $\mathbf{e}_R = (\hat{x}-i\hat{y})/\sqrt{2}$ for the left and right circularly polarized lights, respectively. The difference polarization, which is related to the linear optical activity, is then defined as $\Delta \mathbf{P}(\mathbf{r},t) = \mathbf{P}^L(\mathbf{r},t) - \mathbf{P}^R(\mathbf{r},t)$. Again, assuming that $\mathbf{k} = (\omega_0/c)\hat{z}$, the rotationally averaged x-components of the linear polarizations $\mathbf{P}^L(\mathbf{r},t)$ and $\mathbf{P}^R(\mathbf{r},t)$ are

$$P_x^L(\mathbf{r},t) = \frac{1}{\sqrt{2}}\int_0^\infty d\tau \{\chi_{\mu\mu}(\tau) + \chi_{\mu M}(\tau) + \chi_{M\mu}(\tau)\} E(\mathbf{r},t-\tau) \tag{8.10}$$

$$P_x^R(\mathbf{r},t) = \frac{1}{\sqrt{2}}\int_0^\infty d\tau \{\chi_{\mu\mu}(\tau) - \chi_{\mu M}(\tau) - \chi_{M\mu}(\tau)\} E(\mathbf{r},t-\tau) \tag{8.11}$$

where

$$\chi_{\mu M}(t) \equiv \varrho_0 \left(\frac{i}{\hbar}\right)\langle [\mu_x(t), -iM_x(0)]\varrho_{eq}\rangle \tag{8.12}$$

$$\chi_{M\mu}(t) \equiv \varrho_0 \left(\frac{i}{\hbar}\right) \langle [M_x(t), i\mu_x(0)]\varrho_{eq}\rangle \tag{8.13}$$

Thus, we find that the rotationally averaged difference polarization amplitude is

$$\Delta P_x(t) = \frac{1}{\sqrt{2}} \int_0^\infty d\tau \Delta\chi(t-\tau)E(\tau) \tag{8.14}$$

where the linear optical activity susceptibility $\Delta\chi(t)$ [$= \chi_L(t) - \chi_R(t)$] is defined as

$$\Delta\chi(t) \equiv 2\{\chi_{\mu M}(t) + \chi_{M\mu}(t)\} \tag{8.15}$$

Note that $\Delta P_x(t)$ in this case has no contribution from the electric dipole–electric dipole and the electric dipole–electric quadrupole responses because they all vanish after rotational averaging of the corresponding second- and third-rank tensorial response functions over randomly oriented chiral molecules in solutions. The linear optical activity susceptibility in Equation 8.15 can be rewritten as

$$\begin{aligned}\Delta\chi(\omega) &= \frac{\varrho_0}{\hbar}\int_0^\infty dt e^{i\omega t}\{\langle[\mu_x(t), M_x(0)]\varrho_{eq}\rangle - \langle[M_x(t), \mu_x(0)]\varrho_{eq}\rangle\} \\ &= \frac{\varrho_0}{\hbar}(1-e^{-\beta\hbar\omega})\sum_{a,b} P(a)(\mu_{ab}M_{ba} - M_{ab}\mu_{ba})\int_0^\infty dt e^{i(\omega-\omega_{ba})t}\end{aligned} \tag{8.16}$$

where $\mu_{ab} = \langle a|\mu_x(0)|b\rangle$ and $M_{ab} = \langle a|M_x(0)|b\rangle$ in this case. Using the relationship $\int_{-\infty}^\infty dt e^{i\omega t}\langle A(t)B(0)\rangle = \left(\int_{-\infty}^\infty dt e^{i\omega t}\langle B(t)A(0)\rangle\right)^*$, the line shape function of the circular dichroism in an isotropic medium is then given by

$$\Delta I(\omega) = \frac{3\hbar\Delta\varepsilon''(\omega)}{4\pi^2(1-e^{-\beta\hbar\omega})} = \frac{\varrho_0}{\pi}\text{Im}\int_{-\infty}^\infty dt e^{i\omega t}\langle\boldsymbol{\mu}(t)\cdot\mathbf{M}(0)\rangle \tag{8.17}$$

This result shows that the cross-correlation function of electric dipole and magnetic dipole is directly related to the CD spectrum via Fourier transformation. Consequently, a direct calculation of the cross-correlation function in time is enough to obtain the CD spectrum.

8.3
Direct Time-Domain Calculation with QM/MM MD Simulation Methods

The VCD line shape and intensity are determined by the anisotropy ratio $g = \Delta A/A = 4R/D$ [2]. Here, the rotational strength (R) is the imaginary part of inner product of the electric and the magnetic dipole transition moments, and the dipole strength (D) is the square of the electric dipole transition moment. The rotational strength of a given vibrational transition, which is the critical quantity in predicting VCD spectrum by performing proper quantum chemical computations, can be obtained by calculating the atomic axial and polar tensors. The computed VCD spectra of isolated peptides were shown to predict the specific sign patterns of the representative protein secondary structures in a satisfactory manner [15]. In the cases where the conformational distribution of a given molecule of interest is comparatively narrow at around a well-defined structure, that is, relatively rigid chiral molecule, the optimized geometry of an isolated molecule in gas phase can be used to predict the observed VCD spectra. However, this is not the case in general for polypeptides and complicated polyatomic chiral molecules in solutions. Then, the band broadening and subtle line shape variations due to hydration or conformational heterogeneity require the explicit inclusion of solvent molecules [16–19], and this would be impractical for polypeptides or proteins.

The QM/MM molecular dynamics simulation method combines the mixed QM/MM Hamiltonian approach with the molecular dynamics simulation and is believed to be a promising approach to resolving those issues inherent in studying condensed phase dynamics. With this method, the chiral solute molecule is treated quantum mechanically and the solvent dynamics is explicitly included in a molecular mechanical way. The total effective Hamiltonian is decomposed into QM, MM, and QM/MM Hamiltonians [20]

$$\hat{H}_{\text{eff}} = \hat{H}_{\text{QM}} + \hat{H}_{\text{MM}} + \hat{H}_{\text{QM/MM}} \tag{8.18}$$

and the coupled Hamiltonian is defined as

$$\hat{H}_{\text{QM/MM}} = -\sum_{i,M} \frac{q_M}{r_{iM}} + \sum_{\alpha,M} \frac{Z_\alpha q_M}{R_{\alpha M}} + \sum_{\alpha,M} \left\{ \frac{A_{\alpha M}}{R_{\alpha M}^{12}} - \frac{B_{\alpha M}}{R_{\alpha M}^6} \right\} \tag{8.19}$$

where the lower case index is to denote the QM part (solute) and the capital letter index is to denote the MM part (solvent). The first term in Equation 8.19 is the Coulomb interaction between the electrons in the QM region and the MM atoms with partial charge q_M, while the second term is that between the QM nuclei and the MM atoms. The last term in Equation 8.19 is the van der Waals interaction between the QM nuclei and the MM atoms with optimized parameters A and B. The solute trajectory is directly governed by the QM and the coupled QM/MM Hamiltonians and indirectly by the MM Hamiltonian.

The simulation procedure follows the conventional MD protocol except for some specifics listed below [21], and we use the GAMESS-UK [22] combined with CHARMM [23] to do the QM/MM MD simulations. The solute whose VCD spectrum is of concern is restrained by a weak harmonic potential at the center of a sphere filled with solvent molecules. The bulk solvent effect outside the spherical system is modeled by the spherical solvent boundary potential [24]. The nonbonding interactions between molecules are evaluated without using a cutoff. The MD integration time step should be compromised between the computational cost and the details in high-frequency dynamics information. When the subtle hydrogen bonding dynamics of the solute is the major issue, however, the time step size should be smaller than 1 fs. Apart from the integration time step size, the trajectory saving time step size determines the upper limit of the spectral window. The finite system size inevitably limits the range of the long-range interactions, which could not fully account in the QM/MM MD algorithm as of now, and the total energy of the simulated system is kept constant to ensure the stability of the spherical system. The population analysis is done for every data saving time step and the atomic partial charges (q_i) along with the solute trajectory (atomic positions and velocities, $\mathbf{r}_i$ and $\mathbf{v}_i$) are used in calculating the electric and the magnetic dipole moments of the QM solute molecule,

$$\mu(t) = \sum_i q_i(t)\mathbf{r}_i(t)$$

$$\mathbf{M}(t) = \frac{1}{2c}\sum_i q_i(t)\mathbf{r}_i(t) \times \mathbf{v}_i(t) \qquad (8.20)$$

where the summation index denotes the solute atoms. To compute the VCD spectra using the QM/MM MD trajectories, one should just carry out a Fourier transformation of the cross-correlation function of the electric dipole moment and the magnetic dipole moment of the solute. Then the raw spectrum is smoothed out to have the experimental spectral resolution.

The characteristic features of VCD spectra provide complementary and often critical information used to elucidate the conformation of polypeptides and proteins [25]. As the first application of this time-correlation function calculation method developed recently by us [12], the alanine dipeptide analogue, $(CH_3)CO$-Ala-NH(CH_3), was chosen to compute the VCD spectra of α-helix, β, P_{II}, and random coils [26]. The basis sets of 3–21G, 4–31G, 6–31G, and 6–31G* were used at the restricted HF level and the solvent is modeled with TIP3P [27]. At each level of the QM method, several MD trajectories were independently generated with different initial conformations and velocities, that is, different initial conditions. The atomic partial charges were obtained by the Löwdin population analysis [28]. The dihedral angles φ and ψ of the peptide were evaluated for each MD snapshot and the trajectory that is dominantly populated with a specific secondary motif was selectively used to calculate the corresponding VCD spectra. We were able to assign the amide I, II, and III modes in the computed VCD spectra by calculating the autocorrelation functions of the relevant internal local vibrations [26].

The polyproline II (P_{II}) structure, defined by the dihedral angles φ = −75° and ψ = 145°, has been extensively studied over the last decade because it was believed to be one of the possible structures of unfolded or denatured proteins [29]. The distinction between the random coil and P_{II} structures has been, therefore, one of the main issues paid a great deal of attention and was often not so clear. The amide VCD bands of P_{II} conformer were observed experimentally to be negative couplets, each with a much weaker positive peak on the higher frequency side. The computed VCD spectrum using a P_{II}-dominant trajectory is shown in the uppermost panel of Figure 8.1. All amide VCD bands were found to be negative couplets in the computed spectra and are indeed compatible to the experimental VCD of P_{II} or random coil polypeptides. However, the relative intensity of the amide II in comparison to the amide I VCD band is very large, which is contrary to the observed VCD spectra.

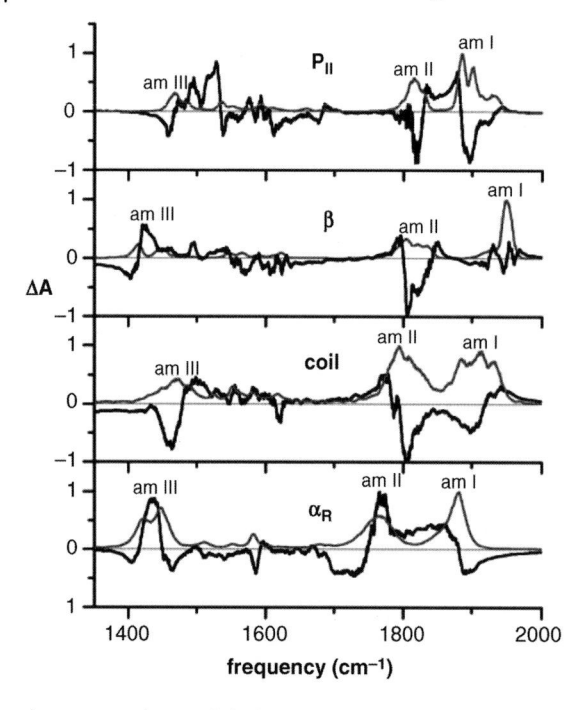

Figure 8.1 Vibrational absorption (gray) and circular dichroism spectra (black) of alanine dipeptide analogue in water using QM/MM MD trajectories highly populated with P_{II}, β, coil, and α-helix conformations, respectively. The trajectories of extended structures, such as P_{II} and β conformers, are obtained with 6-31G* basis and those of α-helix structures with 3-21G basis.

The VCD spectrum of β-structure, including both parallel ($\phi = -119°$, $\psi = 113°$) and antiparallel β-sheet conformers ($\phi = -139°$, $\psi = 135°$), which cannot be distinguished from each other in a single-stranded peptide, is shown in the second panel of Figure 8.1 and is the averaged one over several trajectories generated at the same level of basis set. In the experimental VCD spectra of β-sheet-rich proteins, the amide I VCD is weaker than the amide II VCD. In the computed VCD spectra in Figure 8.1, the amide I band is very weak in intensity and is difficult to assign the band character. Referring to other spectra that are calculated using trajectories generated at different basis level but are represented by β-conformers (not shown here), however, the computed amide I VCD of β-conformers is commonly a weak negative couplet.

The computed coil VCD spectrum is presented in the third panel of Figure 8.1 and this was obtained using a trajectory generated at 6–31G* basis level. Furthermore, we found that the coil VCD spectra obtained at other level of QM method show more or less qualitatively spectral profiles (not shown here) similar to the one presented in Figure 8.1. Comparing the VCD spectra of the three conformers P_{II}, β, and coil, which are shown in the upper three panels in Figure 8.1, one can find that the overall spectral sign patterns are quite similar. The amide I and II VCD bands are negative and the amide III VCD bands are negative couplets, even though the relative peak intensities appear to depend on the specific structures. This indicates that the similar

VCD spectral profile itself does not always guarantee the homogeneous distribution of a specific secondary structure of a peptide in the solution.

It was shown that the VCD spectrum of α-helix ($\phi = -57°$, $(= -47°)$) exhibits the opposite amide I spectral profile in comparison to that of P_{II} VCD and is a positive couplet, while the amide II VCD band was known to be negative and as strong as the amide I VCD band. In the bottom panel in Figure 8.1, we plot the computed VCD spectrum using QM/MM MD trajectories populated dominantly with the α-helical conformation, which is the averaged one over several trajectories generated at the 3–21G basis level. The amide I VCD is a positive couplet that is in agreement with experimental results, while the amide II VCD is a negative couplet with a strong positive peak in the high-frequency region. The amide III VCD is a strong positive peak and the sign pattern is in good agreement with the previous computational and experimental results.

There are issues to be addressed in future studies to improve the agreement between computed and experimental VCD spectra and they include the QM level and basis sets, the population analysis method, the integration time step size in MD simulations, the polarizability of solvent molecules, and so on. The application of the QM/MM MD simulations and the time-correlation function formalism to compute the VCD spectra of alanine dipeptide analogue in water is the very first example showing the promising aspects of the combined method.

8.4
Direct Time-Domain Measurement of VOA Free Induction Decay Field

8.4.1
Conventional Differential Measurement Method

The CD is the result of slightly different responses ($\chi_{L,R}$) of a chiral sample with LCP and RCP lights so that the conventional CD measurement technique is based on this differential measurement scheme, which is depicted in Figure 8.2. Note that the chiral signal is detected at the intensity level not at the amplitude level. Experimentally, an equal amount of LCP or RCP beam (I_0) is alternately created by a polarization modulator (PM) and injected into the chiral sample. Then, the transmitted beam

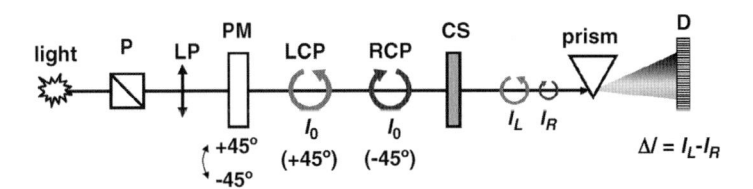

Figure 8.2 Conventional differential CD measurement scheme. P: polarizer, LP: linearly polarized, PM: polarization modulator, CS: chiral sample, D: detector. The PM generates LCP or RCP beam by alternately switching its optic axis to ±45°. The attenuated intensities I_L and I_R by the CS are individually measured and their difference ($\Delta I = I_L - I_R$) is then taken.

intensities I_L and I_R attenuated after traveling through the absorptive medium are separately recorded by detector (D). By taking the difference of their logarithmic scales, the CD spectrum (ΔA) is finally obtained as

$$\Delta A = A_L - A_R = -\log\left(\frac{I_L}{I_0}\right) + \log\left(\frac{I_R}{I_0}\right) = \log\left(\frac{I_R}{I_L}\right) \tag{8.21}$$

With $\Delta I = I_L - I_R$ and $I = (I_L + I_R)/2$, for $\Delta I \ll I$, ΔA is approximately given by $-\Delta I/(2.303 \times I)$. In the case of VCD, $\Delta A/A$ value is typically about 10^{-4}–10^{-5}. Therefore, even if the incident light ($I = \bar{I} + \delta I$) is fairly stable so that intensity fluctuation (δI) is less than 0.1% of its average intensity ($\bar{I}$), it is very difficult to discriminate such a weak chiral signal (ΔI) from largely fluctuating background noise (δI) that is typically one to two orders of magnitude larger than the chiral signal. This is a fundamental and intrinsic problem of the conventional differential measurement method. In the following section, an alternative time-domain method will be discussed.

8.4.2
Femtosecond Spectral Interferometric Approach

8.4.2.1 Cross-Polarization Detection Configuration

Now, let us consider the cross-polarization detection (CPD) configuration shown in Figure 8.3, which was found to be useful to isolate the polarization originating from the chiral response ($\Delta \chi$) as well as to eliminate the strong background signal from the transmitted beam [8–11]. A femtosecond pulse propagates along the z-axis and is polarized in the direction of the y-axis by the first linear polarizer (P1). After passing through the chiral sample, only the x-component of the transmitted field is selected by the second linear polarizer (P2) whose optic axis is perpendicular to the

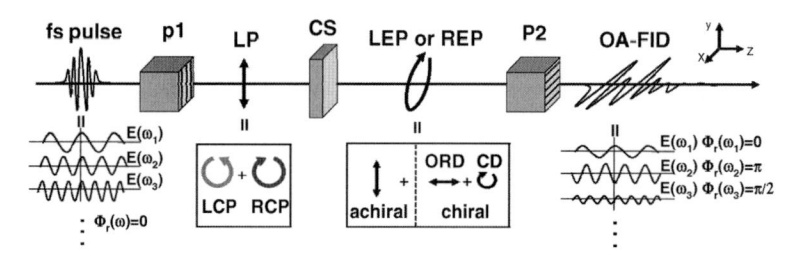

Figure 8.3 Cross-polarization detection configuration for producing OA-FID field. P1–P2: polarizers, LP: linearly polarized, CS: chiral sample, $E(\omega)$: electric field, $\Phi_r(\omega)$: relative spectral phase. The incident femtosecond pulse first becomes vertically polarized by the P1. The CS then transforms it into LEP or REP light whose major axis is inclined to the left or right side. Only the horizontal field component containing chiral information (CD and ORD) is finally selected by the P2. The output field from the P2 can be viewed as a wave packet formed by a superposition of individual frequency components with different phases [$\Phi_r(\omega)$] and amplitudes [|$E(\omega)$|], which are modulated by a chirospecific response $\Delta \chi(\omega)$ of the CS. This time-domain signal field is called OA-FID.

first one. With this CPD configuration, the unit vector of the incident beam is given as $\mathbf{e} = \hat{\mathbf{y}}$ in Equation 8.3 and the rotationally averaged x-component of the linear polarization $P_x^{CPD}(t)$ over the ensemble of randomly oriented chiral molecules in solution is then given as

$$P_x^{CPD}(t) = -i \int_0^\infty d\tau \{\chi_{\mu M}(\tau) + \chi_{M\mu}(\tau)\} E(t-\tau)$$
$$= -\frac{i}{2} \int_0^\infty d\tau \Delta\chi(\tau) E(t-\tau) \tag{8.22}$$

From this, we find that $P_x^{CPD}(t)$ is the same as the difference polarization $\Delta P_x(t)$ in Equation 8.14 except for the constant factor and that the CPD geometry enables a direct measurement of the chiral response without relying on the conventional differential measurement scheme.

The resultant Equation 8.22 includes information on both CD ($\Delta\varkappa = \varkappa_L - \varkappa_R$) and CB ($\Delta n = n_L - n_R$). The incident pulse that is linearly polarized by the first polarizer (P1) is essentially an equal superposition of LCP and RCP components (Figure 8.3). Then, the chiral sample absorbs these opposite components differently due to the CD so that the linearly polarized pulse is transformed into left or right elliptically polarized (LEP or REP) light depending on the handedness of the chiral molecule. At the same time, the major axis of the polarization ellipse becomes rotated due to the CB effect, which in turn gives rise to a phase delay between the LCP and RCP lights. Consequently, the perpendicular polarization component $[P_x^{CPD}(t)]$ to the incident polarization direction ($\mathbf{e} = \hat{\mathbf{y}}$) contains information on the full optical activity of the chiral molecule, whereas the parallel one $[P_y(t)]$ reflects the achiral response. The second polarizer (P2) oriented perpendicular to the polarization direction of the incident beam was thus used to selectively measure the chiral component only.

Often, the parallel component $P_y(t)$ is referred to as optical *free induction decay* (FID), a time-domain observable representing the relaxation of a nonequilibrium ensemble of oscillating electric dipoles created by a short optical pulse whose temporal width is shorter than the corresponding dephasing time. It is related to the electric dipole–electric dipole time-correlation function $\langle \boldsymbol{\mu}(t) \cdot \boldsymbol{\mu}(0) \rangle$ as can be seen in Equations 8.5–8.9. Now, the perpendicular component $P_x^{CPD}(t)$ can thus be named as *optical activity FID* (OA-FID) because it carries detailed information on the chirality of the molecule in its phase and amplitude. In contrast to the conventional FID, the OA-FID is led by the electric dipole–magnetic dipole time-correlation function $\langle \boldsymbol{\mu}(t) \cdot \mathbf{M}(0) \rangle$.

In reality, the experimentally measured quantity is not the polarization itself but the electric field $E(t)$. For the x- and y-components of the emitted signal electric field at position z inside the sample, $E_x(z, t)$ and $E_y(z, t)$, we found that the Maxwell equation is given as

$$\nabla^2 E_x(z,t) - \frac{1}{c^2} \frac{\partial^2}{\partial t^2} E_x(z,t) = \frac{4\pi}{c^2} \frac{\partial^2}{\partial t^2} P_x^{CPD}(z,t) \tag{8.23}$$

where

$$P_x^{CPD}(z,t) = -\frac{i}{2}\int_0^\infty d\tau \Delta\chi(\tau) E_y(z, t-\tau) + \int_0^\infty d\tau \chi_{\mu\mu}(\tau) E_x(z, t-\tau) \qquad (8.24)$$

Note that $P_x^{CPD}(z,t)$, determined by both $E_x(z,t)$ and $E_y(z,t)$, acts as the source of $E_x(z,t)$. By solving this equation in the frequency domain [8], it was found that the x-component of the emitted electric field, that is, OA-FID, after the sample length L is given as

$$E_x(L,\omega) = \frac{\pi\omega L}{cn(\omega)} \Delta\chi(\omega) E_y(L,\omega) \qquad (8.25)$$

where $E_y(\omega)$ is the complex transmitted electric field spectrum given by the interference between the input field and the electric dipole-allowed FID and the complex function $\Delta\chi(\omega)$ is the linear optical activity susceptibility. This suggests that once the phases and amplitudes of both $E_x(\omega)$ and $E_y(\omega)$ are measured in the frequency domain, the linear optical activity susceptibility $\Delta\chi(\omega)$ whose imaginary and real parts correspond to the CD and ORD spectra, respectively, can be fully characterized.

8.4.2.2 Fourier Transform Spectral Interferometry

It is the Fourier transform spectral interferometry (FTSI) that turned out to be quite useful in precisely determining both the phase and amplitude of any unknown weak electric field [30, 31]. Figure 8.4 shows the standard FTSI data conversion procedure. When a reference electric field $E_{ref}(\omega)$ precedes the unknown field $E_{unknown}(\omega)$ by a fixed time delay τ_d, they interfere with each other in the frequency domain. The heterodyne term of the resultant spectral interferogram $S_{het}(\omega)$ can be detected by a spectrometer as

$$S_{het}(\omega) = 2\text{Re}\left[E_{ref}^*(\omega) E_{unknown}(\omega) \exp(i\omega\tau_d)\right] \qquad (8.26)$$

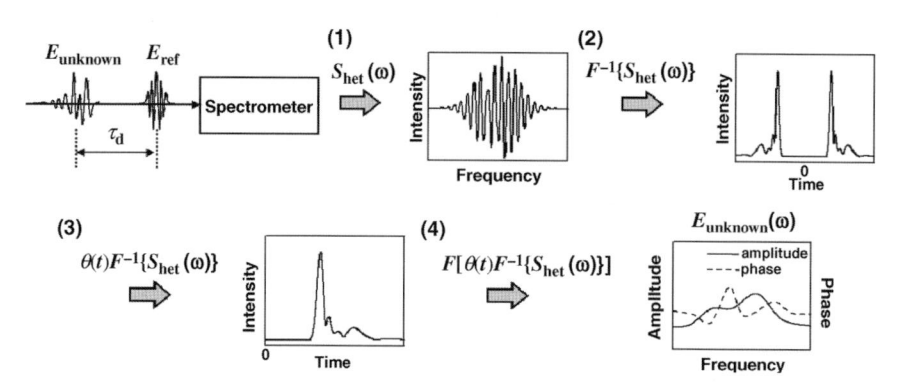

Figure 8.4 Standard FTSI procedure for characterizing an unknown electric field $E_{unknown}$. See the text for details.

Once the real function $S_{het}(\omega)$ is recorded (step 1), it is inverse Fourier transformed into the time-domain signal $F^{-1}\{S_{het}(\omega)\}$ (step 2). Then, by multiplying the Heaviside step function $\theta(t)$ with $F^{-1}\{S_{het}(\omega)\}$ (step 3) and Fourier transforming the $\theta(t) F^{-1}\{S_{het}(\omega)\}$ (step 4), the complex form of Equation 8.26 is obtained as $F[\theta(t) F^{-1}\{S_{het}(\omega)\}]$. Consequently, the complex electric field $E_{unknown}(\omega)$ is now given by

$$E_{unknown}(\omega) = \frac{F[\theta(t)F^{-1}\{S_{het}(\omega)\}]\exp(-i\omega\tau_d)}{2E_{ref}^*(\omega)} \tag{8.27}$$

If one has a well-defined reference field $E_{ref}(\omega)$ with a fixed τ_d, the phase-and-amplitude information of the $E_{unknown}(\omega)$ can be fully retrieved from the measured spectral interferogram.

Similarly, the weak OA-FID field can be measured using the FTSI procedure described above. However, the y-component of the emitted FID (nonchiral FID) in addition to its x-component (OA-FID) should also be measured to completely characterize the optical activity susceptibility $\Delta\chi(\omega)$ because it is given by the ratio of $E_x(\omega)$ to $E_y(\omega)$. Figure 8.5 is a schematic outline of the femtosecond spectral interferometry setup used to measure such FID fields. A linearly polarized femtosecond pulse whose polarization direction is inclined at 45° with respect to the y-axis is injected into a Mach–Zehnder interferometer [13] and is split into two parts, signal and reference arms, by a beam splitter (BS1). In the signal arm, the OA-FID or nonchiral FID signal electric field (E_s) is created through the CPD or parallel polarization detection configuration, respectively. In the reference arm, on the other

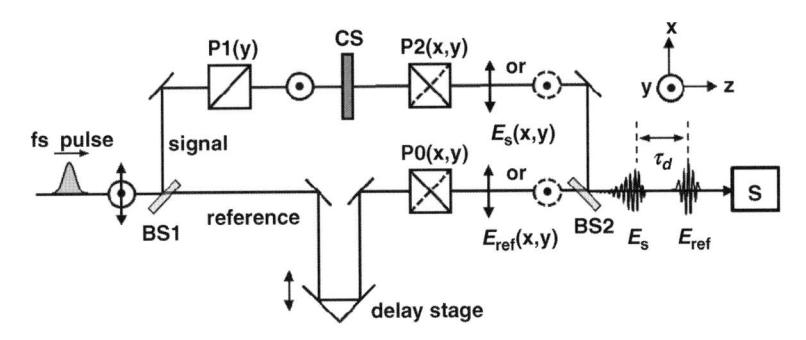

Figure 8.5 Femtosecond spectral interferometric measurement setup. P0–P2: polarizers, BS1–BS2: beam splitters, CS: chiral sample, S: spectrometer. The incident femtosecond pulse is split into the signal and reference arms of Mach–Zehnder interferometer by the BS1. One is used as an excitation pulse generating the FID signal fields (E_s) and the other as a reference (E_{ref}). For the measurement of OA-FID (x-component of E_s), the transmission axis of P1 is aligned to be the Y-axis, whereas those of P2 and P0 to be the X-axis. For the measurement of nonchiral FID (y-component of E_s), on the other hand, those of all the polarizers (P0–P2) are in the Y-axis. For spectral interferometric detection, the delay stage is controlled to make the E_{ref} precede the E_s by a fixed time delay τ_d and they are combined at the BS2. Finally, the resultant spectral interferograms ($S_{(\perp,\parallel)}$) are recorded at spectrometer.

hand, the reference electric field (E_{ref}) is properly delayed such that it precedes the signal field by a fixed time delay τ_d for the spectral interferometric detection. Then, the signal and reference fields are combined by another beam splitter (BS2) and finally recorded as a spectral interferogram by a spectrometer. In the CPD configuration, the transmission axes of the polarizers P1, P2, and P0 lie in the y-, x-, and x-axes, respectively, and the perpendicular detected spectral interferogram $S_\perp(\omega)$ is then measured, whereas for the measurement of the parallel detected spectral interferogram $S_\parallel(\omega)$, those of both P2 and P0 are rotated into the y-axis with the P1 being fixed. Note that the transmission axis of P0 is controlled to be the same as that of P2 for maximal heterodyning efficiency.

Assuming that the phase and amplitude of the reference field $E_{ref}(\omega)$ and the time delay τ_d remain unaltered during the two measurements of $S_{(\perp,\parallel)}(\omega)$, we showed that by combining Equations 8.25 and 8.27 $\Delta\chi(\omega)$ is given as

$$\Delta\chi(\omega) = \frac{cn(\omega)}{\pi\omega L} \frac{F^{-1}[\theta(t)F^{-1}\{S_\perp(\omega)\}]}{F^{-1}[\theta(t)F^{-1}\{S_\parallel(\omega)\}]} \tag{8.28}$$

Note that the reference field $E_{ref}(\omega)$ in the denominator of Equation 8.27 and an additional phase term $\exp(-i\omega\tau_d)$ in the numerator of Equation 8.27 cancel out by taking the ratio of $E_x(\omega)$ to $E_y(\omega)$. From the definition of the differential absorption coefficient $\Delta\varkappa(\omega)$, we find that the absorption difference ΔA (CD) of chiral molecules for LCP and RCP beams is given as

$$\Delta A(\omega) = \frac{\Delta\varkappa(\omega)L}{2.303} = \frac{4}{2.303}\,\text{Im}\left[\frac{F[\theta(t)F^{-1}\{S_\perp^{het}(\omega)\}]}{F[\theta(t)F^{-1}\{S_\parallel^{het}(\omega)\}]}\right] \tag{8.29}$$

where

$$\Delta\varkappa(\omega) \equiv \varkappa_L(\omega) - \varkappa_R(\omega) = \frac{4\pi\omega}{n(\omega)c}\,\text{Im}[\Delta\chi(\omega)] \tag{8.30}$$

On the other hand, for a dilute solution, the ORD $\Delta\varphi(\omega)$ defined by the half of phase difference between LCP and RCP after passing through the sample is given as

$$\Delta\varphi(\omega) \equiv \Delta n(\omega)\frac{\omega}{2c}L = \text{Re}\left[\frac{F[\theta(t)F^{-1}\{S_\perp^{het}(\omega)\}]}{F[\theta(t)F^{-1}\{S_\parallel^{het}(\omega)\}]}\right] \tag{8.31}$$

where

$$\Delta n(\omega) = n_L(\omega) - n_R(\omega) = \frac{2\pi}{n(\omega)}\,\text{Re}[\Delta\chi(\omega)] \tag{8.32}$$

Equations 8.29 and 8.31 are the principal results of the femtosecond spectral interferometric detection method, which demonstrate that both CD and ORD spectra can simultaneously be obtained from the measured spectral interferograms $S_{\perp,\parallel}(\omega)$. Also, it should be emphasized that even though the interferogram $S_\perp(\omega)$ containing chiral information is recorded in the frequency domain via the spectral interferometric detection, the measured quantity is the time-domain OA-FID signal electric field that is directly associated with the electric dipole–magnetic dipole time-correlation function $\langle \boldsymbol{\mu}(t) \cdot \mathbf{M}(0) \rangle$.

8.4.2.3 Vibrational OA-FID Measurement

The theory and the experimental scheme of the femtosecond OA-FID measurement approach presented above is sufficiently general, so this method can be readily applied to a variety of optical activity measurement experiments over electronic (UV-Vis) to vibrational (IR) transition frequency ranges. Experimentally, however, a measurement of vibrational OA-FID (VOA-FID) has been believed to be extremely challenging because of its weak chiral susceptibility and low detectivity of IR detectors. In the CPD configuration, extinction ratio of the crossed polarizer pairs (P1 and P2) is a critical factor for successfully discriminating such a weak VOA-FID signal from comparatively huge achiral FID. For typical chiral molecules with ΔA (VCD) $= 10^{-4}$–10^{-5}, linear polarizers with the exceptionally small extinction ratio of $\sim 10^{-8}$ are required. Dichroic absorptive calcite polarizers meet this rigorous criterion [32], even though its effective working frequency window is restricted within narrow spectral limits (3.35–3.5 μm, 3.9–4.1 μm, and so on), and it can thus be used to examine VCD-active C–H stretching vibrational modes (2800–3000 cm^{-1}) of organic chiral molecules [9, 10].

Figure 8.6 depicts the step-by-step procedure for retrieving the VCD and VORD spectra from the perpendicular and parallel detected spectral interferograms $S_{\perp,\parallel}(\omega)$ of (1S)-β-pinene, which has been considered to be a standard chiral molecule often used to verify performance of VCD spectrometer. The dispersed heterodyned spectral interferograms $S_\perp(\omega)$ (solid) and $S_\parallel(\omega)$ (dashed) are depicted in Figure 8.6a and they exhibit distinct spectral shapes in terms of phases and amplitudes. In particular, the $S_\perp(\omega)$ has highly oscillating features than the $S_\parallel(\omega)$, which implies that the $S_\perp(\omega)$ contains more complicated positive/negative sign information on the optical activity. Each interferogram is then inverse Fourier transformed and multiplied by the Heaviside step function. The resultant time-domain signal amplitudes $|\theta(t) F^{-1}\{S_\perp(\omega)\}|$ (solid) and $|\theta(t) F^{-1}\{S_\parallel(\omega)\}|$ (dashed) are displayed in Figure 8.6b. Here $\theta(t) F^{-1}\{S_\parallel(\omega)\}$ is the convolution product of $E_{\text{ref}}(t)$ and $E_y(t)$ given by the interference

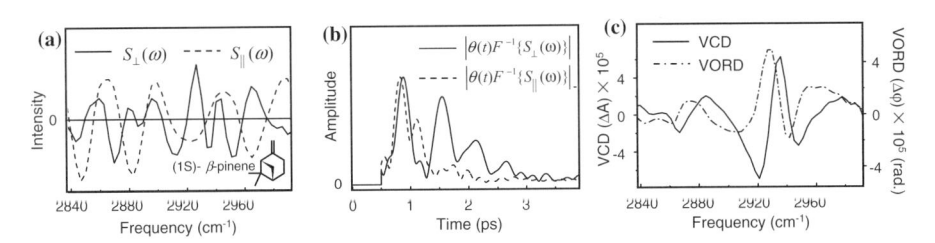

Figure 8.6 VCD and VORD measurements of (1S)-β-pinene in CCl$_4$ using the femtosecond spectral interferometry. (a) Experimentally measured heterodyned spectral interferograms $S_\perp(\omega)$ (solid) and $S_\parallel(\omega)$ (dashed). (b) Inverse Fourier transformed time-domain signal amplitudes $|\theta(t) F^{-1}\{S_\perp(\omega)\}|$ (solid) and $|\theta(t) F^{-1}\{S_\parallel(\omega)\}|$ (dashed). (b) Retrieved VCD (solid) and VORD (dashed-dotted) spectra obtained by using Equations 8.29 and 8.31, respectively.

between the input field and the achiral FID field, whereas the $\theta(t)F^{-1}\{S_\perp(\omega)\}$ is that between $E_{ref}(t)$ and $E_x(t)$, which is in turn given by the convolution between $\Delta\chi(t)$ and $E_y(t)$. Although the time-domain signal in Figure 8.6b is not absolutely identical to the chiral response $\Delta\chi(t)$ itself or the achiral response $\chi_{\mu\mu}(t)$, $\theta(t)F^{-1}\{S_\perp(\omega)\}$ and $\theta(t)F^{-1}\{S_\parallel(\omega)\}$ are governed by $\Delta\chi(t)$ and $\chi_{\mu\mu}(t)$, that is, $\langle \boldsymbol{\mu}(t) \cdot \mathbf{M}(0) \rangle$ and $\langle \boldsymbol{\mu}(t) \cdot \boldsymbol{\mu}(0) \rangle$, respectively. Finally, Figure 8.6c shows the VCD (solid) and VORD (dashed-dotted) spectra retrieved by Fourier transforming the FID signals in Figure 8.6b and then by using Equations 8.29 and 8.31. It is noteworthy that the simultaneous acquisition of both VCD and VORD spectra is allowed by directly measuring the phase and amplitude of the time-domain OA-FID field through the femtosecond spectral interferometry.

8.5
Summary and a Few Concluding Remarks

In this chapter, we presented both time-domain measurement and calculation methods for vibrational CD and ORD spectra of chiral molecules in condensed phases. From the time-correlation function theory for the optical activity, we showed that the direct calculation of electric dipole–magnetic dipole cross-correlation function is possible by using a QM/MM molecular dynamics simulation method. Here, the chiral solute molecule is treated quantum mechanically, whereas the surrounding solvent molecules are treated classically. Since the vibrational dynamics is determined by quantum mechanically calculated potential surface, the anharmonic natures of the potential energy surface are fully taken into consideration. Note that the classical force fields used in conventional molecular mechanical MD simulation methods are often inaccurate in quantitatively describing vibrational dynamics of polyatomic molecule in condensed phases. Furthermore, the fluctuating partial charge effects originating from the intrinsic polarizable nature of molecules in solutions were also automatically included in our direct calculation method utilizing combined QM and MM molecular dynamics simulation methods.

In addition, we have developed experimental method that can also be used to directly measure the optical activity response function in time domain by using femtosecond laser pulses and employing a properly designed Mach–Zehnder spectral interferometry. Detailed procedures on how to obtain the vibrational CD and ORD spectra from the measured spectral interferograms have also been discussed here. Thus, we showed that the time-domain calculation and measurement techniques for simultaneously characterizing vibrational CD and ORD spectra of chiral molecules in condensed phases are not only feasible but also potentially useful. In particular, the present phase- and amplitude-sensitive VCD and VORD measurement technique has a superior time resolution in comparison to the conventional VCD spectroscopy based on the differential absorption measurement. Consequently, one can directly extend this method to study ultrafast dynamics of chiral molecules in solutions with unprecedented achievable time resolution.

References

1 Born, M. and Wolf, E. (1999) *Principles of Optics*, Cambridge University Press, Cambridge.
2 Berova, N., Nakanishi, K., and Woody, R.W. (2000) *Circular Dichroism: Principles and Applications*, John Wiley & Sons, Inc., New York.
3 Kronig, R.deL. (1926) *J. Opt. Soc. Am.*, **12**, 547–557.
4 Kramers, H.A. (1927) *Atti. Congr. Intern. Fisica. Como.*, **2**, 545–557.
5 Nusenzveig, H.M. (1972) *Causality and Dispersion Relations*, Academic Press, New York.
6 Holzwarth, G., Hsu, E.C., Mosher, H.S., Faulkner, T.R., and Moscowitz, A. (1974) *J. Am. Chem. Soc.*, **96**, 251–252.
7 Nafie, N.A., Keiderling, T.A., and Stephens, P.J. (1976) *J. Am. Chem. Soc.*, **98**, 2715–2723.
8 Rhee, H., Ha, J.-H., Jeon, S.-J., and Cho, M. (2008) *J. Chem. Phys.*, **129**, 094507.
9 Rhee, H., June, Y.-G., Lee, J.-S., Lee, K.-K., Ha, J.-H., Kim, Z.H., Jeon, S.-J., and Cho, M. (2009) *Nature*, **458**, 310–313.
10 Rhee, H., June, Y.-G., Kim, Z.H., Jeon, S.-J., and Cho, M. (2009) *J. Opt. Soc. Am. B*, **26**, 1008–1017.
11 Rhee, H., Kim, S.-S., Jeon, S.-J., and Cho, M. (2009) *ChemPhysChem*, **10**, 2209–2211.
12 Jeon, J., Yang, S., Choi, J.-H., and Cho, M. (2009) *Acc. Chem. Res.*, **42**, 1280–1289.
13 Loudon, R. (1983) *The Quantum Theory of Light*, Clarendon, Oxford.
14 Cho, M. (2009) *Two-Dimensional Optical Spectroscopy*, CRC Press, Boca Raton.
15 Bour, P. and Keiderling, T.A. (1993) *J. Am. Chem. Soc.*, **115**, 9602–9607.
16 Han, W.-G., Jalkanen, K.J., Elstner, M., and Suhai, S. (1998) *J. Phys. Chem. B*, **102**, 2587–2602.
17 Kubelka, J., Huang, R., and Keidering, T.A. (2005) *J. Phys. Chem. B*, **109**, 8231–8243.
18 Bour, P. and Keiderling, T.A. (2005) *J. Phys. Chem. B*, **109**, 23687–23697.
19 Turner, D.R. and Kubelka, J. (2007) *J. Phys. Chem. B*, **111**, 1834–1845.
20 Lipkowitz, K.B., Cundari, T.R., and Boyd, D.B. (1996) *Reviews in Computational Chemistry*, Wiley–Interscience, Hoboken, NJ.
21 Allen, M.P. and Tildesley, D.J. (1987) *Computer Simulation of Liquids*, Oxford University Press, New York.
22 Guest, M.F., Bush, I.J., van Dam, H.J.J., Sherwood, P., Thomas, J.M.H., Lenthe, J.H., Havenith, R.W.A., and Kendrick, J. (2005) *Mol. Phys.*, **103**, 719–747.
23 Brooks, B.R., Bruccoleri, R.E., Olafson, B.D., States, D.J., Swaminathan, S., and Karplus, M. (1983) *J. Comput. Chem.*, **4**, 187–217.
24 Beglov, D. and Roux, B. (1994) *J. Chem. Phys.*, **100**, 9050–9063.
25 Keiderling, T.A. (2002) *Curr. Opin. Chem. Biol.*, **6**, 682–688.
26 Yang, S. and Cho, M. (2009) *J. Chem. Phys.*, **131**, 135102.
27 Jorgensen, W.L., Chandrasekhar, J., Madura, J.K., Impey, R.W., and Klein, M.L. (1983) *J. Chem. Phys.*, **79**, 926–935.
28 Löwdin, P.-O. (1950) *J. Chem. Phys.*, **18**, 365–375.
29 Rose, G.D. (2002) *Advances in Protein Chemistry: Unfolded Proteins*, vol. 62, Academic Press, San Diego.
30 Lepetit, L., Chériaux, G., and Joffre, M. (1995) *J. Opt. Soc. Am. B*, **12**, 2467–2474.
31 Fittinghoff, D.N., Bowie, J.L., Sweetser, J.N., Jennings, R.T., Krumbugel, M.A., DeLong, K.W., Trebino, R., and Walmsley, I.A. (1996) *Opt. Lett.*, **21**, 884–886.
32 Bridges, T.J. and Klüver, J.W. (1965) *Appl. Opt.*, **4**, 1121–1125.

9
Electronic Circular Dichroism
Lorenzo Di Bari and Gennaro Pescitelli

9.1
Introduction

Electronic circular dichroism (ECD) is one of the most simple and sensitive spectroscopic methods for structural analysis. It is quite strange that it is looked at with completely different eyes by two classes of researchers: while those dealing mostly with small molecules consider it as a tool for absolute configuration assignments, those working with macromolecules (synthetic or biological polymers) use it to investigate conformations and more in particular regular secondary structures. To our experience, it is uncommon for a "molecular chemist" to think of ECD for conformational analysis or to follow, for example, the adjustments taking place during the course of a reaction. On the other hand, biochemists and polymer chemists tend to forget that 100% enantiomeric purity hardly exists, that racemization and epimerization processes occur, and that there are very relevant examples of violation of what can be regarded as a homochirality dogma.

Let us broadly classify some relevant problems of molecular structure, with reference to the specific example of compound **1** (Blennolide E, Scheme 9.1) [1].

a) Absolute configuration (AC), that is, ($2R,9S,10S,11R$) versus ($2S,9R,10R,11S$);
b) Relative configuration (RC), for example, ($2R,9S,10S,11R$) versus ($2S,9S,10S,11R$); and
c) Conformation.

Points (b) and (c) change the molecular shape and very importantly imply different spatial relations (distances and angles) between atoms and groups. Many – or even most – spectroscopies are sensitive to these parameters, which are scalar quantities. On the contrary, point (a) implies a complete coordinate inversion, which leaves all the scalar quantities unchanged, while inverting pseudoscalars, like dihedral angles. Only chiroptical techniques (ECD, vibrational circular dichroism or VCD, circularly polarized luminescence or CPL, Raman optical activity or ROA, and optical rotation or OR) sense this, and this is one reason why they are regarded as a tool for solving

Computational Spectroscopy: Methods, Experiments and Applications. Edited by Jörg Grunenberg
Copyright © 2010 WILEY-VCH Verlag GmbH & Co. KGaA, Weinheim
ISBN: 978-3-527-32649-5

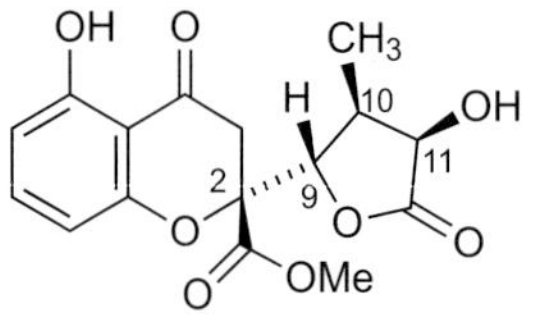

Scheme 9.1

this problem. Nonetheless, all of them strongly depend on scalar quantities as well, as we shall discuss below.

Let us focus on ECD: its fundamental equation was given by Rosenfeld and states that, if we define the molar ECD as the difference between the molar extinction coefficients versus left and right circularly polarized radiation ($\Delta\varepsilon = \varepsilon^{\ell} - \varepsilon^{r}$), then the integral of an ECD band (also called Cotton effect, CE) due to one electronic transition i between the states is proportional to

$$R_i = \text{Im}(\boldsymbol{\mu}_i \cdot \mathbf{m}_i) \qquad (9.1)$$

where $\boldsymbol{\mu}_i$ and $\mathbf{m}_i$ are the electric and magnetic transition dipoles, respectively. The quantity R_i is called *rotational strength* (in an analogy with the oscillator strength) and it is a pseudoscalar (i.e., it changes sign upon coordinate inversion) because it is the result of the scalar product between a proper vector $\boldsymbol{\mu}_i$ and a pseudovector $\mathbf{m}_i$.

Whichever computational tool may provide $\boldsymbol{\mu}_i$ and $\mathbf{m}_i$ for all the relevant transitions i, it can be used in principle to predict the rotational strengths: after applying a suitable band shape to the set of calculated R_i (see Section 9.5), one obtains a calculated ECD spectrum to be compared with experimental data. This idealized situation must take into account temperature effects and calls for Boltzmann averaging over a structural ensemble or even for the existence of hot bands. Both these effects will be discussed in some detail in this chapter.

Coming back to the stereochemical problems outlined above, we may put forward that often carbon stereogenic centers are inert, that is, they do not lead to an appreciable R/S inversion with temperature, possibly leading to racemization or epimerization process. In other words, at the very first approximation, we may consider absolute and relative configurations (points (a) and (b) above) thermally stable. On the contrary, depending on the structure, there may be a considerable amount of temperature-dependent conformational variation (point (c) above). An interesting borderline case is provided by *atropisomers*, that is, frozen conformations that cannot jump from one to the other. This has important fallouts in our context, for example, for axial chirality in biaryls or dimeric compounds. We shall treat these *conformational* stereogenic elements as if they were *configurational*.

Application of Rosenfeld equation requires that we calculate all the relevant magnetic and electric transition dipoles of a molecule provided, with reference to the points a–c above,

- a') an absolute configuration is more or less arbitrarily chosen;
- b') for molecules containing more than one stereogenic element, the relative configurations are known, or a set of all the possible diastereomers must be built and taken into full account; and
- c') for each diastereomer, a reasonable set of conformational isomers is considered.

Because for two enantiomers all the rotational strengths must be exactly opposite, for point (a') either we produce a calculated ECD spectrum or its mirror image. There is no symmetry relation for the isomers to be considered in point (b') and (c'), and each should be separately calculated. Very often, one does have an exact knowledge of the relative configurations (e.g., from other spectroscopies such as NMR or IR, or from X-ray diffraction, XRD), but there are examples when ECD has been used to discriminate among diastereomers [1]. As long as we can consider the configurations of all the elements in the molecule inert, point (b') does not require any temperature averaging.

Conformational manifolds, as implied by point (c'), will be the object of a more thorough discussion in Section 9.3.

9.2
Molecular Anatomy

What we said in Section 9.1 is common about circular dichroism and related techniques, that is, it holds for both ECD and VCD. What makes ECD a particularly interesting tool is that it looks at electronic transitions, which is practically viable only in the presence of specific chemical features called chromophores. A chromophore may be defined as part of a molecule responsible for one or more absorption bands in the UV or in the visible spectrum. In the context of organic chemistry, it is usually a functional group or a combination of several groups with a more or less extended π-electron system.

In fact, it is unlikely that bonding σ-orbitals contribute significantly to absorption bands above 180 nm. This is often regarded as a limitation of ECD versus, for example, VCD, where practically all the bonds can be regarded as potential chromophores. Indeed, UV-vis transparent molecules cannot be used for ECD. At the same time, it is a great advantage, because one may focus only on chromophoric moieties, neglecting the rest of the molecular backbone, which has relevant consequences in computing ECD.

Moreover, the fact that optical activity is essentially centered on chromophores gives chemists the opportunity to change ECD by modifying the chromophores, that is, by inserting groups that extend conjugation or by adding new chromophores [2, 3].

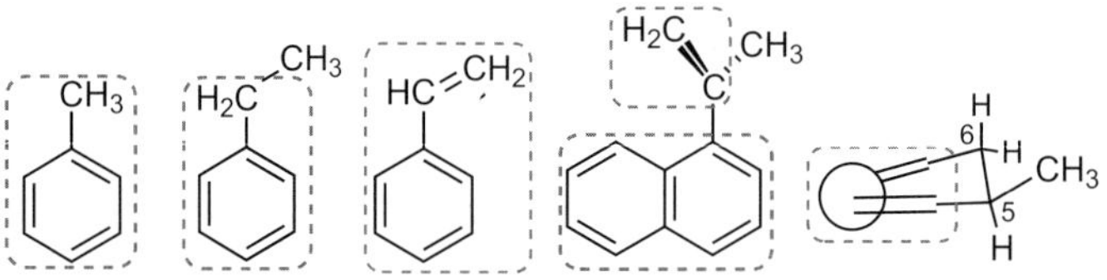

Scheme 9.2

Unfortunately, this fascinating field goes beyond the scope of the present discussion on computational methods.

We wish to introduce here a concept of *molecular dissection*, that is, the separation of a molecule's anatomy into structural and functional parts, like a skeleton and the organs in a living organism. Clearly, there may be parts sharing both functions.

As we mentioned before, organic chromophores must contain π-orbitals and the whole set of atoms, where electron delocalization may extend, must be considered part of the chromophore (see framed portions in Scheme 9.2). Thus, for example methyl benzene must be considered as a whole, but the methyl group in ethyl benzene can be thought as not conjugated to the rest and can be neglected, that is, methyl and ethyl benzene can be considered equivalent from our point of view. Styrene is clearly another whole, at least as long as conjugation is not prevented by, for example, steric effects, as in the case of 1-(1′-methylvinyl)naphthalene where the two unsaturated planes are at right angles and electron delocalization is impossible.

The latter example should tell us that in certain cases the notion of chromophore must be actively treated and may need revision depending on the notion of geometrical features. The degree of overlap between *p*-orbitals depends on the cosine of the dihedral they define; thus, it is very large even for almost orthogonal arrangements ($\cos 80° = 0.17$!). Skewed conjugation constitutes an example of *intrinsic chirality*: the chromophore itself, lacking improper symmetry elements (planes, inversion, or improper axes in general), has electronic transitions endowed with nonvanishing and nonorthogonal $\boldsymbol{\mu}_i$ and $\mathbf{m}_i$, which leads to nonvanishing rotational strength R_i. When perfect conjugation leads to planar chromophores, a fundamental symmetry property of pseudoscalars leads to vanishing R_i (either because one of the two transition dipole moments is zero or because they are orthogonal). In this case, we speak of *intrinsically achiral* chromophores and their optical activity must arise from their coupling with the surroundings.

Spectroscopically silent moieties perform two actions: they contribute to the environment, where the chromophore is hosted, and they participate in giving the molecule a more or less defined shape or flexibility.

Thus, the σ-backbone may cause a twist in the conjugated chromophore and determine its intrinsic chirality. Alternatively, it may provide symmetry-breaking groups, which lower the local symmetry of an intrinsically achiral chromophore and determine the emergence of optical activity in its transitions. Clearly, their effects must be taken into account in calculations.

For example, let us consider 5-methyl cyclohexa-1,3-diene (last molecule in Scheme 9.2). The 6-membered ring prevents the diene from being planar. The methyl group must occupy a pseudoequatorial position, determining one sense of the twist of the diene, which becomes intrinsically chiral. At the same time, the two allylic axial hydrogen atoms in positions 5 and 6 constitute chiral perturbations to the chromophore, and the observed ECD allied to the low-lying $\pi-\pi^*$ transition of this compound around 240 nm is the result of the combination of both effects [4].

From this brief discussion, it should be clear that a correct molecular dissection must take into account the chromophores together with their surroundings, *through space*, or independent of the existence of covalent bonds. This raises an obvious question about the size of the chromophore environment to be taken into account. As we shall further see here, a primary reason of through space interaction is the electric dipole interaction, which not only decays with the third power of the distance (r^{-3}) but also depends on the polarity or the polarizability of atoms and groups. For apolar and weakly polarizable systems, a radius of a few Å from the chromophore may be adequate.

Because of their very strong polarizability, if two or more chromophore are present in the same system, their interaction may extend over very large distances and very often provides a contribution to the observed ECD, which largely overrides other terms and calls for a different approach. Exciton-coupled ECD is a theoretical framework where the spectroscopically silent portions are completely neglected (or are considered only for their structural function) and one focuses on the interaction between electric dipole-allowed transitions. First-order quantum mechanics (QM) perturbation theory predicts states mixing and possibly (if the geometric arrangement allows it) ECD for a pair of chromophores in a rather straightforward way. If one deals with more than two electronic transitions, things become more complicated, and sometimes additivity of pair-wise interactions has been invoked. We believe that there is no need for this simplification because hybrid approaches described below have been shown very successful in accounting for a large number of transitions without the limitation to first order but rather exactly (*all order*) by means of classical electrodynamics.

The biological world offers a useful illustrative example of the involved relationship between the chromophore and its environment. Many proteins crucial for life contain a chromophoric prosthetic group (also indicated as dye or pigment) either covalently or noncovalently attached, endowed with redshifted absorption bands well distinguished from those of the peptide skeleton (below 240 nm) or its side groups (below 300 nm) [5]. Noteworthy examples are myoglobin and hemoglobin, containing a heme, that is, a porphyrin analogue, with absorptions at 400–430 nm and between 500 and 600 nm, and rhodopsin, containing retinal with absorption at 335 and 500 nm. Although the isolated chromophores are achiral, when embedded in the protein, they show moderately intense ECD bands corresponding to their respective absorptions. Where do they arise from? In principle, the following two different mechanisms may be responsible for them:

- When the dye is placed in the protein, it is distorted from its originally planar structure to a nonsymmetric, chiral structure; therefore, it constitutes an intrinsically chiral chromophore whose electronic transitions are ECD-active.
- The electronic transitions of the chromophore couple with those of the protein chromophores (amide bonds and aromatic side groups).

Which of the two mechanisms dominates the observed ECD depends on the case. Matrix-type calculations (see Section 9.4) have, for example, shown that the ECD of myoglobin is due to the exciton coupling between the porphyrin core and the aromatic side groups [6], while a combination of the same calculations and TDDFT (Section 9.5) has demonstrated that the ECD of rhodopsin is dominated by the intrinsic chirality of retinal [7].

9.3
Conformational Manifolds and Molecular Structure

We shall see below that ECD is very sensitive to molecular conformations, which is in strong contrast to most frequent cases of ordinary absorption. Indeed, frequency and transition probability to an excited state are primarily related to an electric transition dipole, which is often only weakly affected by the geometry of the environment of an organic chromophore. The situation may be very different for metal complexes where the coordination geometry can have a strong influence on the absorption spectrum: in this case, however, one should consider at least the whole first coordination sphere as the chromophore. ECD, in contrast, arises from an interference effect (see Rosenfeld equation (9.1)) and depends not only on the electric and magnetic dipole moments but also on their relative orientation (owing to the scalar product), that is, explicitly on the geometry.

Before going into any further discussion, we wish to define what should be regarded as a molecular structure. In a static picture, and for a single molecule, one may consider each conformer as one structure (possibly endowed with more or less large vibrational features), but for an ensemble and in the general case, a full conformer distribution must be taken into account, which amounts to saying that a molecular structure is a set of conformers (i.e., geometrical parameters defining them) associated with temperature-dependent Boltzmann weights. For a relatively rigid molecule, this indeed may reduce to one conformer.

The calculated ECD spectrum over a conformational distribution, that is, over a structure according to our definition, is the linear combination of the contributions of the various conformers χ, with their relative populations a_χ as multipliers

$$\Delta\varepsilon^{calc}(\lambda) = \sum a_\chi \Delta\varepsilon^\chi(\lambda) \tag{9.2}$$

Only this should be compared with the experiment, and not the individual $\Delta\varepsilon^\chi(\lambda)$, unless one postulates that the structure consists in one conformer (sometimes, this can be defined as monorotameric approximation).

Even if the ECD spectrum encodes relevant geometric information, it is usually very hard or impossible to extract it from the direct analysis of experimental data (which is in contrast, for example, with XRD or some NMR data). What is often done is calculate the various $\Delta\varepsilon^\chi(\lambda)$ and assess the Boltzmann weights a_χ at the working temperature T by means of computed free energies E (often approximated to enthalpies, or internal energies, without zero-point vibrational corrections):

$$a_\chi = \exp(-E_\chi/RT) \tag{9.3}$$

Alternatively, one may try and optimize the multipliers a_χ by fitting experimental and calculated ECD spectra. More sophisticated procedures, essentially based on linear algebra, have rarely proved satisfactory.

If the objective of an ECD analysis is primarily the absolute configuration assignment, it may be very useful to reduce the conformational freedom as much as one can. At least two strategies can be envisaged.

First, one can take advantage of the structural homogeneity guaranteed by crystals: if a substance gives rise to one crystal form, whose structure can be solved by XRD, apart from its AC, one is in the ideal situation of carrying out the ECD calculation on one conformer of known structure. This solid-state ECD (ss-ECD) approach will be discussed in Section 9.5.1.1.

Second, several chemical modifications to limit molecular flexibility by the formation of cycles have been proposed [8]. This approach displaces the problem of AC assignment from a flexible target molecule to some conformationally homogeneous derivative, provided substrate AC and RCs are preserved or predictably transformed.

These two approaches are of prime importance for AC assignment, but they lead to overriding conformational richness and overlooking the role ECD may have in revealing it.

9.4
Hybrid Approaches

As we have seen before, provided certain conditions are met, we may partition a molecule into one or more separate chromophores, that is, in chemical groups, which are responsible for one or more electronic transitions giving rise to absorptions in the UV-vis region. Sometimes, we can recognize in one system (molecule, supramolecule or aggregate) several chromophores, which following what we said before must be considered having negligible differential overlap, which can also be expressed by saying that they are *independent* (independent systems approximation or ISA). Although independent, they are not isolated, which means that they sense and perturb each other through space, and we shall provide here a framework to rationalize this coupling and to determine its effects on the ECD.

To derive a complete description of the molecular orbitals (MOs) and the electronic transitions of each isolated chromophore, one may use the quantum mechanical methods described in Chapter 5. By so doing, we depict the molecule as a collection of chromophores held together by an inert (from the spectroscopic point of view) scaffold. Indeed, we do not even need to invoke the existence of covalent bonds connecting the various chromophores, which gives us the liberty to use our arguments for one single molecule, as well as for an aggregate, a compound, a supramolecular structure, or even a crystal. In the following discussion, we shall define our system a *polymer*, consisting of N monomer units, each containing a chromophore. No restriction will be applied to the nature of the linkage between monomers, neither to any similarity nor to a symmetry relation between them.

To a zeroth-order approximation, the various chromophores are independent, but their proximity leads to a mutual perturbation, primarily of Coulombic nature. From a QM point of view, one may treat the polymer system by perturbation theory, by introducing corrections to the individual chromophores wavefunctions, which would lead to a rapidly increasing computational effort.

In contrast, we switch to classical electrodynamics and treat the through-space interactions between monomers in terms of local fields acting on polarizable charge distributions. This is the reason why we define this one as *hybrid approach*: the individual chromophores are dissected from the molecular framework, subject to full MO characterization by suitable QM methods, and then they are made to interact classically. In many literature reports, the spectroscopic parameters of the individual chromophores are derived from the experimental absorption spectra of a model compound (ideally, the dissected portion), like in a sort of *semiempirical* approach. Nowadays, it seems to us more reasonable and correct to use such an experimental information, when it is available, to check the quality of computations. This is because there are two sets of parameters that are either difficult or impossible to determine experimentally but are of prime importance in the ECD prediction, as we shall see below. They are the directions of the transition dipoles and their location. The first one may be accessed, although with several limitations and difficulties, through linear dichroism (LD) measurements; for the second one, there is no chance, apart from symmetry considerations that may occasionally hold.

9.4.1
Coupled Oscillators and the DeVoe Method

The coupled oscillators approximation is especially meant to account for interacting electric dipole transitions located in distinct chromophores. Within a classical framework, it is an exact treatment and has the merit of explicitly introducing line shapes. As we will see in Section 9.4.3, it has been used most often for organic molecules and organometallic species.

The radiation wavelength (typically ≥ 180 nm) is much larger than the system size (at least neglecting very large nanoscopic aggregates), which ensures that we can consider the electric field **E** of the electromagnetic wave as uniform throughout the sample.

9.4 Hybrid Approaches

Each transition of the individual chromophores is characterized by a complex electronic polarizability $\alpha_i^0(\nu)$ and a polarization direction represented by the unit vector $\mathbf{e}_i^\mu$. The real and imaginary components of $\alpha_i^0(\nu)$ depend on one another through Kronig–Kramers relation, and $\text{Im}(\alpha_i^0(\nu))$ is related to the molar extinction coefficient ε_i^0 through

$$\text{Re}(\alpha_i^0(\nu)) = \frac{2C_1}{\pi} \int_0^\infty \frac{\varepsilon_i^0(x)dx}{x^2 - \nu^2}$$

$$\text{Im}(\alpha_i^0(\nu)) = -\frac{C_1 \varepsilon_i^0(\nu)}{\nu} \tag{9.4}$$

where the numerical constant $C_1 = 4.356 \times 10^{-14}$ in SI units.

Upon experiencing a local electric field $\mathbf{E}^{\text{local}}$, each polarizability gives rise to an induced moment $M_i^0(\nu)$

$$M_i^0(\nu) = \alpha_i^0(\nu)\, \mathbf{E}^{\text{local}} \cdot \mathbf{e}_i^\mu = \alpha_i^0(\nu)\left[\mathbf{E}\cdot\mathbf{e}_i^\mu - \sum_j G_{ij} \cdot M_j^0(\nu)\right] \tag{9.5}$$

where the local field is the sum of the electromagnetic radiation field $\mathbf{E}$ and of the contributions arising from all the surrounding $M_j^0(\nu)$, stemming from the polarizabilities $\alpha_j^0(\nu)$ located in the other monomers. In fact, each of them gives rise to a dipolar field, whose effect on the ith transition is quantified through G_{ij} by the point dipole interaction[1]

$$G_{i\ne j} = \frac{1}{r_{ij}^3}\left(\mathbf{e}_i^\mu \cdot \mathbf{e}_j^\mu - 3\frac{(\mathbf{e}_i^\mu \cdot \mathbf{r}_{ij})(\mathbf{e}_j^\mu \cdot \mathbf{r}_{ij})}{r_{ij}^2}\right)$$

$$G_{ii} = 0 \tag{9.6}$$

In the previous equation, $\mathbf{r}_{ij}$ is the vector connecting the two electric dipoles aligned along $\mathbf{e}_i^\mu$ and $\mathbf{e}_j^\mu$. Apparently, Equation 9.5 is recursive, that is, by introducing the initial values relative to the jth transition, one obtains first-order corrections to $M_i^0(\nu)$, which then need to be reinserted in the same equations to improve to second and higher orders. By means of matrix algebra, there is an exact solution, which can be defined correct to all orders.

Let us define the following arrays and matrices, whose generic elements (with $i, j = 1, \ldots, N$) are indicated:

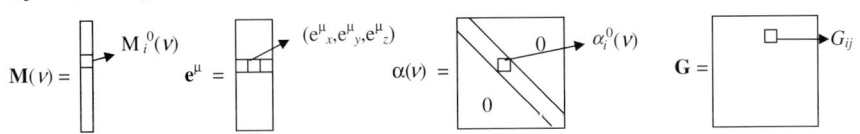

$$\mathbf{M}(\nu) = \begin{bmatrix} M_i^0(\nu) \end{bmatrix} \quad \mathbf{e}^\mu = \begin{bmatrix} (e_x^\mu, e_y^\mu, e_z^\mu) \end{bmatrix} \quad \boldsymbol{\alpha}(\nu) = \begin{bmatrix} \alpha_i^0(\nu) & 0 \\ 0 & \end{bmatrix} \quad \mathbf{G} = \begin{bmatrix} 0 & G_{ij} \\ & 0 \end{bmatrix}$$

---

[1] Especially when the dipole approximation is poor, for example, in the case of closely spaced monomers, one should substitute the above expression for G_{ij} with one taking into account diffuse transition charge densities or charge monopoles, instead.

Then,

$$\mathbf{M}(\nu) = \boldsymbol{\alpha}(\nu) \cdot [\mathbf{e}^{\mu} \cdot \mathbf{E} - \mathbf{G} \cdot \mathbf{M}(\nu)] \qquad (9.7)$$

which can be solved for $\mathbf{M}(\nu)$

$$\mathbf{M}(\nu) = [\boldsymbol{\alpha}(\nu)^{-1} + \mathbf{G}]^{-1} \cdot (\mathbf{e}^{\mu} \cdot \mathbf{E}) = \mathbf{A}(\nu) \cdot (\mathbf{e}^{\mu} \cdot \mathbf{E}) \qquad (9.8)$$

with

$$\mathbf{A}(\nu) = [\boldsymbol{\alpha}(\nu)^{-1} + \mathbf{G}]^{-1} \qquad (9.9)$$

The array $\mathbf{M}(\nu)$ contains the dipole moments $M_i^0(\nu)$ correct to *all orders*, within the point dipole approximation.

We can appreciate that all we need to do is to invert the frequency-dependent matrix $[\boldsymbol{\alpha}(\nu)^{-1} = \mathbf{G}]$, which is Hermitian: it is the sum of a complex diagonal matrix $\boldsymbol{\alpha}(\nu)^{-1}$ and of the real symmetric matrix $\mathbf{G}$. Even for a large number of interacting chromophores, modern computers and software can easily solve this problem.

Here, we account for two contributions to ECD: one due to the interacting electric dipoles and the other one arising from the coupling of electric and magnetic transition dipole moments. Equation 9.10 provides an operational tool for calculating the ECD of an aggregate of chromophores.

$$\Delta\varepsilon(\nu) = 4.76 \times 10^{17} \frac{(n_s^2 + 2)}{3} \nu^2 \sum_{ij} (\mathbf{e}_i^{\mu} \times \mathbf{e}_j^{\mu} \cdot \mathbf{r}_{ij} - b_j \mathbf{e}_i^{\mu} \cdot \mathbf{e}_j^{m}) \mathrm{Im}(A_{ij}(\nu))$$

$$(9.10)$$

where $\mathbf{e}_j^m$ is the direction of a *magnetic* transition dipole, n_s is the medium refractive index, and $b_j = c(\mathrm{Im}|\mathbf{m}_j|)|\boldsymbol{\mu}_j|/2\pi\nu_j$.

The DeVoe approach to ECD was translated into a Fortran routine by Hug [9, 10].

Following the above derivation, input parameters are for all electronic transitions i's:

- Electric and magnetic transition dipole moment amplitudes;
- Half-height line widths;
- Directions of the electric and magnetic transition dipole moments $\mathbf{e}_i^{\mu}$ and $\mathbf{e}_j^{m}$; and
- Locations of the electric and magnetic transition dipoles, $\mathbf{x}_i^{\mu}$ and $\mathbf{x}_j^{m}$.

The above first two points are used to calculate the frequency-dependent complex polarizabilities $\alpha_i^0(\nu)$ upon approximating ε_i^0 in Equation 9.4 with a Lorentzian band.

In the present form of Hug's routine, the two latter sets of input parameters are inserted as their Cartesian components/coordinates or by reference to atoms provided a table of atomic coordinates is also supplied. This latter aspect is particularly convenient when one needs to consider a set of conformers because it is sufficient to change the table defining the molecular structure, while leaving the rest of the input unchanged.

The program outputs absorption and ECD spectra, and the matrix $\mathbf{G}$ is also provided, which measures the strength of couplings of the various transitions. What influences the terms G_{ij} are geometrical parameters arising from the dipolar

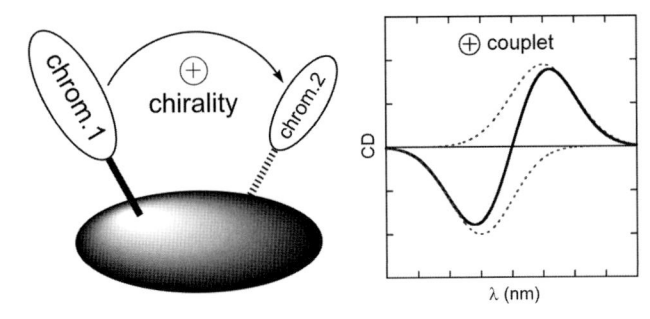

Figure 9.1 Exciton chirality rule. A clockwise twist to bring the chromophore (actually its electric dipole transition moment) in the front onto that in the back is defined as positive chirality, and corresponds to a positive couplet in the ECD spectrum. This consists in a bisignate feature with positive long-wavelength branch [2, 3, 11].

interaction, namely, distance and relative orientations of the dipoles, as well as purely spectroscopic terms. Since we are dealing with coupling between oscillators, it should be clear that their proper frequencies play a role, as well, and in fact the stronger the interaction is, the closer to resonance they are (i.e., the smaller is the difference between their frequencies). The limiting case is provided by what is called *degenerate coupling*, obtained when two oscillators are identical. The consequence of this interaction is the formation of two different *normal oscillation modes* (in-phase and out of phase) with different frequencies (leading to a more or less pronounced splitting or broadening of the absorption band) and opposite rotational strengths. For reviews of the so-called *exciton chirality rule*, depicted in Figure 9.1, see Refs [2, 3, 11].

9.4.2
The Matrix Method

This procedure is in many respects similar to DeVoe approach outlined above, insofar it is based once more on a classical treatment of the interaction of transition dipoles. There are two main differences: on the one hand, it accounts fully and explicitly for all electric and magnetic dipole interactions and, on the other, it is based on strictly localized transitions, that is, the line shape is introduced only *a posteriori*.

In this method, developed by Schellman [12, 13], each chromophore is described by two elements, an electric and a magnetic transition dipole, and for a collection of N chromophores, one deals with dimension $2N$.

For a system of two chromophores, one can build the Hamiltonian matrix $\mathbf{H}$

$$\mathbf{H} = \begin{bmatrix} E_{n_1} & V_{n_1 m_1} & V_{n_1 n_2} & V_{n_1 m_2} \\ V_{n_1 m_1} & E_{m_1} & V_{m_1 n_2} & 0 \\ V_{n_1 n_2} & V_{m_1 n_2} & E_{n_2} & V_{n_2 m_2} \\ V_{n_1 m_2} & 0 & V_{n_2 m_2} & E_{m_2} \end{bmatrix} \quad (9.11)$$

The diagonal terms E_{n_1}, E_{m_1}, E_{n_2}, and E_{m_2} represent the zero-order transition energies for the electric (n) and magnetic (m) dipole transitions of chromophores 1 and 2. The off-diagonal elements contain the interaction terms and specifically:

- $V_{n_1 m_1}$ is a pure term for chromophore 1 (and similarly $V_{n_2 m_2}$ for 2) and represents what is also called one-electron mechanism. In the independent systems approximation, the transitions described as n_1 and m_1 originate from one ground state to two possibly different excited states, one allied with electric and the other with magnetic dipole character. For a local achiral chromophore, either the two do not mix or the two moments are orthogonal, so as to lead to vanishing rotational strength. On the contrary, in the absence of improper symmetry operations, that is, when the chromophore becomes intrinsically chiral neither transition is pure, which for small distortions may be treated by means of perturbation theory, whereby the transition allied to n_1 acquires a character of m_1 and vice versa. This perturbative coupling term is $V_{n_1 m_1}$.
- $V_{n_1 n_2}$ is the electric dipole coupling described in the previous section and named there G_{ij}.
- $V_{n_1 m_2}$ and $V_{m_1 n_2}$ are also called **μ·m** terms because they describe the coupling of electric and magnetic dipoles (as required in Rosenfeld equation) but occur on different chromophores, that is, at different locations.

Generalization to a larger number of chromophores is straightforward because only pair-wise terms as those described above are required. The mixing between two transitions is determined not only by the off-diagonal coupling terms but also by the difference in their transition energies. Corrected eigenstates and transition energies (eigenvalues) are obtained by diagonalizing matrix **H**, which is Hermitian.

The matrix method shows its full potential in the quantitative structural analysis of biopolymers [14]. Particularly for what concerns proteins, many transitions of different nature are responsible for the observed CD in the far- and near-UV, either electric or magnetic dipole allowed (amide $\pi-\pi^*$ and $n-\pi^*$, aromatic $\pi-\pi^*$ and disulfide $n-\sigma^*$ in the side chains). Therefore, considerable effort has been put in studying the relevant transitions at a high level of theory [15, 16], to afford the most reliable parameters for matrix method calculations [17].

Since in this chapter we focus on small molecules, where it has been somewhat less used [13], we shall not deal further with the matrix method.

9.4.3
Applications

Applications of the DeVoe method to stereochemistry have been recently reviewed by Rosini and coworkers [18] and do not really need to be covered here. We shall provide only a couple of examples, from our own work, to illustrate the power of this approach. We wished to investigate the solution conformation of enantioselective catalytic precursors, based on 1,1′-bi(2-naphthol) (BINOL) as the chiral element. In the parent system, the only degree of internal freedom is the dihedral angle θ between the two aromatic planes, which can hardly be assessed through mostly used and reliable NMR techniques. 1,1′-Binaphthalenes are a common example of atropisomeric axial chiral moieties. The sp^2-sp^2 joint, in the absence of other constraints, ensures a large flexibility and the dihedral angle θ can sweep a range as wide as

60–120° without exceeding 1 kcal mol^{-1} above the conformational minimum at 90°. One can easily encounter in the literature compounds including this moiety, often substituted at 2,2'- and possibly at 6,6'-positions. Other examples are somewhat less common. C_2 symmetry makes corresponding nuclei isochronous, which prevents the direct observation of NOEs, for example, between the protons at positions 8 and 8'. On the contrary, ECD provides an excellent alternative by means of DeVoe approach.

In the first place, we need to look at the specific molecule we wish to investigate and correctly recognize the chromophore(s). First, quasi-orthogonal arrangement of the two aromatics ($\theta \approx 90°$) leads to negligible electron delocalization and the two subunits can be considered independent. Second, the presence of substituents that may extend conjugation must be taken into account; thus, for molecules such as **2** and **3** (Scheme 9.3) the active chromophore is simply naphthalene, while **4–7** must be regarded as dimers of 2-naphthol, 6-bromo-2-naphthol, 2-iodonaphthalene, or 2-naphthoate, respectively. Wavelength, dipole strength, transition dipole orientations and center of gravity may be more or less affected by nature and position of the substituents.

In the simplest cases, when naphthalene is the chromophore, calculation of the absorption and ECD spectra as a function of θ with Hug's routine is straightforward and yields the results depicted in Figure 9.2a, where we can appreciate the extreme sensitivity of the computed ECD to conformation [19].

The experimental ECD spectra for (R)-**2** and (R)-**3**, shown in Figure 9.2b, reveal profound differences in intensity and wavelength separation between peak and trough (related to the so-called Davydoff splitting), and a comparison with the calculated curves reveals a good match between compound **2** and $\theta = 90°$ and compound **3** and $\theta = 50–60°$. In fact, this finding is confirmed by computational geometry optimization: **2**, lacking particular constraints, can be expected to assume

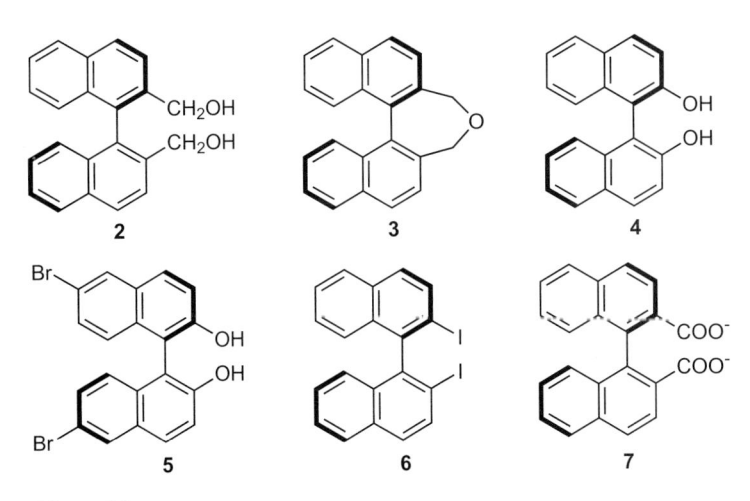

Scheme 9.3

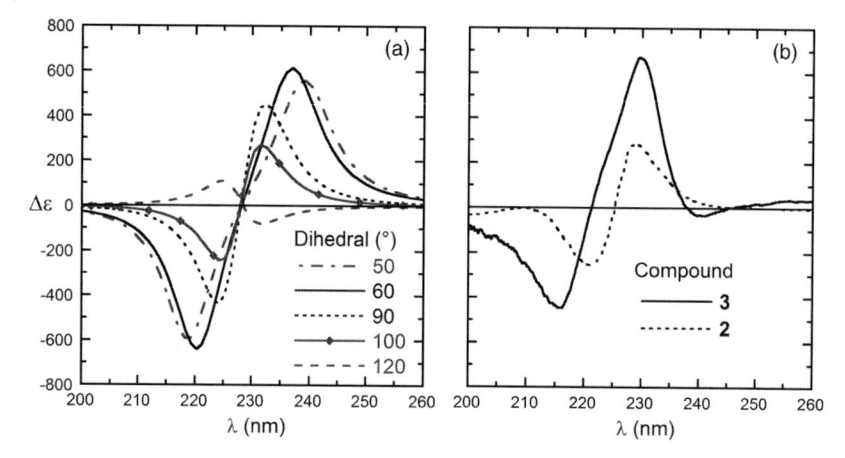

Figure 9.2 Calculated (a) and experimental (b) ECD spectra of (R)-1,1′-binaphthyls as a function of the dihedral angle θ.

an orthogonal arrangement ($\theta \approx 90°$), ensuring minimum steric interaction, while for **3** the seven-membered cycle constraints θ to a smaller value ($\theta \approx 55°$).

BINOL (1,1′-bi-(2-naphthol)) enters the composition of many efficient enantioselective catalysts, where it (or its conjugated base) is used as a ligand for transition metal and lanthanide ions. Owing to its pliancy, BINOL (or its *ate* form) can accommodate any metal ion M^{n+} both as a chelate and as a bridging ligand, namely, the two oxygen atoms are bonded to the same or to different M^{n+}, giving rise to a wide variety of structures. In the recent past, we dealt with many BINOLate complexes, with variable geometries and possibly involving several BINOLate units in the same compound, which introduces the possibility of interactions between transitions centered on aryl moieties that are not directly bonded (*inter*binaphthol as opposed to *intra*binaphtol) [20–23]. These are long-range interactions; although made relatively weaker by the term r_{ij} in Equation 9.6, these nevertheless need to be properly accounted for to correctly calculate the spectra. Once more, Hug's program can easily implement this feature and predict the absorption and ECD spectra starting from tentative structures of the aggregate based on other evidence.

One of the most complex cases treated so far is represented by porphyrins **8** and **9**, which differ in atropisomerism around the two bonds indicated by gray straight arrows depicted in Scheme 9.4 [24]. NMR reveals that the two species are endowed with C_4 and D_2 symmetry, respectively. Accordingly, only two degrees of conformational freedom need to be taken into account: the dihedral ψ between the porphyrin and the directly attached naphthalene and the binaphthyl angle θ.[2] The fourfold symmetry found above reveals that only one pair (ψ, θ) fully describes the conformation of each porphyrin.

---

2) A further angle describes the orientation of the methoxy group, but this is irrelevant to the present summary.

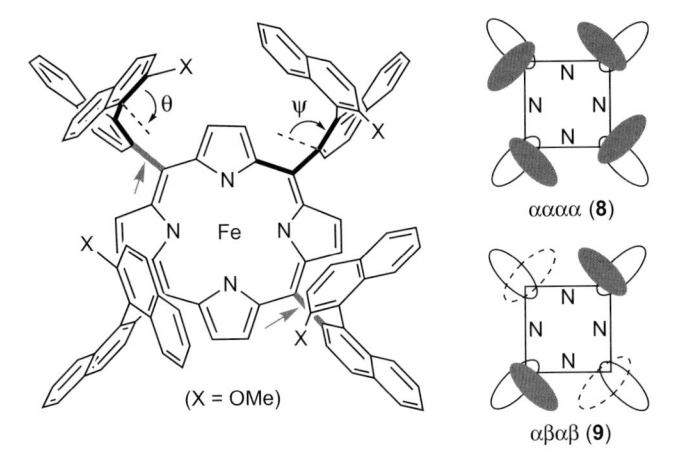

Scheme 9.4

The chromophores can be identified as the porphyrin, described through a circular oscillator for its Soret transition (i.e., a degenerate pair of orthogonal transition dipoles, oriented along the two axes connecting opposite nitrogen atoms); a naphthalene (once more delocalization to the porphyrin is hampered by quasi-orthogonal arrangement); and a 2-naphthol. All of them are well described in the literature and need no further comment; their spectroscopic features can be directly introduced in the program. Two sets of structures, one for the C_4 and one for the D_2 species are produced, by changing the conformational variables (ψ, θ) and ECD, and absorption spectra are calculated as shown in Figure 9.3. The high frequency part of the spectra (below 300 nm), which are dominated by naphthalene-centered transitions, can be used to investigate binaphthyl conformation: the amount of Davydoff splitting

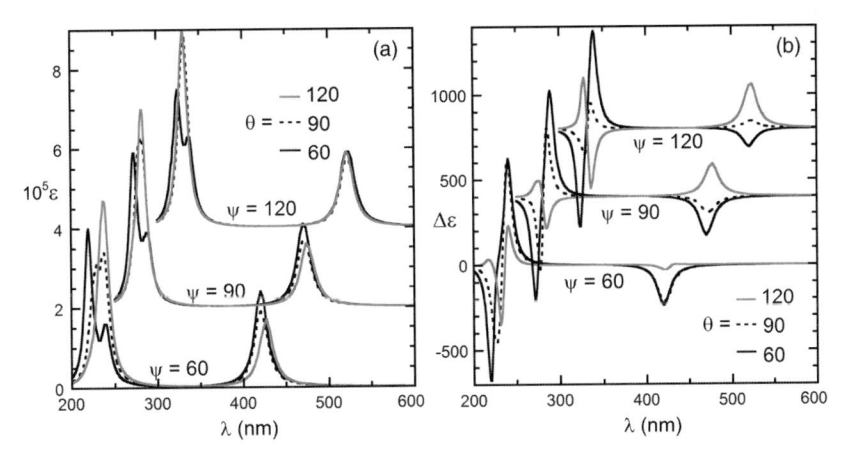

Figure 9.3 Absorption (a) and ECD (b) spectra calculated for porphyrin **8** with DeVoe method on varying dihedral angles θ and ψ. Spectra are stacked and shifted to the right for clarity.

(in some cases, apparent also in the absorption spectrum), the couplet amplitude, and in extreme cases its sign, can also be considered primary reporters of the dihedral θ, mainly owing to degenerate coupling of the ^1B_b transitions.

In these compounds, Soret transition is predicted, and also experimentally found, to be allied with a monosignated ECD band. In the DeVoe approach that we are following, the intensity of this Cotton effect arises from the nondegenerate coupling with the high-energy naphthalene transitions and it is strongly affected by both dihedrals. As a result, for both the C_4 and the D_2 porphyrins, each conformation is uniquely associated with a pair of calculated absorption and ECD spectra. Thus, the comparison of the experimental data with the computed spectra allows one to pick up one single best match. For compound **8**, this is found for $\theta = 75°$ and $\psi = 75°$, and for compound **9**, for $\theta = 90°$ and $\psi = 90°$. Semiempirical geometry optimizations and considerations on the ^{1}H-NMR chemical shifts further support this assignment [24].

9.5
The QM Approach

"Direct" or "full" ECD calculation with quantum mechanical methods or, in other terms, from first-principles theory, is nowadays a fully practicable option for moderately large organic molecules and metal complexes. In principle, any QM method affording excited state description (in terms of wavefunctions, transition energies, and moments) may provide rotational strengths with a little extra computational effort. In practice, however, it is not yet completely understood what level of theory is required to obtain the most accurate predictions of rotational strengths. In general, chiroptical properties are very sensitive to every approximation made in electronic structure calculations. It is, however, not uncommon that a more approximate method may perform more efficiently than more rigorous (and computationally expensive) ones.

Most of the QM approaches described in Chapter 5 used to calculate absorption (UV-vis) spectra have been implemented for both rotational strengths and optical rotation calculations. The relative theoretical formulations may be found in the literature, summarized by several reviews [25–28]. Our intention is to stay on a rather practical ground, and to face with the following basic questions:

- Which are the state-of-the-art and/or most popular QM methods employed to calculate ECD spectra;
- What are their respective applicability, scope, and limitations; and
- What kind of structural information is expected to be gained and how accurate.

First, we shall quickly describe the most important approaches and later we shall give a few examples of applications.

Reliable ECD predictions necessarily require theoretical methods that take electron correlation into account. This also implies the use of large basis sets, often including diffuse functions. Large basis sets (ideally approaching the so-called basis set limit) are also beneficial to circumvent the problem of origin dependence of magnetic properties in finite basis set calculations (see below). Put all together, a post-Hartree–Fock (HF) method including the highest possible level of electron correlation (i.e., approaching a full CI treatment) [29] along with the largest possible basis set, which may be regarded as the most rigorous treatment, would be impractical for any "real" molecule. This is the reason why alternative methods must necessarily be employed. They fall into two main families that can be identified as high-level *ab initio* calculation approaches and semiempirical QM methods (described in Chapter 5).

Ab initio methods imply explicit solution of time-dependent Schrödinger or Kohn–Sham equation.[3] Among the several different high-level methods that have been employed for ECD calculations, three have especially emerged over the years as those leading to the best accuracy/cost compromise and stand out as the methods of choice for practicable and accurate ECD calculations. Time-dependent density functional theory (TDDFT) represents by far the most popular one and will be the main subject of this section. Another DFT-based method, namely, DFT/MRCI (multireference configuration interaction), is computationally more expensive than TDDFT but has one advantage that the detrimental impact of small basis sets is reduced. One step above in terms of theoretical rigor and reliability stands the coupled cluster (CC) theory, which contrary to TDDFT is a convergent approach, that is, it is systematically extensible toward the "exact" limit. However, it is rather computationally expensive limiting its scope. Other *ab initio* methods such as RPA (random-phase approximation, also known as time-dependent HF) were relatively popular in the past to calculate ECD spectra, but are now outperformed by more modern approaches and will not be discussed here [30, 31].

To describe the optical and chiroptical response of a molecule to an applied time-dependent (oscillating) field, the interaction is treated as a perturbation of the initial (ground) state and expressed as a linear dependence on the field strength through the system polarizabilities. This linear response theory avoids the length and slow convergence of the alternative sum-over-states approach, which requires summation over all excited states to derive chiroptical properties at any frequency. In the linear response regime, the chiroptical response is expressed in terms of a frequency-dependent magnetic/electric polarizability tensor β that relates induced electric and magnetic moments to the time-dependent magnetic and electric fields, respectively (in au):

$$\boldsymbol{\mu} = \alpha \mathbf{E} - c^{-1} \beta \frac{\partial}{\partial t} \mathbf{B}$$

$$\mathbf{m} = \chi \mathbf{B} + c^{-1} \beta \frac{\partial}{\partial t} \mathbf{E}$$

(9.12)

---

3) Many DFT functionals such as Becke's B3 exchange functional contain parameters derived from best fits of experimental data; therefore, they are not strictly *ab initio*. Here, however, we will include all DFT functionals in the *ab initio* family.

where **E** and **B** are the applied time-dependent electric and magnetic fields, α and χ the electric polarizability and magnetic susceptibility, respectively. The factor c^{-1} (1/137 in au) explains the relative order of magnitude of the electronic absorption (related to the first terms in the equations) and circular dichroism phenomenon, which is phenomenologically quantified by the factor $g = \Delta\varepsilon/\varepsilon$. For isotropic samples, only the complex quantity β (ν) related to the trace of tensor **β** needs to be considered.

$$\beta(\nu) = \frac{1}{3}\text{Tr}(\boldsymbol{\beta}) = \frac{1}{3}(\beta_{xx} + \beta_{yy} + \beta_{zz}) \tag{9.13}$$

A noteworthy property of parameter β (ν) is that its poles occur in correspondence with excitation frequencies, which is exploited in most linear response approaches (such as TDDFT and CC) to compute vertical excitation frequencies without requiring explicit calculation of excited state wavefunctions or densities.

One problem related to tensor **β** is its origin dependence. For finite basis sets, the computed chiroptical properties to **β** are origin-dependent, that is, they vary upon translation of molecular coordinates, which is of course an unphysical outcome. One way to overcome such a problem is by using gauge-independent atomic orbitals (GIAOs) such as London AOs that directly incorporate magnetic-field dependence [32]. The use of LAOs has, however, effect only when employed with the so-called variational computational methods (i.e., based on energy minimization, such as TDDFT and DFT/MRCI but not CC). Alternatively, the tensor β may be represented through an origin-independent velocity gauge formulation, instead of the origin-dependent length gauge formulation. In practice, it consists in evaluating instead of the electric moment μ its momentum $\mathbf{p} = -i\boldsymbol{\mu}(2\pi\nu)^{-1}$. Dipole velocity (DV) rotational strengths usually lead to poorer results than dipole length (DL) ones when small basis sets are used, that is, DL tends to converge faster toward the basis set limit. However, the use of modern polarized double- or triple-ζ quality basis sets leads to substantially equivalent DL and DV rotational strengths with most high-level methods. On the contrary, with semiempirical calculations large discrepancies are often obtained, and use of DL values is especially questionable.

TDDFT has a very favorable balance of cost and accuracy [28]. Density functional theory, which was practically unexplored in the context of ECD calculations only 10 years ago, has most of the merit of the recent "renaissance" of chiroptical methods [33]. This is surprising in light of the fact that DFT functionals such as the most popular B3LYP have been designed and optimized to reproduce thermochemical data [34], and there is no obvious reason for them to reproduce equally well chiroptical properties. Nonetheless, hybrid functionals such as B3LYP, PBE0 and BH&HLYP [35, 36] do predict transition energies and rotational strength with high accuracy in many cases. Still, the typical DFT drawbacks described in Chapter 5 (self-interaction error and wrong asymptotic behavior) should always be kept in mind when using functionals, such as BP86, employing adiabatic local density approximation. Their practical consequence is that poorly localized states such as charge transfer, diffuse, and Rydberg states are often inadequately described. These problems may be alleviated by increasing the fraction of "exact" HF exchange, that is, using hybrid functionals such as B3LYP and especially PBE0 and BH&HLYP (20, 25 and 50% HF exchange, respectively), or solved using the newest range-separated functionals such as CAM-B3LYP or functionals based on the so-called statistical average of orbital potentials (SAOP) model. In the field of newly developed functionals, double-hybrid perturbatively corrected functionals such as B2PLYP yield very accurate excitation energies and lead to less spurious or "ghost" states than hybrid functionals. Another issue related to TDDFT calculations is that, because of the perturbative nature of the approach employed, they are intrinsically more accurate in predicting low-lying excited states. Strictly speaking, all states falling beyond the computed ionization potential IP should be disregarded [37]. The problem is worsened by the fact that the absolute value of the HOMO energy (eigenvalue), which equals the IP in HF theory, is often underestimated by DFT, so that the number of "fallen" states is artificially increased. Most often, however, this IP threshold is below the observable ECD spectra cutoff ($\approx$ 185 nm), so that the true impact of the described problem in practical applications is only marginal.

Contrary to *ab initio* methods, semiempirical quantum mechanical methods (Chapter 5) rely on strong simplifications [38]. They ignore core electrons and use the neglect of differential overlap approximation (NDO) to skip the explicit calculation of computationally demanding one- and two-electron integrals. These latter are either put to zero or suitably parameterized to atom-related values depending on quantities such as ionization potential and electron affinity. In the context of excited state calculations, electron correlation is taken into account by means of a truncated configuration interaction (CI). When only singly excited states are included, a CI single (CIS) procedure is employed in connection with various semiempirical models developed for spectroscopic purposes, such as CNDO/S (complete NDO) and ZINDO/S (Zerner's intermediate NDO). Both of them have been successfully employed for ECD calculations [13, 39, 40]. The main advantage of such techniques is that they may be simultaneously extremely fast and accurate enough, at least for certain types of chromophores and transitions. In fact, because of the parametric essence of NDO methods, their efficiency depends case by case on the system investigated and their consistency is somewhat limited. Both CNDO/S and ZINDO/S have been especially designed to treat aromatic and heteroaromatic chromophores (including ligands for transition metals, for ZINDO) [41, 42], so that they perform well for compounds such as biaryls whose UV-vis and ECD spectra are dominated by strong π–π^* transitions. On the contrary, their accuracy in predicting high-lying electric dipole forbidden transitions such as n–σ^* is comparatively much poorer [43]. Also, systematic shifts between computed and experimental transition wavelengths are common and must be taken into account by the so-called wavelength correction (see below).

How should one choose from among all available QM methods for ECD predictions? The answer to such a question depends essentially on how complex (large, flexible) is the system and on the nature of the chromophores contributing to the ECD spectra. For very large molecular systems with many degrees of conformational freedom (that is, for which several different low-energy structures must be considered), any *ab initio* method may be unpractical. In such a situation, semiempirical NDO methods may be the answer especially if the ECD is dominated by strong aromatic π–π* transitions [39]. For moderately complex systems, TDDFT is the lead choice in most common cases [25, 26]. The most frequent applications of TDDFT calculations are in the AC assignment of relatively rigid organic molecules, whose conformation is known from other means (in solution or in the solid state), and in the interpretation of chiroptical properties (ECD and OR) of prototypical molecules with established AC. Relatively much less popular is their use in conformational analyses, in a way similar to what is seen in Section 9.4.3. A few specific examples of applications of TDDFT ECD calculations will be presented in the following sections.

Situations where the use of alternative QM methods may be desirable are the simulation of high-lying aromatic transitions with large contribution from double and higher excitations, which need multireference methods such as DFT/MRCI to be treated [39]; and of diffuse or Rydberg transitions of relatively weak chromophores, for which CC methods may be well employed [44, 45].

Focusing on TDDFT calculations, the most essential issue is the choice of the functional and the basis set. As discussed above, hybrid functionals with different HF character usually perform better than pure DFT functionals. In general, B3LYP, PBE0 and BH&HLYP should be employed as the first choices for ECD calculations among standard functionals. On the other hand, range-separated (CAM-B3LYP) [46], double-hybrid functionals (B2PLYP) [47], and SAOP-based functionals [48] will probably represent the future benchmarks in TDDFT calculations. The choice of the basis set is a little more involved decision. Here, in principle, the larger is the better but also the more time consuming. Therefore, one should always wonder how necessary a large basis set is for a given situation. Split-valence double- or triple-ζ quality basis sets with polarization functions usually represent very good compromises for moderately large molecules where ECD is dominated by valence excitations. Our personal favorite is Ahlrichs' TZVP set [49]. Especially when only low-lying excited states need to be evaluated, however, using smaller basis sets such as 6-31G(d) may be sufficient [50]. Smaller molecules with relatively weak chromophores may, on the other hand, require the use of basis sets augmented with diffuse functions, like the "aug-" ones of Dunning's correlation-consistent basis sets (aug-cc-pVDZ and so on) [51].

Apart from the approximations specifically related to the TDDFT method, two further very crude simplifications are usually made in computing ECD, that is, the impact of environment (commonly, a solvent) and vibrations is neglected. Ideally, they should be taken into account both in the generation of input geometries (as seen above) *and* in the prediction of optical and chiroptical spectra. The appearance of ECD spectra may be strongly affected by the presence of vibrational progressions and/or vibronic coupling. This is, however, especially true for a few particular cases, for example, for low-lying electric dipole forbidden transitions of weak chromophores. As for the solvent, the dependence of ECD is usually less pronounced than, for example, of OR and VCD. However, specific interactions such as hydrogen bonds may affect overall ECD spectra, for example, of amino acids and oligopeptides to a large extent and imply the inclusion of explicit solvent molecules [52, 53], although their impact may be minimal for less polar compounds [54]. Whenever possible and desired, vibronic coupling [55–57] and/or solvent models [58] (see also Chapter 5) may be introduced in ECD calculations although they will considerably increase the computational time.

Whatever the QM method employed for ECD calculations, the computational outcome is normally a list of rotational strengths R_i (possibly, both in DL and DV gauges) at discrete transition frequencies v_i, or, in other words, a stick plot. To compare it with an experimental ECD spectrum, it is advisable to apply a broadening or band shape. That is, each rotational strength is associated with a Gaussian or Lorentzian shape function with intensity proportional to the absolute rotational strength value, and then a sum of all bands is evaluated. Such a procedure requires a more or less arbitrary bandwidth σ to be chosen. In principle, band broadening is associated with excited states having finite lifetimes, and related damping constants could be introduced at some point in the OR or ECD calculation formalism. In practice, empirically derived bandwidths are usually employed, that is, those providing the best fit with the experimental spectra. Reasonable values are between $\sigma = 1500$ and 3000 cm^{-1}; alternatively, a frequency-dependent bandwidth (proportional to the frequency or wave number square root, $\sigma_i \propto \sqrt{\tilde{v}_i}$) may be employed. Comparison between experimental and calculated ECD spectra is most properly made on the frequency or wave number scale. Expressing the rotational strengths in 10^{-40} cgs units and $\Delta\varepsilon$ in the common M^{-1} cm^{-1} (where M is molarity, mol l^{-1}) units, the ECD spectrum calculated as sums of Gaussians in the wave numbers ($\tilde{v}$ in cm^{-1}) domain is

$$\Delta\varepsilon(\tilde{v}) = 0.0247 \sum_i \frac{R_i \tilde{v}_i}{\sigma_i} \exp\left[-\left(\frac{\tilde{v}-\tilde{v}_i}{\sigma_i}\right)^2\right] \qquad (9.14)$$

Computed transition frequencies may be systematically shifted with respect to experimental ones. With TDDFT, this shift depends on the functional used and, for hybrid functionals, on the amount of exact exchange included: an increasing HF fraction brings about a blueshift over all estimated transition wavelengths [59]. To better compare computed and experimental spectra, the former ones may be appropriately shifted; in other words, a frequency or wavelength correction may be applied. A common way of doing it is to look for the match between computed and experimental UV-Vis spectra, then the same shift is applied to ECD ones (the so-called UV correction) [39]. After band shape application and shift correction, the computed spectrum in the wave number domain may be translated into that in the wavelengths one ($\lambda = 10^4 \tilde{v}^{-1}$) for a quicker comparison with the experimental ECD, which is usually recorded as a function of wavelength.

9.5.1
Assignments of Absolute Configurations

TDDFT calculations of ECD spectra have emerged over the years as one of the most efficient tools for assigning absolute configurations of organic molecules, in particular, of natural products and, less frequently, of metal complexes. The prerequisites for this kind of application are only a few but of fundamental nature:

a) The compound must show a significant ECD spectrum in the commonly accessible range, that is, above 185 nm.
b) A reliable input structure, or a series of input structures, is available.
c) If the molecule is flexible, the number of populated conformers at the working temperature is not too large.
d) The molecule is not too big (let us say, below 30–35 nonhydrogen atoms).

Point (a) is not a severe limitation in the context of chiral organic molecules and especially of natural products, most of which contain a conjugated moiety leading to nonnegligible ECD spectrum. The problem of a consistent conformational sampling and of the generation of a set of "correct" input structures (points (b) and (c)) has been already discussed in Section 9.3. As for point (d), it is gratifying to note that ECD of increasingly larger and complex molecular systems has been treated by TDDFT in recent years. Notable examples are the photosynthetic reaction center [60] and gold nanoclusters covered with chiral adsorbates [61].

In general, the common simplifications employed in TDDFT calculations (the neglect of environment and vibrational structure), as well as the problems associated with standard functionals discussed above, do not hamper a safe application of the method, provided a proper functional/basis set combination is employed. If the above conditions are fulfilled, a reliable AC assignment is usually achieved by comparing experimental and TDDFT-calculated ECD spectra. This approach represents a valid option to methods based on other chiroptical spectroscopies such as VCD and ROA [62–64] when the method based on X-ray anomalous scattering cannot be applied.

The usual procedure for assigning ACs of a moderately flexible molecule via QM ECD calculations consists of many steps. Recalling what we discussed in previous sections, we may summarize them as follows:

1) Thorough conformational search with MM or semicmpirical methods, possibly based on Monte Carlo or molecular dynamics treatment, to afford a more or less wide ensemble of structures.
2) If applicable, dissection of molecular portions (e.g., flexible chains remote from the chromophore) that seemingly do not contribute to the observed ECD.
3) Geometry optimization of all above structures at a higher level (DFT, MP2, and so on), to afford a reduced set of low-energy minima with the same arbitrarily assumed AC.
4) Experimental check of optimized geometries based on NMR data such as NOEs and J-couplings.
5) TDDFT (or other QM) ECD calculations on all minima within a certain energy threshold (2–3 kcal/mol), that is, with significant population (say, >3–5%) at the working temperature (normally, 298 or 300 K).
6) Weighted average of all computed ECD spectra according to the Boltzmann weights at the working temperature.

7) Comparison with the experimental ECD spectrum: if the match is good with either the average calculated ECD (for the initially assumed AC) or its mirror image (that is, for the opposite AC), the configuration may be assigned.

The most drastic approximations used are the neglect of zero-point vibrational corrections to the free energies, and of solvent and vibronic effects on geometries and calculated ECD. They can be relieved by *ad hoc* methods that are, however, computationally quite demanding.

Of the many examples published in the past 10 years of AC assignments based on the above procedure (referred to as "solution ECD-TDDFT//DFT" if solution ECD spectra, DFT geometry optimizations, and TDDFT ECD calculations are employed), we will discuss only a few illustrative cases.

Compounds **10a–10c** (Scheme 9.5) were synthesized to be screened for anti-inflammatory activity (COX-2 inhibitor). They were prepared in racemic form and resolved through enantioselective HPLC. Their ECD spectra were recorded online stopping the elution in correspondence with the peaks for the two enantiomers of each compound. In Figure 9.4b, the ECD for the two enantiomers of compound **10a** are shown. To assign their ACs, the solution ECD-TDDFT//DFT method was applied [23]. The molecule was first split into two moieties to investigate the conformation around the chiral fragment attached at C-3 and the two aryl–aryl torsions at N-1 and C-5, which were thought to be independent of each other. The relevant dihedral angles for the first moiety (C-3/C-2' and C-2'/C-1') were systematically varied in the 0–360° range by 15° step. From this double torsional energy scan with AM1 method, a 3D plot was generated (energy versus the two angles) whose low-energy points (eight in total) were optimized with DFT, B3LYP/6-31G* method, affording four different structures. The two lowest energy ones were in agreement with observed NOE between pyrrole H-4 and CH_3-3' methyl group protons, and with the measured dependence of OH chemical shift with the temperature, which indicated intramolecular hydrogen bonding with the ester group. NMR experiments gave instead no clear indication about the situation around the two aryl–aryl linkages, thus only computational data were available. DFT geometry optimizations revealed that the two aryl–aryl torsions are correlated and two enantiomeric situations are

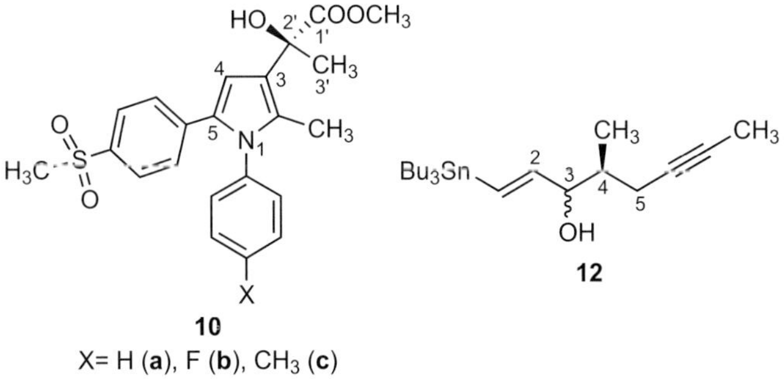

10
X= H (**a**), F (**b**), CH_3 (**c**)

Scheme 9.5

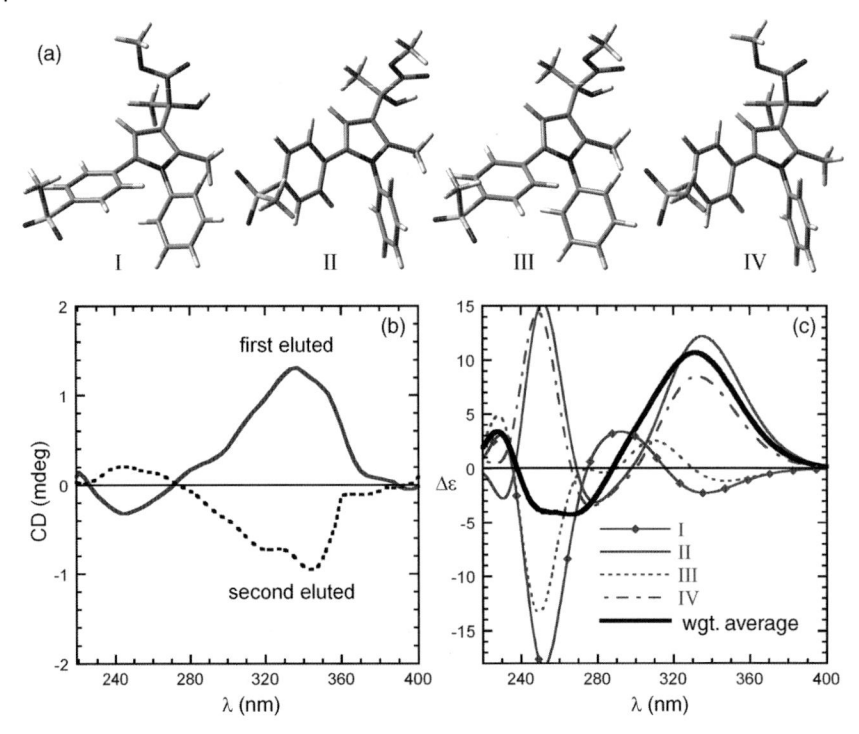

Figure 9.4 (a) DFT-optimized lowest energy structures (I–IV) of **10a**. (b) ECD spectra recorded online for the two enantiomers. (c) ECD spectra calculated with TDDFT method on the four structures (I–IV) and Boltzmann-weighted average for (S)-**10a**.

possible with both positive or both negative values for N-1/phenyl and C-5/aryl dihedrals. When all results were put together and the whole molecule reconstituted and subdued to final DFT geometry optimizations, four energy minima (I–IV) were obtained with sizable population at 300 K (Figure 9.4a). TDDFT ECD calculations with B3LYP/TZVP method led to quite different ECD spectra for the four structures (Figure 9.4c). Such a situation is far from ideal because, in principle, a small change in the relative Boltzmann's factors may strongly affect the weighted average ECD. In this situation, however, the average calculated ECD was in good agreement with the experimental one and the AC could safely be established [23].

One may expect that upon increasing the molecular complexity and flexibility, the computational cost and possible errors associated with the conformational search will increase too. This is well exemplified by the 3,8″-biflavonoid morelloflavone (**11**, Figure 9.5) that contains flavanone (ABC) and flavone (DEF) moieties giving rise to atropisomerism around the C-3/C-8″ bond [65]. The two atropisomers (**11a** and **11b**) have around 7:3 population in methanol found by NMR. Apart from that, several other degrees of freedom are apparent in the geometry of **11**, in particular, the two torsions around rings B–C and E–F. Independent AM1 torsional energy scans around these torsions resulted in 16 conformers (corresponding to

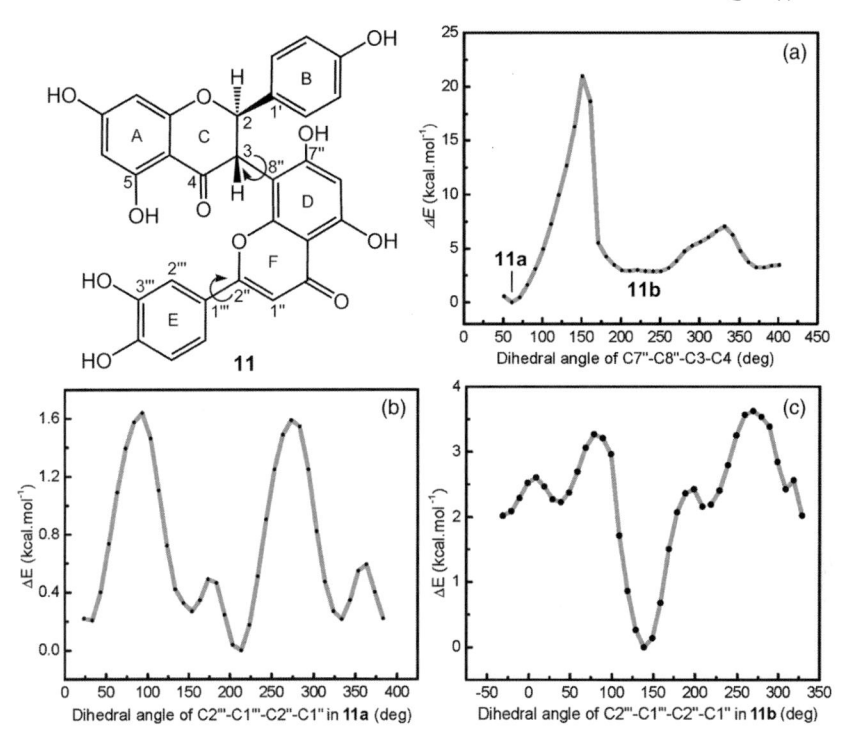

Figure 9.5 Torsional AM1 energy scans for relevant dihedrals of compound **11**. (a) Atropisomers **11a** and **11b** correspond to the two minima found for the torsion between rings C and D. (b and c) Energy scans relative to the torsions between rings E and F for the two atropisomers. Reprinted with permission from Ref. [65]. Copyright 2007, the American Chemical Society.

the various minima in the graphs in Figure 9.5) that were optimized with DFT (B3LYP/6-31G*). The absolute lowest energy structure (with, in principle, almost 100% population) was, however, discarded because its exceptional stabilization is due to a long-range intramolecular hydrogen bond between C-3''' and C-5 hydroxyls. This feature was consistently found even including a solvent model for methanol in DFT calculations; however, it was considered a computational artifact unlikely to survive in the experimental conditions. After removal of that minimum, five low-energy DFT minima were left with population >3% at 300 K. They were subdued to TDDFT ECD calculations (B3LYP/6-31G*) and the weighted average ECD was in agreement with the experimental one at low energies, thus allowing the AC assignment [65].

A quite common situation occurs when a set of important molecular torsions are all correlated with each other, making any independent torsional energy scan meaningless. If so, a thorough conformational search may be run by employing suitable algorithms consisting in a more or less systematic variation of some or all possible torsions to generate a large set of starting structures. Each of them is optimized (usually with MM or semiempirical method) and checked for acceptance on the basis of an energy criterion. Finally, the set of optimized structures is screened for duplicates and an ensemble of distinguishable low-energy structures within a certain energy window (say, 2.5–3 kcal/mol) is obtained. Usually, $C(sp^3)$–$C(sp^3)$ single bonds are varied by 120° in each step, $C(sp^2)$–$C(sp^2)$ single bonds by 180°, and so on; some tricks are used to include endocyclic torsions in the search. When a full systematic search would be too long, a Monte Carlo procedure may be employed where the bonds are randomly rotated. This is a simulated annealing method where the molecule is first placed at high temperature to overcome high rotation barriers, then as an increasing number of minima is found, the temperature is lowered so that the procedure tends to converge. Both methods are usually very effective in exploring conformational ensembles even for very flexible molecules. Usually, the set of minima thus found is refined by higher level calculations. Such a procedure is illustrated, for example, by diastereomeric compounds (3R,4S) and (3S,4S)-**12** (Scheme 9.5) that were investigated as enantiopure starting materials in the synthesis of prostacyclin analogues [66]. Four adjacent torsions in the middle of the molecules are the main degrees of freedom, clearly correlated with each other. A Monte Carlo conformational search with MNDO method led to around 20 low-energy minima (within 3 kcal/mol) for each isomer, which were reoptimized with DFT (BP86/TZVP) and subdued to TDDFT ECD calculations (same level as above). It turned out that for (3R,4S)-**12** the first 5 conformers substantially contributed to the average ECD, while for (3S,4S)-**12** the first 10 did. The average calculated ECD values for the two isomers were in good agreement with experimental ones and their ACs could thus be assigned [66].

9.5.1.1 The Solid-State ECD TDDFT Method

From the discussion and examples given in the above section, it is clear that flexible molecules may represent very difficult cases to handle for ECD QM calculations. First, the conformational analysis step is prone to inaccuracy, the major sources of error lying in the prediction of relative energies and in the possible missing of one or more significant conformers. Second, the ECD calculation must be run on each structure at the same level of theory, which may require a lot of computational time.

One possibility of avoiding large time consumption and uncertainty related to the conformational flexibility is offered by considering the crystal state, where the molecular conformation is fixed and univocal. Moreover, if crystals amenable to X-ray analysis are available, such a structure can be determined with high accuracy. It must be stressed that the availability of good crystals is not sufficient for assigning AC by means of the Bijovet's method because this additionally requires the presence of a strong scatterer, that is, a "heavy" atom (second row and beyond). ECD spectra may be recorded in the solid state with various techniques, the most common of which consists in mixing a microcrystalline sample with an excess of a salt such as KBr or KCl, the mixture is then pressed to obtain a glassy transparent pellet similar to that used for IR measurements. Reproducing such a solid-state ECD spectrum by means of QM calculations would require a unique and, if possible, already determined conformation to be taken into account. This so-called ss-ECD/TDDFT method has been purposely developed for assigning AC of natural products [67] and requires the following steps: (1) isolation and structural identification of the natural compound; (2) growth of suitable crystals; (3) solid-state structure determination by X-ray single-crystal diffraction; (4) generation of an input geometry for ECD calculations with initial arbitrary AC, after optimizing hydrogen atoms of the X-ray structure with DFT; (5) TDDFT ECD calculation on the latter geometry with TDDFT, possibly employing various functional/basis set combinations; (6) measurement of solid-state ECD spectrum as KCl pellet; and (7) comparison between experimental solid-state ECD spectrum and TDDFT-calculated one using the X-ray geometry. The method is computationally very fast and usually leads to a very good agreement between calculated and experimental spectra [67]. Its main limitation is that it is applicable only to samples giving single crystals suitable for X-ray diffraction, although they need not contain heavy elements, which is the situation found for a large majority of natural compounds that are made up of H and C-to-O atoms. The second limitation is that intermolecular interactions between molecules closely packed in the crystals should not make significant contributions to the solid-state ECD spectrum because they cannot be reproduced by single-molecule calculations. Finally, since solid-state ECD spectra may be easily plagued by artifacts, a strict protocol for sample preparation and measurement must be followed. One of the best qualities of the ss-ECD/TDDFT method is that it may be applied to treat difficult cases, such as very flexible molecules or compounds that exist as mixture of tautomers or epimers in solution. It seems especially well suited for natural products because they frequently have rather complicated structures, which justifies the efforts toward obtaining crystals, and are often available in very small amounts preventing manipulations necessary for other ECD options.

As an example of application of the ss-ECD/TDDFT approach, let us consider macropodumine B (**13**, Figure 9.6) a unique zwitterionic natural compound containing a rare cyclopentadienyl anion and an iminio counterion [68]. It was isolated from *Daphniphyllum macropodum*, a Chinese medicinal plant, along with several related metabolites including macropodumine C (**14**), and its solid-state structure was determined by X-ray analysis. The AC of macropodumines B and C was established by comparing their ECD spectra with TDDFT-calculated ones, employing for **13** the solid-state approach and for **14** the more common approach described in Section 9.5.1. For macropodumine B (**13**), the ECD spectrum calculated on the X-ray geometry (B3LYP/TZVP) was in very good agreement with the experimental solid-state ECD (Figure 9.6a), and the AC could be safely established after a moderate computational effort. For macropodumine C (**14**), a Monte Carlo-based conformational search (MMFF method) followed by DFT geometry optimizations (B3LYP/6-31G*) revealed the presence of several minima due to the rotation of $-CH_2CH_2OH$ and $-COOCH_3$ substituents accompanied by fluctuations in the ring system. When used as input structures in B3LYP/TZVP calculations, the first four low-energy DFT minima (with Boltzmann populations >4% at 300 K) led to quite different ECD spectra. Their weighted average was in agreement with the ECD of **14** measured in solution in the low-energy region, while some discrepancy was obtained at high energies (Figure 9.6b). Moreover, the AC assignment required a considerable computational effort, about 4.5 times longer than for **13** [68].

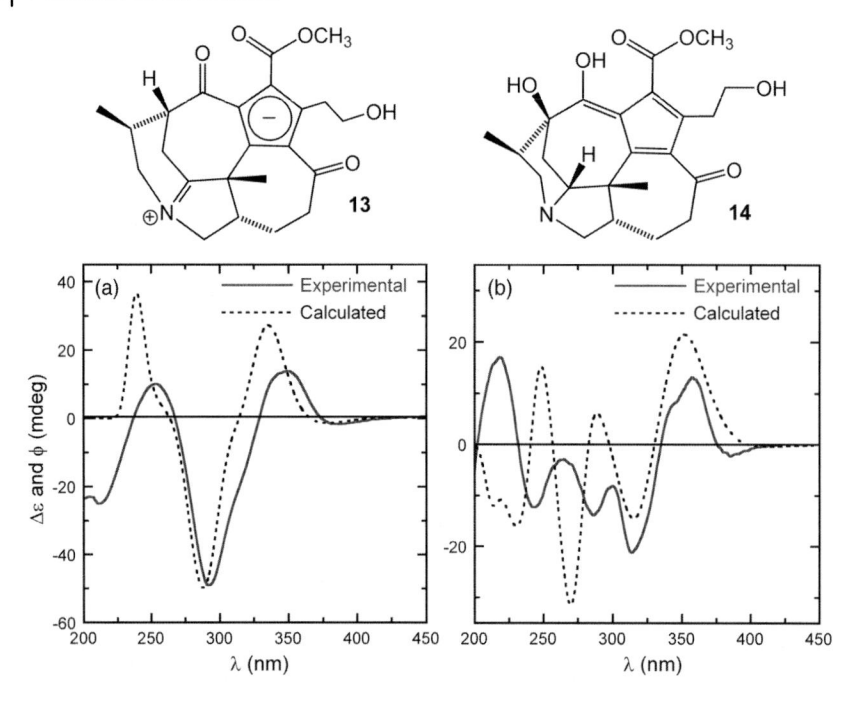

Figure 9.6 Experimental and TDDFT-calculated CD spectra for compounds **13** and **14**. For **13**, the solid-state CD spectrum (KCl pellet) is compared with that calculated using the X-ray geometry as input (a). For **14**, the spectrum in CH_3CN solution is compared with the Boltzmann-weighted average over four DFT-optimized lowest energy minima (b).

9.5.2
Interpretations of ECD Spectra

Having reliable calculations methods in hand may have some unexpected inconveniences. A possible one is the tendency to use them as "black box" that just provides the result we sought for without the need for any control of the underlying operations. Although such "superficial" use of computational packages may well solve everyday problems, it will not add much to our knowledge of physicochemical processes and properties such as those responsible for an ECD signal. On the contrary, one can fully benefit from all the possibilities offered by a reliable ECD calculation.

A deeper analysis of an excited state calculation result than a mere spectral comparison will first of all provide indication of the chromophores and the specific transitions involved in the computed ECD bands and allow one to rationalize the observed ECD. A classical way of analyzing UV-vis and ECD spectra of molecules is to assign each band to a certain transition, in terms of one-electron excitation or configuration, centered on a certain chromophore, as seen in Section 9.2. However, the typical output of QM calculations often describes most transitions as the combination of a more or less large number of configurations. Moreover, the "canonical" MOs are usually delocalized on large molecular portions, rather than on single chromophores. It must also be stressed that, strictly speaking, DFT or Kohn–Sham orbitals (KSO), defined starting from the electron density, are *not* the same as classical MOs, although they are known to be qualitatively similar and convey the same kind of information [69]. Overall, describing any ECD band as primarily due to a specific chromophore transition, although desirable, is always a little simplistic even in the most favorable cases. Let us consider the case of phospholene oxide **15** (Figure 9.7). The first transition computed with TDDFT (B3LYP/6-31++G(d,p)) is made up of eight different contributions, that is, single excitations involving four occupied and two virtual orbitals, most of which are delocalized on a large molecular portion [70]. In such a circumstance, it is possible to adopt a mathematical trick known as singular value decomposition (SVD) that transforms the space of canonical orbitals into a new one, that of "natural" orbitals, which are more strictly localized on individual chromophores. In addition, the number of configurations contributing to each transition is heavily reduced, making description of interesting transitions, and therefore of ECD bands, much simpler. After SVD, the first transition of compound **15** clearly emerges to be due to solely two π–π^* excitations with the four involved orbitals well localized on the benzene ring (Figure 9.7).

One of the most important qualities of QM calculations is that they may provide the chiroptical response of purposely designed molecular geometries that would not be experimentally accessible. In addition, one can also disentangle the contributions due to different conformers that sum up to yield the experimental average. The combination of the two pieces of information – the structural and the spectroscopic one – would provide a basis for understanding and/or establishing structure-to-spectra relationships. Put on a ground familiar to ECD users, these relationships correspond to the so-called helicity or sector rules. They relate the sign of a specific ECD band to the inherent chirality of the chromophore responsible for it (*helicity rules* like the diene one) or to the spatial disposition of a certain perturber with respect to an achiral chromophore responsible for it (*sector rules* like the ketone octant rule) [2, 71, 72]. These rules have represented in the past popular means of interpreting ECD spectra and especially of assigning ACs, but nowadays their use is greatly downsized. However, QM calculations offer a unique possibility to recheck the validity of old sector and helicity rules, put them on a firmer theoretical basis, and establish new ones. This is because, as stressed before, any calculated ECD spectrum

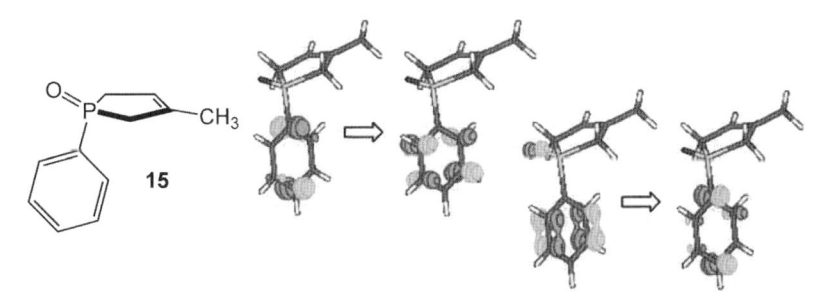

Figure 9.7 Principal components of the main electronic transition for **15** obtained after SVD treatment. Adapted from Ref. [70], with permission from Elsevier.

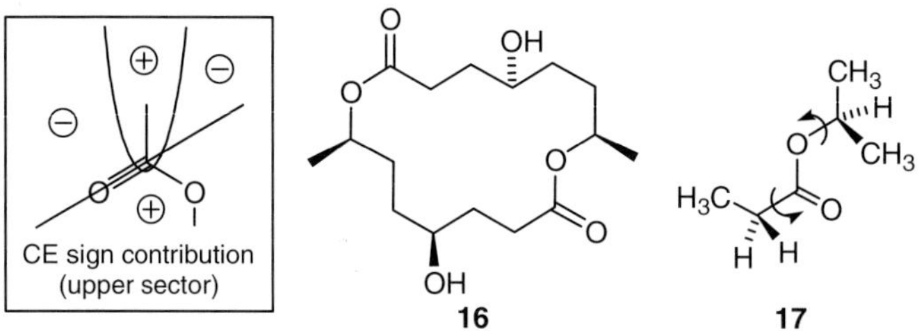

Figure 9.8 Empirical comet rule for ester n–π^* transition (framed) and compounds tested: a real one (**16**) and a hypothetical one (**17**).

is directly associated with its input structure, and calculation results may be easily analyzed in terms of involved transitions and orbitals. The case of ester chromophore will exemplify what stated. Various semiempirical rules were developed in the 1960s for esters and lactones, relating the sign of the observed n–π^* band around 220 nm to the geometry of the groups surrounding the chromophore [72]. In particular, Snatzke proposed a sector (or "comet") rule for acyclic esters valid only for a planar (achiral) ester chromophore (Figure 9.8).

The AC of a compound of natural origin, the diester pyrenophorol (**16**), was assigned through the solid-state ECD/TDDFT approach described in Section 9.5.1.1 [73]. In addition, Snatzke's comet rule was considered with the advantages provided by the solid state. In fact, the geometry found by X-ray analysis (showing a planar ester moiety) was subdued to TDDFT (B3LYP/TZVP) calculations and the sign calculated for the n–π^* band at 210 nm was in keeping with the rule's prediction, based on the same structure. Moreover, the hypothetical "chirogenic" fragment **17** was used in a series of calculations where the two relevant dihedral angles (curved arrows in Figure 9.8) were systematically varied and the ECD calculated. In this way, the consistency between the sign calculated for n–π^* band and the geometry around the ester chromophore could be more precisely ascertained. When analyzed through a sector scheme, the sector surfaces thus found turned out to be very similar to those suggested by Snatzke. The second rule established by Legrand and Bucourt for five–seven-membered ring lactones has recently been similarly rechecked [74, 75].

Analyses of this kind may help rejuvenating old semiempirical rules, as demonstrated by other examples where the very popular sector rules for the benzene chromophore have been investigated [76, 77].

9.5.3
Other Applications

Because of relatively large time consumption of high-level QM methods, ECD calculations with these methods are poorly employed as a source of conformational information. In particular, the approach described in Section 9.4.3 for DeVoe-type calculations, consisting in a systematic conformational sampling followed by ECD calculations on each single structure to find the best-matching one with the experiment, has seldom been explored with TDDFT calculations [54, 78, 79]. However, thanks to the advances in computer technologies we expect this kind of treatment to become more widespread in the future.

Thus far, we have not mentioned a rather fundamental application of TDDFT or other QM ECD calculations, which actually represents a fundamental prerequisite of everything discussed in this section. After a certain computational method is theoretically established, it needs to be validated for everyday uses. Normally, a set of "real" molecules, usually both structurally simple and conformationally rigid, is considered for calculations and the results compared with the experiment in order to assess the performance of the method tested. To check the applicability and scope of TDDFT ECD calculations, for example, various series of prototypical test molecules have been considered, either with very diverse skeletons [45, 80, 81] or belonging to certain classes, for example, alkenes [82]. Alternatively, single small molecules are investigated upon systematic variation of functionals and basis sets, to explore their respective capacities for specific problems [83, 84]. Finally, critical evaluation of new developments such as new functionals, solvent models, and so on, usually requires consideration of extensive sets of test molecules.

9.6
Conclusions and Perspectives

Nowadays, direct calculations of ECD spectra of increasingly large and complex molecules is possible with high-level QM approaches being both efficient and practical. It is amazing to notice that in a review about ECD calculations published in 2000, the authors observed that TDDFT "method has not been used for optical activity calculations" [31]. Less than 10 years later, TDDFT has become the most popular tool in the field of ECD calculations, and the development of efficient DFT functionals is still one of the hottest topics in theoretical chemistry.

Excited state calculations of medium-sized organic molecules are presently workable on a common personal computer. Clearly, a limitation is still posed by molecular flexibility because taking into account many different populated conformations may still be computationally very demanding. The whole procedure is complicated by the fact that, most often, computational packages for conformational searches that perform efficient conformational searches are not implemented for ECD calculations, and vice versa. An expected development for the near future is the appearance of packages capable of an automatic procedure consisting in conformational search, selection of relevant energy minima, computation of spectroscopic properties (including ECD), and averaging to afford a response immediately comparable to experimental data. Computational software is likely to further spread among nonspecialized users, although this "black box usage" may have its inconveniences, as discussed above.

Ab initio or TDDFT prediction of the ECD spectrum of a biopolymer is still not routine, though filling this gap will probably be just a matter of time. However, one should always use common sense to select the right means for the right purpose. The fact that TDDFT calculations are possibly applicable for a very complex and large molecular system does not necessarily mean that they *have* to be employed. Provided certain assumptions are met, hybrid approaches such as DeVoe's may afford the desired answer with a comparably negligible computational effort. Therefore, some space will always be left for simplified calculation tools in the present – and future – computational era.

Abbreviations

AC	absolute configuration
CE	Cotton effect
CI	configuration interaction
DFT	density functional theory
DL	dipole length
DV	dipole velocity
ECD	electronic circular dichroism
HF	Hartree–Fock
HOMO	highest occupied MO
IP	ionization potential
ISA	independent systems approximation
KSO	Kohn–Sham orbital
MO	molecular orbitals
NDO	neglect of differential overlap
QM	quantum mechanics
RC	relative configuration
ss-ECD	solid-state ECD
SVD	singular value decomposition
TDDFT	time-dependent DFT
UV-Vis	ultraviolet visible
XRD	X-ray diffractometry

References

1 Zhang, W., Krohn, K., Ullah, Z., Flörke, U., Pescitelli, G., Di Bari, L., Antus, S., Kurtán, T., Rheinheimer, J., Draeger, S., and Schulz, B. (2008) New mono- and dimeric members of the secalonic acid family: blennolides A-G isolated from the fungus *Blennoria* sp. *Chem. Eur. J.*, **14** (16), 4913–4923.

2 Berova, N., Di Bari, L., and Pescitelli, G. (2007) Application of electronic circular dichroism in configurational and conformational analysis of organic compounds. *Chem. Soc. Rev.*, **36** (6), 914–931.

3 Berova, N. and Nakanishi, K. (2000) Exciton chirality method: principles and applications, in *Circular Dichroism*, 2nd edn (eds N. Berova, K. Nakanishi, and R.W. Woody), John Wiley & Sons, Inc., New York, pp. 601–620.

4 Salvadori, P., Rosini, C., and Di Bari, L. (1997) Conformation and chiroptical

properties of dienes and polyenes, in *The Chemistry of Dienes and Polyenes* (ed. Z. Rappoport), John Wiley, Chichester, pp. 111–147.

5 Sreerama, N. and Woody, R.W. (2000) Circular dichroism of peptides and proteins, in *Circular Dichroism*, 2nd edn (eds N. Berova, K. Nakanishi, and R.W. Woody), John Wiley & Sons, Inc., New York, pp. 601–620.

6 Woody, R.W. and Hsu, M.-C. (1971) Origin of the heme Cotton effects in myoglobin and hemoglobin. *J. Am. Chem. Soc.*, **93** (14), 3515–3525.

7 Pescitelli, G., Sreerama, N., Salvadori, P., Nakanishi, K., Berova, N., and Woody, R.W. (2008) Inherent chirality dominates the visible/near-ultraviolet CD spectrum of rhodopsin. *J. Am. Chem. Soc.*, **130** (19), 6170–6181.

8 Rosini, C., Scamuzzi, S., Focati, M.P., and Salvadori, P. (1995) A general, multitechnique approach to the stereochemical characterization of 1,2-diarylethane-1,2-diols. *J. Org. Chem.*, **60** (25), 8289–8293.

9 Hug, W., Ciardelli, F., and Tinoco, I. (1974) Optically active hydrocarbon polymers with aromatic side chains. VI. Chiroptical properties of helical copolymers with aromatic side chains. *J. Am. Chem. Soc.*, **96** (11), 3407–3410.

10 Cech, C.L., Hug, W., and Tinoco, I., Jr. (1976) Polynucleotide circular dichroism calculations: use of an all-order classical coupled oscillator polarizability theory. *Biopolymers*, **15** (1), 131–152.

11 Harada, N. and Nakanishi, K. (1983) *Circular Dichroic Spectroscopy: Exciton Coupling in Organic Stereochemistry*, University Science Books, Mill Valley, CA.

12 Bayley, P.M., Nielsen, E.B., and Schellman, J.A. (1969) Rotatory properties of molecules containing two peptide groups: theory. *J. Phys. Chem.*, **73** (1), 228–243.

13 Sandstrom, J. (2000) Determination of absolute configurations and conformations of organic compounds by theoretical calculations of CD spectra. *Chirality*, **12** (4), 162–171.

14 Rizzo, V. and Schellman, J.A. (1984) Matrix-method calculation of linear and circular dichroism spectra of nucleic acids and polynucleotides. *Biopolymers*, **23** (3), 435–470.

15 Woody, R.W. and Koslowski, A. (2002) Recent developments in the electronic spectroscopy of amides and α-helical polypeptides. *Biophys. Chem.*, **101–102**, 535–551.

16 Oakley, M.T., Bulheller, B.M., and Hirst, J.D. (2006) First-principles calculations of protein circular dichroism in the far-ultraviolet and beyond. *Chirality*, **18** (5), 340–347.

17 Woody, R.W. (2005) The exciton model and the circular dichroism of polypeptides. *Monatsh. Chem.*, **136** (3), 347–366.

18 Superchi, S., Giorgio, E., and Rosini, C. (2004) Structural determinations by circular dichroism spectra analysis using coupled oscillator methods: an update of the applications of the DeVoe polarizability model. *Chirality*, **16** (7), 422–451.

19 Di Bari, L., Pescitelli, G., and Salvadori, P. (1999) Conformational study of 2,2′-homosubstituted 1,1′-binaphthyls by means of UV and CD spectroscopy. *J. Am. Chem. Soc.*, **121** (35), 7998–8004.

20 Di Bari, L., Lelli, M., Pintacuda, G., Pescitelli, G., Marchetti, F., and Salvadori, P. (2003) Solution versus solid-state structure of ytterbium heterobimetallic catalysts. *J. Am. Chem. Soc.*, **125** (18), 5549–5558.

21 Pescitelli, G., Di Bari, L., and Salvadori, P. (2004) Multiple solution species of titanium(IV) 1,1′-bi-2-naphtholate elucidated by NMR and CD spectroscopy. *Organometallics*, **23** (18), 4223–4229.

22 Pescitelli, G., Di Bari, L., and Salvadori, P. (2006) Effect of water on BINOL/Ti((OPr)-Pr-i)(4) solution mixtures: the nature of a catalytic precursor of enantioselective sulfoxidation. *J. Organomet. Chem.*, **691** (10), 2311–2318.

23 Di Bari, L., Pescitelli, G., Salvadori, P., Rovini, M., Anzini, M., Cappelli, A., and Vomero, S. (2007) Synthesis, resolution, and absolute configuration of two novel and selective cyclooxygenase-2 inhibitors based on the 1,5-diarylpyrrole structure.

Tetrahedron: Asymmetr., **17** (24), 3430–3436.

24 Di Bari, L., Pescitelli, G., Reginato, G., and Salvadori, P. (2001) Conformational investigation of two isomeric chiral porphyrins: a convergent approach with different techniques. *Chirality*, **13** (9), 548–555.

25 Autschbach, J. (2010) Computing chiroptical properties with first–principles theoretical methods: background and illustrative examples. *Chirality*, **21** (1E), E116–E152.

26 Crawford, T.D. (2006) Ab *initio* calculation of molecular chiroptical properties. *Theor. Chem. Acc.*, **115** (4), 227–245.

27 Pecul, M. and Ruud, K. (2005) The *ab initio* calculation of optical rotation and electronic circular dichroism. *Adv. Quantum Chem.*, **50**, 185–212.

28 Elliott, P., Burke, K., and Furche, F. (2008) Excited states from time-dependent density functional theory, in *Reviews in Computational Chemistry* (eds K.B. Lipkowitz and T.R. Cundari), John Wiley & Sons. Inc., New York, pp. 91–166.

29 Cramer, C.J. (2004) *Essentials of Computational Chemistry*, 2nd edn, John Wiley & Sons Ltd., Chichester.

30 Hansen, A.E. and Bouman, T.D. (1980) Natural chiroptical spectroscopy: theory and computations, in *Advances in Chemical Physics* (eds I. Prigogine and S.A. Rice), Wiley–Interscience, pp. 545–644.

31 Koslowski, A., Sreerama, N., and Woody, R.W. (2000) Theoretical approach to electronic optical activity, in *Circular Dichroism: Principles and Applications*, 2nd edn (eds N. Berova, K., Nakanishi, and R.W. Woody), John Wiley & Sons. Inc., New York, pp. 55–95.

32 Pecul, M., Ruud, K., and Helgaker, T. (2004) Density functional theory calculation of electronic circular dichroism using London orbitals. *Chem. Phys. Lett.*, **388** (1–3), 110–119.

33 Polavarapu, P.L. (2007) Renaissance in chiroptical spectroscopic methods for molecular structure determination. *Chem. Rec.*, **7** (2), 125–136.

34 Becke, A.D. (1993) Density-functional thermochemistry. III. The role of exact exchange. *J. Chem. Phys.*, **98** (7), 5648–5652.

35 (a) Becke, A.D. (1988) Density-functional exchange-energy approximation with correct asymptotic behavior. *Phys. Rev. A*, **38**, 3098–3100. (b) Becke, A.D. (1993) A new mixing of Hartree–Fock and local density-functional theories. *J. Chem. Phys.*, **98** (2), 1372–1377.

36 Adamo, C., and Barone V. (1999) Toward reliable density functional methods without adjustable parameters: the PBE0 model. *J. Chem. Phys.*, **110** (13), 6158–6170.

37 Casida, M.E., Jamorski, C., Casida, K.C., and Salahub, D.R. (1998) Molecular excitation energies to high-lying bound states from time-dependent density-functional response theory: characterization and correction of the time-dependent local density approximation ionization threshold. *J. Chem. Phys.*, **108** (11), 4439–4449.

38 Zerner, M.C. (1991) Semiempirical molecular orbital methods. *Rev. Comput. Chem.*, **2**, 313–365.

39 Bringmann, G., Bruhn, T., Maksimenka, K., and Hemberger, Y. (2009) The assignment of absolute stereostructures through quantum chemical circular dichroism calculations. *Eur. J. Org. Chem.*, (17) 2717–2727.

40 Bringmann, G., Gulder, T.A.M., Reichert, M., and Gulder, T. (2008) The online assignment of the absolute configuration of natural products: HPLC-CD in combination with quantum chemical CD calculations. *Chirality*, **20** (5), 628–642.

41 Telfer, S.G., Tajima, N., and Kuroda, R. (2004) CD spectra of polynuclear complexes of diimine ligands: theoretical and experimental evidence for the importance of internuclear exciton coupling. *J. Am. Chem. Soc.*, **126** (5), 1408–1418.

42 Telfer, S.G., Tajima, N., Kuroda, R., Cantuel, M., and Piguet, C. (2004) CD spectra of d-f heterobimetallic helicates with segmental di-imine ligands. *Inorg. Chem.*, **43** (17), 5302–5310.

43 Fabian, J., Diaz, L.A., Seifert, G., and Niehaus, T. (2002) Calculation of excitation energies of organic

chromophores: a critical evaluation. *THEOCHEM*, **594** (1–2), 41–53.

44 Crawford, T.D., Tam, M.C., and Abrams, M.l. (2007) The current state of *ab initio* calculations of optical rotation and electronic circular dichroism spectra. *J. Phys. Chem. A*, **111** (48), 12057–12068.

45 Diedrich, C. and Grimme, S. (2003) Systematic investigation of modern quantum chemical methods to predict electronic circular dichroism spectra. *J. Phys. Chem. A*, **107** (14), 2524–2539.

46 Shcherbin, D. and Ruud, K. (2008) The use of Coulomb-attenuated methods for the calculation of electronic circular dichroism spectra. *Chem. Phys.*, **349**, 234–243.

47 Goerigk, L. and Grimme, S. (2009) Calculation of electronic circular dichroism spectra with time-dependent double-hybrid density functional theory. *J. Phys. Chem. A*, **113** (4), 767–776.

48 Trindle, C. and Altun, Z. (2009) Circular dichroism of some high-symmetry chiral molecules: B3LYP and SAOP calculations. *Theor. Chem. Acc.*, **122** (3–4), 145–155.

49 Schafer, A., Huber, C., and Ahlrichs, R. (1994) Fully optimized contracted Gaussian basis sets of triple zeta valence quality for atoms Li to Kr. *J. Chem. Phys.*, **100** (8), 5829–5835.

50 Giorgio, E., Tanaka, K., Verrotta, L., Nakanishi, K., Berova, N., and Rosini, C. (2007) Determination of the absolute configurations of flexible molecules: synthesis and theoretical simulation of electronic circular dichroism/optical rotation of some pyrrolo[2,3-b]indoline alkaloids: a case study. *Chirality*, **19** (6), 434–445.

51 Kendall, R.A., Dunning, T.H., Jr., and Harrison, R.J. (1992) Electron affinities of the first-row atoms revisited. Systematic basis sets and wave functions. *J. Chem. Phys.*, **96** (9), 6796–6806.

52 Tanaka, T., Kodama, T.S., Morita, H.E., and Ohno, T. (2006) Electronic and vibrational circular dichroism of aromatic amino acids by density functional theory. *Chirality*, **18** (8), 652–661.

53 Sebek, J., Gyurcsik, B., Sebestik, J., Kejik, Z., Bednarova, L., and Bour, P. (2007) Interpretation of synchrotron radiation circular dichroism spectra of anionic, cationic, and zwitterionic dialanine forms. *J. Phys. Chem. A*, **111** (14), 2750–2760.

54 Cappelli, C., Bronco, S., and Monti, S. (2005) Computational study of conformational and chiroptical properties of $(2R,3S,4R)$-(+)-3,3′,4,4′,7-flavanpentol. *Chirality*, **17** (9), 577–589.

55 Lin, N., Santoro, F., Zhao, X., Rizzo, A., and Barone, V. (2008) Vibronically resolved electronic circular dichroism spectra of (R)-(+)-3-methylcyclopentanone: a theoretical study. *J. Phys. Chem. A*, **112** (48), 12401–12411.

56 Dierksen, M. and Grimme, S. (2006) A theoretical study of the chiroptical properties of molecules with isotopically engendered chirality. *J. Chem. Phys.*, **124** (17), 174301–174312.

57 Nooijen, M. (2006) Investigation of Herzberg–Teller Franck–Condon approaches and classical simulations to include effects due to vibronic coupling in circular dichroism spectra: the case of dimethyloxirane continued. *Int. J. Quantum. Chem.*, **106** (12), 2489–2510.

58 Tomasi, J., Mennucci, B., and Cammi, R. (eds) (2007) *Continuum Solvation Models in Chemical Physics: From Theory to Applications*, John Wiley & Sons Ltd., Chichester.

59 Jacquemin, D., Perpete, E.A., Scuseria, G.E., Ciofini, I., and Adamo, C. (2008) TD-DFT performance for the visible absorption spectra of organic dyes: conventional versus long-range hybrids. *J. Chem. Theory Comput.*, **4** (1), 123–135.

60 Ren, Y., Ke, W., Li, Y., Feng, L., Wan, J., and Xu, X. (2009) Understanding the spectroscopic properties of the photosynthetic reaction center of *Rhodobacter sphaeroides* by a combined theoretical study of absorption and circular dichroism spectra. *J. Phys. Chem. B*, **113** (30), 10055–10058.

61 Goldsmith, M.-R., George, C.B., Zuber, G., Naaman, R., Waldeck, D.H., Wipf, P., and Beratan, D.N. (2006) The chiroptical signature of achiral metal clusters induced by dissymmetric adsorbates. *Phys. Chem. Chem. Phys.*, **8**, 63–67.

62 Freedman, T.B., Cao, X., Dukor, R.K., and Nafie, L.A. (2003) Absolute configuration determination of chiral molecules in the solution state using vibrational circular dichroism. *Chirality*, **15** (9), 743–758.

63 Nafie, L.A. (2008) Vibrational circular dichroism: a new tool for the solution-state determination of the structure and absolute configuration of chiral natural product molecules. *Nat. Prod. Commun.*, **3** (3), 451–466.

64 Barron, L.D., Zhu, F., Hecht, L., Tranter, G.E., and Isaacs, N.W. (2007) Raman optical activity: an incisive probe of molecular chirality and biomolecular structure. *J. Mol. Struct.*, **834–836**, 7–16.

65 Ding, Y., Li, X.-C., and Ferreira, D. (2007) Theoretical calculation of electronic circular dichroism of the rotationally restricted 3,8″-biflavonoid morelloflavone. *J. Org. Chem.*, **72** (24), 9010–9017.

66 Voloshina, E.N., Raabe, G., Fleischhauer, J., Kramp, G.J., and Gais, H.-J. (2004) (*E*)-4-Methyl-1-tributylstannyl-oct-1-en-6-yn-3-ol: circular dichroism measurement and determination of the absolute configuration by quantum-chemical CD calculations. *Zeits. Naturfor.*, **59a** (3), 124–132.

67 Pescitelli, G., Kurtán, T., Flörke, U., and Krohn, K. (2010) Absolute structural elucidation of natural products: a focus on quantum-mechanical calculations of solid-state CD spectra. *Chirality*, **21** (1E), E181–E201

68 Guo, Y.-W., Kurtán, T., Krohn, K., Pescitelli, G., and Zhang, W. (2009) Assignment of the absolute configuration of zwitterionic and neutral macropodumines by means of TDDFT CD calculations. *Chirality*, **21** (6), 561–568.

69 Stowasser, R. and Hoffmann, R. (1999) What do the Kohn–Sham orbitals and eigenvalues mean? *J. Am. Chem. Soc.*, **121** (14), 3414–3420.

70 Mayer, Z.A., Kállay, M., Kubinyi, M., Keglevich, G., Ujj, V., and Fogassy, E. (2009) Assignment of absolute configurations of chiral phospholene oxides by UV/CD spectroscopy using TD-DFT quantum chemical calculations and singular value decomposition approach for the analysis of the spectra. *J. Mol. Struct. THEOCHEM*, **906** (1–3), 94–99.

71 Eliel, E.L. and Wilen, S.H. (1994) *Stereochemistry of Organic Compounds*, John Wiley & Sons, Inc., New York.

72 Ciardelli, F. and Salvadori, P. (1973) *Fundamental Aspects and Recent Developments in Optical Rotatory Dispersion and Circular Dichroism*, Heyden, London.

73 Krohn, K., Farooq, U., Flörke, U., Schulz, B., Draeger, S., Pescitelli, G., Salvadori, P., Antus, S., and Kurtán, T. (2007) Secondary metabolites isolated from an endophytic *Phoma* sp.: absolute configuration of tetrahydropyrenophorol using the solid-state TDDFT CD methodology. *Eur. J. Org. Chem.*, (19), 3206–3211.

74 Bertin, V., Hussain, H., Kouam, S.F., Dongo, E., Pescitelli, G., Salvadori, P., Kurtán, T., and Krohn, K. (2008) Antialactone: a new γ-lactone from *Antiaris africana*, and its absolute configuration determined from TDDFT CD calculations. *Nat. Prod. Commun.*, **3** (2), 215–218.

75 Stecko, S., Frelek, J., and Chmielewski, M. (2009) Chiraloptical properties of adducts derived from cyclic nitrones and 2(5*H*)-furanones. Combined experimental and theoretical studies. *Tetrahedron: Asymmetry*, **20** (15), 1778–1790.

76 Pescitelli, G., Di Bari, L., Caporusso, A.M., and Salvadori, P. (2008) The prediction of the circular dichroism of the benzene chromophore: TDDFT calculations and sector rules. *Chirality*, **20** (3/4), 393–399.

77 Kundrat, M.D. and Autschbach, J. (2009) Modeling of the chiroptical response of chiral amino acids in solution using explicit solvation and molecular dynamics. *J. Chem. Theory Comput.*, **5** (4), 1051–1060.

78 Fukuyama, T., Matsuo, K., and Gekko, K. (2005) Vacuum-ultraviolet electronic circular dichroism of L-alanine in aqueous solution investigated by time-dependent density functional theory. *J. Phys. Chem. A*, **109** (31), 6928–6933.

79 Mori, T., Inoue, Y., and Grimme, S. (2007) Experimental and theoretical study of the CD spectra and conformational properties of axially chiral 2,2′-, 3, 3′-, and 4,4′-biphenol ethers. *J. Phys. Chem. A*, **111** (20), 4222–4234.

80 Autschbach, J., Ziegler, T., van Gisbergen, S.J.A., and Baerends, E.J. (2002) Chiroptical properties from time-dependent density functional theory. I. Circular dichroism spectra of organic molecules. *J. Chem. Phys.*, **116** (16), 6930–6940.

81 Autschbach, J., Jorge, F.E., and Ziegler, T. (2003) Density functional calculations on electronic circular dichroism spectra of chiral transition metal complexes. *Inorg. Chem.*, **42** (9), 2867–2877.

82 McCann, D.M. and Stephens, P.J. (2006) Determination of absolute configuration using density functional theory calculations of optical rotation and electronic circular dichroism: chiral alkenes. *J. Org. Chem.*, **71** (16), 6074–6098.

83 Skomorowski, W., Pecul, M., Sałek, P., and Helgaker, T. (2007) Electronic circular dichroism of disulphide bridge: *ab initio* quantum-chemical calculations. *J. Chem. Phys.*, **127** (8), 085102 085101-085108.

84 Kowalczyk, T.D., Abrams, M.L., and Crawford, T.D. (2006) Ab *initio* optical rotatory dispersion and electronic circular dichroism spectra of (*S*)-2-chloropropionitrile. *J. Phys. Chem. A*, **110** (24), 7649–7654.

10
Computational Dielectric Spectroscopy of Charged, Dipolar Systems

Christian Schröder and Othmar Steinhauser

10.1
Methods

10.1.1
Dielectric Field Equation

Dielectric spectroscopy obtains information of a molecular system by exposing the sample to an external, spatially homogeneous electric field $\vec{E}_{ext}(\omega)$. For not too strong fields, the response of the system, that is, the total dielectric polarization $\vec{P}_{tot}(\omega)$, is a linear function of the Maxwell field $\vec{E}(\omega)$:

$$\vec{P}_{tot}(\omega) = \frac{\langle \vec{M}_{tot} \rangle_{\vec{E}_{ext}}}{V} = \frac{\Sigma^*(\omega)}{4\pi} \vec{E}(\omega) \tag{10.1}$$

with the proportionality factor $\Sigma^*(\omega)/4\pi$. Hence, $\Sigma^*(\omega)$ is commonly called dielectric "susceptibility" or generalized dielectric constant (GDC) (although being a function of ω). The polarization itself represents the averaged total collective dipole moment $\vec{M}_{tot}$ per volume V in the presence of the external field $\vec{E}_{ext}$. The constitutive relation (10.1) always involves the Maxwell or internal field $\vec{E}$, that is, the average field acting in the sample. It is composed of the external field $\vec{E}_{ext}$ (which is assumed to be spatially homogeneous and hence does not depend on $\vec{r}$) and the average molecular fields inside:

$$\vec{E}(\vec{r}) = \vec{E}_{ext} + \int_V \vec{S}(\vec{r}-\vec{r}') \cdot \varrho(\vec{r}')d\vec{r}' + \int_V \overleftrightarrow{T}(\vec{r}-\vec{r}') \cdot \vec{P}_D(\vec{r}')d\vec{r}' \tag{10.2}$$

The coupling vector $\vec{S}(\vec{r})$ and the tensor $\overleftrightarrow{T}(\vec{r})$ are the gradients of the Green function $G(\vec{r})$:

$$G(\vec{r}) = \frac{1}{|\vec{r}|} \tag{10.3}$$

Computational Spectroscopy: Methods, Experiments and Applications. Edited by Jörg Grunenberg
Copyright © 2010 WILEY-VCH Verlag GmbH & Co. KGaA, Weinheim
ISBN: 978-3-527-32649-5

$$\vec{S}(\vec{r}) = -\vec{\nabla}\frac{1}{|\vec{r}|} \tag{10.4}$$

$$\overleftrightarrow{T}(\vec{r}) = \vec{\nabla}\vec{\nabla}\frac{1}{|\vec{r}|} \tag{10.5}$$

The molecular fields in Equation 10.2 consist of a contribution from the rotation of the dipoles (represented by the dipolar density $\vec{P}_D(\vec{r})$) and a contribution from the charges in the system formulated in terms of the charge density $\varrho(\vec{r})$. This charge density is connected to the translational or current polarization via the equation of continuity

$$\int \vec{\nabla} \cdot \frac{\vec{J}(t)}{V} dt + \varrho = 0$$
$$\vec{\nabla} \cdot \vec{P}_J + \varrho = 0 \tag{10.6}$$

keeping in mind that the current $\vec{J}(t)$ is the time derivative of the translational dipole moment $\vec{M}_J(t) = \int \vec{J}(t)dt$, which divided by the volume gives the current polarization $\vec{P}_J = \vec{M}_J/V$. The sum of the rotational polarization $\vec{P}_D(\vec{r})$ and the current polarization $\vec{P}_J(\vec{r})$ yields the total polarization $\vec{P}_{tot}(\vec{r})$:

$$\vec{P}_{tot}(\vec{r}) = \vec{P}_J(\vec{r}) + \vec{P}_D(\vec{r}) \tag{10.7}$$

The inhomogeneous, internal field in Equation 10.2 is a particular solution of the Poisson equation (10.8) (see Equation 4.33 of Ref. [30]):

$$\vec{\nabla} \cdot \vec{E} = 4\pi(\varrho - \vec{\nabla} \cdot \vec{P}_D)$$
$$= -4\pi(\vec{\nabla} \cdot \vec{P}_J + \vec{\nabla} \cdot \vec{P}_D) \tag{10.8}$$

$$= -4\pi \vec{\nabla} \cdot \vec{P}_{tot} \tag{10.9}$$

which can be reformulated by means of Equations 10.6 and 10.7 to show the dependence of the internal field $\vec{E}$ on the total polarization $\vec{P}_{tot}$. The fact that at least parts of the system consist of simultaneously charged and dipolar species manifests itself in the appearance of translational and rotational dipole densities $\vec{P}_J$ and $\vec{P}_D$. Replacing the total polarization by the constitutive relation (10.1) gives

$$\vec{\nabla} \cdot (\vec{E} + \Sigma^*(\omega) \cdot \vec{E}) = 0$$
$$\vec{\nabla} \cdot ((\Sigma^*(\omega) + 1)\vec{E}) = 0 \tag{10.10}$$

This last equation can be used to calculate the local internal field $\vec{E}(\vec{r})$, which is the negative gradient of the potential $\phi(\vec{r})$. First, the universal, spatially independent GDC is made space dependent

$$\Sigma^*(\omega) + 1 \to \Sigma^1(\vec{r}, \omega) \tag{10.11}$$

transforming Equation 10.10 into

$$\vec{\nabla} \cdot (\Sigma^1(\vec{r}, \omega)\vec{\nabla}\phi(\vec{r})) = 0 \tag{10.12}$$

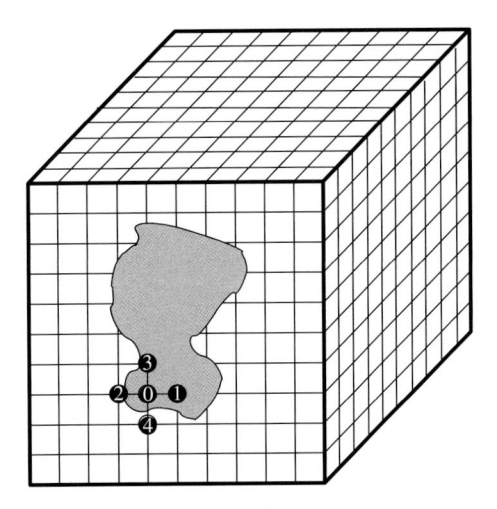

Figure 10.1 Computation of the dielectric constant $\Sigma^1(\vec{r}, \omega)$ on a basis of a r-dependent potential ϕ. In order to calculate the potential ϕ_0, the adjacent potentials $\phi_{1\ldots 6}$ are used. Due to this schematic 2D visualization, ϕ_5 and ϕ_6 cannot be displayed. For the evaluation (Equation 10.15) of Σ_0^1, Σ_1^1, and Σ_3^1, the dielectric constant ε and the conductivity σ of the gray-shaded body are used, whereas the dielectric constant of the conductivity of the "white" solvent determine Σ_2^1 and Σ_4^1.

For an approximate numerical solution, the volume of the sample is subdivided into finite elements, in the simplest case into cubes of edge length h, as depicted in Figure 10.1. Integrating the equation over a cube and applying the Gaussian theorem, we get the surface integral

$$0 = \iint_A \Sigma^1(\vec{r}, \omega) \vec{\nabla} \phi(\vec{r}) d\vec{A}$$
$$= \sum_{k=1}^{6} \Sigma_k^1 \frac{\phi_k - \phi_0}{h} h^2 \qquad (10.13)$$

from which the potential ϕ_0 at the center of each cube is finally given by

$$\phi_0 = \frac{\sum_{k=1}^{6} \Sigma_k^1 \phi_k}{\sum_{k=1}^{6} \Sigma_k^1} \qquad (10.14)$$

This formula has a simple interpretation. The potential at the central cube ϕ_0 is determined by the potential ϕ_k at the six nearest neighbors multiplied by the respective GDC Σ_k^1. Practically, ϕ_k is obtained in a self-consistent procedure starting from appropriate estimates of ϕ_k. So far we have followed the traditional concept of the widely used Poisson–Boltzmann method [24]. The essential difference is the replacement of the simple dielectric permittivity ε by the GDC. We emphasize that the crucial point is the choice of the GDC of each grid point. For example, in Figure 10.1, Σ_0^1, Σ_1^1, and Σ_3^1 are equal and correspond to the GDC of the gray-shaded body. Σ_2^1 and Σ_4^1 are computed from the dielectric permittivity ε and the conductivity σ of the "white" solute according to

$$\Sigma_k^1 = \varepsilon_k + \frac{4\pi i \sigma_k}{\omega} \tag{10.15}$$

In this approximation of Ref. [3], ε_k and σ_k are the static values of the dielectric permittivity $\varepsilon(\omega = 0)$ and the conductivity $\sigma(\omega = 0)$. Actually, these quantities are not varied at each cube, but are kept constant in a complete region representing a biomolecule, the solvent, the membrane, and so on. The estimate (10.15) is inspired by the general relation

$$\Sigma^*(\omega) = \varepsilon(\omega) - 1 + \frac{4\pi i \sigma(\omega)}{\omega} \tag{10.16}$$

derived in the subsequent section. Upon making the drastic approximation to replace both $\varepsilon(\omega)$ and $\sigma(\omega)$ by their static values leads to the estimate in Equation 10.15 since the frequency dependence is solely introduced by the denominator ω.

10.1.2
Molecular Resolution of the Total Collective Dipole Moment

In principle, the computation of the GDC in a molecular dynamics (MD) simulation is directly possible from the constitutive relation (10.1) by applying an additional external field [48, 63, 64, 71]. The molecular forces $\vec{F}_{i,\alpha}$ acting on each atom α of the molecule i with their velocities $\vec{v}_{i,\alpha}$ are augmented by charge-weighted term $q_{i,\alpha} \cdot \vec{E}_{ext}$:

$$m_{i,\alpha} \frac{d\vec{v}_{i,\alpha}}{dt} = \vec{F}_{i,\alpha} + q_{i,\alpha} \cdot \vec{E}_{ext} \tag{10.17}$$

The external field influences the individual atoms that are located off-center at $\vec{r}_{i,\alpha}$ and simultaneously generates a translation of the center-of-mass as well as a rotation about that center. As we are dealing with charged species, their collective translation creates a current $\vec{J}(t)$. Without an external field, that is, for a system in equilibrium, the direction of $\vec{J}(t)$ fluctuates and finally averages to 0. In the presence of the field, however, a preferred direction remains and produces a continuous current whose molecular contributions obey the equation of motion:

$$\frac{d\vec{j}_i}{dt} = q_i \frac{d\vec{v}_{cm,i}}{dt} = \frac{q_i}{m_i} \left(\sum_\alpha \vec{F}_{i,\alpha} + \left(\sum_\alpha q_{i,\alpha} \right) \cdot \vec{E}_{ext} \right)$$
$$= \frac{q_i}{m_i} \vec{F}_i + \frac{q_i^2}{m_i} \cdot \vec{E}_{ext} \tag{10.18}$$

If the external field is strong enough to overrule the molecular forces $\vec{F}_i$, the second term in the last equation dominates the time evolution of the molecular current $\vec{j}_i$. The sign of the molecular charge carrier q_i becomes insignificant and the molecular current follows the direction of the applied field. In Equation 10.18, $m_i = \sum_\alpha m_{i,\alpha}$, $q_i = \sum_\alpha q_{i,\alpha}$, and $\vec{F}_i = \sum_\alpha \vec{F}_{i,\alpha}$ denote the net mass, net charge, and net force acting on molecule i. These quantities govern the time evolution of the center-of-mass velocity $\vec{v}_{\text{cm},i}$. In other words, the translation of the covalently bound charge set $\{q_{i,\alpha}\}$ may be considered as the movement of the net charge q_i located at the center-of-mass. Consequently, the molecular contribution to the current is $\vec{j}_i(t) = q_i \cdot \vec{v}_{\text{cm},i}(t)$ [14]. The total current $\vec{J}(t)$ is given as the sum over all molecules:

$$\vec{J}(t) = \sum_i \vec{j}_i(t) = \sum_i q_i \cdot \vec{v}_{\text{cm},i}(t) \tag{10.19}$$

This relation holds irrespective of the presence or absence of an external field. Integrating this current over time yields the corresponding collective translational dipole moment:

$$\vec{M}_J(t) = \int \vec{J}(t) dt = \sum_i q_i \cdot \vec{r}_{\text{cm},i}(t) \tag{10.20}$$

With respect to actual simulations, the evaluation of $\vec{M}_J(t)$ as the integral of $\vec{J}(t)$ is to be preferred over a sum involving the center-of-mass positions $\vec{r}_{\text{cm},i}(t)$, because the latter experiences discontinuities through toroidal shifts due to minimum image convention (for details, see Section 10.1.4). However, $\vec{M}_J(t)$ is only the translational part of the total collective dipole moment $\vec{M}_{\text{tot}}(t)$,

$$\vec{M}_{\text{tot}}(t) = \sum_i \sum_\alpha q_{i,\alpha} \cdot \vec{r}_{i,\alpha}(t) \tag{10.21}$$

entering the constitutive Equation 10.1. The nontranslational complement of $\vec{M}_J(t)$

$$\vec{M}_D(t) = \vec{M}_{\text{tot}}(t) - \vec{M}_J(t) \tag{10.22}$$

$$= \sum_i \sum_\alpha q_{i,\alpha} \vec{r}_{i,\alpha}(t) - \sum_i q_i \vec{r}_{\text{cm},i}(t) \tag{10.23}$$

$$= \sum_i \sum_\alpha q_{i,\alpha} (\vec{r}_{i,\alpha}(t) - \vec{r}_{\text{cm},i}(t)) \tag{10.24}$$

$$= \sum_i \vec{\mu}_{D,i}(t) \tag{10.25}$$

is not only essentially determined by overall rotation but also includes bond and angle vibrations as well as torsional motions. For the sake of simplicity, however, we use the term collective rotational dipole moment for $\vec{M}_D(t)$.

At the molecular level, we map the molecular $\vec{\mu}_{D,i}$ to the ith contribution in Equation 10.24:

$$\vec{\mu}_{D,i} = \sum_\alpha q_{i,\alpha} (\vec{r}_{i,\alpha}(t) - \vec{r}_{\text{cm},i}(t)) \tag{10.26}$$

$$= \vec{\mu}_i - q_i \cdot \vec{r}_{\text{cm},i} \tag{10.27}$$

$$= \sum_\alpha \left(q_{i,\alpha} - q_i \frac{m_{i,\alpha}}{m_i} \right) \vec{r}_{i,\alpha}(t) \tag{10.28}$$

$$= \sum_\alpha q'_{i,\alpha} \cdot \vec{r}_{i,\alpha}(t) \tag{10.29}$$

The last equation shows that the so-defined molecular dipole moment $\vec{\mu}_{D,i}$ is independent of the choice of origin, because the sum over $q'_{i,\alpha}$ is zero. However, this independence neither applies to $\vec{\mu}_i$ nor to its translational counterpart $q_i \cdot \vec{r}_{cm,i}(t)$ in the case of charged molecules. Taking the difference of both quantities, the dependence cancels out.

So far, the molecular formulas given are valid for charged and neutral molecules likewise, as demonstrated by Equation 10.29. In the case of neutral molecules, $\vec{\mu}_{D,i}$ equals $\vec{\mu}_i$, because the net charge q_i does not exist. Consequently, $q'_{i,\alpha}$ is $q_{i,\alpha}$. In the case of charged molecules, the role of the modified charges $q'_{i,\alpha}$ is merely to show that $\vec{\mu}_{D,i}$ is still a well-defined dipole moment. They are never used to compute physical quantities, for example, forces, in simulations. Moreover, $\vec{\mu}_{D,i}$ for charged species can be alternatively computed by

$$\vec{\mu}_{D,i} = q_i(\vec{r}_{cq,i}(t) - \vec{r}_{cm,i}(t)) \tag{10.30}$$

with the center of charge $\vec{r}_{cq,i}(t)$ defined analogously to the center-of-mass, but using the partial charges as weight factors. In Figure 10.2, the center-of-mass and the center of charge are shown for the organic cation 1-butyl-3-methyl-imidazolium (BMIM$^+$). The difference vector $\vec{r}_{cq,i}(t) - \vec{r}_{cm,i}(t)$ is the so-called charge arm [33]. The net charge of the molecule is located at $\vec{r}_{cq,i}(t)$, that is, off to the center-of-mass. This implies that the action of the external field on the charge center creates both a translation of the center-of-mass of the molecule and a rotation about this center. Therefore, $\vec{\mu}_{D,i}$ is best suited to describe the molecular rotation. In fact, it was shown in Ref. [49] that the equation of motion for the angular momentum is governed by the torque $\vec{\mu}_{D,i} \times \vec{E}_{ext}$. Furthermore, $\vec{\mu}_{D,i}$ is also important for the static structure. It is essential in constructing the so-called g-coefficient expansion [28, 62].

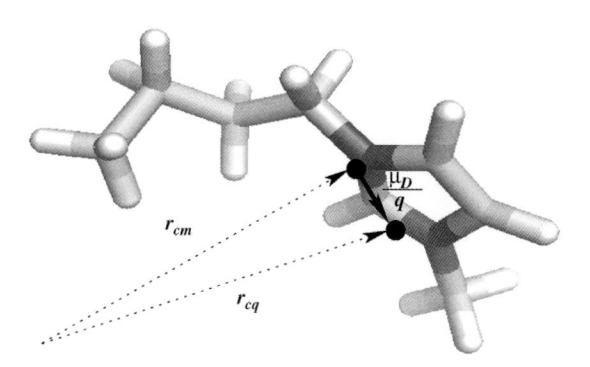

Figure 10.2 Schematic view of a 1-butyl-3-methylimidazolium cation. The center-of-mass $\vec{r}_{cm}$ and the center of charge $\vec{r}_{cq}$ does not coincide for a molecular ion. The distance vector between these two centers is called "charge arm" and equals the dipole vector $\vec{\mu}_D$ divided by the net charge q of the molecule.

If the whole system exclusively consists of charged species, the concept of charge center can also be used to compute the total collective dipole moment:

$$\vec{M}_{tot}(t) = \sum_i q_i \cdot \vec{r}_{cq,i} \tag{10.31}$$

In this sense, the asymmetry of the molecular charge distribution $\{q_{i,\alpha}\}$ determines the dielectric properties via the position of the charge center in the respective molecular ions. For atomic ions, the center-of-mass and the center of charge coincide, thus giving no charge arm descriptive of rotation. As a result, $\vec{M}_{tot}(t)$ is only given by $\vec{M}_J(t)$.

Figure 10.3 summarizes the considerations made in this chapter. The central quantity $\vec{M}_{tot}$ is on top of the figure and stands for the macroscopic regime. This is the definite level of experiments measuring the complex spectrum of the GDC. This is also the level where experiment and simulation meet each other. The step down to the mesoscopic level $\vec{M}_{tot}$ can be decomposed into its rotational ($\vec{M}_D$) and translational ($\vec{M}_J$) components. This decomposition is in a strict sense an interpretation of the well-defined collective property $\vec{M}_{tot}$. In other words, this decomposition is not necessary but useful. From the rotational component $\vec{M}_D$, the dielectric permittivity $\varepsilon(\omega)$ can be evaluated and the translational component $\vec{M}_J$ can be used to compute the conductivity $\sigma(\omega)$. Both quantities contain a contribution from the cross-correlation of the rotational and translational dipolar components, but this contribution is very small [17, 34, 53]. In a complex mixture of positively, negatively, and uncharged species, a further subdivision into collective moments $\vec{M}_D^+, \vec{M}_D^-, \vec{M}_D^0, \vec{M}_J^+$, and $\vec{M}_J^-$ can be made for a finer grained analysis. At this point, we anticipate that the rotational components affect the low-frequency regime of the GDC spectrum, while the translational ones show up at higher frequencies.

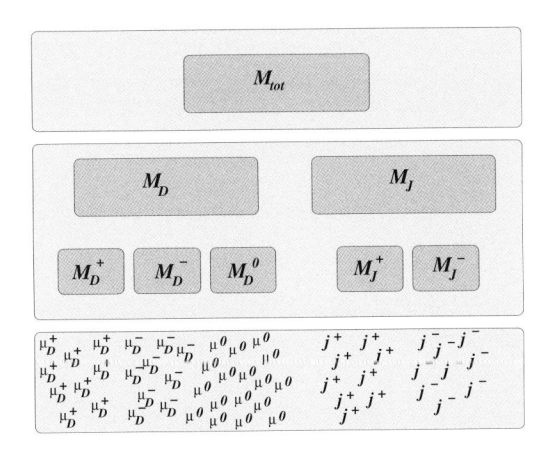

Figure 10.3 Decomposition of the total collective dipole moment $\vec{M}_{tot}$: The rotational part $\vec{M}_D$ can be calculated for each species since it is made up by the corresponding molecular dipole moments $\vec{\mu}_D$. The translational collective dipole moment $\vec{M}_J$ is basically determined by the current j of each molecule.

At the finest level of graining, a resolution into molecular dipole moments $\vec{\mu}_D^+$, $\vec{\mu}_D^-$, and $\vec{\mu}^0$ as well as molecular currents $\vec{j}^+$ and $\vec{j}^-$ is possible.

10.1.3
Computing the Generalized Dielectric Constant in Equilibrium

In the previous section, we applied an external field to the sample in order to create a total collective dipole moment $\vec{M}_{tot}$. Simulations of this kind are seldom found in literature [48, 63, 64, 71]. The reason is that one needs strong external fields to overrule the molecular forces as visible in Equation 10.18. In this case, however, the average total dipole moment $\langle \vec{M}_{tot} \rangle$ is no more a linear function of the applied field as quadratic and higher terms enter Equation 10.1. On the contrary, for weak fields, the response of the system is too noisy to be seriously analyzed. An alternative way is the equilibrium MD simulation.

Although the averages of the collective dipole moments are zero in equilibrium, they nevertheless have a memory. The traditional measures of temporal memory are time correlation functions, such as

$$\Phi(t) = \langle \vec{A}(0) \cdot \vec{B}(t) \rangle = \lim_{T \to \infty} \frac{1}{T} \int_0^T \vec{A}(t') \cdot \vec{B}(t+t') dt' \tag{10.32}$$

Quite generally, the initial value $\Phi(t=0)$ is the average of the product $\langle \vec{A} \cdot \vec{B} \rangle$, while the asymptotic value is the product of the averages $\langle \vec{A} \rangle \cdot \langle \vec{B} \rangle$. In equilibrium, however, the last value should be zero. The special case is autocorrelation functions with $\vec{A} = \vec{B}$. Frequently, they decay monotonically from the initial value $\langle \vec{A}^2 \rangle$ to the asymptotic one $\langle \vec{A} \rangle^2$. In contrast, the current autocorrelation functions show a typical behavior of a damped oscillator. Furthermore, time correlation functions have a translational symmetry in time, that is, they are independent of the initial reference time $\langle \vec{A}(0) \cdot \vec{B}(t) \rangle = \langle \vec{A}(t') \cdot \vec{B}(t+t') \rangle$. The first derivative $d\Phi(t)/dt$ is given by

$$\frac{d\Phi(t)}{dt} = \frac{d}{dt} \langle \vec{A}(0) \cdot \vec{B}(t) \rangle = \left\langle \vec{A}(0) \cdot \frac{d\vec{B}(t)}{dt} \right\rangle \tag{10.33}$$

$$= \langle \vec{A}(0) \cdot \dot{\vec{B}}(t) \rangle \tag{10.34}$$

Furthermore, the second derivative $d^2\Phi(t)/dt^2$ is equal to negative of the correlation function of the first derivatives:

$$\frac{d^2\Phi(t)}{dt^2} = \frac{d^2}{dt^2} \langle \vec{A}(0) \cdot \vec{B}(t) \rangle = -\left\langle \frac{d\vec{A}(0)}{dt} \cdot \frac{d\vec{B}(t)}{dt} \right\rangle \tag{10.35}$$

$$= -\langle \dot{\vec{A}}(0) \cdot \dot{\vec{B}}(t) \rangle \tag{10.36}$$

Instead of simulating the situation of an applied field, one can use linear response theory that includes the external field in the Liouville equation [26]. Under the assumption of not too strong fields, the augmented Liouville equation is linearized such that the system's response can be expressed in terms of time correlation functions computed in equilibrium. For the concrete case of dielectric spectroscopy, one gets

$$\vec{P}_{tot}(\omega) = \frac{\langle \vec{M}_{tot} \rangle_{\vec{E}_{ext}}}{V} = \frac{1}{3Vk_BT} \mathcal{L}\left[-\frac{d}{dt}\langle \vec{M}_{tot}(0) \cdot \vec{M}_{tot}(t)\rangle_{eq}\right] \vec{E}_{ext}(\omega) \quad (10.37)$$

Since the left-hand side of Equations 10.1 and 10.37 are equal, the right-hand side must also be equal:

$$\Sigma(\omega) = \Sigma^*(\omega) - (\varepsilon_\infty - 1) = \frac{4\pi}{3Vk_BT} \mathcal{L}\left[-\frac{d}{dt}\langle \vec{M}_{tot}(0) \cdot \vec{M}_{tot}(t)\rangle_{eq}\right] \cdot \frac{E_{ext}}{E} \quad (10.38)$$

In principle, the dielectric polarization contains an electronic contribution that shows up in the GDC as ε_∞ visible in the high-frequency limit. In theoretical studies, ε_∞ is often set to unity corresponding to a neglect of electronic contributions [38]. The field ratio E_{ext}/E depends in a characteristic way on the boundary conditions used. In Section 10.1.4, this dependence is analyzed. For the so-called conducting boundary conditions [40, 42], the ratio is unity [49, 54, 59]. $\mathcal{L}[\ldots]$ denotes the Fourier–Laplace transform of a temporal function $f(t)$:

$$\mathcal{L}[f(t)] = \int_0^\infty f(t)e^{i\omega t}dt \quad (10.39)$$

As already discussed at the end of the previous section, the interpretation of the dielectric spectrum proceeds via the decomposition of the total collective dipole moment $\vec{M}_{tot}$. Therefore, its autocorrelation function $\Phi_{tot} = \langle \vec{M}_{tot}(0) \cdot \vec{M}_{tot}(t) \rangle$ may be split into two correlation functions:

$$\Phi_{tot}(t) = \Phi_D(t) + \Phi_J(t) \quad (10.40)$$

$$\Phi_D(t) = \langle \vec{M}_D(0) \cdot \vec{M}_D(t) \rangle + \langle \vec{M}_D(0) \cdot \vec{M}_J(t) \rangle \quad (10.41)$$

$$\Phi_J(t) = \langle \vec{M}_J(0) \cdot \vec{M}_J(t) \rangle + \langle \vec{M}_D(0) \cdot \vec{M}_J(t) \rangle \quad (10.42)$$

In this first step of decomposition, the appearance of cross-correlations $\langle \vec{M}_D(0) \cdot \vec{M}_J(t) \rangle$ becomes already obvious. They represent the coupling of rotation and translation at the mesoscopic level that cannot be described independently by autocorrelation functions alone. Fortunately, the cross-terms are much smaller than the self-terms, which enables a *de facto* decoupling of translation and rotation for numerical reasons [17, 34, 53].

We evaluate the rotational contribution $\Phi_D(t)$ first:

$$\mathcal{L}\left[-\frac{d}{dt}\Phi_D(t)\right] = \mathcal{L}\left[-\frac{d}{dt}\langle \vec{M}_D(0)\cdot \vec{M}_D(t)\rangle\right] + \mathcal{L}\left[-\frac{d}{dt}\langle \vec{M}_D(0)\cdot \vec{M}_J(t)\rangle\right] \quad (10.43)$$

The first autocorrelation term in the equation above can be reformulated using partial integration

$$\mathcal{L}\left[-\frac{d}{dt}\langle \vec{M}_D(0)\cdot \vec{M}_D(t)\rangle\right] = -\int_0^\infty e^{i\omega t}\frac{d}{dt}\langle \vec{M}_D(0)\cdot \vec{M}_D(t)\rangle dt \quad (10.44)$$

$$= -\left[\langle \vec{M}_D(0)\cdot \vec{M}_D(t)\rangle e^{i\omega t}\right]_0^\infty$$

$$+ i\omega \int_0^\infty e^{i\omega t}\langle \vec{M}_D(0)\cdot \vec{M}_D(t)\rangle dt \quad (10.45)$$

$$= \langle \vec{M}_D^2\rangle + i\omega \mathcal{L}\left[\langle \vec{M}_D(0)\cdot \vec{M}_D(t)\rangle\right] \quad (10.46)$$

while the second cross-term may be transformed using relation (10.34):

$$\mathcal{L}\left[-\frac{d}{dt}\langle \vec{M}_D(0)\cdot \vec{M}_J(t)\rangle\right] = -\int_0^\infty \langle \vec{M}_D(0)\cdot \vec{J}(t)\rangle e^{i\omega t}dt \quad (10.47)$$

$$= -\mathcal{L}\left[\langle \vec{M}_D(0)\cdot \vec{J}(t)\rangle\right]$$

Fortunately, the contribution from the cross-correlation of the rotational and translational collective dipole moments can be computed by the Fourier–Laplace transform of $\langle \vec{M}_D(0)\cdot \vec{J}(t)\rangle$ since $\vec{J}(t)$ (as opposed to $\vec{M}_J(t)$) does not suffer from the jumps of individual molecules due to the periodic boundary conditions. This is important to note because it increases the statistical quality of the correlation function and thus allows computing this very small contribution. Altogether, the rotational contribution is given by

$$\mathcal{L}\left[-\frac{d}{dt}\Phi_D(t)\right] = \langle \vec{M}_D^2\rangle + i\omega \mathcal{L}\left[\langle \vec{M}_D(0)\cdot \vec{M}_D(t)\rangle\right] - \mathcal{L}\left[\langle \vec{M}_D(0)\cdot \vec{J}(t)\rangle\right] \quad (10.48)$$

The general form of the translational contribution is

$$\mathcal{L}\left[-\frac{d}{dt}\Phi_J(t)\right] = \mathcal{L}\left[-\frac{d}{dt}\langle \vec{M}_J(0)\cdot \vec{M}_J(t)\rangle\right] - \mathcal{L}\left[\langle \vec{M}_D(0)\cdot \vec{J}(t)\rangle\right] \quad (10.49)$$

While the correlation function of the rotational dipole moment $\langle \vec{M}_D(0)\cdot \vec{M}_D(t)\rangle$ exhibits the expected classical behavior, its translational counterpart $\langle \vec{M}_J(0)\cdot \vec{M}_J(t)\rangle$ behaves rather peculiar and needs more careful analysis. In order to elucidate this behavior, we first investigate the so-called dipolar displacement:

$$\langle \Delta \vec{M}_J^2(t)\rangle = \langle (\vec{M}_J(0)-\vec{M}_J(t))^2\rangle \quad (10.50)$$

An alternative expression can be given following the approach of Berne and Pecora [6] for the mean-squared displacement in terms of the velocity correlation function. The pair $\vec{M}_J$ and $\vec{J}$ can be treated formally equivalent to the pair $\vec{r}$ and $\vec{v}$.

$$\langle \Delta \vec{M}_J^2(t) \rangle = 2 \left(\int_0^t \langle \vec{J}(0) \cdot \vec{J}(t') \rangle dt' \right) \cdot t - 2 \int_0^t t' \langle \vec{J}(0) \cdot \vec{J}(t') \rangle dt' \quad (10.51)$$

In the asymptotic limit at sufficiently long times, this formula gives a linear relation for the dipolar displacement:

$$\lim_{t \to \infty} \langle \Delta \vec{M}_J^2(t) \rangle = 6 V k_B T \sigma(0) t + 2 \langle \vec{M}_J^2 \rangle \quad (10.52)$$

The first term on the right-hand side follows from the Green–Kubo approach to compute the static conductivity:

$$\sigma(0) = \frac{1}{3 V k_B T} \int_0^\infty \langle \vec{J}(0) \cdot \vec{J}(t) \rangle dt \quad (10.53)$$

Using Equation 10.34, the second term on the right-hand side is

$$-\int_0^\infty t \cdot \langle \vec{J}(0) \cdot \vec{J}(t) \rangle dt = \underbrace{[t \cdot \langle \vec{M}_J(0) \cdot \vec{J}(t) \rangle]_0^\infty}_{= 0}$$
$$- \int_0^\infty \langle \vec{M}_J(0) \cdot \vec{J}(t) \rangle dt \quad (10.54)$$

$$= -[\langle \vec{M}_J(0) \cdot \vec{M}_J(t) \rangle]_0^\infty \quad (10.55)$$

$$= \langle \vec{M}_J^2 \rangle \quad (10.56)$$

A typical example of the dipolar displacement for the molecular ionic liquid 1-ethyl-3-methyl-imidazolium dicyanoamide is shown in Figure 10.4 as dashed line. Evaluating the binomial in Equation 10.50, a formulation of the correlation function $\langle \vec{M}_J(0) \cdot \vec{M}_J(t) \rangle$ in terms of the dipolar displacement is found.

$$\langle \vec{M}_J(0) \cdot \vec{M}_J(t) \rangle = -\frac{1}{2} \langle \Delta \vec{M}_J^2(t) \rangle + \frac{1}{2} \left(\langle \vec{M}_J^2(0) \rangle + \langle \vec{M}_J^2(t) \rangle \right) \quad (10.57)$$

Since t is a reference point along the trajectory in the same way as the initial point 0, the two averages $\langle \vec{M}_J^2(0) \rangle$ and $\langle \vec{M}_J^2(t) \rangle$ become equal for sufficiently long times. This should not be confused with $\langle \Delta \vec{M}_J^2(t) \rangle$ that increases linearly in time. Now, the peculiar behavior of $\langle \vec{M}_J(0) \cdot \vec{M}_J(t) \rangle$ becomes obvious: it had to be a linear function of time over a long range of time. For example, this linear behavior of $\langle \vec{M}_J(0) \cdot \vec{M}_J(t) \rangle$ is shown in Figure 10.4 (solid line). According to Equation 10.52, its derivative or slope is determined by the static conductivity after an initial regime:

$$\lim_{t \to \infty} \frac{d}{dt} \langle \vec{M}_J(0) \cdot \vec{M}_J(t) \rangle = 3 V k_B T \sigma(0) \quad (10.58)$$

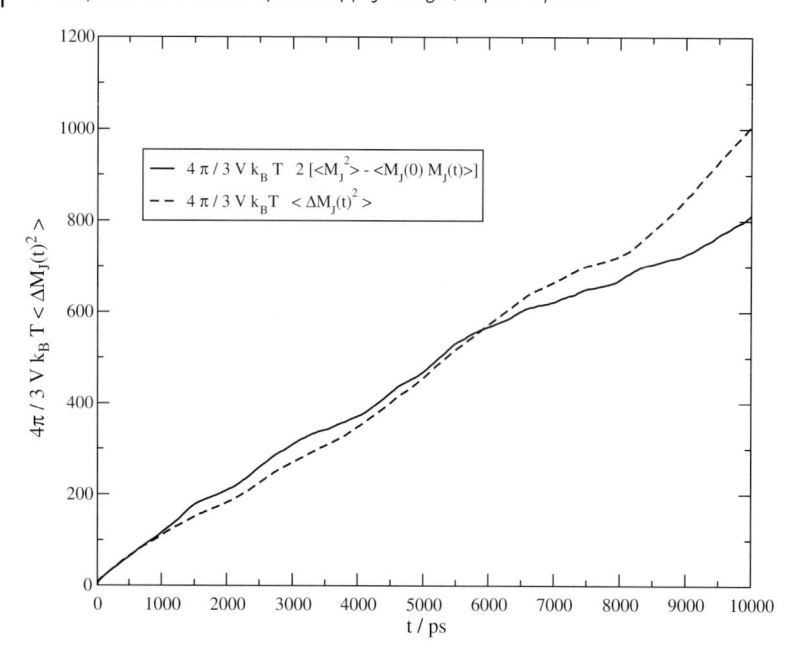

Figure 10.4 Mean-squared displacement of the translational collective dipole moment $\langle \Delta \vec{M}_J(t)^2 \rangle$ (dashed line) of the simulated 1-ethyl-3-methyl-imidazolium dicyanoamide that can be used to calculate the static conductivity $\sigma(0)$. The solid line shows the linear behavior of the autocorrelation function $\langle \vec{M}_J(0) \cdot \vec{M}_J(t) \rangle$ within the first nanoseconds.

As the time dependence in the initial region is quadratic, the derivative $(d/dt)\langle \vec{M}_J(0) \cdot \vec{M}_J(t) \rangle$ vanishes for $t \to 0$.

Now, all tools for the evaluation of $\mathcal{L}\left[-(d/dt)\langle \vec{M}_J(0) \cdot \vec{M}_J(t) \rangle\right]$ are available. Upon partial integration, it transforms to

$$\mathcal{L}\left[-\frac{d}{dt}\langle \vec{M}_J(0) \cdot \vec{M}_J(t) \rangle\right] = \int_0^\infty \frac{e^{i\omega t}}{i\omega} \frac{d^2}{dt^2} \langle \vec{M}_J(0) \cdot \vec{M}_J(t) \rangle dt$$

$$- \left[\frac{e^{i\omega t}}{i\omega} \frac{d}{dt} \langle \vec{M}_J(0) \cdot \vec{M}_J(t) \rangle\right]_0^\infty \quad (10.59)$$

$$= \frac{i}{\omega} \mathcal{L}[\langle \vec{J}(0) \cdot \vec{J}(t) \rangle]$$

where in the second step we have used Equation 10.36 with $\vec{A} = \vec{B} = \vec{M}_J$.

Assembling Equations 10.48 and 10.49, multiplying all these terms by $4\pi/3 V k_B T$, and inserting them into Equation 10.38, we get the final expression for the GDC:

$$\Sigma(\omega) = \frac{4\pi}{3Vk_BT}\left(\langle \vec{M}_D^2\rangle + i\omega\mathcal{L}[\langle \vec{M}_D(0)\cdot\vec{M}_D(t)\rangle] - \mathcal{L}[\langle \vec{M}_D(0)\cdot\vec{J}(t)\rangle]\right.$$
$$\left. + \frac{i}{\omega}\{\mathcal{L}[\langle \vec{J}(0)\cdot\vec{J}(t)\rangle] + i\omega\mathcal{L}[\langle \vec{M}_D(0)\cdot\vec{J}(t)\rangle]\}\right) \quad (10.60)$$

In the zero-frequency limit the real part of Fourier-Laplace transform of the current auto-correlation yields the static conductivity σ(0). Hence, the ratio σ(0)/ω represents a parabola in the imaginary part of the dielectric spectrum.

$$\Sigma_0(\omega) = \Sigma(\omega) - \frac{4\pi i\sigma(0)}{\omega} \quad (10.61)$$

$$= \varepsilon(\omega) - 1 + \frac{4\pi i(\sigma(\omega)-\sigma(0))}{\omega} \quad (10.62)$$

While removing the singularity in the imaginary part, this does not affect the real part of the spectrum. The different contributions to the GDC in Equation 10.60 may be arranged into two groups: one with the prefactor i/ω and one without. Accordingly, the different contributions to the GDC in Equation 10.60 can be assigned to $\varepsilon(\omega)$ as well as to the conductivity $\sigma(\omega)$ in Equation 10.62:

$$\varepsilon(\omega)-1 = \frac{4\pi}{3Vk_BT}\left(\langle\vec{M}_D^2\rangle + i\omega\mathcal{L}[\langle\vec{M}_D(0)\cdot\vec{M}_D(t)\rangle] - \mathcal{L}[\langle\vec{M}_D(0)\cdot\vec{J}(t)\rangle]\right) \quad (10.63)$$

$$\sigma(\omega) = \frac{1}{3Vk_BT}\left(\mathcal{L}[\langle\vec{J}(0)\cdot\vec{J}(t)\rangle] + i\omega\mathcal{L}[\langle\vec{M}_D(0)\cdot\vec{J}(t)\rangle]\right) \quad (10.64)$$

The last two equations can be seen as the motivation for the "arbitrary" decomposition of $\vec{M}_{tot}$: Despite the marginal cross-term $\mathcal{L}[\langle\vec{M}_D(0)\cdot\vec{J}(t)\rangle]$, the dielectric permittivity $\varepsilon(\omega)$ is built up by the rotational collective dipole moment $\vec{M}_D(t)$. The time derivative of the translational collective dipole moment $\vec{M}_J(t)$ almost exclusively determines the conductivity $\sigma(\omega)$. However, the influence of the conductivity $\sigma(\omega)$ on the GDC is indirect since the ratio $\sigma(\omega)/\omega$ enters the spectrum of the GDC. Therefore, we introduce the dielectric conductivity:

$$\vartheta_0(\omega) = 4\pi i\frac{\sigma(\omega)-\sigma(0)}{\omega} \quad (10.65)$$

The difference between the dielectric conductivity $\vartheta(\omega)$ and the conductivity $\sigma(\omega)$ is more than a mere downscaling by the frequency ω: First, it shifts the contribution from $\sigma(\omega)$ to smaller frequencies. Second, and even more important, it is the nonvanishing static limit $\lim_{\omega\to 0}\vartheta_0(\omega)$. Both, the numerator and the denominator tend to zero for zero frequency. So, the static limit of the dielectric conductivity is the result of a delicate balance of the ratio of two zero terms. Neglecting for the moment the cross-term $\langle\vec{M}_D(0)\cdot\vec{J}(t)\rangle$, Equation 10.64 gives the explicit expression for the zero-frequency limit of $\vartheta_0(\omega)$:

$$\lim_{\omega\to 0} \vartheta_0(\omega) \simeq \frac{4\pi i}{3Vk_B T} \int_0^\infty \frac{e^{i\omega t}-1}{\omega} \langle \vec{J}(0)\cdot\vec{J}(t)\rangle dt \tag{10.66}$$

As the limit

$$\lim_{\omega\to 0} \frac{e^{i\omega t}-1}{\omega} = i\cdot t \tag{10.67}$$

the static limit of the dielectric conductivity is given by

$$\lim_{\omega\to 0} \vartheta_0(\omega) \simeq \frac{4\pi}{3Vk_B T} \int_0^\infty -t\cdot \langle \vec{J}(0)\cdot\vec{J}(t)\rangle dt \tag{10.68}$$

$$\simeq \frac{4\pi}{3Vk_B T} \langle \vec{M}_J^2\rangle \tag{10.69}$$

where in the last step we have used Equation 10.56. The zero-frequency limit of the cross-term $\langle \vec{M}_D(0)\cdot\vec{J}(t)\rangle$ yields

$$\lim_{\omega\to 0} \mathcal{L}[\langle \vec{M}_D(0)\cdot\vec{J}(t)\rangle] = \int_0^\infty \langle \vec{M}_D(0)\cdot\vec{J}(t)\rangle dt \tag{10.70}$$

$$= [\langle \vec{M}_D(0)\cdot\vec{M}_J(t)\rangle]_0^\infty \tag{10.71}$$

$$= -\langle \vec{M}_D(0)\cdot\vec{M}_J(0)\rangle \tag{10.72}$$

by means of Equation 10.34. Augmenting Equation 10.69 by the static limit of the cross-term in Equation 10.72, the static limit of the dielectric conductivity finally reads

$$\lim_{\omega\to 0} \vartheta_0(\omega) = \frac{4\pi}{3Vk_B T}\left(\langle \vec{M}_J^2\rangle + \langle \vec{M}_D(0)\cdot\vec{M}_J(0)\rangle\right) \tag{10.73}$$

From Equation 10.63, the static limit of the dielectric permittivity can be easily read:

$$\lim_{\omega\to 0} \varepsilon(\omega)-1 = \frac{4\pi}{3Vk_B T}\left(\langle \vec{M}_D^2\rangle + \langle \vec{M}_D(0)\cdot\vec{M}_J(0)\rangle\right) \tag{10.74}$$

Uniting the last two equations, the generalized static dielectric constant $\varepsilon_{\text{stat}}$ is given by

$$\varepsilon_{\text{stat}}-1 = \lim_{\omega\to 0} \Sigma_0(\omega) = \frac{4\pi}{3Vk_B T}\left(\langle \vec{M}_D^2\rangle + 2\langle \vec{M}_D(0)\cdot\vec{M}_J(0)\rangle + \langle \vec{M}_J^2\rangle\right)$$

$$= \frac{4\pi}{3Vk_B T}\langle \vec{M}_{\text{tot}}^2\rangle \tag{10.75}$$

The derivation of this expression shows that the conductivity although residing in the high-frequency regime finally makes a knowledgeable contribution to the static GDC ε_{stat}. This demonstrates that in charged, dipolar systems, the conductivity is an equal partner of the permittivity.

So far, we have split the macroscopic total dipole moment $\vec{M}_{tot}$ into its rotational $\vec{M}_D$ and translational $\vec{M}_J$ contributions in order to get separate equations for the dielectric permittivity and the dielectric conductivity. At the mesoscopic level, $\vec{M}_D(t)$ as well as $\vec{J}(t)$ may be further decomposed into contributions from the cations, the anions, and the neutral solvent. Formally, we can write

$$\vec{M}_D(t) = \sum_k \vec{M}_D^k(t) \tag{10.76}$$

$$\vec{J}(t) = \sum_k \vec{J}^k(t) \tag{10.77}$$

where $\vec{M}_D^k(t) \in \{\vec{M}_D^+(t), \vec{M}_D^-(t), \vec{M}_D^0(t)\}$ and $\vec{J}^k(t) \in \{\vec{J}^+(t), \vec{J}^-(t)\}$ as indicated in Figure 10.3. At this level of decomposition, we have contracted all species with respect to their charge. In a finer grained decomposition, a resolution into species can be done. Even more, the species itself may be further decomposed: The solvent may be divided into several solvation shells of the solute and the remaining bulk. The solute, for example, a protein, may be split into regions of helical structure, β-sheets, or loops. In DNA molecules, the contribution from the sugar and the phosphate group may be even separated.

Since $\langle \vec{M}_D(0) \cdot \vec{M}_D(t) \rangle$ correlates the sum over several species or their moieties, one gets autocorrelations $\langle \vec{M}_D^k(0) \cdot \vec{M}_D^k(t) \rangle$ and cross-terms $2\langle \vec{M}_D^k(0) \cdot \vec{M}_D^l(t) \rangle$ with $k \neq l$. These cross-correlations are the mesoscopic analogue of intermolecular or intermoiety coupling. They always appear by a factor of 2 such that one cross-term can be attributed to each partner in the sense of an equal sharing of the coupling. In other words, one can define correlation functions

$$\Phi_{D,k}(t) = \langle \vec{M}_D^k(0) \cdot \vec{M}_D^k(t) \rangle + \sum_{l \neq k} \langle \vec{M}_D^k(0) \cdot \vec{M}_D^l(t) \rangle \tag{10.78}$$

specific for species, moieties, or regions. Thus, the sum of all individual $\Phi_{D,k}(t)$ equals the total $\langle \vec{M}_D(0) \cdot \vec{M}_D(t) \rangle$. Since the Fourier–Laplace transform is linear, each $\Phi_{D,k}(t)$ can be transformed separately. In this sense, each Fourier–Laplace transform stands for the contribution of a species, moiety, or region to the total dielectric permittivity ε. This contribution, however, should not be mixed up with the term "local dielectric constant" of the respective species, moiety, or region. This is already obvious from the fact that all terms are weighted by the same factor $1/V$ and not by their individual volumes. In other words, if a species is present at infinite dilution, its contribution to the whole sample vanishes despite its individual dielectric properties. An analogous decomposition of the current leads to $\Phi_{J,k}(t)$ and, consequently, to contributions to the conductivity from individual charged species, moieties, or regions. We emphasize that such a decomposition is much more appropriate to the collective nature of the current than a description in terms of ion pairs.

10.1.4
Finite System Electrostatics

In a mesoscopic description of dielectrics, the Poisson equation plays a central role as shown in the first section. At the molecular level, the equations of motion involving the electrostatic forces are central. Both descriptions rest upon the multiple application of Coulomb's law

$$\vec{F}_{ij} = q_i \cdot \vec{E} = -\vec{\nabla} \frac{q_i \cdot q_j}{r_{ij}} \tag{10.79}$$

acting between a pair of charges. Unfortunately, Coulomb's law is the force law of longest range found in nature. Therefore, the finiteness or boundary of the system under consideration is of importance. So far, we have solved equations with no respect to the finiteness or boundary conditions. In simulation studies, this question of finiteness is even more urgent because of the limitations in memory and storage. Therefore, the correct treatment of finite size effects, the so-called finite system electrostatics, is a prerequisite for the interpretation of dielectric spectra.

The field Equation 10.2 is of dual nature as it can be used for both, a molecular and a mesoscopic description of finiteness of the system: At the molecular level the charge and dipole density describes the atomistic or molecular set of point charges $\{q_i\}$ and point dipoles $\{\vec{\mu}_{D,i}\}$.

$$\varrho(\vec{r}) = \frac{1}{V} \sum_i q_i \cdot \delta(\vec{r}-\vec{r}_i) \tag{10.80}$$

$$\vec{P}_D(\vec{r}) = \frac{1}{V} \sum_i \vec{\mu}_{D,i} \cdot \delta(\vec{r}-\vec{r}_i) \tag{10.81}$$

If these expressions are inserted into the field equation (10.2), one gets electrostatic part of the molecular forces used in MD simulations, for example, Equation 10.17.

At the mesoscopic level, averaging of the field equation yields the internal or Maxwell field entering the constitutive relation (10.1):

$$\vec{E} = \frac{1}{V} \int_V \vec{E}(\vec{r}) \mathrm{d}\vec{r} \tag{10.82}$$

$$= \vec{E}_{\text{ext}} + \int_V \vec{S}(\vec{r}) \mathrm{d}\vec{r} \cdot \int_V \varrho(\vec{r}) \mathrm{d}\vec{r} + \int_V \overleftrightarrow{T}(\vec{r}) \mathrm{d}\vec{r} \cdot \int_V \vec{P}_D(\vec{r}) \mathrm{d}\vec{r} \tag{10.83}$$

Thus, we have used the convolution property of the integrals appearing in the field equation (10.2). The integral $\int_V \varrho(\vec{r}) \mathrm{d}\vec{r}$ equals the net charge q_{tot} of the sample. As a result, the second contribution in Equation 10.83 vanishes for a neutral system $q_{\text{tot}} = 0$. If the integral of the dipole–dipole tensor $\overleftrightarrow{T}(\vec{r})$ also vanish, Maxwell and external field would be identical. Consequently, one always tries to accomplish overall charge neutrality and a vanishing integral of the dipole–dipole tensor.

The integral of $\overleftrightarrow{T}(\vec{r})$ is essential for the dielectric boundary conditions that we use here as a synonym for a complex situation. On the one hand, this term comprises the finite range of the Coulomb interaction. On the other hand, it refers to the finite size and shape of the samples volume. The geometrical part is usually handled by the introduction of the so-called toroidal boundary conditions based on the minimum image convention: If a particle crosses the face of the simulation volume, it is reinserted with the same velocity but via the opposite face. The pair distances $\vec{r}_{ij}$ connecting particles are also subject to this toroidal shift. Thus, toroidal boundary conditions simulate a finite system but without surface. One however has to pay for this advantage. While all quantities involving the velocities remain unchanged, those based on the coordinates experience a jump. For example, this has severe consequence on the translational collective dipole moment $\vec{M}_J(t)$, while its derivative the current $\vec{J}(t)$ is unaffected since it is built up from the velocities. One remedy to cope with the jumps in $\vec{M}_J(t)$ is the unfolding of the trajectory post simulation. In other words, the toroidal moves performed during simulation are undone for the analysis. One has to be aware, however, that the mechanism of unfolding always depends on the initial geometry. Computation of the dipolar displacement $\langle \Delta \vec{M}_J^2(t) \rangle$ essentially removes this dependence. Therefore, the average static value of $\langle \vec{M}_J^2 \rangle$ is taken from the axis intercept of $\langle \Delta \vec{M}_J^2(t) \rangle$ instead of the initial value of $\langle \vec{M}_J(0) \cdot \vec{M}_J(t) \rangle$.

The other aspect of the boundary conditions is the truncation of Coulomb interaction. Subsequently, we will discuss the most important types of truncation or modification, namely, truncation at the cutoff, inclusion of the reaction field, and the Ewald summation over periodic replica. Of course, truncation or modification of the Coulomb law is not independent of the system size or geometry. Indeed, the combination of both makes what we call dielectric boundary conditions. Fortunately, the toroidal shifts are included in Equation 10.2 because it is of convolution type that means that the relative distance $\vec{r} - \vec{r}'$ is always kept in the simulation volume V.

The simplest and most time-saving method to treat electrostatic forces is the cutoff principle. Beyond a typical distance of $r_{\text{cut}} = 10\text{-}12$, all interactions with more distant neighbors are neglected. Formulated in terms of the $\overleftrightarrow{T}(\vec{r})$ tensor, one has

$$\overleftrightarrow{T}_{\text{cut}}(\vec{r}) = \begin{cases} \overleftrightarrow{T}(\vec{r}) & : r \leq r_{\text{cut}} \\ 0 & : r > r_{\text{cut}} \end{cases} \tag{10.84}$$

The integral of this truncated dipole tensor is

$$\int_V \overleftrightarrow{T}(\vec{r}) d\vec{r} = \int_{K_{\text{cut}}} \overleftrightarrow{T}(\vec{r}) d\vec{r} \tag{10.85}$$

$$= \frac{-4\pi}{3} \int_{K_{\text{cut}}} \delta(\vec{r}) \overleftrightarrow{I} d\vec{r} \tag{10.86}$$

$$= \frac{-4\pi}{3} \overleftrightarrow{I} \tag{10.87}$$

where K_{cut} is a sphere of radius r_{cut} and $\overleftrightarrow{I}$ is the unit tensor. Thus, we have used Equation 10.5. While the off-diagonal elements of $\overleftrightarrow{T}(\vec{r})$ vanish and the diagonal elements become equal due to the isotropy of the simulation volume, the above integral actually calculates $1/3$ of the Laplacian of the original Green function $1/r$ that gives $\Delta G(r) = -4\pi\delta(\vec{r})$ [15]. These special relations are lost for an anisotropic volume, for example, a cuboid. As a result, the Maxwell field becomes the well-known Lorentz field [13, 31, 35]

$$\vec{E} = \vec{E}_{ext} - \frac{4}{3}\pi \vec{P}_D \tag{10.88}$$

and deviates considerably from the applied external field $\vec{E}_{ext}$. While appropriate for the short-range interaction, for example, Lennard–Jones, the cutoff principle has tremendous consequences on electrostatic forces. A critical measure is the so-called r-dependent Kirkwood $g_K(r)$ factor [10, 11, 15]. The neighborhood of a reference dipole $\vec{\mu}_D$ is decomposed into spherical shells of increasing radius r and the rotational collective dipole moment $\vec{M}_D(r)$ is projected onto $\vec{\mu}_D$

$$g_K(r) = \frac{\vec{\mu}_D \cdot \vec{M}_D(r)}{\vec{\mu}_D^2} \tag{10.89}$$

It measures the collective orientational order of dipoles. Alternatively, $g_K(r)$ may be written as integral over spherical shells:

$$g_K(r) = 1 + 4\pi \int_0^r g_{\mu\mu}(r') r'^2 dr' \tag{10.90}$$

where the integrand $g_{\mu\mu}(r)$ is the radial distribution function weighted by the direction cosine of pair of dipoles μ_D separated by a distance r [15, 28, 37, 57]. The higher sensitivity of $g_K(r)$ as compared to the integrand $g_{\mu\mu}(r)$ comes from the r^2 amplifier at larger distances. In an infinite, unperturbed system, one would expect that $g_K(r)$ – after some initial oscillations over coordination shells – approaches a plateau value when $g_{\mu\mu}(r)$ has died off. Within the cutoff scheme, such a plateau value is never reached. After the initial oscillations, $g_K(r)$ drops sharply and even reaches negative values at $r = r_{cut}$ [61]. Afterward it recovers, but the final positive value is very small, because truncation of interaction does not automatically imply the truncation of spatial correlations. An extension to solvated biomolecules shows that the cutoff principle distorts the helical structure of peptides [50, 51].

A first remedy against the cutoff artifacts is to approximate the neglected interactions beyond the cutoff by a reaction field (RF) with dielectric permittivity ε_{RF} illustrated in Figure 10.5a. When embedding the cutoff sphere in this dielectric, the reaction field contributes an additional term

$$\overleftrightarrow{T}_{RF} = \frac{1}{r_{cut}^3} \cdot \frac{2\varepsilon_{RF} - 2}{2\varepsilon_{RF} + 1} \overleftrightarrow{I} \tag{10.91}$$

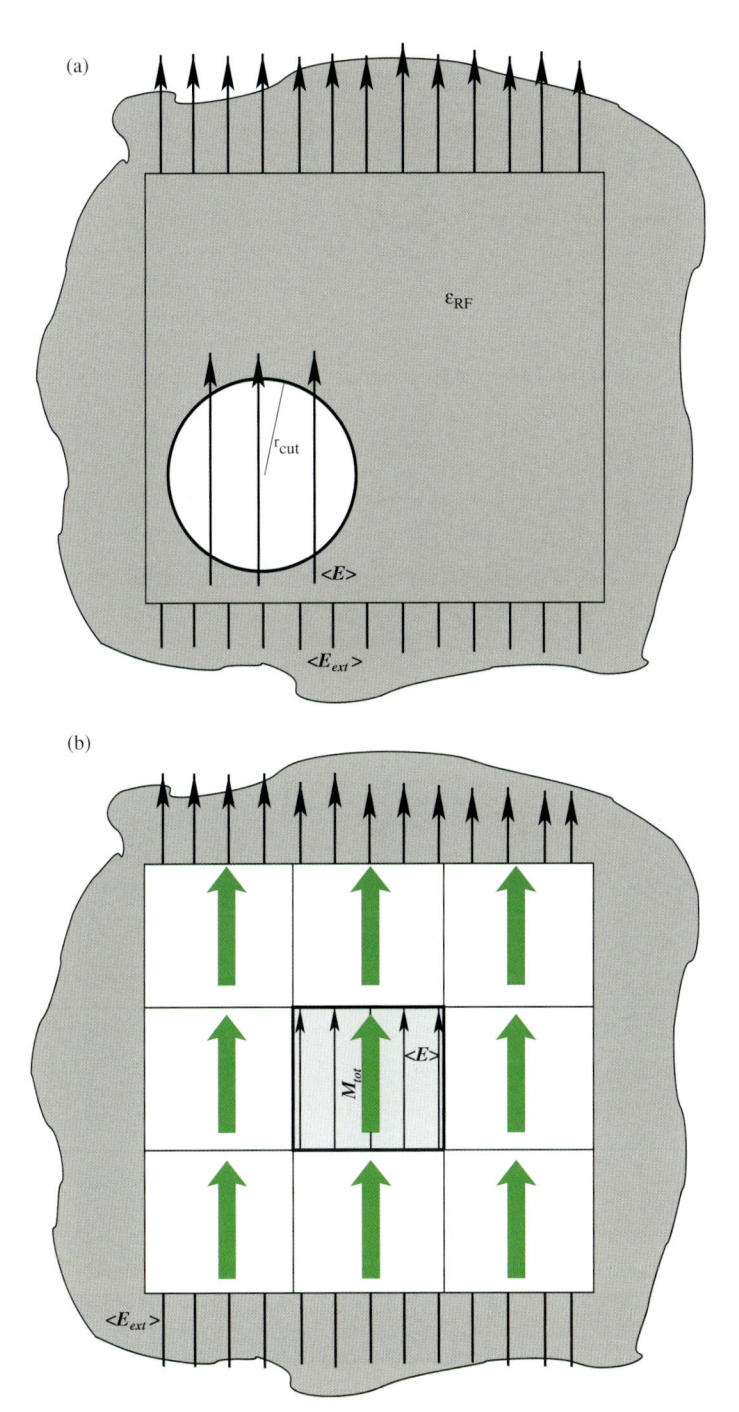

Figure 10.5 (a) Schematic view of the reaction field method. The dielectricum outside the cutoff sphere (white area) is treated with a dielectric constant ε_{RF}. With the increasing ε_{RF}, the local field $\langle \vec{E} \rangle$ becomes more similar to the external field $\langle \vec{E}_{ext} \rangle$. (b) In case of the Ewald summation technique, the local field $\langle \vec{E} \rangle$ equals the external field $\langle \vec{E}_{ext} \rangle$.

to the original $\overleftrightarrow{T}(\vec{r})$ in Equation 10.5. The integral of the combined tensors $\overleftrightarrow{T}(\vec{r}) + \overleftrightarrow{T}_{RF}$ now results in a field relation

$$\vec{E} = \vec{E}_{ext} - \frac{4}{3}\pi \cdot \frac{3}{2\varepsilon_{RF}+1} \vec{P}_D \tag{10.92}$$

For the isolated sample, that is, in the absence of the reaction field corresponding to $\varepsilon_{RF} = 1$, we obtain the Lorentz field again. With increasing ε_{RF}, however, internal and external field come closer to each other. For a surrounding dielectric of infinite dielectric permittivity ε_{RF}, the so-called conducting boundary conditions, both fields are equal. While $g_K(r)$ now gives the correct asymptotic value, its spatial resolution is still not satisfying. Although being now positive, there is still a minimum at $r = r_{cut}$ [62]. It is interesting to note that the inclusion of the reaction field tensor in Equation 10.91 corresponds to a change of the original Green function $G(r)$ to

$$G_{RF} = \frac{1}{r} + \frac{1}{2} \cdot \frac{1}{r_{cut}^3} \cdot \frac{2\varepsilon_{RF}-2}{2\varepsilon_{RF}+1} \cdot r^2 \tag{10.93}$$

This additional harmonic term counteracts the original Coulomb law: The attraction between unlike charges and the repulsion between like charges is reduced.

Lattice sum techniques replace the neglected interactions by a sum over translational replica, as illustrated in Figure 10.5b. In other words, an effective Green function implicitly containing the interaction with all images of the interaction partner in the primary cell is used. In the Ewald scheme, this effective Green function is

$$G_{EW} = \frac{1}{r} - \frac{\text{erf}(\lambda r)}{r} + \frac{4\pi}{V} \sum_k \frac{e^{-k^2/4\lambda^2}}{k^2} e^{i\vec{k}\vec{r}} \tag{10.94}$$

where in the last term, the sum over all relevant $\vec{k}$ vectors in reciprocal space has to be performed. The Ewald tensor $\overleftrightarrow{T}_{EW}$ is the double gradient of G_{EW} and its integral is given by [39]

$$\overleftrightarrow{T}_{EW} = \frac{4\pi}{3V} \cdot Q \overleftrightarrow{I} \tag{10.95}$$

with Q given by the integral

$$Q = \left(\frac{\lambda}{\sqrt{\pi}}\right)^3 \int_0^{r_{EW}} 4\pi r^2 e^{-\lambda^2 r^2} dr \tag{10.96}$$

Here r_{EW} is the threshold beyond which the screened Coulomb potential $(1-\text{erf}(\lambda r))/r$ acting in real space can be neglected for numerical reasons. The screening parameter λ governs the balance of real space and reciprocal space Coulombic lattice sums. Equation 10.96 represents the integral of a nonnegative function and therefore increases monotonically with λ. For sufficiently large λ, where the Gaussian integrand has become sufficiently small, Q tends to unity as it represents the integral over a normalized Gaussian function. In this case, the Ewald tensor completely compensates the Lorentz field created by the original $\overleftrightarrow{T}$ tensor such that the internal Maxwell field and the external field become equal. This is only valid, however, for sufficiently large screening parameter λ, otherwise both fields deviate and the field ratio E_{ext}/E is not unity. Therefore, all expressions for the GDC have to be corrected by this field ratio E_{ext}/E. It is true that the reaction field method and the Ewald sum may result in identical asymptotic values of r-dependent g_K factor. So one may question the high effort of the Ewald sum as compared to that of the reaction field being almost identical to the cutoff scheme. At a radial scale, however, the improvement of the Ewald method is impressing. $g_K(r)$ reaches the expected plateau value and lacks any further artifacts [10]. Nowadays, the Ewald method is the standard technique to handle the long-range Coulomb forces in finite systems. In order to reduce computational effort, however, mesh-based methods project the position of the fluctuating charges on a fixed grid. This is known as "particle mesh–Ewald" (PME) method [20].

10.2
Applications and Experiments

Although the previous theory sections have established the principal connection between simulation data and the dielectric spectrum, the application of these analytical expressions to the real simulation data is not straightforward. The problem of finite system electrostatics makes one select the size and shape of the simulation box carefully. The box shape has to be spatially isotropic and the size is even more important in a twofold sense: First, we have learnt that Coulomb's law has to be modified in order to handle the long-range Coulomb forces in finite systems. We have presented three possibilities: The cutoff method, the reaction field, and the Ewald sum. Each of these has its own size dependence. While the cutoff method would require unrealistically large systems in order to give results free from artifacts, the reaction field and the Ewald sum yield reliable results for practically feasible system sizes. The computational efforts for the reaction field are almost identical to those of the cutoff methods, but they correctly give the global properties only. In order to be correct on a local scale too, the more expensive Ewald sum has to be implemented. Therefore, the Ewald sum exhibits the least size dependence of all three methods and its higher numerical effort is thus partially paid for. Apart from this consideration concerning the undisturbed buildup of electrostatics, there is a second demand on system size. As we need collective properties to compute the spectrum, the system must be large enough to behave like a piece of dielectric material. In practice, several hundred molecules are necessary to fulfill this requirement.

Provided a reasonable system size has been chosen, the preparation or equilibration of the system prior to the actual simulation is the next critical issue. The traditional method to start from crystal structure usually implies unreasonably high collective rotational dipole moments. In principle, long equilibration runs at constant pressure could bring down $\vec{M}_D(t)$, however, at extreme computational cost. Therefore, one is forced to create a starting configuration of moderate dipole alignment. This usually consumes several nanoseconds of equilibration time.

For a well-prepared system of reasonable size, the so-called production period where the system's trajectory is generated has to last sufficiently long in time. As a rule of thumb, the simulation time must cover several dielectric relaxation times being defined as the time necessary for the time correlation functions of the collective dipole moments to get leveled off. Bearing in mind that for a collective property, every point along the trajectory provides a single value only, because averaging over particles, residue, and so on is not possible, these extreme demand on simulation length can be easily understood. This explains why time correlation functions of collective dipole moments are still noisy for simulation length of 100 ns. In order to smooth this statistical roughness, time correlation functions are often fitted to analytical expressions. For $\langle \vec{M}_D(0) \cdot \vec{M}_D(t) \rangle$ and its mesoscopic decompositions, multiexponential fits are frequently used:

$$\langle \vec{M}_D(0) \cdot \vec{M}_D(t) \rangle \simeq \sum_k A_k \cdot e^{-t/\tau_k} \tag{10.97}$$

The number of exponential terms depends on the correlation function under investigation and the precision desired. This may require up to four terms. As the amplitudes A_k determine the height of the peaks in the imaginary part of the dielectric spectrum, the careful determination of the amplitudes and their correct sum is crucial. As similar relaxation times result in a single broaden peak in the spectrum, they are not a real problem. However, an increasing number of exponentials impedes a reasonable interpretation. On the other hand, a Fourier–Laplace transform of the raw data of $\langle \vec{M}_D(0) \cdot \vec{M}_D(t) \rangle$ is not desirable since the integration of statistical errors of the correlation function at very long times leads to blurred spectra. Therefore, one has to restrict the upper integration limit of $\langle \vec{M}_D(0) \cdot \vec{M}_D(t) \rangle$. In Figure 10.6, the gray line represents the direct Fourier–Laplace transform of the raw data up to six times the longest time constant τ_k, which can be fairly reproduced by the Fourier–Laplace transform of the fit function (black solid line). Due to the linearity of the Fourier–Laplace transform, the multiexponential fit results in a superposition of Debye processes:

$$\mathcal{L}\left[\sum_k A_k \cdot e^{-t/\tau_k}\right] = \sum_k \frac{A_k \tau_k}{1 - i\omega\tau_k} \tag{10.98}$$

In order to keep the number of fit parameters at the minimum, one could alternatively use the so-called Kohlrausch–Williams–Watt (KWW) functions [68], which model the spread of exponentials by a single parameter β:

$$\langle \vec{M}_D(0) \cdot \vec{M}_D(t) \rangle \simeq A \cdot e^{-(t/\tau)^\beta} \tag{10.99}$$

The corresponding Fourier–Laplace transform can only be approximated by a series [23, 27]:

$$\mathcal{L} = \left[A \cdot e^{-(t/\tau)^\beta}\right] = \frac{A\tau}{\pi} \sum_{k=0} \frac{\Gamma(\beta k + 1)}{k!} \frac{(-1)^{k+1}}{(\omega\tau)^{\beta k + 1}} \left(\cos\left(\frac{\pi}{2}k\beta\right) + i\sin\left(\frac{\pi}{2}k\beta\right)\right)$$

$$\tag{10.100}$$

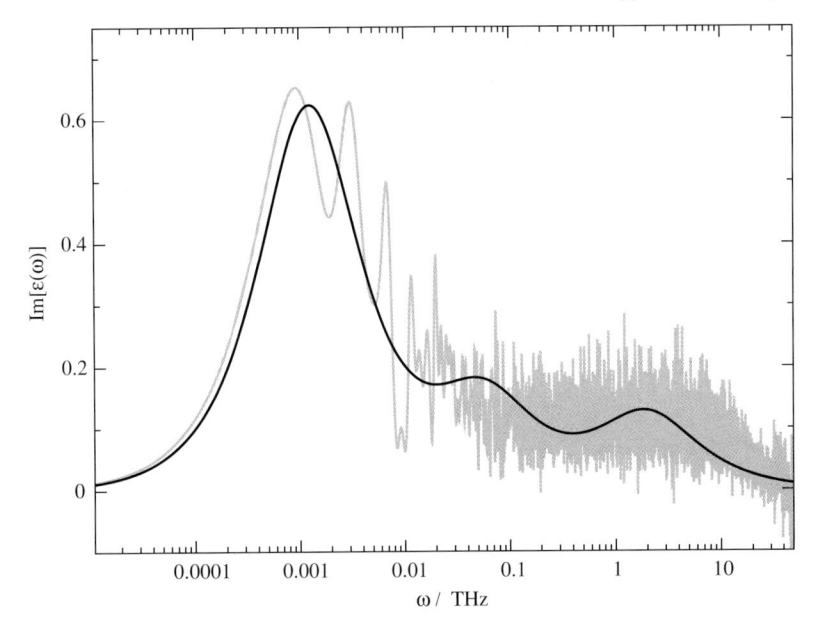

Figure 10.6 Imaginary part of the dielectric constant $\varepsilon(\omega)$ of simulated 1-ethyl-3-methyl-imidazolium dicyanoamide. The solid line represents the spectrum evaluated by the fit function $f(t) = \sum_k A_k e^{-t/\tau_k}$ and the gray line represents the result from the direct Fourier–Laplace transform of $\langle \vec{M}_D(0) \cdot \vec{M}_D(t) \rangle$ up to 5 ns.

Unfortunately, this series has problems to converge for very low frequencies $\omega\tau \ll 1$. Therefore, one has to switch in this case to the asymptotic series:

$$\lim_{\omega\tau \ll 1} \mathcal{L}\left[A \cdot e^{-(t/\tau)^\beta}\right] \simeq \frac{A\tau}{\beta} \sum_{k=0} (\omega\tau)^k \frac{\Gamma\left(\frac{k+1}{\beta}\right)}{k!} \left(\cos\left(\frac{\pi}{2}k\right) + i\sin\left(\frac{\pi}{2}k\right)\right) \tag{10.101}$$

In order to avoid the switching between these two series, we represent the dielectric spectrum originating from the KWW function by Havriliak–Negami function

$$\mathcal{L}\left[-\frac{d}{dt} A \cdot e^{-(t/\tau)^\beta}\right] \simeq \frac{1}{(1-(i\omega\tau)^\alpha)^\gamma} \tag{10.102}$$

with fitted parameters α and γ [1, 2].

As a general rule, the rotational relaxation times scale with the viscosity of the underlying system. This may be considered as a remnant feature of hydrodynamics that provides formulas involving the product of the volume and the viscosity [21].

$$\tau = \xi \frac{3V_i}{k_B T} \eta \tag{10.103}$$

with ξ being the shape factor. In case of ionic liquids, the evaluation of the prefactor usually gives molecular volumes V_i that are unreasonably small [59]. For neutral molecular liquids, however, one gets reasonable molecular volumes. In both cases, a linear scaling of relaxation times τ with the viscosity η is valid.

As opposed to the monotonic behavior of the collective rotational dipole functions, the correlation functions involve the current relax within few picoseconds and are oscillatory in nature. Therefore, we have generalized the above multiexponential fit $f(t)$ including phase-shifted cosine functions [58]:

$$\langle \vec{J}(0) \cdot \vec{J}(t) \rangle \simeq f(t) = \sum_k A_k \cos(\omega_k \cdot t + \delta_k) e^{-t/\tau_k} \quad (10.104)$$

The relaxation times τ_k are found to be much shorter than those of $\langle \vec{M}_D(0) \cdot \vec{M}_D(t) \rangle$. It is important to note that this fit function does not only mimic the overall behavior of $\langle \vec{J}(0) \cdot \vec{J}(t) \rangle$, but its Fourier–Laplace transform

$$\mathrm{Re}[\mathcal{L}[f(t)]] = \sum_k \frac{A_k \tau_k}{2} \left(\frac{\cos(\delta_k) - \tau_k(\omega_k - \omega) \sin(\delta_k)}{1 + \tau_k^2(\omega_k - \omega)^2} \right. \\ \left. + \frac{\cos(\delta_k) - \tau_k(\omega_k + \omega) \sin(\delta_k)}{1 + \tau_k^2(\omega_k + \omega)^2} \right) \quad (10.105)$$

$$\mathrm{Im}[\mathcal{L}[f(t)]] = \sum_k \frac{A_k \tau_k}{2} \left(\frac{-\sin(\delta_k) - \tau_k(\omega_k - \omega) \cos(\delta_k)}{1 + \tau_k^2(\omega_k - \omega)^2} \right. \\ \left. + \frac{\sin(\delta_k) + \tau_k(\omega_k + \omega) \cos(\delta_k)}{1 + \tau_k^2(\omega_k + \omega)^2} \right) \quad (10.106)$$

also gives the correct limiting behavior of the dielectric conductivity $\vartheta(\omega)$. In particular, the fit function describes the divergence of imaginary part of $\vartheta(\omega)$ that is represented by $\mathrm{Re}[\mathcal{L}[f(t)]]$ at zero frequency. After correcting for the static conductivity, that is, considering $\vartheta_0(\omega)$,

$$\mathrm{Im}[\vartheta_0(\omega)] = \frac{4\pi}{3 V k_B T} \frac{\mathrm{Re}[\mathcal{L}[f(t)]] - \sigma(0)}{\omega} \quad (10.107)$$

$$= \omega \sum_k \frac{A_k \tau_k^3 \left[\left(\tau_k^2(3\omega_k^2 - \omega^2) - 1\right) \cos(\delta_k) + \tau_k \omega_k \sin(\delta_k) \left(3 - \tau_k^2(\omega_k^2 - \omega^2)\right) \right]}{\left(1 + \tau_k^2 \omega_k^2\right)\left(1 + \tau_k^2(\omega_k - \omega)^2\right)\left(1 + \tau_k^2(\omega_k + \omega)^2\right)} \quad (10.108)$$

the limit $\omega \to 0$ is zero. The application of the set of fit functions for the rotational and translational correlation functions is, however, not limited to $\langle \vec{M}_D(0) \cdot \vec{M}_D(t) \rangle$ and $\langle \vec{J}(0) \cdot \vec{J}(t) \rangle$, but can be applied to the correlation functions of their subcomponents.

Having developed the necessary theoretical and practical tools for the interpretation of dielectric spectra, we now turn to a discussion of concrete charged, dipolar systems. They may be grouped into two classes: solvated biomolecules and molecular ionic liquids.

10.2.1
Solvated Biomolecules

Starting from the building block of single amino acid in its zwitterionic form, we already encounter two dipoles: the dipole generated by the dislocation of the terminal charges of the amino and the carboxyl group and the side chain dipole in case of polar amino acids. Assembling the individual building blocks to peptides or proteins creates a third type of dipole, namely, that of the peptide unit $-\mathrm{NH-CO}-$. These single peptide dipoles may be summed up to a collective peptide dipole moment. It depends on the secondary structure whether the dipoles cooperate and enhance the collective value or they compensate each other decreasing the collective value. For example, α-helices are characteristic of a parallel alignment of peptide dipoles. Therefore, in this case, the collective value reaches its maximum value. In contrast, β-sheets are typical of dipole compensation. Loop structures exhibit a certain residual correlation of dipoles in the sense of an intramolecular Kirkwood g_K factor. Since a protein is characterized by a diversity of rather different structural elements, the compensation of dipoles is larger compared to peptides, which in their turn are more compensatory than the single amino acids. In other words a protein with a smaller number of residues has a higher dipole density or polarization $\vec{P}_D$ per residue. This order refers to a mixture of structural elements. If a protein consists almost exclusively of α-helices, its collective dipole moment and consequently the dielectric increment of the protein solution may be exceedingly high. As this polarization of the solute replaces the dipole density of the solvent, for example, water, it is decisive for the dielectric increment representing the excess value of the biomolecular solution compared to the pure solvent.

One should not overlook, however, the influence of ions that are usually present in a protein solution. As atomic ions occupy space but contribute nothing to the collective rotational dipole moment, they counteract the protein dipoles. In extreme cases, they may even generate a dielectric decrement. However, one has to bear in mind that the ions contribute to the static dielectric properties via the current density $\vec{P}_J$. In other words, the influence of the ions is a delicate balance between the lack of dipole density $\vec{P}_D$ and the contribution of current density $\vec{P}_J$. Charged protein residues unite both features: They contribute to the dielectric permittivity via $\vec{P}_D$ and to the dielectric conductivity $\vec{P}_J$. However, the number of charged residues in a protein is typically 10% of the total number of residues. Furthermore, the difference in sign leads to a compensation of the overall net charge that is usually a few charge units only. Not to forget that roughly half of the proteins possess a metallic ionic cofactor enhancing the charge. A typical motif of such a charge-stabilized structure is the zinc finger, which is extensively used by nature as a tandem sequence. In this sense, peptides and proteins may be considered as charged macroions moving through the solution. DNA molecules are extreme in this respect as each residue carries a negative charge. Therefore, one has to add an appropriate number of counterions to keep the whole system electrically neutral.

In the following, we discuss the concrete applications of "computational dielectric spectroscopy" and compare the results with corresponding experiments. While the experiments provide the total spectrum, the computational analysis enables the decomposition down to the molecular level. This helps to interpret the experimental spectra. It is a long tradition that the dielectric properties either experimental or computational are given in dimensionless units. In order to subject the above dielectric theory to this criterion, all relevant formulas have to be divided by $4\pi\varepsilon_0$.

10.2.1.1 Peptides

The importance of finite system electrostatics for a peptide dissolved in an ionic solution of NaCl and water at room temperature was analyzed by Smith and Pettitt 20 years ago [60]. The truncation of electrostatic interactions using a switching function to smooth the discontinuous drop at several cutoffs was compared to an implementation of the Ewald summation technique in a bundle of simulation studies. Thus, the size of the cutoff, the water content, the inclusion or exclusion of ions, and the initial configurations of the ions were varied. As criteria for judgment of all these variations, the single-particle relaxation time of water molecules and $g_K(r)$ are used. Without NaCl the relaxation time is retarded by a factor of 2 when using the cutoff instead of the Ewald summation. In addition, g_K drops by a factor of 30 to an unphysical low value of 0.2 that can also be derived from the $g_{\mu\mu}$ functions of Equation 10.90 shown in Figure 14 of Ref. [60]. In the presence of NaCl, the deviation becomes less but it is still remarkable. Reaction field simulations of pure water included from the literature yield relaxation times close to those gained by the Ewald simulations [41, 42]. All these findings nicely demonstrate the extreme sensitivity of dielectric properties to boundary conditions. Subsequently, the influence of finite system electrostatics on the solvated peptides was systematically investigated in a series of papers [50–52].

Based on this analysis of boundary conditions, Pettitt and coworkers [69] provided first the dielectric permittivities for a similar peptide solvated in an ionic solution. The system was decomposed into neutral and charged peptide moieties, water, and ions. Although restricted to static values, the paper reports averages of the collective dipoles themselves, of their square, and of the cross-correlation with others for all four components. Since the cross-correlation are found to be rather small, the averaged mean-squared total dipole moment $\langle \vec{M}_D^2 \rangle$ can be approximated by the sum of the individual contributions $\langle (\vec{M}_D^k)^2 \rangle$. Among the latter, the charged components make the major contribution as expected from the theory. Subsequently, the $\langle (\vec{M}_D^k)^2 \rangle$ are converted to dielectric permittivities ε_k of the respective component. This is achieved by using a component-specific molar volume V_K instead of the total volume V. This is somewhat misleading as it suggests a "molecular dielectric permittivity" that has no theoretical basis. In order to be consistent with Section 10.1, one could multiply these ε_k by the ratio V_k/V and then call it the contribution of the component to the dielectric permittivity of the system.

A complete spectrum of solvated peptides consisting of alanine subunits in roughly 400 TIP3P water is given in Ref. [12]. All electrostatic forces were exclusively calculated by the Ewald method. Several systems containing different concentrations of alanine and dialanine were simulated over a period of 10 ns. Increasing the concentration of alanines and dialanines leads to a linearly rising dielectric increment. Only when the zwitterionic nature is suppressed, a decrement is observed. The autocorrelation function of the total collective dipole moment $\langle \vec{M}_{tot}(0) \cdot \vec{M}_{tot}(t) \rangle$ was fitted mono- and biexponentially as well as by a KWW function. While the monoexponential fit turned out to be too crude, the biexponential and KWW fits yield similar spectra. Water molecules showed a different behavior as a function of the distance to the nearest peptide. Consequently, the collective dipole moment of the water was decomposed into a contribution from the first shell around alanine and the bulk. The imaginary part of the dielectric spectrum is determined by three contributions: The slowest contribution at lowest frequency stems from correlations of the alanine with alanine, first shell water, and bulk water. The cross-term between the shell and bulk water holds the medium position. The self-terms of shell and bulk water comprise the fast contribution. The frequency spread in case of two solvated dialanines is larger as compared to four solvated alanines. The fast contributions do not change their position in frequency, but the slow contribution is shifted to lower frequency because the larger volume of the dipeptide slows down molecular motion.

The model of bulk, solvation shell, and peptide was also analyzed for hydrophilic NAGMA and amphiphilic NALMA dissolved in TIP4P-Ew water at different concentrations [36]. The molecular dynamics simulation with the Ewald summation was carried out over a period from 6 to 21 ns. The collective dipole correlation functions $\langle \vec{M}_D^k(0) \cdot \vec{M}_D^k(t) \rangle$ were fitted biexponentially (refer to Equation 10.97). The decomposition was done at three levels: the whole system, peptide and water, and peptide, first hydration shell, and water. The fit parameters are given explicitly for all three levels of decomposition. For the whole and the two-component systems, the dielectric permittivity spectra are given graphically. While the hydrophilic NAGMA shows a dielectric decrement for both concentrations, the amphiphilic NALMA yields a decrement at lower concentration but shows a dielectric increment at higher concentrations. Even more interesting, the highly concentrated NALMA system displays a doublet shape in the imaginary part, while the other three systems show a single peak structure. This double-peak structure is generated by the downshift of the peptide–peptide and peptide–water peaks at the higher NALMA concentration.

10.2.1.2 Proteins

A comparative study of the static dielectric properties of a suite of the four proteins (P) hen egg white lysozyme, α-lactalbumin, rat fatty acid binding protein, and llama antibody heavy-chain variable domain was performed over a simulation period of 5 ns within the framework of the reaction field method [45]. Instead of conducting boundary conditions ($\varepsilon_{RF} = \infty$), the strength of the reaction field was set to $\varepsilon_{RF} = 68$ for the simulation in SPC water and $\varepsilon_{RF} = 5$ for the simulation in chloroform. The values given correspond to static dielectric constants of the pure solvents. This implies that the field ratio

$$E_{ext}/E = \frac{2\varepsilon_{RF} + \varepsilon}{2\varepsilon_{RF} + 1} \tag{10.109}$$

has to be included in the expression for the evaluation of the simulated dielectric constant. A central issue of the paper was the careful analysis of the convergence of the mean-squared total dipole moment $\langle(\vec{M}_D^P)^2\rangle$ of the respective protein. In order to achieve a better convergence, the square of the average $\langle\vec{M}_D^P\rangle^2$ was subtracted from $\langle(\vec{M}_D^P)^2\rangle$. So, the static dielectric permittivity was finally calculated from

$$\varepsilon - 1 = \frac{4\pi}{3Vk_BT} \frac{\langle(\vec{M}_D^P)^2\rangle - \langle\vec{M}_D^P\rangle^2}{V_P} \frac{2\varepsilon_{RF} + \varepsilon}{2\varepsilon_{RF} + 1} \tag{10.110}$$

It is important to note that the volume V_P is not the volume of the sample V, but the volume of the respective protein calculated from its mass and mass density. Therefore, the values given for the static dielectric permittivity ε_P should be scaled by the ratio V_P/V in order to get the contribution of the protein to the dielectric constant of the whole system. The convergence of the corrected dipole fluctuations of the protein within the simulation period was carefully analyzed for each protein. The time for convergence was 1–4 ns depending on the system. The contribution from the water and the ions was ignored. The dipole fluctuations of the protein were further subdivided into the contribution of the backbone and the contribution when leaving out the charged residues and the NH_3^+–COO^- rear-head dipole. One clearly sees that the charged components give the major contribution. The compensation of the peptide dipoles along the backbone already anticipated at the beginning of this section leads to a small dielectric contribution.

The first simulation providing the complete dielectric spectrum of a solvated protein is given in Ref. [8]. The 76-residue protein ubiquitin (P) dissolved in 5523 TIP3P water molecules was simulated over 5 ns with the PME method. Since ubiquitin contains the same number of positively and negatively charged amino acids at neutral pH, the total collective dipole moment $\vec{M}_{tot}$ is only made up by the rotational part $\vec{M}_D$ because $\vec{M}_J$ is zero due to the zero net charge $q_{tot} = 0$. The cumulative dipole fluctuations of the protein $\langle\vec{M}_{tot}^{P}{}^2\rangle$ gradually decline during the simulation. Nevertheless, it needs 5 ns to bring this value in the vicinity of zero. In other words, in this case, 5 ns is sufficiently long to calculate a well-behaved $\langle\vec{M}_{tot}^P(0) \cdot \vec{M}_{tot}^P(t)\rangle$. The water–water self-term $\langle\vec{M}_{tot}^W(0) \cdot \vec{M}_{tot}^W(t)\rangle$ and the cross-term $\langle\vec{M}_{tot}^P(0) \cdot \vec{M}_{tot}^W(t)\rangle$ relax faster to zero. All these three correlation functions were fitted biexponentially in order to compute the dielectric contributions. The bimodal structure of the imaginary part of $\varepsilon(\omega)$ comes from the superposition of the protein–protein and protein–water contributions in the first peak at lower frequencies and the water–water self-term at the second higher peak, as shown in Figure 10.8a. It should be noted that the protein self-term and the cross-term contribute almost equally to the first peak. This structure was also found experimentally [32] and attributed to the so-called β- and γ-processes. The experimental investigation of the δ-process was difficult due to the small amplitude. In the computational spectrum, however, this δ-process may be explained by the protein–water cross-term. This contribution is not directly visible as a single peak in the spectrum because the protein–protein (β-process) and the water–water (γ-process) peaks overlap in the dielectric spectrum. This interpretation is not restricted to ubiquitin,

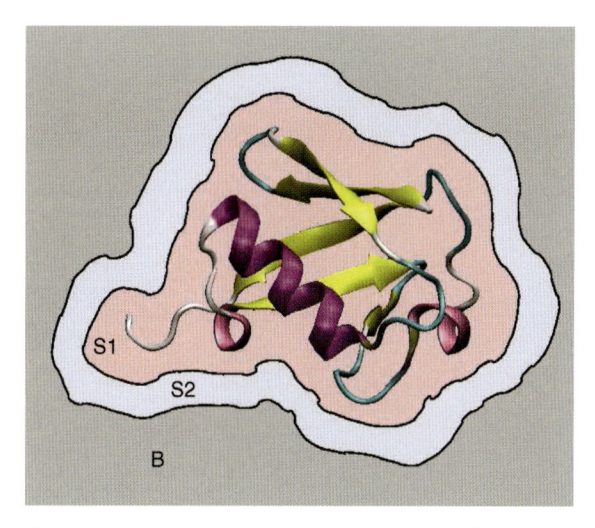

Figure 10.7 Ubiquitin and its solvation layers. The first shell (S1) consists of water molecules directly attached to the protein. The second shell (S2) water molecules are neighbors of the S1 water molecules. All water that do not belong to S1 or S2 are considered as bulk water (B).

but also applies to the spectrum of Myoglobin [18]. Another advantage of the simulation is the classification of water as a function of the distance to the protein. Based on the Voronoi decomposition [5, 43, 65] that enables a parameter-free construction of the first, second, and subsequent solvation shells, the dielectric contribution from these shells (depicted in Figure 10.7) can be analyzed separately. It turns out that the first solvation shell (S1) is anticorrelated with the protein (red dashed–dotted line in Figure 10.8b) supporting the picture of a suprasolute with a quenched dipole moment. The correlation of the suprasolute with the second shell (S2) is very small and demonstrates the importance of the first solvent layer of a protein. For subsequent shells, the correlation with the suprasolute is still there and even stronger compared to the second shell.

While the previous study dealt with a single protein ubiquitin and was restricted to the static dielectric properties of the solvation shell, a comparative study of three proteins, namely, ubiquitin, apo-calbindin D_{9K}, and the C-terminal SH2 domain of phospholipase C-γ1 provided a decomposition of the dielectric spectrum into contributions from the protein, the first and second shells, and bulk water [49]. The number of water molecules surrounding the protein in a truncated octahedron was 8500 for ubiquitin and apo-calbindin and 10 000 TIP3P water for the SH2 domain. The simulations were performed under Ewald boundary conditions and covered 20–35 ns. The imaginary part of the dielectric spectrum shows the typical bimodality, but the relative weight of the β- and γ-processes differ among the proteins. While for ubiquitin, the low-frequency protein peak is lower in height as compared to the water peak, it is of equal height for apo-calbindin, and even higher than the water peak for the SH2 domain. In order to elucidate the complete fine structure of the spectrum, the sample was split into four components: the

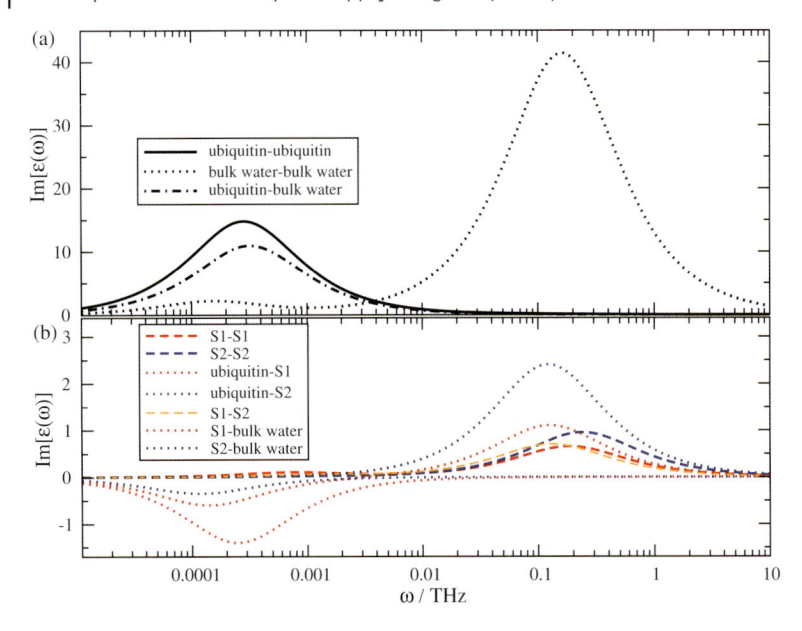

Figure 10.8 (a) Imaginary part of the dielectric constant $\varepsilon(\omega)$ of ubiquitin in water. The protein–protein self-term (β-process) is represented as solid line, the bulk–bulk interaction (γ-process) as dotted line. The coupling between the bulk and ubiquitin is shown as dashed–dotted line. (b) The interactions of the first and second solvation shells.

protein, the first and second hydration shells, and the bulk water. It is interesting to analyze the static dielectric properties of these components. The contribution from the protein–protein self-term does not follow the sequence of the height of the first peak in the spectrum. Rather, we have now the sequence ubiquitin < SH2 domain < apo-calbindin. It is the additional cross-term between protein and the bulk that enhances the first peak in case of the solvated SH2-domain dielectric spectrum. The three proteins also differ with respect to the correlation of the protein and the first hydration shell. While ubiquitin and apo-calbindin are anticorrelated with their first water layer, the SH2 domain is positively correlated. The correlation with the second shell is negative for all three proteins, but the actual value of −0.4 is almost negligible and leaves the protein–water correlation mainly to the bulk. In principle, the decomposition into protein (P), first (S1) and second (S2) hydration shells, and bulk (B) water leads to 10 dipolar correlation functions, 4 self-terms, and 6 cross-terms. It turns out that the spectrum is dominated by the P–P and B–B self-terms with the P–B playing an important role for the height of the first low-frequency peak, as shown in Figure 10.8b. The remaining seven contributions, two self-terms S1–S1 and S2–S2, as well as all other cross-terms involving S1 and S2 altogether give a small contribution, but their bimodal structure is also instructive: however, ubiquitin and apo-calbindin show a bimodal shape with a first negative peak, which is reversed to a positive peak for the SH2 domain. The superposition of all minor contributions involving self-terms and cross-terms with S1 and S2 may be viewed as a difference spectrum when subtracting P–P, P–B, and B–B from the total spectrum. As the first two are essentially the β-process and B–B is almost exclusively the γ-process, the difference spectrum must reflect what experimenter refer to as the δ-process. The present analysis shows that this δ-process is not a single relaxation, but a rather complicated superposition of shell-based contributions.

All studies discussed so far dealt with the dielectric permittivity spectrum. The zinc finger as the classical model for a charged protein introduces a "protein current" and its cross-term with the proteins a rotational dipole moment. Thus, all terms developed in Section 10.1 come into action in this model system. The charge of the protein was focused on Zn^{2+} and two chloride anions in 2872 SPC/E water molecules guaranteed the total charge neutrality. The simulation was performed over 13 ns with Ewald conditions [34]. The system was split into overall neutral body of the zinc finger and the neutral water as well as the ionic component Zn^{2+} and two Cl^-. For the two neutral components, the collective dipole correlation functions $\langle \vec{M}_D^P(0) \cdot \vec{M}_D^P(t) \rangle$, $\langle \vec{M}_D^W(0) \cdot \vec{M}_D^W(t) \rangle$, and $\langle \vec{M}_D^P(0) \cdot \vec{M}_D^W(t) \rangle$ were fitted biexponentially to compute the two contributions $\varepsilon_P(\omega)$ and $\varepsilon_W(\omega)$ being determined by the Fourier–Laplace transform of the respective self-term and the cross-term. Again, the γ-process is almost exclusively made up by the water self-term, while the β-process is a superposition of the protein self-term and the protein–water cross-term. In contrast to the larger protein discussed above, the contribution of the cross-term is not of equal weight but considerably smaller. The current–current correlation function $\langle \vec{J}(0) \cdot \vec{J}(t) \rangle$ of the ionic component was Fourier–Laplace transformed to give a conductivity spectrum. The cross-correlation function $\langle \vec{M}_D^W(0) \cdot \vec{J}(t) \rangle$ appears more or less as numerical noise. In other words, there is no coupling between water and the ionic component. In a subsequent study [9], the water component was decomposed into two hydration shells S1 and S2 and the bulk B. From the contribution to the static dielectric permittivity, we learnt again that the first shell is anticorrelated with the protein, although the contribution is rather small. This is also true for the coupling with S2 that is just opposite in sign to S1. The coupling of the protein with the bulk is rather small in this case too. This is also visible in the dielectric spectrum where the β-process does not play the dominant role as for the larger proteins discussed above. This brings into action the self-terms S1–S1 and S2–S2 that now contribute to the γ-process.

10.2.1.3 DNA

As already stated in the general introduction, solvated DNA molecules are the most complicated molecular systems to study since each nucleotide carries a net negative charge. Therefore, a DNA oligomer is a macroion with a negative charge proportional to the number of nucleotides. To ensure a neutral system, positive counterions are usually added. If doing so, the total dipole moment of the system $\vec{M}_{tot}(t)$ is a well-defined quantity. Any decomposition, however, into charged fragments automatically creates the problem how to compute the fragment dipole. As the fragments are covalently bonded, they share a common center-of-mass. Therefore, the splitting of the fragment dipole into a contribution to current of its net charge and a neutral charge set referring to fragment's individual center-of-mass is possible, but the so obtained rotational dipole moments of each fragment do not add up to the DNA's total rotational dipole moment $\vec{M}_D(t)$ as they refer to individual center-of-mass. This general principle has been overlooked in a paper by Yang et al. [70] simulating a triple helical DNA d(CG·G)$_7$ with 37 Na$^+$ and 16 Cl$^-$ and 837 SPC/E water molecules over a period of 1155 ps. The total collective dipole moment is split up into water component, ions, and DNA, which in its turn is decomposed into its bases, sugar moieties, and phosphate groups. Therefore, we have a lot of charged fragments with the problem how to compute their dipole moments. Nevertheless, this was done in this study, and from the dipole moment fluctuations of the self- and cross-terms, static dielectric contributions were calculated.

A more recent study by Ikeda et al. [29] tackles the dielectric spectrum of a solvated DNA again. This time, simulating a octamer random sequence ds-DNA with 16 Na$^+$ and 2 Cl$^-$ dissolved in 1239 water molecules over a time period of 7.3 ns. The self-terms $\langle \vec{M}_{DNA}(0) \cdot \vec{M}_{DNA}(t) \rangle$ and $\langle \vec{M}_W(0) \cdot \vec{M}_W(t) \rangle$ as well as the mixed term were computed. The ions and their current seem to be completely ignored. It should be noted that the correlation functions refer to the net dipole moment of the respective component. No splitting into a rotational and a translational part was performed. This may explain why the fit function contains an exponential and a linear function that may be assigned to these two components. The nonintuitive linear behavior of $\langle \vec{M}_J(0) \cdot \vec{M}_J(t) \rangle$ was already discussed in Section 10.1 and graphically displayed in Figure 10.4. All three correlation functions are directly Fourier–Laplace transformed to give the real and imaginary part of the frequency-dependent dielectric function.

10.2.1.4 Biological Cells

So far we have discussed applications at the molecular resolution. The numerical solution of Poisson equation (10.10) outlined at the end of Section 10.1 offers the possibility to compute a dielectric spectrum even at the macroscopic level. This was done by Asami [3, 4]. As a test case, he first studied the method for a water-in-oil and oil-in-water emulsion, that is, for a mixture with components of rather different dielectric properties [3]. The real-world application is now the computation of the dielectric spectra of cells during cell division. The single cell is modeled by a sphere, the membrane by an attached spherical shell, and the cytosol by the rest of the volume. Each of the three regions is assigned a different but constant value of ε and σ. Intermediate state of cells are modeled by penetrating spheres.

An alternative and more elaborate method to describe the biological cells was developed by Prodan and Prodan [46]. First, they augment the Poisson equation by a diffusion term and then they convert the solution to an integral over the surface. The electric field is represented by a surface integral over dipoles oriented in the direction of the external field. These surface dipoles are assumed to be polarizable and their polarizability is related to the generalized dielectric constant of the two media contacting at the surface. The method was applied by di Biaso et al. [22]

to the numerical simulation of dielectric spectra of aqueous suspensions of non-spheroidal differently shaped biological cells. For the transition from a sphere over two penetrating spheres to a dumbbell, the changes in the dielectric spectrum are computed.

10.2.2
Molecular Ionic Liquids

So far we have discussed solutions of charged, dipolar species. A biomolecular solute is considerably larger than the water solvent molecules and the added monoatomic ions. Thus, charge polarity is heterogeneous in an otherwise essentially uncharged solvent. A homogeneous system of charged, dipolar species is realized by ionic liquids consisting of molecular ions where the charge and the dipole reside on the very same molecule. It is the anisotropy in molecular shape that makes this system liquid at ambient temperature. Isotropic monoatomic ions are ionic melts with melting points of hundreds degrees of Celsius. For comparison, liquid NaCl melts at 801 °C. Replacing the atomic cation by imidazolium brings the melting point down to 80 °C. Further replacement of the anion by slightly anisotropic species like BF_4^- or PF_6^- reduces the melting point to ambient temperature. Therefore, these systems are liquid, but one should not forget that solidification is retarded by the molecular anisotropy. As a consequence of this remnant feature of the solid state, the viscosity of molecular ionic liquids is quite high compared to the viscosity of water of roughly 1 mPa·s. The viscosity of ionic liquids ranges from 20 mPa·s in case of 1-ethyl-3-methyl-imidazolium [$EMIM^+$] dicyanoamide [$N(CN)_2^-$] up to 100 mPa·s and more. For ionic liquids composed of substituted amino acid viscosities up to 1000 mPa s have been measured. This high viscosity has important implications for simulation studies as the molecular dynamics usually scales linearly with viscosity (see Equation 10.103). This means that for water, a simulation period of 1 ns might be quite sufficient, but has to cover 100 ns for ionic liquids. Bearing in mind that both systems proceed in the same time step of 1 fs, the largely enhanced computational effort in simulating ionic liquids becomes obvious. The situation is even more acuminated when calculating collective properties. As an averaging over individual molecules is not possible in this case, the statistical demands on the simulation length are very high. As this study exclusively deals with collective properties, their evaluation requires a maximum of computational effort.

Nevertheless, it is worthwhile to undertake simulations of ionic liquids as their fascinating properties can be varied in a wide range because of the vast amount of combinations of cations and anions. In some sense, one may speak of "task-specific" ionic liquids designed for the special purposes needed. As the properties of ionic liquids are mainly governed by the type of the anion, the set of applied cations concentrates to a few prominent examples, for example, imidazolium cations. The plethora of anions ranges from simple atomic chloride ions over slightly anisotropic molecular ions like BF_4^- or PF_6^- to dipolar species like $CF_3SO_3^-$ to hydrophobic species derived from perfluorated hydrocarbons.

10.2.2.1 Conductivity and Dielectric Conductivity

When dealing with ionic liquids, the conductivity is one of the first macroscopic properties to be used for their characterization. As outlined in Section 10.1, the frequency-dependent conductivity is determined by the Fourier–Laplace transform of the current–current correlation function $\langle \vec{J}(0) \cdot \vec{J}(t) \rangle$ in Equation 10.53. Its zero-frequency limit is the static or "DC" conductivity. This value is needed to correct the imaginary part of the generalized dielectric constant, as outlined in Section 10.1 (refer to Equation 10.62). A rough estimate for the static conductivity frequently used is given by the Nernst–Einstein relation:

$$\sigma_{NE} = \frac{\varrho q^2}{k_B T}(D^+ + D^-)(1-\Delta) \tag{10.111}$$

Thus, the static conductivity can be computed from the single-particle diffusion coefficients of the cations D^+ and the anions D^-. Besides, the coupling between cations and anions as well as the coupling between different cations or anions is neglected. The Δ-parameter is a measure for the deviation caused by this simplification. A typical value of Δ is 0.35, but in extreme cases much higher Δ-values up to 0.87 may be found [7, 44], as given in Table 10.1.

In an interesting comparative study, Rey-Castro and Vega have compared the diffusion coefficients, the viscosity, and the conductivity of molten NaCl and liquid EMIM$^+$Cl$^-$ in order to show the influence of the anisotropy of the cation [47]. The Ewald method was used to calculate the electrostatic forces and the simulation covered 8 ns. Studying 64, 125, and 216 ion pairs, they tried to investigate the effect of system size. While the diffusion turned out to be rather insensitive, as expected, a speculative trend for a lower viscosity and higher conductivity for the larger system was observed. In order to figure out the cooperativity of the current correlation function, the self-terms of the EMIM$^+$ cation and the chloride anion were subtracted to get the cross-term between cations and anions. The integral over this cross-term turned out to be very small, such that the conductivity estimated from the

Table 10.1 Collective properties for several common ionic liquids.

Ionic liquid	Δ	$\varepsilon(0)$	$\varepsilon_{MDJ}(0)$
EMIM$^+$Cl$^-$	0.87		
EMIM$^+$PF$_6^-$	0.32		
EMIM$^+$N(CN)$_2^-$	0.18–0.35	2.2	−0.2
EMIM$^+$CF$_3$SO$_3^-$	0.61	7.6	−0.2
BMIM$^+$PF$_6^-$	0.34–0.38	9.5	−0.8
BMIM$^+$BF$_4^-$	0.36	8.2	−0.3
BMIM$^+$CF$_3$COO$^-$	0.14	26.2	−0.5

Δ describes the deviation of static conductivity $\sigma(0)$ from the Nernst–Einstein relation (10.110). $\varepsilon(0)$ is the static dielectric permittivity [53, 57]. Its contribution from the cross-correlation of the collective rotational dipole moment and the current is $\varepsilon_{MDJ}(0)$. These two properties are in units of $4\pi\varepsilon_0$.

Nernst–Einstein equation and the integral of the current correlation function were found to be pretty close. The linear dependence of the conductivity on the fluidity or inverse viscosity, the so-called Walden rule, was found to be fulfilled over a considerable temperature range of 378–489 K.

In a molecular dynamics study, Pićalek and Kolafa [44] computed the conductivity from the linear slope of the dipolar displacement according to Equation 10.52. The ionic liquids $BMIM^+PF_6^-$, $BMIM^+BF_4^-$, and $EMIM^+PF_6^-$ were simulated for 1.2 ns. Two hundred, 400, and 800 ion pairs were investigated in order to find out some influence of the system size. The cationic diffusion coefficient turned out to be insensitive, while the anionic diffusion was slowed down in the larger systems. The conductivities derived from the dipolar displacements, including all collective effects, were compared to the estimates from the Nernst–Einstein equation retaining the self-terms only. Raising the temperature from 360 to 400 K, the Δ-values decreased from 0.46 to 0.19. In other words, the higher the temperature, the smaller the effect of cooperativity.

While the previous studies were restricted to the static conductivity at zero frequency, the first simulated conductivity spectrum was given in Ref. [53]. Three ionic liquids $BMIM^+BF_4^-$, $BMIM^+CF_3COO^-$, and $EMIM^+N(CN)_2^-$ were simulated at room temperature over a period of 84, 100, and 66 ns, respectively. While below 7 THz, all spectra looked rather similar to the divergence for higher frequencies with the peak ranking $EMIM^+N(CN)_2^- > BMIM^+BF_4^- > BMIM^+CF_3COO^-$. While the real part is always positive with a broad maximum, the imaginary part displays a typical sigmoidal shape passing zero at 11.9 THz ($BMIM^+BF_4^-$), 9.4 THz ($BMIM^+CF_3COO^-$), and 15.5 THz ($EMIM^+N(CN)_2^-$). While previous studies derived the static conductivity either from the integral of the current correlation function or the linear slope of the dipolar displacement, this study applies both methods and thus permits a critical comparison. It turns out that the dipolar displacement method is more robust and reliable.

In principle, the frequency-dependent conductivity $\sigma(\omega)$ is a macroscopic property to characterize the collective movement of ions and can thus be discussed stand-alone. When considering the generalized dielectric constant, however, one has to consider the dielectric conductivity $\vartheta_0(\omega)$ as defined in Equation 10.65. In the first presentation of $\vartheta_0(\omega)$ in Ref. [53], this quantity was computed directly from its definition as the ratio $\sigma(\omega)/\omega$. As we have already learnt from the computation of the static value $\sigma(0)$, the upper integration limit of $\langle \vec{J}(0) \cdot \vec{J}(t) \rangle$ is of major importance. This numerical problem can be circumvented by using the fit function $f(t)$ of $\langle \vec{J}(0) \cdot \vec{J}(t) \rangle$ presented in Section 10.2 [58]. However, the parameters of $f(t)$ cannot be determined by a simple least square fit. Rather, they have to be optimized according to the correct behavior of the running integral of $\langle \vec{J}(0) \cdot \vec{J}(t) \rangle$. Thus, it is found that the oscillatory components compensate each other almost perfectly. Therefore, an additional nonoscillatory ($\omega_k = \delta_k = 0$) fit component plays an important role. Although of low-amplitude A_k, it is essential for the static value $\sigma(0)$ as well as for the computational spectrum of $\vartheta_0(\omega)$. The imaginary part of $\vartheta_0(\omega)$ typically resides in the regime from $\omega = 0.01$ to 10 THz. Its shape is not only determined by the time correlation of individual ion currents $\langle \vec{j}^+(0) \cdot \vec{j}^+(t) \rangle$ and

$\langle \vec{j}^-(0) \cdot \vec{j}^-(t) \rangle$ (refer to Figure 10.3) but it also involves all possible intermolecular couplings. The fact that $\vartheta_0(\omega)$ is a dielectric conductivity, that is, it behaves like a dielectric constant, becomes obvious when considering its real part. Starting from a static value $\vartheta_0(0)$, it remains constant with increasing ω until the imaginary part $\mathrm{Im}[\vartheta_0(\omega)]$ shows its peak structure. From this frequency on, the real part $\mathrm{Re}[\vartheta_0(\omega)]$ declines.

10.2.2.2 Dielectric Permittivity

So far for ionic liquids, our focus was on the dielectric behavior resulting from translational motion. Its rotational counterpart, the classical dielectric permittivity ε, was first compared to experiment for $BMIM^+BF_4^-$ [59]. The static value $\varepsilon(0)$ was found to be in fair agreement with Refs [19, 57, 67]. The frequency dependence, however, differed because the computed rotational relaxation times were much higher than the experimental ones. Both, computational and experimental results are not in line with hydrodynamic theory because the extracted molecular volumes of $BMIM^+$ are too small: In simulation, they differ by a factor of 2 and in experiment up to an order of magnitude [59]. This rather different behavior of computational and experimental motions is restricted to rotation and does not affect translation. In simulation, the extracted molecular volume is too small but consistent between rotation and translation. The large discrepancy between these two types of motion in experiment may be explained by a possible spatial heterogeneity or segregation: The butyl chains may form hydrophobic islands [16, 66], which in their turn favor anisotropic rotation of $BMIM^+$. The fastest of these anisotropic rotations seems to be those observed in experiment. In recent studies [66], it was found that a potential segregation occurs for a chain length $n \geq 4$, that is, for imidazolium substituents with butyl and larger alkyl chains. Apart from all these considerations, the linear dependence of relaxation times on viscosity seems to be a general feature of motions in ionic liquids. The only remnant feature of hydrodynamic theory was analyzed in detail in a recent study [56].

While dielectric experiments always yield the complete spectrum, a computational approach enables the decomposition into various contributions. In particular, the cationic and anionic contribution can be separated [54, 55]. In this way, the first main peak in the imaginary part of the spectrum of $BMIM^+CF_3COO^-$ can be assigned to the cation, while the high-frequency shoulder comes from the anion [59]. However, the sum of those two does not give the complete spectrum of $\varepsilon(\omega)$, which also includes a cross-term between the ions.

The appearance of cross-terms is not restricted to the rotational motion of ionic species. It can also refer to the coupling of translational and rotational motions of the very same species. In dielectric terms, there exists a cross-correlation function $\langle \vec{M}_D(0) \cdot \vec{J}(t) \rangle$ whose Fourier–Laplace transform contributes to the dielectric spectrum. As a measure of their magnitudes, the static values ε_{MDJ} for some ionic liquids are tabulated in Table 10.1 together with the total $\varepsilon(0)$. As can be seen, the contribution of the cross-terms are small [53, 57] and therefore often neglected.

10.2.2.3 Generalized Dielectric Constant

In the last two sections, we have discussed the dielectric permittivity $\varepsilon(\omega)$, that is, the rotational part of the dielectric spectrum, and its translational counterpart, the dielectric conductivity $\vartheta_0(\omega)$, separately. Experimentally accessible, however, is only the sum $\Sigma(\omega)$ [19, 67]. Therefore, computational dielectric spectroscopy offers the possibility for an interpretation of experimental spectra via a decomposition of $\Sigma(\omega)$ into $\varepsilon(\omega)$ and $\vartheta_0(\omega)$ as done for EMIM$^+$N(CN)$_2^-$ in Figure 10.9. The real part of these three spectra is given in Figure 10.9a, c, and e, while Figure 10.9b, d, and f present the corresponding imaginary part. Figure 10.9d and f also contain the black dotted parabola $4\pi i \sigma(0)/\omega$ resulting from the static conductivity. The dielectric absorption Re[$\varepsilon(\omega)-1$] in Figure 10.9a is essentially determined by the autocorrelation function $\langle \vec{M}_D(0) \cdot \vec{M}_D(t) \rangle$ of the collective rotational dipole moment corresponding to the first line of Equation 10.63 shown as dashed line. The additional cross-term $\mathcal{L}[\langle \vec{M}_D(0) \cdot \vec{J}(t) \rangle]$ is depicted as solid gray line in Figure 10.9a–d.

In Figure 10.9b, the first peak of Im[$\varepsilon(\omega)$] located at 1.3 GHz originates from the cations EMIM$^+$. Because of the low dipole moment of N(CN)$_2^-$, the

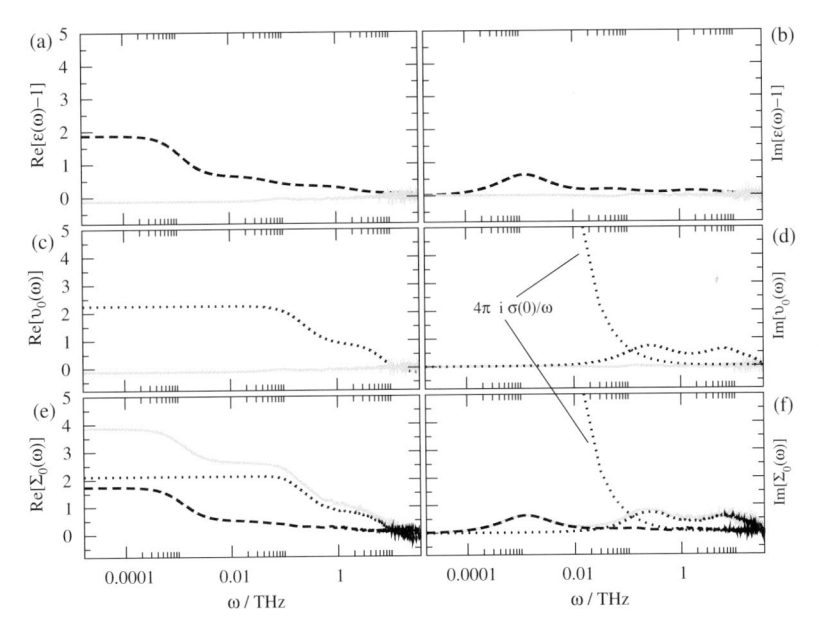

Figure 10.9 Spectral decomposition of the generalized dielectric constant $\Sigma_0(\omega)$: (a) The black dashed line represents the contribution of $\langle \vec{M}_D(0) \cdot \vec{M}_D(t) \rangle$ to Re[$\varepsilon(\omega)-1$] and the gray line stands for the contribution of $\langle \vec{M}_D(0) \cdot \vec{J}(t) \rangle$. (b) Imaginary counterpart Im[$\varepsilon(\omega)-1$] to (a). (c) The black dashed–dotted line stems from the contribution of $\langle \vec{J}(0) \cdot \vec{J}(t) \rangle$ to Re[$\vartheta_0(\omega)$]. (d) Same as (c) but imaginary part of $\vartheta_0(\omega)$. (e) The black dashed line, the dashed–dotted line, and the gray solid line represent Re[$\varepsilon(\omega)-1$], Re[$\vartheta_0(\omega)$], and Re[$\Sigma_0(\omega)$], respectively. (f) Same as (e) but imaginary parts. The black dotted line stands for the conductivity parabola $4\pi i\sigma(0)/\omega$. Note that the experimental frequency ν is $\omega/2\pi$.

anionic contributions is hidden in the peak shoulder up to 5 THz. In principle, Figure 10.9a and b may be interconverted using the Kramers–Kronig relation [25]. The black dashed–dotted line in Figure 10.9c shows the contribution from the collective translational dipole moment $\vec{M}_J(t)$. Surprisingly, for this system, the static value $\vartheta_0(\omega = 0)$ has the same magnitude as its rotational counterpart $\varepsilon(0)$. This leads to the counterintuitive result that the translational motion of the ions, that is, their conductivity, residing in the high-frequency regime makes a considerable contribution to the static dielectric constant. For other ionic liquids, however, the weighting between these two static values may be quite different favoring $\varepsilon(0)$. Nevertheless, the dielectric conductivity $\vartheta_0(\omega)$ behaves like a dielectric constant. It merely differs from the dielectric permittivity by a shift to higher frequencies. As opposed to $\varepsilon(\omega)$, where the respective molecular dipole moment $\vec{\mu}$ of cations and anions determine the peak structure, one might argue that the equal strength of the charges leads to comparable peak heights in $\vartheta_0(\omega)$. This conclusion is wrong, however, the double-peak structure in Figure 10.9d is made up by almost coincidence of cation and anion peaks at high frequencies, while the peak at lower frequencies stems from the interaction of cations and anions. Even more, this interaction also reduces the height of the second peak. In the computational analysis, the conductivity parabola (dotted line) is not necessary to compute $\varepsilon(\omega)$ and $\vartheta_0(\omega)$. On the contrary, it has to be subtracted from the experimental spectrum in order to get the sum $\text{Im}[\varepsilon(\omega) + \vartheta_0(\omega)]$. This procedure is delicate because at low frequencies ω, the parabola dominates the spectrum by far. Consequently, the residual spectrum left after the subtraction of $4\pi i\sigma(0)/\omega$ is plagued with considerable numerical errors.

The generalized dielectric constant $\Sigma_0(\omega) = \varepsilon(\omega) - \varepsilon_\infty + \vartheta_0(\omega)$ for the simulated EMIM$^+$N(CN)$_2^-$ is displayed as gray solid line in Figure 10.9e and f. ε_∞ represents the high-frequency limit of the electronic contribution. Since these effects are usually neglected in simulation, ε_∞ is unity. Therefore, experimental and computational spectra have to be matched by subtracting the respective ε_∞ [54, 59].

One can see that the multiple steps in the real part come from the superposition of rotational and translational contributions. In the imaginary part, the interplay of these two contributions becomes even more obvious, as they populate a common frequency region around 10 GHz in case of simulated EMIM$^+$N(CN)$_2^-$. This region profits from the overlay of the decaying $\text{Im}[\varepsilon(\omega)]$ and the rising $\text{Im}[\vartheta_0(\omega)]$ that is shifted to lower frequencies because of the $1/\omega$ downscaling of $\sigma(\omega)$. On the other hand, the decay of $\text{Im}[\varepsilon(\omega)]$ can be shifted to higher frequencies by lowering the viscosity of that ionic liquid. Altogether, one can imagine that in ionic liquids where these two contributions are closer in frequency, the common region will profit even more such that a very broad plateau-like spectrum emerges [54].

It is interesting to see that the cross-term $\langle \vec{M}_D(0) \cdot \vec{J}(t) \rangle$ between the rotational and translational motions is also located at the intersection of $\text{Im}[\varepsilon(\omega)]$ and $\text{Im}[\vartheta_0(\omega)]$. However, its contribution in this transition region is too small compared to the rotational $\langle \vec{M}_D(0) \cdot \vec{M}_D(t) \rangle$ and the translational $\langle \vec{J}(0) \cdot \vec{J}(t) \rangle$ self-terms. In other words, it is the overlap of the rotational and translational self-terms and not the cross-correlation between them.

10.3
Summary and Outlook

Quite generally, dielectric spectroscopy may be seen from two viewpoints. On the one hand, the dielectric spectrum of a system may be seen as the signature of its collective relaxation phenomena involving rotational as well as translational motions in the case of charged, dipolar systems. On the other hand, dielectric properties pave the way toward a better understanding of solvation processes. The extent a solvent can influence a solute depends on the magnitude and frequency distribution of its generalized dielectric constant.

As the system size governs the computational effort, it also determines the method applied and hence the level of resolution. The solvation characteristics may be already analyzed at the macroscopic level prior to any molecular analysis. The largest systems studied so far are biological cells whose dielectric heterogeneity is described by regions or bodies differing with respect to their generalized dielectric constant. This concept of a regional dielectric constant is compatible with the dielectric field equation mentioned in Section 10.1.1. At this very crude level of resolution, one only gets the complete dielectric spectrum.

Decomposition of the dielectric spectrum is only possible at the higher level of molecular resolution that necessitates molecular dynamics simulations. These simulations are usually performed at several dozens of nanoseconds without applying an external field and cover the dynamic range of the underlying collective motions of the molecules. As raw data, these simulations yield time correlation functions of collective dipole moment of the whole sample and their constitutive species (see Sections 10.1.2 and 10.1.3). Selecting a subset of these time correlation functions corresponds to a decomposition of the dielectric spectrum. This cannot be done only with respect to the different species but also with the different modes of motion, that is, rotation and translation. Rotational relaxation typically appears multiexponentially, while translation has to be described by damped oscillatory functions. By using fit functions of the appropriate type, one can cope with the disturbing noise of the time correlation functions because of the limitation of the elapsed simulation period. Finiteness in space is an even more critical point. Therefore, a prerequisite for all these calculations is the correct treatment of the "finite system electrostatics" described in Section 10.1.4. This rules out any attempt to emulate bulk properties by clusters of several dozens of molecules or equivalently a biomolecule with one or two solvation layers.

In the last two decades, the theoretical framework of computing dielectric spectra was applied to charged, dipolar systems, in particular to biomolecules in aqueous solution described in Section 10.2.1. The computational decomposition of corresponding dielectric spectra revealed the importance of cross-correlation between the biomolecule and the water bulk. The first and second hydration layers of the biomolecule contribute to a minor degree to the dielectric properties. Furthermore, all correlation functions involving the coupling between the biomolecule and any fraction of water, hydration layer or bulk, falls in the same time range as the relaxation of the biomolecule itself. Thus, the protein peak in the dielectric spectrum is usually enhanced by these interactions, which in certain cases may comprise one half of the overall peak.

Quite recently, dielectric spectra of the prospering field of molecular ionic liquids (see Section 10.2.2) were analyzed computationally. These molecular ionic liquids are real examples that combine a net charge and a molecular dipole on the very same molecule. Starting from the pure ionic liquids, the range of applicability has extended even to hydrated ionic liquids. Thus, it was found that relaxation times scale with the viscosity of the system. Furthermore, the appearance of a translational contribution to dielectric spectrum at high frequencies was demonstrated for the first time. Even more, translational motion of molecular ions contributes to the static dielectric constant.

From the dielectric point of view, solvated biomolecules are prestigious objects, while the molecular ionic liquids are outstanding model systems. This makes the combination of both, namely, solvation of biomolecules in hydrated ionic liquids, a charming future perspective.

References

1 Alvarez, F., Alegria, A., and Colmenero, J. (1991) Relationship between the time-domain Kohlrausch–Williams–Watts and frequency-domain Havriliak–Negami relaxation functions. *Phys. Rev. B*, **44**, 7306.

2 Alvarez, F., Alegria, A., and Colmenero, J. (1993) Interconnection between frequency-domain Havriliak–Negami and time-domain Kohlrausch–Williams–Watts relaxation functions. *Phys. Rev. B*, **47**, 125.

3 Asami, K. (2005) Simulation of dielectric relaxation in periodic binary systems of complex geometry. *J. Colloid Interface Sci.*, **292**, 228.

4 Asami, K. (2006) Dielectric dispersion in biological cells of complex geometry simulated by the three-dimensional finite difference method. *J. Phys. D: Appl. Phys.*, **39**, 492.

5 Aurenhammer, F. (1991) Voronoi diagrams: a survey of a fundamental geometric data structure. *ACM Comput. Surv.*, **23**, 345.

6 Berne, B.J. and Pecora, R. (2000) *Dynamic Light Scattering*, Dover Publications, New York.

7 Bhargava, B.L. and Balasubramanian, S. (2005) Dynamics in a room-temperature ionic liquid: a computer simulation study of 1,3-dimethylimidazolium chloride. *J. Chem. Phys.*, **123**, 144505.

8 Boresch, S., Höchtl, P., and Steinhauser, O. (2000) Studying the dielectric properties of a protein solution by computer simulation. *J. Phys. Chem. B*, **104**, 8743.

9 Boresch, S., Ringhofer, S., Höchtl, P., and Steinhauser, O. (1999) Towards a better description and understanding of biomolecular solvation. *Biophys. Chem.*, **78**, 43.

10 Boresch, S. and Steinhauser, O. (1997) Presumed and real artifacts of the Ewald summation technique: the importance of dielectric boundary conditions. *Ber. Bunsenges. Phys. Chem.*, **101**, 1019.

11 Boresch, S. and Steinhauser, O. (1999) Rationalizing the effects of modified electrostatic interactions in computer simulations: the dielectric self-consistent field method. *J. Chem. Phys.*, **111**, 8271.

12 Boresch, S., Willensdorfer, M., and Steinhauser, O. (1994) A molecular dynamics study of the dielectric properties of aqueous solutions of alanine and alanine dipeptide. *J. Chem. Phys.*, **120**, 3333.

13 Böttcher, C.J.F. and Bordewijk, P. (1978) *Theory of Electric Polarization*, vol. 2, Elsevier, Amsterdam.

14 Brandt, S. and Dahmen, H.D. (2005) *Elektrodynamik*, vol. 4, Springer, Berlin.
15 Caillol, J.M. (1992) Asymptotic behavior of the pair-correlation function of a polar liquid. *J. Chem. Phys.*, **96**, 7039.
16 Canongia Lopes, J.N.A. and Padua, A.A.H. (2006) *J. Phys. Chem.*, **110**, 3330.
17 Chandra, A., Wei, D., and Patey, G.N. (1992) Dielectric relaxation of electrolyte solutions: is there really a kinetic dielectric decrement. *J. Chem. Phys.*, **98**, 4959.
18 Dachwitz, E., Parak, F., and Stockhausen, M. (1989) On the dielectric relaxation of aqueous myoglobin solutions. *Ber. Bunsenges. Phys. Chem.*, **93**, 1454.
19 Daguenet, C., Dyson, P.J., Krossing, I., Oleinikova, A., Slattery, J., Wakai, C., and Weingärtner, H. (2006) Dielectric response of imidazolium-based room-temperature ionic liquids. *J. Phys. Chem. B*, **110**, 12682.
20 Darden, T., York, D., and Pedersen, L. (1993) Particle mesh Ewald: an $n \cdot \log(n)$ method for Ewald sums in large systems. *J. Chem. Phys.*, **98**, 10089.
21 Debye, P. (1954) *Polar Molecules*, Chemical Catalog, Dover, p. 72.
22 di Biasio, A., Ambrosone, L., and Cametti, C. (2009) Numerical simulation of dielectric spectra of aqueous suspensions of non-spheroidal differently shaped biological cells. *J. Phys. D: Appl. Phys.*, **42**, 025401.
23 Ferguson, R., Arrighi, V., McEven, I.J., Gagliardi, S., and Triolo, A. (2006) An improved algorithm for the Fourier integral of the KWW function and its application to neutro scattering and dielectric data. *J. Macromol. Sci. Part B*, **45**, 1065.
24 Fogolari, F., Brigo, A., and Molinari, H. (2002) The Poisson–Boltzmann equation for biomolecular electrostatics: a tool for structural biology. *J. Mol. Recognit*, **15** (6), 377.
25 Halle, B., Johannesson, H., and Venu, K. (1998) Model-free analysis of stretched relaxation dispersions. *J. Magnet. Reson.*, **135**, 1.
26 Hansen, J.P. and McDonald, I.R. (2006) *Theory of Simple Liquids*, 3rd edn, Elsevier.
27 Hilfer, R. (2002) H-function representations for stretched exponential relaxation and non-Debye susceptibilities in glassy systems. *Phys. Rev. E*, **65**, 061510.
28 Höchtl, P., Boresch, S., Bitomsky, W., and Steinhauser, O. (1998) Rationalization of the dielectric properties of common three-site water models in terms of their force field parameters. *J. Chem. Phys.*, **109**, 4927.
29 Ikeda, M., Nakazato, K., Mizuta, H., Green, M., Hasko, D., and Ahmed, H. (2003) Frequency-dependent electrical characteristics of DNA using molecular dynamics simulation. *Nanotechnology*, **14**, 123.
30 Jackson, J.D. (1975) *Classical Electrodynamics*, John Wiley & Sons, Inc., New York.
31 Kirkwood, J.G. (1936) On the theory of dielectric polarization. *J. Chem. Phys.*, **4**, 592.
32 Knocks, A. and Weingärtner, H. (2001) The dielectric spectrum of ubiquitin in aqueous solution. *J. Phys. Chem. B*, **105**, 3635.
33 Kobrak, M.N. and Sandalow, N. (2004) *Molten Salts XIV*, Electrochemical Society, Pennington, NJ.
34 Löffler, G., Schreiber, H., and Steinhauser, O. (1997) Calculation of the dielectric properties of a protein and its solvent: theory and a case study. *J. Mol. Biol.*, **270**, 520.
35 Lorentz, H.A. (1909) *The Theory of Electrons*, Teubner, Leipzig.
36 Murarka, R.K. and Head-Gordon, T. (2008) Dielectric relaxation of aqueous solutions of hydrophilic versus amphiphilic peptides. *J. Phys. Chem. B*, **112**, 179.
37 Neumann, M. and Steinhauser, O. (1983) On the calculation of the frequency-dependent dielectric constant in computer simulations. *Chem. Phys. Lett.*, **102**, 508.
38 Neumann, M. and Steinhauser, O. (1984) Computer simulation and the dielectric constant of polarizable polar systems. *Chem. Phys. Lett.*, **106**, 563.
39 Neumann, M. and Steinhauser, O. (1984) On the calculation of the dielectric constant using the Ewald–Kornfeld tensor. *Chem. Phys. Lett.*, **95**, 417.
40 Neumann, M. (1983) Dipole moment fluctuation formulas in computer simulations of polar systems. *Mol. Phys.*, **50**, 841.

41 Neumann, M. (1985) The dielectric constant of water. Computer simulations with the MCY potential. *J. Chem. Phys.*, **82**, 5663.

42 Neumann, M. (1986) Dielectric relaxation in water. Computer simulations with the TIP4P potential. *J. Chem. Phys.*, **85**, 1567.

43 Okabe, A. (2000) *Spatial Tesselations: Concepts and Applications of Voronoi Diagrams*, John Wiley & Sons, Inc., New York.

44 Pićalek, J. and Kolafa, J. (2007) Molecular dynamics study of conductivity of ionic liquids: the Kohlrausch law. *J. Mol. Liq.*, **134**, 29.

45 Pitera, J.W., Falta, M., and van Gunsteren, W.F. (2001) Dielectric properties of proteins from simulation: the effects of solvents, ligands, pH and temperature. *Biophys. J.*, **80**, 2546.

46 Prodan, C. and Prodan, E. (1999) The dielectric behaviour of living cell suspensions. *J. Phys. D: Appl. Phys.*, **32**, 335.

47 Rey-Castro, C. and Vega, L.F. (2006) Transport properties of the ionic liquid 1-ethyl-3-methyl-imidazolium chloride from equilibrium molecular dynamics simulation: the effect of temperature. *J. Phys. Chem. B*, **110**, 14426.

48 Rotenberg, B., Difrêche, J.-F., and Turq, P. (2005) Frequency-dependent dielectric permittivity of salt-free charged lamellar systems. *J. Chem. Phys.*, **123**, 154902.

49 Rudas, T., Schröder, C., Boresch, S., and Steinhauser, O. (2006) Simulation studies of the protein–water interface. II. Properties at the mesoscopic resolution. *J. Chem. Phys.*, **124**, 234908.

50 Schreiber, H. and Steinhauser, O. (1992) Cutoff size does strongly influence molecular dynamics results on solvated polypeptides. *Biochemistry*, **31**, 5856.

51 Schreiber, H. and Steinhauser, O. (1992) Molecular dynamics studies of solvated polypeptides: why the cutoff scheme does not work. *Chem. Phys.*, **168**, 75.

52 Schreiber, H. and Steinhauser, O. (1992) Taming cut-off induced artifacts in molecular dynamics studies of solvated polypeptides: the reaction field method. *J. Mol. Biol.*, **228**, 909.

53 Schröder, C., Haberler, M., and Steinhauser, O. (2008) On the computation and contribution of conductivity in molecular ionic liquids. *J. Chem. Phys.*, **128**, 134501.

54 Schröder, C., Hunger, J., Stoppa, A., Buchner, R., and Steinhauser, O. (2008) On the collective network of ionic liquid/water mixtures. II. Decomposition and interpretation of dielectric spectra. *J. Chem. Phys.*, **129**, 184501.

55 Schröder, C., Rudas, T., Neumayr, G., Gansterer, W., and Steinhauser, O. (2007) Impact of anisotropy on the structure and dynamics of ionic liquids: a computational study of 1-butyl-3-methyl-imidazolium trifluoroacetate. *J. Chem. Phys.*, **127**, 044505.

56 Schröder, C. and Steinhauser, O. (2008) The influence of electrostatic forces on the structure and dynamics of molecular ionic liquids. *J. Chem. Phys.*, **128**, 224503.

57 Schröder, C., Rudas, T., and Steinhauser, O. (2006) Simulation studies of ionic liquids: orientational correlations and static dielectric properties. *J. Chem. Phys.*, **125**, 244506.

58 Schröder, C. and Steinhauser, O. (2009) On the dielectric conductivity of molecular ionic liquids. *J. Chem. Phys.*, **131**, 114504.

59 Schröder, C., Wakai, C., Weingärtner, H., and Steinhauser, O. (2007) Collective rotational dynamics in ionic liquids: a computational and experimental study of 1-butyl-3-methyl-imidazolium tetrafluoroborate. *J. Chem. Phys.*, **126**, 084511.

60 Smith, P.E. and Pettitt, B.M. (1991) Peptides in ionic solutions: a comparison of the Ewald and switching function techniques. *J. Chem. Phys.*, **95**, 8430.

61 Steinhauser, O. (1983) On the dielectric theory and computer simulation of water. *Chem. Phys.*, **79**, 465.

62 Steinhauser, O. (1983) On the orientational structure and dielectric properties of water. *Ber. Bunsenges. Phys. Chem.*, **87**, 128.

63 Stern, H.A. and Feller, S.E. (2003) Calculation of the dielectric permittivity profile for a nonuniform system: application to a lipid bilayer simulations. *J. Chem. Phys.*, **118**, 3401.

64 Suresh, S.J. and Satish, A.V. (2006) Influence of electric field on the hydrogen bond network of water. *J. Chem. Phys.*, **124**, 074506.

65 Thompson, K.E. (2002) Fast and robust Delaunay tessellation in periodic domains. *Int. J. Numer. Methods Eng.*, **55**, 1345.

66 Triolo, A., Russina, O., Bleif, H.J., and Di Cola, E. (2007) Nanoscale segregation in room-temperature ionic liquids. *J. Phys. Chem. B*, **111** (18), 4641.

67 Wakai, C., Oleinikova, A., Ott, M., and Weingärtner, H. (2005) How polar are ionic liquids? Determination of the static dielectric constant of an imidazolium-based ionic liquid by microwave dielectric spectroscopy. *J. Phys. Chem. B*, **109**, 17028.

68 Williams, G. and Watts, D. (1970) Non-symmetrical dielectric relaxation behaviour arising from a simple empirical decay function. *Trans. Faraday Soc.*, **66**, 80.

69 Yang, L., Valdeavella, C.V., Blatt, H.D., and Pettitt, B.M. (1996) Salt effects on peptide conformers: a dielectric study of tuftsin. *Biophys. J.*, **71**, 3022.

70 Yang, L., Weerasinghe, S., Smith, P.E., and Pettitt, B.M. (1995) Dielectric response of triplex DNA in ionic solution from simulations. *Biophys. J.*, **69** (4), 1519.

71 Yeh, I. and Berkowitz, M.L. (1999) Dielectric constant of water at high electric fields: molecular dynamics study. *J. Chem. Phys.*, **110**, 7935.

11
Computational Spectroscopy in Environmental Chemistry
James D. Kubicki and Karl T. Mueller

11.1
Introduction

11.1.1
Need for Computational Spectroscopy

11.1.1.1 Speciation

Early in the development of environmental chemistry and toxicology, chemicals such as metals and organic contaminants were analyzed and regulated according to their total concentrations. Although this approach made sense in the absence of previous studies on the complex interactions of contaminants with other compounds present in the environment, researchers observed significant deviations between toxicities predicted based on total concentrations and actual effects in the real environment. For example, aqueous Cu concentrations in the laboratory were found to have deleterious effects on organisms at extremely low concentrations (e.g., 10^{-11} M), but the same organisms were found to suffer no toxic effects due to Cu at concentrations that were orders of magnitude higher in nature [1]. The realization that certain chemical species may be responsible for toxic effects, in this case "free" Cu^{2+} ions, and that complexation with naturally occurring organic matter could dramatically decrease the bioavailability and toxicity of contaminants both diminished the risk due to exposure and complicated the process of predicting risk. Environmental scientists could no longer simply collect a water sample and perform a quantitative analysis, such as atomic absorption spectroscopy, to judge whether the biota living in the water were at risk. Instead, matrix effects needed to be considered as the level of dissolved organic matter (DOM) could diminish environmental impacts of the total contaminant concentration [2].

The need to quantify speciation effects led environment chemists to spectroscopic analysis of their samples in order to determine the chemical form of a given contaminant in the environment. Generally, spectroscopic methods are successfully applied to determine chemical structures of compounds. In environmental chemistry, however, the process of collecting and interpreting spectra is complicated by several factors. First, the concentration levels at which a chemical is considered toxic

Computational Spectroscopy: Methods, Experiments and Applications. Edited by Jörg Grunenberg
Copyright © 2010 WILEY-VCH Verlag GmbH & Co. KGaA, Weinheim
ISBN: 978-3-527-32649-5

may be orders of magnitude lower than the detection limits of available spectroscopic techniques. To overcome this problem, chemicals of interest were studied at concentrations above the detection limit of the spectroscopic technique, and it was then assumed that the speciation determined under high concentrations (and/or low pH where solubilities were considerably higher) was applicable to lower concentration conditions in the environment [3]. More sensitive techniques have shown that this extrapolation can be problematic [4]. In addition, interpretation of spectra was commonly guided by previous potentiometric work and thermodynamic databases that rely on model speciation diagrams [5], but these thermodynamic models are generally not sensitive enough to distinguish among species of the same stoichiometry, especially under low concentration conditions [6]. For example, Mueller et al. [7] have demonstrated that aqueous U speciation at lower concentrations more typical of environmental conditions deviates significantly from the speciation predicted by standard thermodynamic models.

Second, naturally occurring compounds and phases may be complex and disordered. Thus, it is difficult to find simple analogue models with which to interpret spectra. Naturally occurring organic matter, such as humic acids, and many solid phases (e.g., nanoparticulate ferric hydroxides) have considerable ranges in structure and composition and hence spectroscopic signatures and reactivities. Consequently, there are examples in the literature where different groups have applied spectroscopic techniques to the same problem and concluded different speciation models. For example, the seemingly simple problem of phosphate adsorption onto the mineral goethite (α-FeOOH) has been the subject of numerous papers and is still debated today [8–12]. Thus, an objective method for analyzing the accuracy of a given speciation model is extremely useful in environmental chemistry. Computational chemistry provides such a tool because one can build the various hypothesized models and predict spectroscopic parameters based on each model independent of empirical data (i.e., no fitting to observed spectra).

11.1.1.2 Surface Reactions

The roles of solid surfaces in environmental chemistry are particularly important. Adsorption, catalysis, and redox reactions all can affect the fate and transport of contaminants (both organic and inorganic) in natural systems. As mentioned above, not all naturally occurring solids have a well-defined crystal structure. Amorphous, poorly crystalline, and nanocrystalline structures are all common and often have greater reactivities than well-crystallized minerals, so they play a disproportionate role in environmental chemistry [13, 14]. Even when a mineral has a well-defined crystal structure, several crystal faces may be exposed in varying proportions depending on the crystal growth kinetics in a given system. Because each face of a mineral may have different functional groups exposed, the reactivity of each face toward a given compound varies. One example is phyllosilicate minerals such as clay or mica. These platy materials are dominantly Si, Al oxides, but their crystal form results in a basal (001) surface that can be dominated by Si—O—Si (siloxane) groups that are hydrophobic. The edge surfaces, (100) and (010), are terminated by Si—OH (silanol) and Al—OH (aluminol) groups that are more hydrophilic with exchangeable OH^- ligands [15, 16]. This leads to the fact that hydrophobic organic contaminants, such as polychlorinated biphenyls (PCBs) and polycyclic aromatic hydrocarbons (PAHs), would be more strongly adsorbed onto the basal surfaces, whereas oxyanions, such as arsenate, are more likely to bond to edge surfaces. This issue is well described in the recent work of Villalobos and coworkers who demonstrated that nucleation and growth kinetics of goethite lead to changes in the relative surface areas of various faces and that the faces terminating the mineral are more reactive than the prismatic faces [17, 18]. Because these terminated faces comprise a minor proportion of the total mineral surface area, the ability to normalize adsorption based on total surface area of a mineral is confounded.

In summary, the problem has been to describe adsorption in the field when experimental studies have measured adsorption isotherms based on a total surface area without information on the specific reactions occurring. The most common method of predicting adsorption thermodynamics has been surface complexation modeling. These approaches previously fit potentiometric and adsorption isotherm data with assumed surface complexes while adjusting model parameters such as the mineral-water interface capacitance. A number of issues hindered predictive capabilities: fitting to high surface coverages while environmental concentrations are typically much lower [19], assuming specific surface complex stoichiometries, and lack of competitive effects in most cases (e.g., carbonate and natural organic matter competing with arsenate for adsorption sites).

These issues have been addressed recently by combining spectroscopy, computational chemistry, and surface complexation modeling. One particularly successful model has been the MUlti-SIte Complexation (MUSIC) model [20]. MUSIC uses surface bonding and H-bonding structures to predict the pK_a values of titratable surface functional groups, which in turn are used to predict adsorption isotherms. Thus, the focus is on particular sites with specific concentrations on a given surface rather than on a total surface area. By collecting spectroscopic data in a given surface, interpreting the spectra with computational chemistry, estimating surface bond distances, and constraining the surface speciation before using the MUSIC model, excellent predictions can be made about adsorption behavior [21–24]. Other models are being developed as well [25], but the combination of the above three techniques is the key to linking molecular level information with field-scale behavior of contaminants.

11.1.2
Types of Spectra Calculated

11.1.2.1 IR/Raman
Vibrational spectroscopy is a commonly used technique to study aqueous solutions and surfaces in environmental chemistry. Infrared, especially attenuated total reflectance Fourier-transform infrared (ATR FTIR), and Raman spectroscopies reflect bond (including H-bond) energies because the vibrations of molecules are directly related to their bond strengths. Thus, as a compound forms an aqueous or

surface complex, the bond strengths change and are reflected in the collected vibrational spectra. Interpretation of these changes is not straightforward because one can have frequency shifts due to a number of possible configuration changes. In addition, as the frequencies change, the IR intensities can also change, complicating spectral interpretation without an estimate of how both frequency and intensity should shift for a given complex.

Computational chemistry addresses both of these spectral parameters, so for a given proposed complex, one can calculate the spectra of the reactants and products to see whether the original frequencies and the frequency shifts in the model match observation. Typically, frequencies are computed more accurately than intensities because frequencies are a function of the square root of the force constant, whereas IR and Raman intensities are higher power functions of the dipole moment and polarizability derivatives, respectively. However, quantum (QM) methods are capable of identifying the relative IR and Raman intensities effectively enough to be useful in assigning structures to vibrational spectra. Although frequencies can be calculated to typically within $\pm 10\,\text{cm}^{-1}$ for the type of complexes studied in environmental chemistry [26–29], frequencies are generally scaled by a predetermined factor to account for basis set, electron correlation, and anharmonic effects [30, 31]. Larger basis sets can lead to relatively small corrections ($\approx 2\%$), so the accuracy of frequency prediction is limited by the available computational resources for a problem. Less precise calculations are still valuable in many instances, however, because the differences in correlations between observed and calculated frequencies can be larger than the effects mentioned above when one model is correct and another is incorrect.

The vibrational frequencies are calculated analytically with the following equations:

$$V_{ij} = \frac{1}{\sqrt{m_i m_j}} \left(\frac{\partial^2 V}{\partial q_i \partial q_j} \right) \quad (11.1)$$

where V_{ij} is the Hessian matrix, m_i refers to the mass of atom i, and ∂q_i refers to a displacement of atom i in the x-, y-, or z-direction,

$$VU = \lambda U \quad (11.2)$$

where U is a matrix of eigenvectors and λ is a vector of eigenvalues, and

$$\lambda_k = (2\pi v_k)^2 \quad (11.3)$$

where λ_k is the kth eigenvalue and v_k is the kth vibrational frequency. The vibrational modes calculated are commonly more complex than assignments made based on observed spectra because more atoms are involved in the computed eigenvectors than typical assignments based on single functional groups. This can complicate comparisons of observed and calculated vibrational spectra. For example, H_2O molecules solvating a molecule can be vibrationally coupled to the solute's vibrations especially if H-bonding is strong [32]. Thus, there can be questions as to what is responsible for changes in the observed spectra with addition of a solute. Is it just the vibration of the compound in question or do changes in the solvent vibrational spectra contribute to the "background-corrected" spectra between the solution of interest and the background solution?

Infrared intensities can be computed with the equation [33]

$$(\partial E_{SCF}/\partial f \partial a) = 2 \sum_i^{d.o.} h_{ii}^{fa} + 4 \sum_i^{d.o.} \sum_j^{all} U_{ji}^a h_{ij}^f \quad (11.4)$$

where

$$h_{ii}^{fa} = \sum_{\mu\nu}^{AO} C_\mu^{i0} C_\nu^{i0} \left(\frac{\partial^2 h_{\mu\nu}}{\partial f \partial a}\right) \quad (11.5)$$

E_{SCF} is the self-consistent field energy, f is the electric field, a is a nuclear coordinate, $h_{\mu\nu}$ is the one-electron atomic orbital integral, U^a is related to the derivative of the molecular orbital coefficients with respect to a by

$$\left(\frac{\partial C_\mu^i}{\partial a}\right) = \sum_m^{all} U_{mi}^a C_\mu^{m0} \quad (11.6)$$

The term "all" in the above summations refers to all occupied and virtual molecular orbitals and "d.o." refers to doubly occupied orbitals such as those found in the ground state of a closed-shell system. Terms such as C_μ^{i0} refer to the coefficients of the atomic orbital m in the ith unperturbed molecular orbital. (See Ref. [33] for a derivation and further explanation).

Raman intensities can be found estimated based on [34]

$$\frac{\partial \alpha_{fg}}{\partial x} = \frac{\partial^3 E}{\partial f \partial g \partial x} = \left\langle \frac{\partial D^f}{\partial x} \frac{\partial P}{\partial g} \right\rangle$$

$$= \left\langle \frac{\partial P}{\partial f} G^x \left(\frac{\partial P}{\partial g}\right) \right\rangle + \left\langle \frac{\partial^2 P}{\partial f \partial g} G^x (P^0) \right\rangle - \left\langle \frac{\partial S}{\partial x} \frac{\partial^2 W}{\partial f \partial g} \right\rangle \quad (11.7)$$

$$+ \left\langle \frac{\partial^2 P}{\partial f \partial g} \frac{\partial h}{\partial x} \right\rangle + \left\langle \frac{\partial P}{\partial f} \frac{\partial D^g}{\partial x} \right\rangle$$

where f and g are directions of the electric field, "$\langle \cdots \rangle$" is the trace of a matrix, D^f is a dipole integral, P is the electron density matrix, G^x is a contraction of the integral derivatives, S is the overlap matrix, W is the energy-weighted density matrix, and h is the Hamiltonian for the atomic core (see Hehre et al. [35], for a detailed explanation of each of these terms).

Both IR and Raman calculated intensities are generally less accurate than the computed frequencies, but the relative estimates of IR and Raman intensities are usually reliable enough to identify which peaks will be observed. Although this chapter will not discuss isotopic fractionations, a number of recent papers have been using calculated vibrational frequencies to predict isotopic fractionation factors that are difficult to measure experimentally [36–38].

11.1.2.2 NMR

Nuclear magnetic resonance (NMR) spectroscopy has been utilized less than a number of other analytical tools in environmental chemistry because it is not a sensitive technique in most cases. However, excellent work has been produced on organic [39], inorganic [40–43], and metal–organic complexation [3, 44, 45]. Because NMR is an element-specific technique, one is allowed to see into the bonding environment surrounding a selected element within this system. Cross-polarization techniques allow one to further target pairs of elements close in space, which is especially helpful when studying surface reactions because surface atoms near to H species that are localized on the surface can be readily distinguished from atoms within the nonprotonated bulk [46]. Two- and three-dimensional techniques allow further refinement of structures, which is particularly important for determining structures of organic molecules relevant to environmental chemistry [47–49].

The most typical NMR parameter calculated is the chemical shift of a given nucleus within the sample of interest. Experimentally, one measures the chemical shielding of the element within the sample referenced to the chemical shielding of a standard reference material and the difference is the chemical shift. Hence, calculating the chemical shift is accomplished in a similar manner – create models of the standard reference and the unknown for the nucleus of interest, calculate the isotropic chemical shielding of each, and then subtract the unknown chemical shielding from the reference compound chemical shielding to obtain the isotropic chemical shift [50]. The gauge-including atomic orbital (GIAO) method ([51] and references therein) is most commonly employed and has been successful for a number of elements in environmental chemistry studies [52, 53].

The basic equation for calculating NMR shielding with the gauge-including atomic orbital method [54] is

$$\sigma_{\alpha\beta}^{N} = \sum_{\mu\nu} D_{\mu\nu} \left[\frac{\delta^2 h_{\mu\nu}}{\delta B_\alpha \delta m_{N\beta}} \right] + \sum_{\mu\nu} \frac{\delta D_{\mu\nu}}{\delta B_\alpha} \times \frac{\delta h_{\mu\nu}}{\delta m_{N\beta}} \quad (11.8)$$

where $\sigma_{\alpha\beta}^{N}$ is the $\alpha\beta$-tensor component of the derivative of the molecular energy for α-nucleus N, D represents the one-electron density matrices, B_α is the magnetic field in the α direction, h is a one-electron Hamiltonian, and m is the magnetic moment of a nucleus. The chemical shieldings of nuclei of interest are then calculated and subtracted from the shielding in a reference compound to give the chemical shift:

$$\delta_{\alpha\beta} = \sigma_{\alpha\beta}(\text{ref}) - \sigma_{\alpha\beta} \quad (11.9)$$

Broad NMR spectra from molecules in the solid state arise from a number of orientation-dependent internal spin interactions, including the well-known anisotropy of the chemical shift. This anisotropy is reflected in the fact that the chemical shift is represented by a tensor rather than a scalar quantity. The isotropic chemical shift observed in the liquid-state NMR arises from the spatial averaging of the full chemical shift tensor, and is equivalent to one-third of the trace of the tensor (which is independent of the axis system chosen to represent the tensor). Under magic-angle spinning (MAS) conditions, the so-called powder pattern arising from the study of a disordered solid breaks up into a series of spinning side bands (SSBs) accompanied by a peak at the isotropic chemical shift. The chemical shift tensor parameters in the principal axis system (that where the tensor is diagonal) can be extracted from the magic-angle spinning spectrum via the method developed by Herzfeld and Berger [55]. This analysis relies on an accurate measurement of the intensity and frequency of the isotropic resonance and the corresponding SSBs, and the MAS spectrum therefore provides the chemical shift tensor components of the corresponding nucleus. These components are conveniently expressed in terms of the anisotropy ($\Delta\delta$) and asymmetry (η) parameters. Mathematically, for a chemical shift anisotropy (CSA) tensor with principal shift components (δ_{xx}, δ_{yy}, and δ_{zz}) defined so that

$$|\delta_{zz}-\delta_{iso}| \geq |\delta_{xx}-\delta_{iso}| \geq |\delta_{yy}-\delta_{iso}| \tag{11.10}$$

where

$$\delta_{iso} = (\delta_{xx}+\delta_{yy}+\delta_{zz})/3 \tag{11.11}$$

the anisotropy ($\Delta\delta$) is

$$\Delta\delta = \delta_{zz}-(\delta_{xx}+\delta_{yy})/2 \tag{11.12}$$

while the asymmetry (η) corresponds to

$$\eta = (\delta_{yy}-\delta_{xx})/(\delta_{zz}-\delta_{iso}) \tag{11.13}$$

While most of the NMR nuclides studied in environmental chemistry have nuclear spin quantum numbers of 1/2 (including ^1H, ^{13}C, ^{15}N, ^{19}F, ^{29}Si, and ^{31}P), other nuclei such as ^{17}O and ^{27}Al have spin greater than 1/2 and therefore are subject to additional internal interactions that can be anisotropic in the solid state, or lead to fast spin relaxation in the liquid state. In particular, resonances in the solid-state NMR spectra of ^{27}Al will exhibit quadrupolar broadening and an additional shift due to the quadrupolar interaction, which depends on the coupling of the nuclear electric quadrupole moment to local gradients of the electric field at the nucleus [56–58]. Most important, for chemical analysis, these effects will vary as a function of bonding environment, providing insight into local structure with isotopic selectivity. Fortunately, the effects of the isotropic shift components for the chemical shift and quadrupolar interaction are separable, because the chemical shift is field independent (in parts per million from the reference frequency), while the observed quadrupolar shifts for ^{27}Al depend inversely on the square of the strength of the magnetic field used in the experiments.

11.1.2.3 EXAFS + CTR + XSW
Extended X-ray absorption fine structure (EXAFS) spectroscopy, crystal truncation rod (CTR), and X-ray standing wave (XSW) techniques have all been used to

characterize mineral–water interface and surface adsorption complex structures (see Ref. [59] where all three are combined). The techniques have been able to provide detailed positions of atoms in a solid–water interface and test previously held assumptions about the structure of the electric double layer (see Ref. [60] for a review). Surface complexation modeling as mentioned above relies upon the theoretical picture of the mineral-water interface, so predicting macroscopic adsorption behavior in environmental systems depends upon having a clear picture of where H_2O and adsorbate molecules reside with respect to the surface.

Although these techniques have rapidly advanced our understanding of the mineral–water interface, CTR and XSW depend upon having a relatively large crystalline sample in order to scatter X-rays from an atomically flat surface. Thus, analysis of many types of environmental substrates such as nanoparticles or poorly crystalline solids is not possible. EXAFS is less dependent upon the substrate and can be used for aqueous species, but spectra can be difficult to interpret when more than one surface species exists, when the surface is disordered, and when a significant fraction occurs as outer-sphere (i.e., an adsorbate separated from the surface by a solvation shell of H_2O molecules) species. Furthermore, EXAFS does not see light elements (especially H), so the determining of the protonation state of the adsorbate and surface can be inhibited.

In this case, computational chemistry becomes a valuable tool for predicting mineral–water interface behavior. Parallel simulations can be performed on well-understood systems such as single faces of TiO_2 [59, 61–67] and benchmarked against high-quality EXAFS, CTR, and XSW data. Once the modeling techniques have proven accurate for these types of systems, one can more confidently perform simulations on substrates that are more difficult to analyze experimentally. In general, computational methods are currently passing through the former stage and beginning to work on the latter types of systems where obtaining and/or interpreting these kinds of analytical data are not possible.

11.1.2.4 QENS and INS

Neutron scattering techniques are not formally a type of spectroscopy, but they do probe similar dynamical behaviors as IR and Raman such as O—H bond stretching, H—O—H angle bending, and so on, as well as slower translational dynamics. Combined with CTR studies, these techniques can provide a detailed picture of the structure and dynamics of water at the interface with a solid [68]. Neutron scattering is highly sensitive to H-atoms, which makes it a good complement to the X-ray techniques that cannot detect light elements. Furthermore, substitution of D for H can dramatically alter the observed spectrum in order to gain further insight into the system of interest.

Because neutron scattering probes both the pico- and nanosecond time frames, both quantum and classical computational methods are applicable. For example, inelastic neutron scattering (INS) can detect changes in the O—H stretching and H—O—H angle bending modes of water interacting with a surface [69]. Density functional theory–molecular dynamics (DFT–MD) can then be used to model these same dynamics with simulations on the order of tens of picoseconds in order to assign the observed shifts to particular OH-bearing species [70]. For longer timescale diffusional dynamics, classical MD simulations can be performed on the order of tens or hundreds of nanoseconds that reproduce H_2O translational motions. Such simulations have proven useful in fitting observed quasi-elastic neutron scattering (QENS) spectra [71]. When the classical force field is based on DFT calculations [72], a self-consistent multiscale model that reproduces numerous experimental observables is the result [59]. Recent success on surfaces has led to a new research challenge into studying the behavior of water in nanopores with the goal of understanding how chemical reactions such as dissolution change with the development of etch pits (see Ref. [73] for a discussion of the importance of etch pit formation on mineral dissolution).

11.2 Methods

The techniques employed for computational spectroscopy in environmental chemistry are no different from other areas of computational chemistry described in this book and elsewhere. Consequently, this chapter will focus on discussing the pitfalls and solutions that are common in the complex chemistry of natural systems.

11.2.1 Model Building

The single most important aspect of applying computational chemistry techniques to environmental chemistry is to have an accurate model of the chemistry that actually occurs. One could solve the Schrodinger equation exactly for some simplified system, but if the model does not include a critical component of interest, then the results are of limited use to the environmental chemist. A realistic approximate model is generally much better than an unrealistic highly accurate model. Of course, high-level calculations are useful in their own way and this will be discussed below.

One important component that is commonly neglected is water. Since most environmental chemistry occurs in the presence of water, including H_2O molecules to explicitly solvate the reaction can be critical. Even in heterogeneous atmospheric reactions where bulk water is not present, water vapor can adsorb onto particles and alter chemical reactions significantly (see Ref. [74] for a recent review). Polarized continuum models (PCMs) of solvation are helpful in many instances, but the strong H-bonding to specific atoms within the system of interest is not reliably modeled through PCM methods. This seems like an obvious statement to make, but the literature is full of examples where explicit solvation is ignored in the interest of saving computational time. An example of the necessity for explicit solvation can be found in Bargar *et al.* [75]. IR spectra of carbonate adsorbed onto hematite (α-Fe_2O_3) were collected and various models of the carbonate surface complex constructed. These spectra were collected "dry" (i.e., in air rather than under solution), so the initial models used to search for surface complexes capable of giving rise to the observed spectra did not include H_2O molecules of hydration. One likely candidate was a monodentate binuclear configuration (i.e., two of the O atoms of the carbonate were bonded to one of the Fe surface atoms). The resulting IR frequencies of the carbonate group did not match the observed spectra, and the Δv between the $C-O_s$ and $C-O_{as}$ modes was 542 cm^{-1} compared to the observed Δv around 200 cm^{-1}. Addition of two H_2O molecules to this complex shifted the calculated frequencies into agreement with observation and the model Δv value decreased to 195 cm^{-1} in good agreement with observation. This is an extreme example because anionic species such as carbonate can form very strong H-bonds with H_2O such that the C–O vibrations are dramatically affected, but the principal lesson should be considered with most environmental chemistry reactions.

How does one determine the minimum number of necessary components to include in a model? Incorporating all possible components is not practical for most studies, so assumptions and approximations are necessary. There are two main reasons for the model results to be wrong: inappropriate computational methodology and neglect of an important component. To address the first, a test model should be selected that represents the quantity to be computed in the system of interest. This test model structure should be well known and have an accurate experimental value for the property of interest (e.g., vibrational spectra, NMR chemical shift, and enthalpy of adsorption). It should also represent the basic chemistry in the more complex system; attempting to model metal complexation with a complex biomolecule would not be wise if one could not model the metal of interest interacting with a simple organic compound. Once a reasonable test model is generated, increasing basis set size and levels of electron correlation can be employed to determine the most computationally efficient method for reproducing the experimental observable with the desired accuracy. Depending on the property to be reproduced, the level of theory can vary considerably as structural information and vibrational frequencies are easier to reproduce than NMR chemical shifts or reaction enthalpies.

Second, the computations should not be performed in a vacuum (sorry for the bad pun). Often discussions will arise concerning how you can prove a simulation result is at a global minimum. This is an exceedingly difficult task for most natural systems; and, if it needed to be answered solely on the basis of computational results, most studies would fall short of succeeding. However, experimental results are commonly available that constrain the possible choices, and comparison of model and observation can distinguish among several possibilities rather than attempting to calculate the properties of all possible configurations. A fortunate result of this is that high accuracy is not always necessary. Some may wince at the idea that ΔG is only accurate to ± 20 kJ/mol or an NMR chemical shift to within ± 5 ppm. Granted, neither of these would be sufficient accuracy to answer many questions, but if there are two proposed configurations for a complex that differ in calculated G by 100 kJ/mol or 20 ppm, then the absolute accuracy of the computational methodology is not as critical. When there are two or more properties for comparison, better relative agreement by one model over the others is even more comforting. Consequently, collaboration with experimental and analytical is imperative for defining the question to be answered *a priori* and for an iterative approach to solving the problem. This is by no means a foolproof solution, but it does build a higher level of confidence in the results, allows an estimation of what the size of the error is in the calculation, and the communication can bring a deeper understanding as what is clear from analytical work may be murky in the simulation and vice versa.

11.2.2
Selecting a Methodology

Details of computational chemistry methods are way beyond the scope of this chapter and are covered in the *Encyclopedia of Computational Chemistry* [76]. An excellent book that provides an overview of many methods is *Essentials of Computational Chemistry: Theories and Models* [77]. However, a few words about the selection of methods are presented here to help the novice get started. The first choice is whether to use a quantum or classical molecular mechanical (MM) approach. When necessary, these two can be combined in the QM/MM methodology. A number of considerations factor into this decision, but a couple of the more important are the size of the system and the process to be modeled. QM methods are inherently more computationally demanding, so systems of thousands of atoms are generally treated with MM. For the same reason, simulations that require a large number (i.e., millions) of steps such as a nanosecond molecular dynamics (MD) simulation tend toward MM rather than QM.

The composition and process to be modeled are, however, equally important. Force fields are typically parameterized to be accurate for certain compounds and physical states. If an appropriate force field is not available for the system of interest, then one should consider QM or creating/modifying a force field for the situation at hand. Furthermore, most force fields do not handle reactivity such that processes involving bond making and bond breaking cannot be simulated. An exception is the reactive force field – ReaxFF [78]. Hence, in some cases, the system size, duration of process, and the chemistry can conspire to make computational chemistry inappropriate.

Even when QM is possible, some types of spectroscopy such as electron spin resonance (ESR) and X-ray photoelectron spectroscopy (XPS) can require large basis sets and high levels of theory such that computational demands become impractical for the common user even for systems of moderate size. Situations that require multiconfigurational approaches such as the complete active space multiconfiguration self-consistent field (MC-SCF or CASSCF) method for examining excited states quickly become impractical for environmental chemistry. The time-dependent density functional theory (TD-DFT) approach can provide a relatively quick shortcut to calculate UV-visible spectra and can be accurate in some instances, but it is not universally applicable.

Once one has decided to use QM or MM, the type of calculation should be selected. For simple systems where one wishes to calculate vibrational (IR and Raman) or NMR parameters, energy minimizations should be performed first. This is especially true for vibrational frequencies because calculated frequencies are valid only at a potential energy minimum for a given methodology. For more complex systems, one may wish to perform MD or Monte Carlo (MC) simulations in order to explore the potential energy surface of the system before trying to find an energy minimum. Complex systems can have multiple minima and performing an energy minimization only on an initial guess for the structure could lead to a local rather than global minimum. MD simulations can also be used to predict vibrational frequencies directly from the velocity autocorrelation functions of the atoms [77].

The last topic discussed here with regard to methodology is the choice of using isolated molecular clusters versus a periodic system. Examples are discussed within this chapter where simple molecules can adequately represent the system of interest in an aqueous solution or on a mineral surface. For vibrational frequencies, simplifying the system to essential components can allow more accurate methodologies that quantitatively reproduce observed frequencies. However, one should test for system size effects and the role of solvation and compare known structures to experiment before concluding one as a realistic representation of the chemistry of interest. Periodic model systems often more realistically represent real systems, but calculations within periodic models may not allow estimation of IR or Raman intensities or they may use less accurate numerical methods for calculating frequencies. Recent developments in periodic calculations of NMR properties are promising, but have not been extensively applied in environmental chemistry to date.

11.3
Examples

11.3.1
IR/Raman Phosphate on Goethite

As mentioned above, vibrational spectroscopy has been used in environmental chemistry for decades, but the application of computational chemistry to help interpret these spectra is a relatively new enterprise [79]. One example where computational methods are particularly useful is the adsorption of phosphate onto the mineral goethite (α-FeOOH). Phosphate is a key nutrient in the environment, so its adsorption to mineral surfaces is an important factor in its transport and bioavailability. Fe-oxyhydroxide minerals tend to have a strong affinity for phosphate [80], and goethite is one of the most common Fe-oxyhydroxide minerals. Furthermore, adsorption of negatively charged oxyanions to a positively charged oxyhydroxide mineral surface can reverse the surface charge and affect the behavior of other solutes in pore waters [81]. For these reasons, numerous studies have been performed on the adsorption of phosphate to goethite ([9] and references therein) and many of these have been IR studies [8, 82–85].

At pHs around 4.2–4.5, observed IR frequencies are found in the ranges 939–945, 1001–1008, 1044–1049, and 1122–1123 cm^{-1} (first value from Ref. [8]; second from Luengo et al. [85]). Persson et al. [8] also observed peaks at 876 and 1178 cm^{-1} and

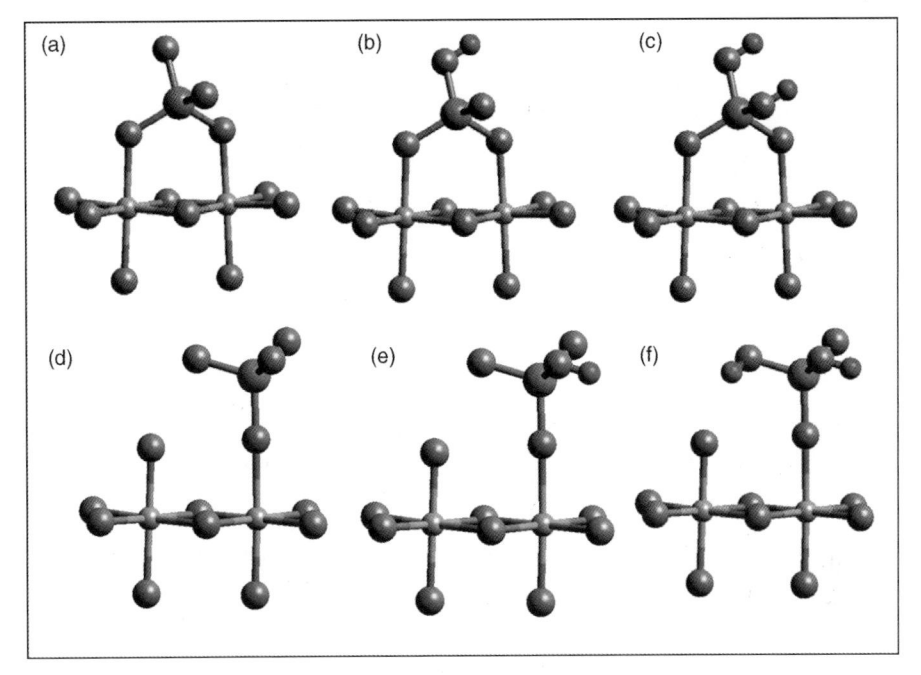

Figure 11.1 Possible surface complex structures of phosphate on goethite as suggested by Elzinga and Sparks [12].

Tejedor-Tejedor and Anderson [84] predicted a weak band at 982 cm^{-1}. Note the range of values from these observed spectra as the broadness of the IR bands and perhaps the variability of the goethite surfaces affect the collected spectra. Possible surface complex structures of phosphate on goethite as suggested by Elzinga and Sparks [12] are shown in Figure 11.1.

Kwon and Kubicki [10] used a simple cluster model approach using a Fe-hydroxide dimer model for the goethite surface and a small number of H_2O molecules to hydrate the phosphate group. All six models suggested by Elzinga and Sparks [12] were subjected to energy minimization using Gaussian 03 [86], the B3LYP density functional method [87, 88], and the 6-31G(d) basis set [89]. The doubly protonated bidentate configuration (Figure 11.1a) resulted in IR-active frequencies of 876, 940, 993, 1080, 1120, and 1178 cm^{-1} that correlate very strongly with the observed IR frequencies of Persson et al. [8]. The presence of the 876 and 1178 cm^{-1} peaks in the Persson et al. [8] spectra that were not reported in other studies are critical in this assignment because no other model predicts a phosphate-related IR-active vibration in the vicinity of 1178 cm^{-1}.

Under moderate pH conditions, it was not possible to distinguish between the deprotonated bidentate and the monoprotonated monodentate configurations based on correlations of observed and calculated frequencies. Estimation of adsorption energies to form these two complexes favored the monoprotonated monodentate

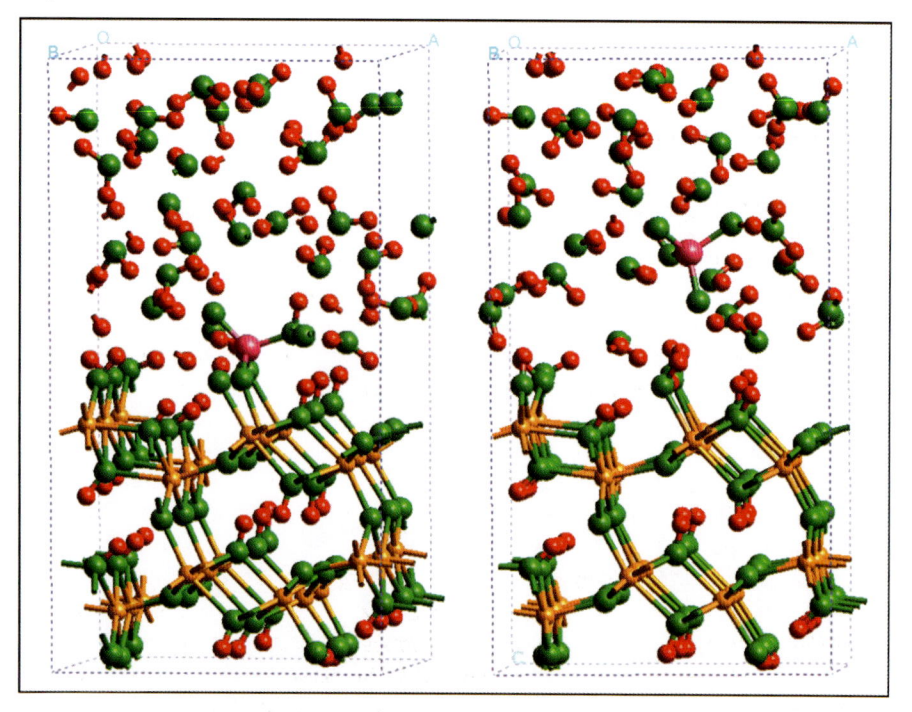

Figure 11.2 Periodic DFT MD simulations using VASP [90] of phosphate on goethite (010) surface. Energies of phosphate in water layer and adsorbed onto surface can be directly compared as the stoichiometry is the same in each model.

species, but this result is questionable due to the approximations used in calculating the reaction energies with a cluster approach. A more accurate method would be to use periodic DFT MD simulations to calculate the energy of the phosphate in solution and on the surface for various protonation states and surface complexes (Figure 11.2). Preliminary results for HPO_4^{2-} adsorption on the (010) surface (*Pbnm* space group) of α-FeOOH (goethite) suggest that the enthalpy of adsorption should be on the order of -200 kJ/mol for the bidentate configuration and that the monodentate configuration is tens of kJ/mol higher in energy.

For alkaline solution conditions, the deprotonated monodentate model resulted in the best match with observed frequencies 939 (944), 970 (973), and 1057 (1050) cm^{-1} (calculated values in parentheses). These assignments are not only consistent with observed IR spectra, they also make sense with the chemistry of increasing pH conditions. At low pH, the surface OH^- groups become protonated to form OH_2 groups that are better leaving groups, favoring a bidentate binuclear configuration such as Figure 11.1c. As pH increases to circumneutral values where OH^- groups dominate the surface, monodentate configurations can become more common because the metal–OH^- bonds are harder to break. Under alkaline conditions, the adsorbed HPO_4^{2-} deprotonates to form PO_4^{3-}.

11.3.2
Solution-State NMR of Al–Organic Complexes

The interaction of metals and hydrophobic organic contaminants with natural organic matter (NOM) is critical for two reasons. First, the solubility of these toxic components can be enhanced dramatically over the solubility in pure water due to associations with NOM. Second, interaction of toxic components with NOM can reduce the toxicological effects and bioavailability of the contaminants. Indeed, it is thought that organisms produce chelating agents to sequester metals such as copper in natural waters in order to survive. Recognition of these two phenomena has caused the US Environmental Protection Agency to re-evaluate how it handles allowable concentration limits of pollutants in natural waters. From the first observation above, it becomes exceedingly difficult to remediate natural waters to extremely low levels because of dissolved NOM such as fulvic and humic acids (i.e., products of partial degradation of biological compounds). However, due to the lowered environmental risk associated with pollutants associated with NOM, there is not as great a need to lower contaminant levels below threshold levels found in laboratory studies utilizing solutions that did not contain these complexing agents.

Fulvic and humic acids mentioned above are not specific compounds. Because they are derived from a variety of metabolic processes on a variety of precursor materials and are affected by the minerals, climate, and other factors, dissolved NOM exhibits extreme complexity. One approach to dealing with this complexity has been to study metal–NOM interactions with various functional groups that are common in NOM that are likely to interact strongly with various metals. For example, carboxylate, sulfhydryl groups, and heterocyclic N-atoms can bond to different metals, and the assumption is that the short-range chemistry of model compounds containing these functional groups will reflect the bonding that occurs in the fulvic and humic acids containing the same functional groups. One common functionality thought to be responsible for binding important metals such as Al^{3+} and Fe^{3+} is the carboxylate + phenol moiety as found in salicylic acid ($C_6H_4OHCOOH$). Thus, ^{27}Al NMR spectra of Al–salicylate solutions were collected by Thomas et al. [3] in order to determine the structure of possible Al–NOM complexes. A $\delta^{27}Al$ value of approximately 3 ppm was observed and assigned to the bidentate salicylate–Al^{3+} complex (Figure 11.3a) based on a thermodynamic speciation model of Rakotonarivo et al. [91].

Kubicki et al. [92] created molecular models of the various possible complexes, performed energy minimizations and frequency calculations, and then calculated ^{27}Al chemical shifts relative to a model standard $Al^{3+} \cdot 6(H_2O)$. Both the monodentate $[C_6H_4OHCOO^-]Al^{3+} \cdot 5(H_2O)$ and the bidentate $[C_6H_4OCOO^{2-}]Al^{3+} \cdot 4(H_2O)$ complex resulted in predicted $\delta^{27}Al$ values consistent with observation within computational accuracy (i.e., 6 and 7 ppm, respectively). However, the latter complex with the phenol group protonated resulted in much poorer correlations to the observed vibrational frequencies of this complex [93]. Consequently, it was proposed that the monodentate configuration was the most probable at the low pH of the observed

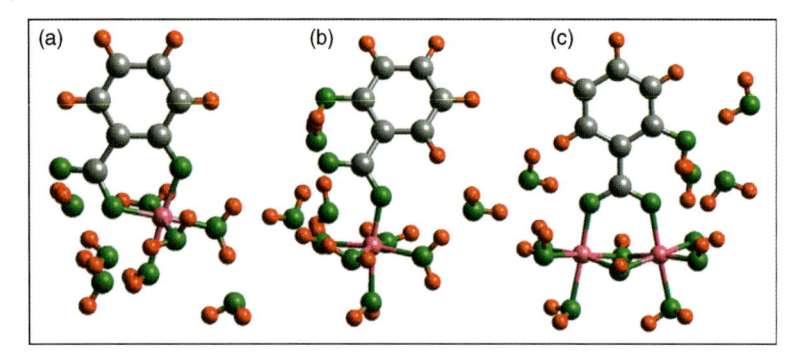

Figure 11.3 Possible Al–salicylic complexes in aqueous solution are represented by (left to right) bidentate (or bidentate mononuclear), monodentate, and bidentate bridging (or bidentate binuclear) models.

NMR spectra even though a bidentate configuration could form at higher pH where the phenol group could deprotonate if salicylate were complexed with Al^{3+} [94]. This example reiterates the points made above that experimental conditions necessary for spectroscopic analysis are not always consistent with environmental conditions and that multiple spectroscopic techniques should be modeled simultaneously in order to distinguish among possible models and verify the accuracy of the theoretical prediction.

In this chapter, this issue is revisited. With the last decade of advances in computational efficiency, it is possible to use larger basis sets (i.e., 6-311G(2d,2p) versus 3-21G(d,p)) and explicit hydration with H_2O molecules to predict the structures of the aqueous Al–salicylate complexes. In addition, the B3LYP [87, 88] density functional methods are included in both the energy minimization and NMR calculations. Although the current methods should be much more accurate than those employed in Kubicki et al. [92], the results for the calculated $\delta^{27}Al$ values are similar. For example, the previous $\delta^{27}Al$ values for the monodentate and bidentate complexes were 6 and 16 ppm, respectively. B3LYP/6-311G(2d,2p) values are 4 and 13 ppm, respectively. Hence, the monodentate configuration is still consistent with the observed value. In addition, a mononuclear bidentate complex with the phenol still protonated (Figure 11.3a with a H^+ on the phenol group connected to the Al^{3+}) results in a ^{27}Al chemical shift of 3 ppm and a binuclear bidentate configuration (Figure 11.3c) has a 6 ppm shift. Three potential complexes are possible by comparing the observed and calculated $\delta^{27}Al$ values.

To resolve this issue, the $\delta^{13}C$ chemical shifts were also calculated relative to tetramethylsilane (TMS) and compared to experimental values. The B3LYP/6-311G(2d,2p) method was chosen for these new calculations because Anandan et al. [95] have shown that this method results in accurate $\delta^{13}C$ values for salicylic acid. As a test, $\delta^{13}C$ values for salicylic acid were taken from the Biological Magnetic Data Bank (www.bmrb.wisc.edu) and the calculated values were correlated against them. Using salicylic acid in the configuration with the phenol OH H-bonded to the O atom in the C=O group, the best correlation with observed values was found. The R^2 value is 0.992 with a slope of 1.07 and intercept of −6.3 ppm compared to ideal values of 1.000, 1.00, and 0.0 ppm. The standard deviation is only ±2.1 ppm. These results are fairly accurate, so the same methodology was applied to predict $\delta^{13}C$ values for the Al–salicylate complex.

Of the three Al–salicylate complexes that gave $\delta^{27}Al$ values reasonably close to observation, the monodentate complex resulted in the best correlation with observed $\delta^{13}C$ values. For example, the monodentate linear fit parameters were $R^2 = 0.964$, slope = 1.04, intercept = −0.44 ppm and standard deviation of ±5.1 ppm. In comparison, the bidentate, phenol-protonated complex gave fit parameters of $R^2 = 0.850$, slope = 0.76, intercept = 35.5 ppm, and standard deviation of ±8.2 ppm. The monodentate match to observation is superior for all parameters. Combined with the superior correlation of the observed and calculated vibrational frequencies at the B3LYP/6-311G(2d,2p) level (i.e., monodentate $R^2 = 0.996$, slope = 1.02, intercept = −36 cm^{-1}, and standard deviation of ±15 cm^{-1} versus bidentate $R^2 = 0.994$, slope = 0.96, intercept = 62 cm^{-1}, and standard deviation of ±18 cm^{-1}), the model results are most consistent with a monodentate complex at pH 3.

11.3.3
Solid-State NMR of Phosphate Binding on Alumina

Understanding the interactions of nucleic acids and related compounds with metal oxides is important for describing many agricultural, environmental, and geochemical problems. In a combined approach aimed at measuring bonding interactions of nucleic acids at the molecular scale, Fry et al. [96] combined the solid-state MAS NMR measurements of the principal components of the full chemical shift anisotropy tensor with *ab initio* calculations. They interrogated the structure of the bound phosphate group of the mononucleotide 2′-deoxyadenosine 5′-monophosphate (dAMP) interacting with the octahedrally coordinated aluminum species on the surface of a mesoporous alumina sample. dAMP forms an inner-sphere complex at the alumina surface that can bind in either a monodentate or bidentate configuration. After using rotational-echo double resonance (REDOR) NMR methods [97] to confirm that the phosphate group is predominantly interacting with aluminum species in octahedral coordination on the surface, a ^{31}P magic-angle spinning NMR spectrum was acquired. The spectrum, shown in Figure 11.4, contains a peak at the isotropic chemical shift value ($\delta_{iso} = -3.2$ ppm) as well as a set of spinning sidebands. Analysis of the sideband intensities using a method introduced by Herzfeld and Berger [55] provides values for the shift anisotropy ($\Delta\delta = -117.2$ ppm) and asymmetry ($\eta = 0.64$) parameters

The surface complexation of dAMP with mesoporous alumina was modeled using the full dAMP molecule and edge-sharing octahedral aluminum clusters, represented in Figure 11.4 as the monodentate complex, [Al$_2$(OH)$_4$(OH$_2$)$_5$dAMP·(H$_2$O)$_6$], and the bidentate complex, [Al$_2$(OH)$_4$(OH$_2$)$_4$dAMP·(H$_2$O)$_6$]. The H$_2$O molecules were added to the complexes as the incorporation of explicit hydrogen bonding to

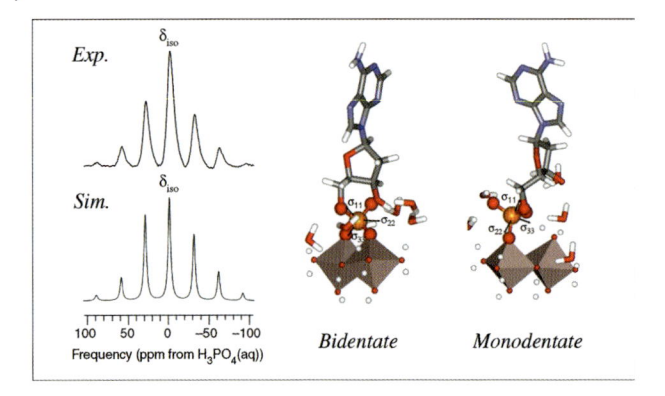

Figure 11.4 The ^{31}P solid-state MAS NMR experimental and simulated spectra (left) provide discrimination between two different inner-sphere binding models for dAMP adsorbed onto an alumina surface. The simulated spectrum is based on calculated chemical shift tensor components for the monodentate complex ($\Delta\delta = -116.5$ ppm and $\eta = 0.79$) and centered at the experimentally determined isotropic chemical shift.

a model system has been shown to improve the agreement between simulated and experimental ^{31}P CSA tensor components [98, 99]. The structures of these complexes were energy-minimized with B3LYP functionals and the 6-31G(d) basis set using Gaussian 03. NMR calculations were also performed in Gaussian 03, using the GIAO method with HF/6-311++G(2d,2p) basis sets and the structures obtained with the B3LYP calculation.

The isotropic chemical shift is first considered as a parameter for discrimination between these two possible binding conformations. However, perturbations such as changes in bond lengths and angles noticeably affect the ^{31}P CSA tensor [100, 101], and changes in one of the tensor components are commonly compensated by equally large changes in another component. Therefore, only minor variations are found in the isotropic chemical shift [102], so the interpretation solely based on the changes in the isotropic chemical shift between the two surface complexation environments is unreliable and complicated by the difficulty and uncertainty in calculating a chemical shift standard for ^{31}P. However, measurements of the remaining CSA parameters and their comparison to those found through computational methods for these two bonding complexes (calculated as $\Delta\delta = -116.5$ ppm and $\eta = 0.79$ for the monodentate complex and $\Delta\delta = 99.9$ ppm and $\eta = 0.58$ for the bidentate complex) indicate that the binding of dAMP to the alumina surface is indeed through a monodentate, inner-sphere complex. These results indicate that for more complex systems where nuclides are present with wide chemical shift ranges and where large variations in principal components of the CSA tensor are present, experimental and computational investigations of additional chemical shift tensor components in addition to the isotropic chemical shift provide an improved and discriminatory molecular scale description of the bonding environments.

11.3.4
Solid-State NMR of Aluminum Species at Mineral and Glass Surfaces

The release of metal species to solution during the aqueous dissolution of aluminosilicate minerals and glasses is tied to soil development and fertility, as well as the global carbon budget. Criscenti et al. [103] used *ab initio* modeling of feldspar dissolution to understand the mechanism of this process and, in particular, the change of coordination state of aluminum from tetrahedral coordination ($^{[4]}$Al) in the bulk structure to octahedral coordination ($^{[6]}$Al) in solution. The formation of octahedrally coordinated aluminum species on the surface itself during dissolution, and not as the result of reprecipitation from solution, has implications for the formation of silicon-rich leached layers on these surfaces as well as the sorption of natural organic matter. Hypothesizing that this transformation takes place at the oxide/water interface, molecular orbital calculations were undertaken on aluminosilicate clusters with discrete water molecules present to account for solvation effects at the interface.

Energy calculations on both fully relaxed and partially constrained clusters demonstrated that the energy difference between $^{[4]}$Al and $^{[6]}$Al on the surface of a feldspar (or similar aluminosilicate material) is small enough to allow the conversion of $^{[4]}$Al to $^{[6]}$Al (each connected to three Si atoms through bridging oxygen species) at or near the surface of the mineral, even before the release of Al ions to the aqueous solution. Experimental NMR results confirm that $^{[6]}$Al is present in the leached layer of a model aluminosilicate glass with a nepheline composition through the observed isotropic chemical shift (8 ppm) of the ^{27}Al NMR resonance acquired using cross-polarization MAS NMR techniques. Calculated isotropic shifts for $Q^{1[6]}$Al, $Q^{2[6]}$Al, and $Q^{3[6]}$Al – containing one, two, or three oxygen bridges, respectively, to silicon atoms – suggest that the $^{[6]}$Al observed on aluminosilicate glass surfaces is $Q^{1[6]}$Al and therefore formed as part of the dissolution process. Importantly, the *observed* shift of the resonance in the NMR spectrum falls at a position that is the sum of both the isotropic chemical shift and the isotropic second-order quadrupolar shift, the former being magnetic field independent while the latter shift is scaled (in ppm units) by the square of the inverse of the field strength. Therefore, performing experiments at two magnetic field strengths (here, 9.4 and 11.7 T) allows the separation of these two isotropic components to the overall shifts and the determination of the true isotropic chemical shift for comparison to calculated values.

11.3.5
Water and Zn(II) on TiO$_2$

Adsorption onto mineral surfaces is a significant process affecting the transport and fate of metals in the environment (see Ref. [60] for a review). Myriad studies have been performed using different approaches (i.e., potentiometric and adsorption isotherm studies with surface complexation modeling, spectroscopic studies, and molecular simulations). Zhang et al. [66] combined these approaches to examine the atomic structure of ions adsorbed on the (110) surface of α-TiO$_2$ (rutile). The combined approach is the key because each technique has its own strengths and weaknesses; a model of the mineral–water interface that describes all the results simultaneously should be superior to attempting to explain each type of result separately. Another reason to utilize various approaches is that the molecular level model can inform macroscopic scale models that are more practical in environmental applications such as predicting adsorption in large-scale (i.e., meters to kilometers) fluid flow simulations.

It is also imperative to note that this paper focused on a specific surface of the mineral. When comparing results from computational chemistry to experimental observations, discrepancies can arise due to inadequacy of the theoretical methodology or due to the model system. Often experiments are performed on minerals that exhibit a number of different crystal faces with various reactivities toward a given species. Thus, the observed value is a weighted average of all the surfaces present. Modeling studies are generally performed on one surface at a time, so even if the results are correct, they can disagree with observations on bulk materials.

The experimental components of this study examined Na^+, Rb^+, Ca^{2+}, Sr^{2+}, Zn^{2+}, Y^{3+}, and Nd^{3+}. All these cations except for Zn^{2+} were found to adsorb at similar sites via X-ray reflectivity and X-ray standing wave measurements (Zhang et al., [66]). This discovery was significant in itself because the assumption that ion such as Na^+ would reside in the "diffuse layer" and not specifically adsorb to the surface had guided much of the experimental design and interpretation up to this point. Periodic DFT calculations on Sr^{2+} and classical MD simulations (see also Ref. [104]) on Na^+, Rb^+, Ca^{2+}, and Sr^{2+} were consistent with the X-ray results. Furthermore, the surface structural information from the simulations was used as inputs to surface complexation models, and the results were consistent with the adsorption isotherms indicating that the spectroscopic, computational, and macroscopic data were all consistent – except for Zn^{2+}.

Why should the Zn^{2+} results be at odds when the other ions resulted in a consistent picture for the state of the adsorbed species? The problem was that the X-ray results suggested that Zn^{2+} adsorbed at two different sites from Sr^{2+} and that it could reside closer to the TiO$_2$ surface. The hypothesis to explain these differences was that Zn^{2+} was undergoing hydrolysis on the surface in much the same manner as it undergoes hydrolysis in solution, except that hydrolysis was occurring at a lower pH. As Zn^{2+} undergoes hydrolysis, the coordination state can change from six- to fourfold. The tetrahedral coordination state would stabilize Zn^{2+} at different sites from the other octahedrally coordinated cations in the study and allow it to approach the TiO$_2$ surface more closely because tetrahedral Zn–O bonds are shorter than octahedral Zn–O bonds (Figure 11.5).

A subsequent paper by Zhang et al. [59] added extended X-ray absorption fine structure spectroscopy to the mix of techniques employed in order to better constrain Zn–O bond lengths and Zn coordination numbers. This study found better agreement between the observed and DFT-predicted Zn positions on the surface when the Zn^{2+} was hydrolyzed compared to the case of the Zn^{2+} coordination sphere being completed by H$_2$O molecules. This hydrolysis led to spontaneous

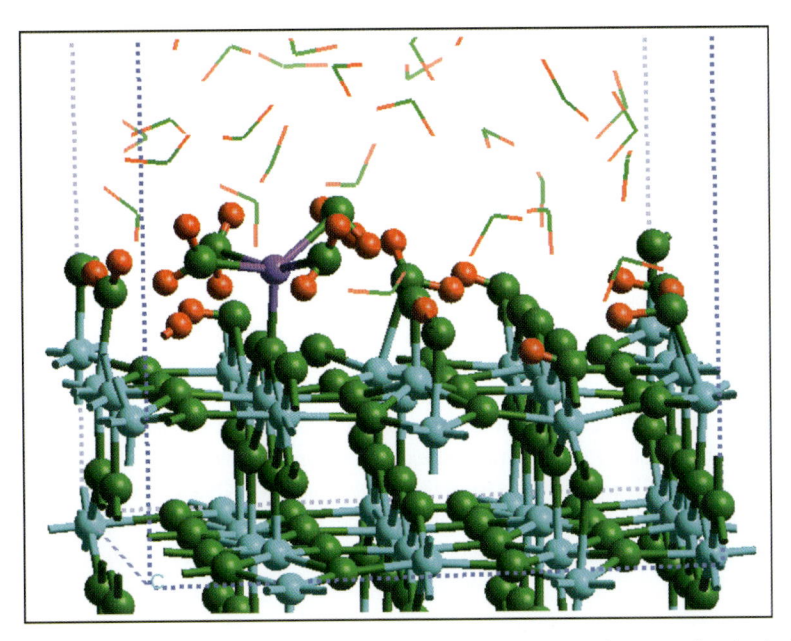

Figure 11.5 Zn(OH) on α-TiO$_2$ (110) from DFT using the VASP code [90] predict that hydrolysis leads to a lower coordination state and closer approach of the Zn^{2+} ion to the surface as observed [59].

lowering of the coordination number in the DFT energy minimizations. Minor discrepancies for the lateral positions of the Zn^{2+} adsorption sites remained among the XSW, EXAFS, and DFT results; however, it is interesting to note that the difference between the XSW and EXAFS positions was greater than the difference between these techniques and the DFT results.

These papers served as an important benchmark and verification of DFT methods in this type of study. A similar study was performed to predict the acid–base behavior of the α-TiO$_2$ (110) surface that combined the second harmonic generation, surface complexation modeling, and DFT calculations [21]. These studies demonstrated the ability of DFT calculations to predict accurately the structures of mineral–water interfaces and adsorption of ions at these surfaces. These studies also served as a starting point for simulations of the dynamic behavior of the mineral–water interface.

11.3.6
Water Dynamics on TiO$_2$ and SnO$_2$

Static pictures of the mineral–water interface atomic structure are useful, but ultimately the environmental chemist is interested in the rates of chemical reactions such as adsorption and desorption. A step toward this goal is to move from the static structures generated from X-ray analyses and DFT energy minimizations to dynamic movies of atomic motions in the interfacial region. A particularly excellent tool for collecting these data is neutron scattering. Quasi-elastic neutron scattering and inelastic neutron scattering probe the translational, rotational, and vibrational movements of molecules and are sensitive to H-atoms of the water molecules near the surface.

Mamontov et al. [68] performed QENS experiments on α-TiO_2 and SnO_2 (cassiterite) nanocrystals to study the structure and dynamics of water around these two minerals. Rutile and cassiterite were chosen because they have the same crystal structure (although cassiterite has lattice parameters approximately 10% larger than rutile), but extremely different bulk dielectric properties. James and Healy [105–107] and Sverjensky [108] had hypothesized that the dielectric constant of the mineral would control the nature of the mineral–water interface, so comparing these two minerals would be a test of that hypothesis.

The structure and dynamics of water around the two types of nanoparticles were found to be significantly different. Both exhibited three types of H_2O molecular behavior as defined by the distance from the mineral surface. Of the designated L1, L2, and L3 layers, the L1 layer closest to the surface was restricted in its diffusional behavior compared to the bulk, whereas L3 was similar to bulk water. The inference of this observation is that the mineral–water interface extends only three molecular

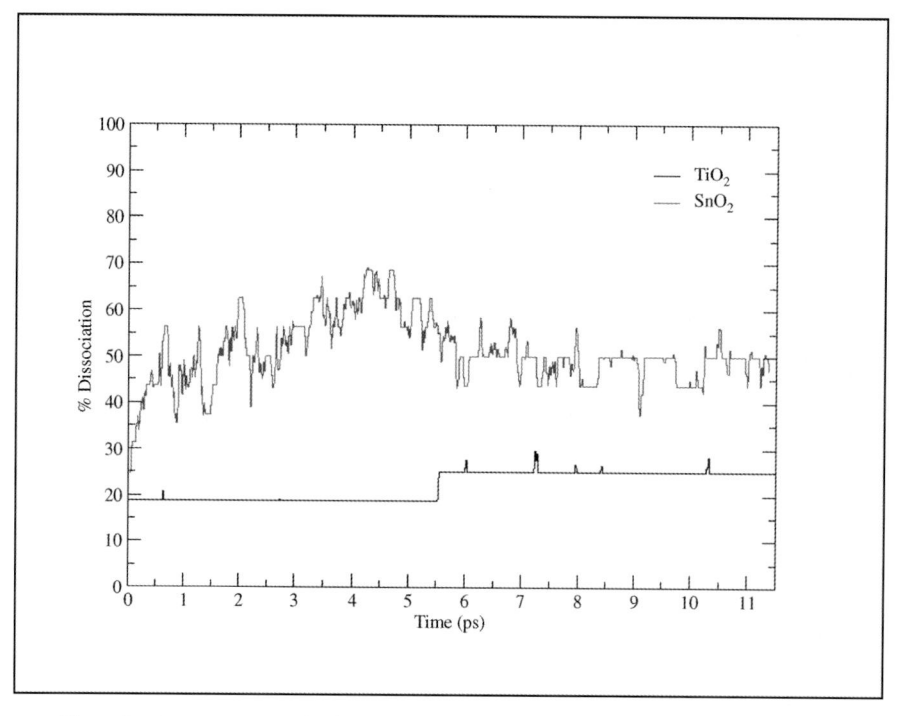

Figure 11.6 Comparison of H-bonding and dynamics of H_2O on TiO_2 and SnO_2 shows significantly more rapid exchange on SnO_2 than TiO_2 (Kumar and Sofo, personal communication).

layers (approximately 1 nm) in terms of its structural and dynamical behavior. Much wider interfacial regions have been calculated in the past based on force measurements [109]. Classical MD simulations (see also Vlcek et al. [71]) based on the SPC/E water force field and modified to provide TiO_2–H_2O and SnO_2–H_2O interactions [72, 110] provided both structural and energy transfer spectra consistent with the QENS data. Both the structural and dynamical behaviors of water was different on the rutile and cassiterite surfaces [68]; however, the attribution of this behavior to the bulk dielectric constant of the mineral is a matter of current research [61, 63, 67, 111, 112].

A similar type of study using DFT-based MD simulations of the TiO_2–H_2O interface produced vibrational densities of state consistent with INS data (Figure 11.6; [70]). The combination of DFT calculations, force field parameterization based on the DFT results, classical MD simulations, and a variety of spectroscopies allows one to probe various spatial and temporal scales in a self-consistent manner.

11.4
Summary and Future

Although the last decade has seen computational chemistry become successful and more common in environmental chemistry, there are a number of challenges that the community is now poised to overcome. Most published papers to date have dealt with identifying aqueous and surface speciation. This complementary role for computational spectroscopy is valuable, but computational chemistry can go beyond what can be observed experimentally. Thus, one can study reaction mechanisms and probe transient phenomenon. Based on numerous examples where the computational methods are consistent with observation, we can begin to trust the simulation results to explore chemical reactions with a level of detail not attainable with analytical instrumentation.

Another area where computational chemistry has lagged in environmental applications is the simulation of defects. Surface defects and disordered materials can play a huge role in environmental chemistry [13], but surface modeling typically is based on an idealized surface. As computer power and code parallelization increase, it becomes possible to build larger models capable of incorporating defects and disorder. These more energetic and reactive sites could be disproportionately important in reaction kinetics and difficult to identify experimentally. Of course, continued collaboration with analytical chemists will be the key in identifying and quantifying the reactive surface area of materials in the environment [113].

As already mentioned, the increasing computational power and parallelization of software allow larger numbers of atoms to be modeled with quantum methods. These quantum-based calculations also serve as benchmarks for the development of classical force fields. Classical MD and Monte Carlo simulations should always be useful, especially if these force fields can handle chemical reactivity [78]. Expansion of the system size in both quantum and classical simulations will allow more detailed and realistic modeling of biogeochemical phenomena, such as the interaction of bacteria with mineral surfaces [114]. As the biotic factors influencing environmental chemistry are increasingly recognized as being critically important, the ability to model a complex inorganic-organic–biotic system will be invaluable in understanding how nature works on a molecular level.

Acknowledgments

The authors appreciate the critical reviews of Heath D. Watts and Christin P. Morrow. Support of JDK's research by the US National Science Foundation (grants CHE-043132 and CHE-0714183) and the US Department of Energy (Office of Basic Energy Sciences – Nanoscale Complexity at the Water/Oxide Interface) is gratefully acknowledged. KTM acknowledges support by the National Science Foundation (grants CHE-0535656 and CHE-0809657) and the US Department of Energy (DE-FG02-06ER64191 and DE-FG02-08ER64615).

References

1 Sunda, W.G. and Huntsman, S.A. (1998) Processes regulating cellular metal accumulation and physiological effects: phytoplankton as model systems. *Sci. Total Environ.*, **219** (2–3), 165–181.

2 Nogueira, P.F.M., Melao, M.G.G. et al. (2009) Natural DOM affects copper speciation and bioavailability to bacteria and ciliate. *Arch. Environ. Con. Tox.*, **57** (2), 274–281.

3 Thomas, F., Masion, A. et al. (1993) Aluminum(III) speciation with hydroxy carboxylic-acids: Al-27-NMR study. *Environ. Sci. Technol.*, **27** (12), 2511–2516.

4 Ainsworth, C.C., Friedrich, D.M. et al. (1998) Characterization of salicylate–alumina surface complexes by polarized fluorescence spectroscopy. *Geochim. Cosmochim. Acta*, **62** (4), 595–612.

5 Eliet, V., Bidoglio, G. et al. (1995) Characterization of hydroxide complexes of uranium(VI) by time-resolved fluorescence spectroscopy. *J. Chem. Soc. Faraday Trans.*, **91** (15), 2275–2285.

6 Mosselmans, J.F.W., Helz, G.R. et al. (2000) A study of speciation of Sb in bisulfide solutions by X-ray absorption spectroscopy. *Appl. Geochem.*, **15** (6), 879–889.

7 Müller, K., Brendler, V., and Foerstendorf, H. (2008) Aqueous uranium(VI) hydrolysis species characterized by Attenuated total reflection Fourier-transform infrared spectroscopy. *Inorg. Chem.*, **47**, 10127–10134.

8 Persson, P., Nilsson, N. et al. (1996) Structure and bonding of orthophosphate ions at the iron oxide aqueous interface. *J. Colloid Interface Sci.*, **177** (1), 263–275.

9 Arai, Y. and Sparks, D.L. (2007) Phosphate reaction dynamics in soils and soil components: a multiscale approach. *Adv. Agron.*, **94**, 135–179.

10 Kwon, K.D. and Kubicki, J.D. (2004) Molecular orbital theory study on complexation structures of phosphates to iron hydroxides: Calculation of vibrational frequencies and adsorption energies. *Langmuir*, **20**, 9249–9254.

11 Rahnemaie, R., Hiemstra, T. et al. (2007) Geometry, charge distribution, and surface speciation of phosphate on goethite. *Langmuir*, **23** (7), 3680–3689.

12 Elzinga, E.J. and Sparks, D.L. (2007) Phosphate adsorption onto hematite: an *in situ* ATR-FTIR investigation of the effects of pH and loading level on the mode of phosphate surface

complexation. *J. Colloid Interface Sci.*, **308** (1), 53–70.

13 Torn, M.S., Trumbore, S.E. et al. (1997) Mineral control of soil organic carbon storage and turnover. *Nature*, **389** (6647), 170–173.

14 Masiello, C.A., Chadwick, O.A. et al. (2004) Weathering controls on mechanisms of carbon storage in grassland soils. *Global Biogeochem. Cycles*, **18** (4).

15 Avena, M.J., Mariscal, M.M. et al. (2003) Proton binding at clay surfaces in water. *Appl. Clay Sci.*, **24** (1–2), 3–9.

16 Tournassat, C., Ferrage, E. et al. (2004) The titration of clay minerals II. Structure-based model and implications for clay reactivity. *J. Colloid Interface Sci.*, **273** (1), 234–246.

17 Villalobos, M. and Perez-Gallegos, A. (2008) Goethite surface reactivity: a macroscopic investigation unifying proton, chromate, carbonate, and lead(II) adsorption. *J. Colloid Interface Sci.*, **326** (2), 307–323.

18 Villalobos, M., Cheney, M.A. et al. (2009) Goethite surface reactivity: II. A microscopic site-density model that describes its surface area-normalized variability. *J. Colloid Interface Sci.*, **336** (2), 412–422.

19 Criscenti, L.J. and Sverjensky, D.A. (2002) A single-site model for divalent transition and heavy metal adsorption over a range of metal concentrations. *J. Colloid Interface Sci.*, **253** (2), 329–352.

20 Hiemstra, T. and Van Riemsdijk, W.H. (2006) On the relationship between charge distribution, surface hydration, and the structure of the interface of metal hydroxides. *J. Colloid Interface Sci.*, **301** (1), 1–18.

21 Fitts, J.P., Machesky, M.L. et al. (2005) Second-harmonic generation and theoretical studies of protonation at the water/alpha-TiO_2 (110) interface. *Chem. Phys. Lett.*, **411** (4–6), 399–403.

22 Fukushi, K. and Sverjensky, D.A. (2007) A predictive model (ETLM) for arsenate adsorption and surface speciation on oxides consistent with spectroscopic and theoretical molecular evidence. *Geochim. Cosmochim. Acta*, **71** (15), 3717–3745.

23 Machesky, M.L., Predota, M. et al. (2008) Surface protonation at the rutile (110) interface: explicit incorporation of solvation structure within the refined MUSIC model framework. *Langmuir*, **24** (21), 12331–12339.

24 Ridley, M.K., Hiemstra, T. et al. (2009) Inner-sphere complexation of cations at the rutile–water interface: a concise surface structural interpretation with the CD and MUSIC model. *Geochim. Cosmochim. Acta*, **73** (7), 1841–1856.

25 Bickmore, B.R., Rosso, K.M. et al. (2006) Bond-valence methods for pK_a prediction. II. Bond-valence, electrostatic, molecular geometry, and solvation effects. *Geochim. Cosmochim. Acta*, **70** (16), 4057–4071.

26 Goyne, K.W., Chorover, J. et al. (2005) Sorption of the antibiotic ofloxacin to mesoporous and nonporous alumina and silica. *J. Colloid Interface Sci.*, **283** (1), 160–170.

27 de Jong, W.A., Apra, E. et al. (2005) Complexation of the carbonate, nitrate, and acetate anions with the uranyl dication: density functional studies with relativistic effective core potentials. *J. Phys. Chem. A*, **109** (50), 11568–11577.

28 Usher, C.R., Paul, K.W. et al. (2005) Mechanistic aspects of pyrite oxidation in an oxidizing gaseous environment: an in situ HATR-IR isotope study. *Environ. Sci. Technol.*, **39** (19), 7576–7584.

29 Tribe, L., Kwon, K.D. et al. (2006) Molecular orbital theory study on surface complex structures of glyphosate on goethite: calculation of vibrational frequencies. *Environ. Sci. Technol.*, **40** (12), 3836–3841.

30 Wong, M.W. (1996) Vibrational frequency prediction using density functional theory. *Chem. Phys. Lett.*, **256** (4–5), 391–399.

31 Andersson, M.P. and Uvdal, P. (2005) New scale factors for harmonic vibrational frequencies using the B3LYP density functional method with the triple-ζ basis set 6-311 + G(d,p). *J. Phys. Chem. A*, **109** (12), 2937–2941.

32 Wander, M.C.F., Kubicki, J.D. et al. (2006) The role of structured water in the calibration and interpretation of

theoretical IR spectra. *Spectrochim. Acta A*, **65** (2), 324–332.

33 Yamaguchi, Y., Frisch, M. *et al.* (1986) Analytic evaluation and basis set dependence of intensities of infrared-spectra. *J. Chem. Phys.*, **84** (4), 2262–2278.

34 Frisch, M.J., Yamaguchi, Y. *et al.* (1986) Analytic Raman intensities from molecular electronic wave-functions. *J. Chem. Phys.*, **84** (1), 531–532.

35 Hehre, W.J., Radom, L., Schleyer, P.R., and Pople, J.A. (1986) *Ab Initio* Molecular Orbital Theory. John Wiley & Sons, NY, New York.

36 Tossell, J.A. (2005) Boric acid, carbonic acid, and N-containing oxyacids in aqueous solution: *ab initio* studies of structure, pK_a, NMR shifts, and isotopic fractionations. *Geochim. Cosmochim. Acta*, **69** (24), 5647–5658.

37 Schauble, E.A., Ghosh, P. *et al.* (2006) Preferential formation of C-13-O-18 bonds in carbonate minerals, estimated using first-principles lattice dynamics. *Geochim. Cosmochim. Acta*, **70** (10), 2510–2529.

38 Domagal-Goldman, S.D. and Kubicki, J.D. (2008) Density functional theory predictions of equilibrium isotope fractionation of iron due to redox changes and organic complexation. *Geochim. Cosmochim. Acta*, **72** (21), 5201–5216.

39 Hsu, P.H. and Hatcher, P.G. (2006) Covalent coupling of peptides to humic acids: structural effects investigated using 2D NMR spectroscopy. *Org. Geochem.*, **37** (12), 1694–1704.

40 Cole, K.E., Paik, Y. *et al.* (2004) H-2 MAS NMR studies of deuterated goethite (alpha-FeOOD). *J. Phys. Chem. B*, **108** (22), 6938–6940.

41 Bowers, G.M. and Kirkpatrick, R.J. (2007) High-field As-75 NMR study of arsenic oxysalts. *J. Magn. Reson.*, **188** (2), 311–321.

42 Cochiara, S.G. and Phillips, B.L. (2008) NMR spectroscopy of naturally occurring surface-adsorbed fluoride on Georgia kaolinite. *Clay Clay Miner.*, **56** (1), 90–99.

43 Davis, M.C., Brouwer, W.J. *et al.* (2009) Magnesium silicate dissolution investigated by Si-29 MAS, H-1-Si-29 CPMAS, Mg-25 QCPMG, and H-1-Mg-25 CP QCPMG NMR. *Phys. Chem. Chem. Phys.*, **11** (32), 7013–7021.

44 Basile-Doelsch, I., Amundson, R. *et al.* (2007) Mineral control of carbon pools in a volcanic soil horizon. *Geoderma*, **137** (3–4), 477–489.

45 Nebbioso, A. and Piccolo, A. (2009) Molecular rigidity and diffusivity of Al^{3+} and Ca^{2+} humates as revealed by NMR spectroscopy. *Environ. Sci. Technol.*, **43** (7), 2417–2424.

46 Tsomaia, N., Brantley, S.L. *et al.* (2003) NMR evidence for formation of octahedral and tetrahedral Al and repolymerization of the Si network during dissolution of aluminosilicate glass and crystal. *Am. Mineral.*, **88** (1), 54–67.

47 Kaiser, E., Simpson, A.J. *et al.* (2003) Solid-state and multidimensional solution-state NMR of solid phase extracted and ultrafiltered riverine dissolved organic matter. *Environ. Sci. Technol.*, **37** (13), 2929–2935.

48 Diallo, M.S., Simpson, A. *et al.* (2003) 3-D structural modeling of humic acids through experimental characterization, computer assisted structure elucidation and atomistic simulations. 1. Chelsea soil humic acid. *Environ. Sci. Technol.*, **37** (9), 1783–1793.

49 Simpson, A.J., Kingery, W.L. *et al.* (2003) The identification of plant derived structures in humic materials using three-dimensional NMR spectroscopy. *Environ. Sci. Technol.*, **37** (2), 337–342.

50 Sykes, D., Kubicki, J.D. *et al.* (1997) Molecular orbital calculation of Al-27 and Si-29 NMR parameters in Q(3) and Q(4) aluminosilicate molecules and implications for the interpretation of hydrous aluminosilicate glass NMR spectra. *J. Phys. Chem. A*, **101** (14), 2715–2722.

51 Wolinski, K., Hinton, J.F. *et al.* (1990) Efficient implementation of the gauge-independent atomic orbital method for NMR chemical-shift calculations. *J. Am. Chem. Soc.*, **112** (23), 8251–8260.

52 Cody, G.D., Mysen, B. *et al.* (2001) Silicate–phosphate interactions in silicate

glasses and melts: I. A multinuclear (Al-27, Si-29, P-31) MAS NMR and *ab initio* chemical shielding (P-31) study of phosphorous speciation in silicate glasses. *Geochim. Cosmochim. Acta*, **65** (14), 2395–2411.

53 Tossell, J.A. (2006) Boric acid adsorption on humic acids: *ab initio* calculation of structures, stabilities, B-11 NMR and B-11, B-10 isotopic fractionations of surface complexes. *Geochim. Cosmochim. Acta*, **70** (20), 5089–5103.

54 Wolinski, K., Haacke, R. *et al.* (1997) Methods for parallel computation of SCF NMR chemical shifts by GIAO method: efficient integral calculation, multi-Fock algorithm, and pseudodiagonalization. *J. Comput. Chem.*, **18** (6), 816–825.

55 Herzfeld, J. and Berger, A.E. (1980) Sideband intensities in NMR spectra of sample spinning at the magic angle. *J. Chem. Phys.*, **73** (12), 6021–6030.

56 Cohen, M.H. and Reif, F. (eds) (1957) Quadrupole effects in nuclear magnetic resonance studies of solids, in *Solid State Physics*, Academic Press, New York.

57 Abragam, A. (1983) *Principles of Nuclear Magnetism*, Oxford University Press, Oxford.

58 Slichter, C.P. (1990) *Principles of Magnetic Resonance*, Springer-Verlag, New York.

59 Zhang, Z., Fenter, P. *et al.* (2006) Structure of hydrated Zn^{2+} at the rutile TiO_2 (110)-aqueous solution interface: comparison of X-ray standing wave, X-ray absorption spectroscopy, and density functional theory results. *Geochim. Cosmochim. Acta*, **70**, 4039–4056.

60 Brown, G.E., Henrich, V.E. *et al.* (1999) Metal oxide surfaces and their interactions with aqueous solutions and microbial organisms. *Chem. Rev.*, **99** (1), 77–174.

61 Goniakowski, J. and Gillan, M.J. (1996) The adsorption of H_2O on TiO_2 and SnO_2(110) studied by first-principles calculations. *Surf. Sci.*, **350** (1–3), 145–158.

62 Lindan, P.J.D., Harrison, N.M. *et al.* (1998) Mixed dissociative and molecular adsorption of water on the rutile (110) surface. *Phys. Rev. Lett.*, **80** (4), 762–765.

63 Lindan, P.J.D. (2000) Water chemistry at the SnO_2(110) surface: the role of intermolecular interactions and surface geometry. *Chem. Phys. Lett.*, **328** (4–6), 325–329.

64 Langel, W. (2002) Car–Parrinello simulation of H_2O dissociation on rutile. *Surf. Sci.*, **496** (1–2), 141–150.

65 Zhang, C. and Lindan, P.J.D. (2003) Towards a first-principles picture of the oxide–water interface. *J. Chem. Phys.*, **119** (17), 9183–9190.

66 Zhang, Z., Fenter, P., Cheng, L. *et al.* (2004) Ion adsorption at the rutile-water interface: Linking molecular and macroscopic properties. *Langmuir*, **20**, 4954–4969.

67 Evarestov, R.A., Bandura, A.V. *et al.* (2006) Plain DFT and hybrid HF-DFT LCAO calculations of SnO2 (110) and (100) bare and hydroxylated surfaces. *Phys. Status Solidi B*, **243** (8), 1823–1834.

68 Mamontov, E., Vlcek, L., and Wesolowski, D.J. (2007) Dynamics and structure of hydration water on rutile and cassiterite nanopowders studied by quasielastic neutron scattering and molecular dynamics simulations. *J. Phys. Chem. C*, **111** (11), 4328–4341.

69 Levchenko, A.A., Kolesnikov, A.I. *et al.* (2007) Dynamics of water confined on a TiO_2 (anatase) surface. *J. Phys. Chem. A*, **111** (49), 12584–12588.

70 Kumar, N., Neogi, S., Kent, P., Bandura, A.V., Kubicki, J., Wesolowski D., Cole D., and Sofo J. (2009) Hydrogen bonds and vibrations of water on (110) rutile. *J. Phys. Chem. C*, **113**, 13732–13740.

71 Vlcek, L., Zhang, Z., Machesky, M.L., *et al.* (2007) Electric double layer at metal oxide surfaces: static properties of the cassiterite–water interface. *Langmuir*, **23** (9), 4925–4937.

72 Bandura, A.V. and Kubicki, J.D. (2003) Derivation of force field parameters for TiO_2–H_2O systems from *ab initio* calculations. *J. Phys. Chem. B*, **107** (40), 11072–11081.

73 Dove, P.M., Han, N.Z. *et al.* (2005) Mechanisms of classical crystal growth theory explain quartz and silicate dissolution behavior. *Proc. Natl. Acad. Sci. USA*, **102** (43), 15357–15362.

74 Bluhm, H. and Siegmann, H.C. (2009) Surface science with aerosols. *Surf. Sci.*, **603** (10–12), 1969–1978.

75 Bargar, J.R., Kubicki, J.D., Reitmeyer, R.L., and J. A. Davis (2005) ATR-FTIR characterization of inner-sphere and outer-sphere carbonate surface complexes on hematite. *Geochim. Cosmochim. Acta*, **69**, 1527–1542.

76 Schleyer, P.vonR. (1998) *Encyclopedia of Computational Chemistry*, John Wiley & Sons, Inc., New York, p. 3429.

77 Cramer, C.J. (2002) *Essentials of Computational Chemistry: Theories and Models*, John Wiley & Sons, Ltd., Chichester.

78 van Duin, A.C.T., Dasgupta, S. et al. (2001) ReaxFF: a reactive force field for hydrocarbons. *J. Phys. Chem. A*, **105** (41), 9396–9409.

79 Kubicki, J.D. (2001) Interpretation of vibrational spectra using molecular orbital theory calculations. *Rev. Mineral. Geochem.*, **42**, 459–483.

80 Nilsson, N., Lovgren, L. et al. (1992) Phosphate complexation at the surface of goethite. *Chem. Spec. Bioavail.*, **4** (4), 121–130.

81 Weng, L.P., Van Riemsdijk, W.H. et al. (2008) Humic nanoparticles at the oxide–water interface: interactions with phosphate ion adsorption. *Environ. Sci. Technol.*, **42** (23), 8747–8752.

82 Parfitt, R.L., Russell, J.D., and Farmer, V.C. (1976) Confirmation of surface-structures of goethite (α-FeOOH) and phosphated goethite by infrared spectroscopy. *J. Chem. Soc.*, **72**, 1082–1087.

83 Hug, S.J. (1997) In situ Fourier transform infrared measurements of sulfate adsorption on haematite in aqueous solutions. *J. Colloid Interface Sci.*, **188**, 415–422.

84 Tejedor-Tejedor, M.I. and Anderson, M.A. (1990) Protonation of phosphate on the surface of goethite as studied by CIR-FTIR and electrophoretic mobility. *Langmuir*, **6**, 602–611.

85 Luengo, C., Brigante, M., Antelo, J., et al. (2006) Kinetics of phosphate adsorption on goethite: comparing batch adsorption and ATR-IR measurements. *J. Colloid Interface Sci.*, **300**, 511–518.

86 Frisch MJT, et al., Gaussian 03, Revision C.02. Wallingford, 2004.

87 Becke, A.D. (1997) Density-functional thermochemistry 5. Systematic optimization of exchange-correlation functionals. *J. Chem. Phys.*, **107**, 8554–8560.

88 Lee, C.T., Yang, W.T., Parr, R.G. (1988) Development of the Colle-Salvetti correlation-energy formula into a functional of the electron-density. *Phys. Rev. B*, **37**, 785–789.

89 Hehre, W., Ditchfield, R., and Pople, J. (1972) Self-consistent molecular-orbital methods 12. Further extensions of gaussian-type basis sets for use in molecular-orbital studies of organic-molecules. *J. Chem. Phys.*, **56**, 2257–2261.

90 Kresse, G. and Furthmuller, J. (1996) Efficient iterative schemes for *ab initio* total-energy calculations using a plane-wave basis set. *Phys. Rev. B*, **54** (16), 11169–11186.

91 Rakotonarivo, E., Bottero, J.Y. et al. (1988) Electrochemical modeling of freshly precipitated aluminum hydroxide: electrolyte interface. *Colloids Surf.*, **33** (3–4), 191–207.

92 Kubicki, J.D., Sykes, D., and Apitz, S.E. (1999) Ab initio calculation of aqueous aluminum and aluminum-carboxylate NMR chemical shifts. *J. Phys. Chem. A*, **103**, 903–915.

93 Biber, M.V. and Stumm, W. (1994) An in-situ ATR-FTIR study – the surface coordination of salicylic-acid on aluminum and iron(III) oxides. *Environ. Sci. Technol.*, **28**, 763–768.

94 Kummert, R. and Stumm, W. (1980) The surface complexation of organic-acids on hydrous gamma-Al_2O_3. *J. Colloid Interface Sci.*, **7**, 373–385.

95 Anandan, K., Kolandaivel, P. et al. (2005) Molecular structural conformations and hydration of internally hydrogen-bonded salicylic acid: ab initio and DFT studies. *Int. J. Quantum Chem.*, **103** (2), 127–139.

96 Fry, R.A., Kwon, K.D. et al. (2006) Solid-state NMR and computational chemistry study of mononucleotides adsorbed to alumina. *Langmuir*, **22**, 9281–9286.

97 Gullion, T. and Schaefer, J. (1989) Rotational-echo double-resonance. *J. Magn. Reson.*, **81**, 196–200.

98 Kriz, J. and Dybal, J. (2004) Simple and cooperative electrostatic binding of ammonium ions to phosphate polyions: NMR, infrared, and theoretical study. *J. Phys. Chem.*, **108** (26), 9306–9314.

99 Potrzebowski, M.J., Assfeld, X. *et al.* (2003) An experimental and theoretical study of the C-13 and P-31 chemical shielding tensors in solid O-phosphorylated amino acids. *J. Am. Chem. Soc.*, **125** (14), 4223–4232.

100 Alam, T.M. (1999) *Modeling NMR Chemical Shifts: Gaining Insights into Structure and Environment, ACS Symposium Series 732*, American Chemical Society, Washington, DC, pp. 320–334.

101 Un, S. and Klein, M.P. (1989) Study of P-31 NMR chemical-shift tensors and their correlation to molecular-structure. *J. Am. Chem. Soc.*, **111** (14), 5119–5124.

102 Gorenstein, D.G. (1994) Conformation and dynamics of DNA and protein–DNA complexes by P-31 NMR. *Chem. Rev.*, **94** (5), 1315–1338.

103 Criscenti, L.J., Brantley, S.L. *et al.* (2005) Theoretical and ^{27}Al CPMAS NMR investigation of aluminum coordination changes during aluminosilicate dissolution. *Geochim. Cosmochim. Acta*, **69** (9), 2205–2220.

104 Predota, M., Zhang, Z. *et al.* (2004) Electric double layer at the rutile (110) surface. 2. Adsorption of ions from molecular dynamics and X-ray experiments. *J. Phys. Chem. B*, **108** (32), 12061–12072.

105 James, R.O. and Healy, T.W. (1972) Adsorption of hydrolyzable metal ions at oxide–water interface. I. Co(II) adsorption on SiO_2 and TiO_2 as model systems. *J. Colloid Interface Sci.*, **40** (1), 42–52.

106 James, R.O. and Healy, T.W. (1972) Adsorption of hydrolyzable metal ions at oxide–water interface. II. Charge reversal of SiO_2 and TiO_2 colloids by adsorbed Co(II), La(III), and Th(IV) as model systems. *J. Colloid Interface Sci.*, **40** (1), 53–64.

107 James, R.O. and Healy, T.W. (1972) Adsorption of hydrolyzable metal ions at oxide–water interface. III. Thermodynamic model of adsorption. *J. Colloid Interface Sci.*, **40** (1), 65–81.

108 Sverjensky, D.A. (2001) Interpretation and prediction of triple-layer model capacitances and the structure of the oxide–electrolyte–water interface. *Geochim. Cosmochim. Acta*, **65** (21), 3643–3655.

109 Ducker, W.A., Senden, T.J. *et al.* (1992) Measurement of forces in liquids using a force microscope. *Langmuir*, **8** (7), 1831–1836.

110 Bandura, A.V., Sofo, J.O. *et al.* (2006) Derivation of force field parameters for SnO_2–H_2O surface systems from plane-wave density functional theory calculations. *J. Phys. Chem. B*, **110** (16), 8386–8397.

111 Bates, S.P. (2002) Full-coverage adsorption of water on SnO_2(110): the stabilisation of the molecular species. *Surf. Sci.*, **512** (1–2), 29–36.

112 Bandura, A.V., Sykes, D.G. *et al.* (2004) Adsorption of water on the TiO_2 (rutile) (110) surface: a comparison of periodic and embedded cluster calculations. *J. Phys. Chem. B*, **108** (23), 7844–7853.

113 Washton, N.M., Brantley, S.L., and Mueller, K.T. (2008) Probing the molecular-level control of aluminosilicate dissolution: a sensitive solid-state NMR proxy for reactive surface area. *Geochimica et Cosmochimica Acta*, **72**, 5949–5961.

114 Lins, R.D. and Straatsma, T.P. (2001) Computer simulation of the rough lipopolysaccharide membrane of *Pseudomonas aeruginosa*. *Biophys. J.*, **81** (2), 1037–1046.

12
Comparison of Calculated and Observed Vibrational Frequencies of New Molecules from an Experimental Perspective

Lester Andrews

12.1
Introduction

Reactions of metal atoms with small molecules have provided a productive route to many new molecules of fundamental importance for their contributions to our understanding of chemical bonding, and the matrix isolation technique has contributed to this large body of information over the past half century [1–3]. Investigations of this type with more volatile metal atoms employed thermal methods for evaporation, but less volatile metal atoms required more challenging higher temperature experimental methods [4, 5]. Laser ablation provides a very efficient and effective means of heating a very small volume element of solid material to very high temperature sufficient for vaporization (and even ionization) [2, 6–9]. The key point here is that the laser is pulsed and focused on a spot less than 0.1 mm in diameter so that a large amount of energy is deposited into a small volume element of sample in a very short time interval. Accordingly, the most refractory metals such as tungsten and uranium can be evaporated for simple atom–molecule reactions [3].

This chapter will describe some examples of the use of laser ablation to generate metal atoms for reactions with small molecules to make interesting new subject molecules and the comparison of calculated and observed vibrational frequencies for the identification of these new molecules using different theoretical methods.

12.2
Experimental and Theoretical Methods

The matrix isolation laser ablation apparatus employed at the University of Virginia and sketched in Figure 12.1 has been described in more detail previously [3]. Closed cycle cryogenic refrigerators provide very good 4–7 K refrigeration when properly configured. A focused, pulsed Nd-YAG laser (1–20 mJ/pulse, 10 Hz) evaporates metal atoms toward the cryogenically cooled window (4 K for hydrogen and neon or 7 K for argon matrix experiments) for codeposition and reaction with pure hydrogen, Ne/reagent, or Ar/reagent gas mixtures. The laser ablation plume not only contains visible light that

Computational Spectroscopy: Methods, Experiments and Applications. Edited by Jörg Grunenberg
Copyright © 2010 WILEY-VCH Verlag GmbH & Co. KGaA, Weinheim
ISBN: 978-3-527-32649-5

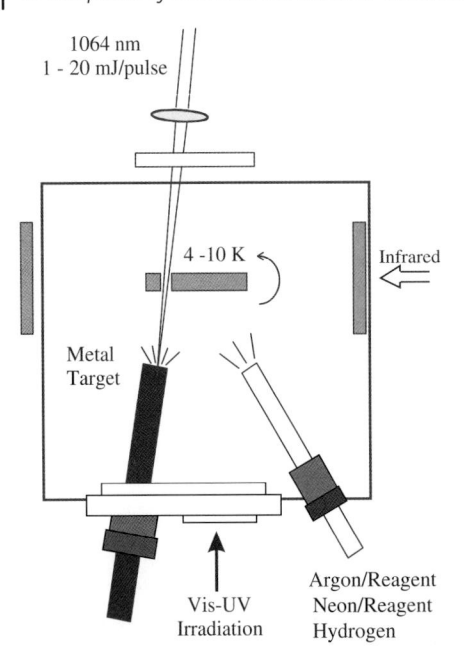

Figure 12.1 Diagram of laser ablation matrix isolation apparatus used for reacting metal atoms with small molecules and trapping the products in solid argon, neon, or hydrogen for infrared spectroscopic and photochemical investigations.

is useful to focus the laser and to direct the ablated materials but also provides ultraviolet and vacuum ultraviolet radiation, which can perform photochemistry on the reagent molecule as well [3]. Infrared spectra are recorded by Nicolet Fourier transform instruments at 0.5 cm^{-1} resolution and 0.1 cm^{-1} frequency accuracy after deposition, after subsequent irradiation by a mercury arc street lamp (Sylvania, 175 watt, outer globe removed) in combination with glass filters, and after sequential annealing of the matrix sample to allow diffusion and further reaction of trapped species.

We routinely compute frequencies, energies, and structures for all molecular products anticipated from the laser evaporated metal atom reactions under investigation. The straightforward method for such calculations is to use the Gaussian program system and the hybrid B3LYP and pure BPW91 density functionals [10–12]. We employ the large Gaussian basis set 6-311++G(3df,3pd) for lighter atoms and the pseudopotentials supported by Gaussian for transition and actinide metal atoms [13]. With selected small systems, we use the CCSD(T) wavefunction-based method for comparison [14].

12.2.1
The LiO$_2$ Ionic Molecule

One of the first and most important new molecules to be prepared in a solid matrix at the University of Virginia was the ionic LiO$_2$ molecule [15]. The bonding model

Table 12.1 Observed neon matrix and calculated harmonic frequencies (cm^{-1}) and infrared intensities (km/mol) for the LiO$_2$ molecule in the $^2A_2$ ground electronic states with the C$_{2v}$ structure[a].

7(16-16) observed	7(16-16) B3LYP	7(18-18) B3LYP	6(16-16) B3LYP	6(18-18) B3LYP	Mode identification
508.9	527.9 (31)	510.1 (32)	545.4 (38)	528.1 (40)	B$_2$, antisym. Li–O
719.7	753.3 (117)	745.4 (114)	802.6 (132)	795.0 (129)	A$_1$, sym. Li–O
1093.9	1175.4 (9)	1108.4 (9)	1176.0 (10)	1109.2 (11)	A$_1$, sym. O–OO
	BPW91	BPW91	BPW91	BPW91	
508.9	515.2 (35)	497.8 (36)	532.2 (43)	515.4 (44)	B$_2$, antisym. Li–O
719.7	724.2 (106)	716.6 (104)	771.5 (120)	764.2 (117)	A$_1$, sym. Li–O
1093.9	1111.2 (12)	1048.0 (11)	1111.9 (13)	1048.9 (14)	A$_1$, sym. O–O
	CCSD(T)	CCSD(T)	CCSD(T)	CCSD(T)	
508.9	533.1	515.0	550.5	532.9	B$_2$, antisym. Li–O
719.7	734.7	727.1	782.9	775.6	A$_1$, sym. Li–O
1093.9	1130.4	1066.0	1130.9	1066.6	A$_1$, sym. O–O

a) See Refs [10–14] for the methods of calculation.

deduced from infrared and Raman spectra was reaffirmed by subsequent theoretical calculations [15–17]. New observations of neon matrix bands at 1093.9, 719.7, and 508.9 cm^{-1} correlate with the argon matrix 1096.9, 698.8, and 492.4 cm^{-1} absorptions, respectively, assigned to the ionic LiO$_2$ molecule [15, 18] and with the frequencies recently computed for this molecule (Table 12.1). The neon matrix frequencies for the interionic modes are higher owing to less interaction with the less polarizable neon matrix and closer to the still yet to be observed gas-phase vibrational frequencies [1]. The near agreement of frequencies calculated with the hybrid B3LYP and the pure BPW91 density functionals and the CCSD(T) wavefunction-based methods is reassuring. Frequencies computed at the SCF level are of course higher [17]. We typically find B3LYP values higher than BPW91 values and both slightly higher than observed values owing in part to the neglect of anharmonic corrections [19–21]. What is important here is the coverage of experimental observations by several theoretical methods to confirm the experimental identification and assignment and the application of theoretical computations to this problem.

The ratios of harmonic lithium-6/lithium-7 isotopic frequencies with oxygen-16 for the strong diagnostic interionic modes are 1.06 545, 1.06 531, and 1.06 561 for the three methods, respectively, which show that the normal modes are described similarly by the three theoretical methods. Also, notice the oxygen-16/oxygen-18 isotopic frequency ratios with lithium 7 computed as 1.0106 and 1.0349 for the symmetric and antisymmetric Li–O stretching modes, respectively, (B3LYP, for example) illustrate different oxygen participations in these modes, as do the experimental values (1.0100 and 1.0343) [18]. Furthermore, the structural parameters computed in Figure 12.2 are within the range of values presented earlier using a variety of theoretical methods [17], and the O–Li–O angle of 44° deduced from isotopic vibrational analysis of the LiO$_2$ molecule [15] is within approximately a degree of the values calculated here. And the symmetrical structure based on the

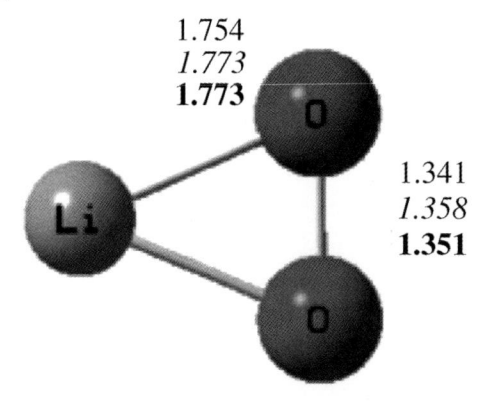

Figure 12.2 Optimized structures calculated for the lithium and oxygen reaction product LiO$_2$. Parameters given for B3LYP, *BPW91*, **CCSD(T)**. Bond lengths in angstroms.

triplet pattern in $^{16}O_2$, $^{16}O^{18}O$, $^{18}O_2$ isotopic spectra [15] has been reaffirmed by all of these calculations. The Mulliken charges on Li are $+0.55, 0.53$, and 0.64 for the three above methods, respectively. In addition, the calculated Mulliken atomic spin densities -0.012 for Li, $+0.506$ for O using B3LYP, -0.009 for Li, $+0.505$ for O using BPW91, and -0.026 for Li, $+0.513$ for O using SCF based on CCSD(T) also substantiate the ionic model for the bonding in LiO$_2$.

12.3
Aluminum and Hydrogen: First Preparation of Dibridged Dialane, Al$_2$H$_6$

The chemistry of boron hydrides has been investigated for over a century, and a large number of boron hydride compounds have been identified and characterized; however, aluminum hydride chemistry under normal conditions is limited to the nonvolatile polymeric solid trihydride (AlH$_3$)$_n$ [22, 23]. The diborane molecule is fundamentally important as the textbook example of hydrogen (μ-hydrido) bridge bonding. Although the isostructural dialane molecule is calculated to be stable, molecular Al$_2$H$_6$ was not isolated until our work with solid hydrogen [24, 25]. The failure to observe dialane earlier is even more surprising in view of the synthesis of the isostructural Ga$_2$H$_6$ molecule [26]. Alane (AlH$_3$) has been observed by three groups from reactions of energetic aluminum atoms with hydrogen in solid argon and characterized by infrared spectroscopy [27]; however, the concentration of AlH$_3$ was not sufficient to form Al$_2$H$_6$ in the rigid argon matrix. Our successful synthesis of Al$_2$H$_6$ for the first time involved pure hydrogen as the matrix. This ensured the selective formation of the highest monohydride AlH$_3$, and diffusion in the soft hydrogen matrix upon annealing allows dimerization to dialane Al$_2$H$_6$, here called DA. The DA bands are produced in much greater yield in the softer hydrogen matrix solid than in the harder argon matrix.

The spectrum of our first deposited Al and H$_2$ sample (Figure 12.3) is dominated by the strong AlH absorption at 1598.7 cm^{-1}, which is intermediate between the gas

12.3 Aluminum and Hydrogen: First Preparation of Dibridged Dialane, Al_2H_6

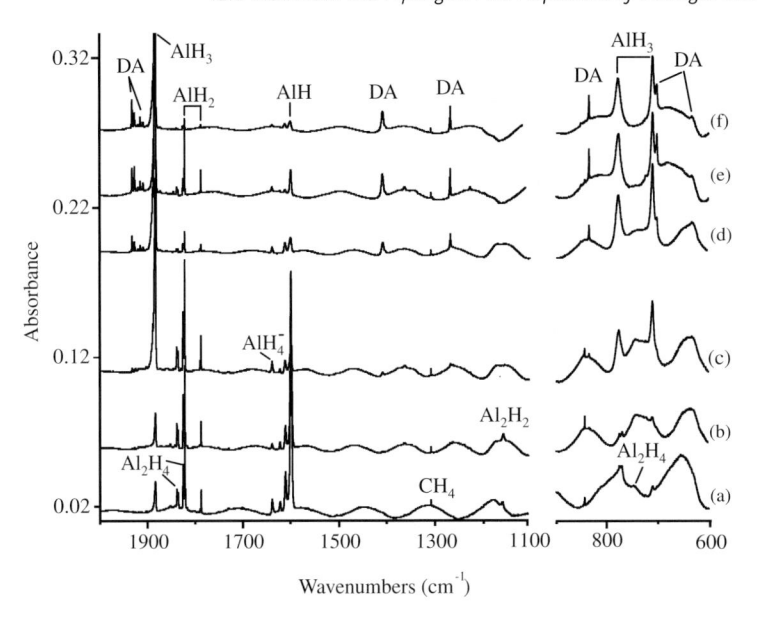

Figure 12.3 Infrared spectra in the 2000–600 cm^{-1} region for laser-ablated Al codeposited with normal hydrogen at 4 K. (a) Spectrum of sample deposited for 25 min, (b) after annealing at 6.2 K, (c) after >290 nm photolysis, (d) after >220 nm photolysis, (e) after annealing at 6.5 K, and (f) after second >220 nm photolysis.

phase (1624.4 cm^{-1}) and the argon matrix (1590.7 cm^{-1}) absorptions for diatomic AlH. Absorptions are also observed for AlH$_3$ (1883.7, 777.9, 711.3 cm^{-1}) and AlH$_2$, which are for the most part higher than the argon matrix values. New absorptions are observed for Al$_2$H$_4$ with the hint of very weak absorptions at 1932.3, 1915.1, 1408.1, 1268.2, 835.6, 702.4, 631.9 cm^{-1} (labeled DA) (Figure 12.3a). Annealing to 6.2 K increased these bands except for AlH (Figure 12.3b). Photolysis stepwise at >290 nm and then at >220 nm decreased the AlH band and increased the AlH$_3$ absorptions and the DA band set (Figure 12.3c,d). A subsequent annealing at 6.5 K increased all but the 1638 cm^{-1} band (Figure 12.3e), and a final >220 nm irradiation increased the DA bands by another 25% (Figure 12.3f). Al atoms and D$_2$ were also codeposited, and the spectra containing all but the weakest DA counterpart bands are assigned in our research papers [24, 25].

The seven-band DA set has a straightforward chemistry. The strongest three bands at 1932.3, 1408.1, and 1268.2 cm^{-1} are detected on laser-ablated Al deposition with normal hydrogen (absorbance 0.001). These bands double upon annealing to 6.0 K and double again upon >290 nm photolysis while AlH is reduced by 90%, AlH$_3$ increases fourfold, and weaker DA bands appear at 1915.1, 835.6, 702.4, and 631.9 cm^{-1}. Subsequent >220 nm photolysis increases the DA bands fourfold in concert and the AlH$_3$ bands twofold, while the next 6.5 K annealing increases DA bands by 25% at the expense of AlH$_3$ (Figure 12.3). Thus, AlH$_3$ is produced upon photoexcitation of AlH with excess hydrogen, and DA is formed upon ultraviolet photolysis along with AlH$_3$ and upon annealing from diffusion and reaction of AlH$_3$.

Table 12.2 Comparison of calculated and observed infrared active frequencies (cm^{-1}) for dibridged Al$_2$H$_6$ in solid hydrogen.

Sym.[a]	Mode	SCF/TZP[b]	CCSD/DZP[c]	B3LYP[d]	BPW91[d]	Observed[e]	H/D[f]
b_{1u}	8	2062 (518)	2047 (344)	1989 (419)	1934 (379)	1932 (0.069)	1.366
	9	977 (393)	954 (317)	866 (244)	808 (199)	836 (0.037)	1.376
	10	249 (18)	235 (15)	223 (13)	205 (10)	—	
b_{2u}	13	1350 (544)	1368 (463)	1292 (352)	1275 (291)	1268 (0.053)	1.379
	14	694 (353)	664 (328)	634 (263)	607 (230)	632 (0.039)	—
b_{3u}	16	2051 (101)	2024 (88)	1966 (126)	1908 (130)	1915 (0.023)	1.364
	17	1603 (1399)	1589 (1162)	1483 (1096)	1431 (968)	1408 (0.128)	1.370
	18	766 (890)	744 (684)	712 (648)	683 (575)	702 (0.048)	1.376

a) Symmetry (D$_{2h}$ point group) and mode description from Liang, C., Davy, R.D., and Schaeffer, H.F., III. Chem. Phys. Lett. 1989, 159, 393.
b) From Liang, C., Davy, R.D., and Schaeffer, H.F., III. Chem. Phys. Lett. 1989, 159, 393 with calculated intensities (km/mol) in parentheses.
c) From Shen, M. and Schaefer, H.F., III. J. Chem. Phys. 1992, 96, 2868.
d) Ref. [24, 25]: 6-311++G** basis set.
e) In solid hydrogen, integrated intensities at maximum yield in parentheses.
f) Hydrogen/deuterium isotopic frequency ratio.

The seven DA bands are therefore assigned to dialane, Al$_2$H$_6$, which is confirmed by comparison to extensive vibrational frequency calculations at different levels of quantum theory, which are compared in Table 12.2 [24, 25]. This table shows how the different levels of computational theory for harmonic frequencies tend to overestimate observed anharmonic frequencies. The two density functionals produce calculated frequencies that are very close to the observed values. Extrapolation through these sets of calculated frequencies to the observed values confirms the assignments to dialane and thus the first preparation of this novel molecule. Notice that the computed and observed infrared intensities are in very good qualitative agreement as well. Thus, the correlation between calculated and observed frequencies makes the case for the identification of this important molecule.

When the hydrogen matrix samples are annealed at 6.8 K, H$_2$ evaporates, molecular aluminum hydrides diffuse, aggregate, and their absorptions decrease, and broad absorptions appear at 1720 (20) and 720 (20) cm^{-1}. These broad bands remain on the CsI window with decreasing absorbance until room temperature is reached. The deuterium matrix samples evaporate D$_2$ at about 10 K and aluminum deuterides produce a broad 1260 (20) cm^{-1} absorption, which remains on the CsI window. The frequency ratio 1720/1260 = 1.365 demonstrates that these bands are due to Al-H/Al-D vibrations. The spectrum of pure solid (AlH$_3$)$_n$ gives strong broad bands at 1760 and 680 cm^{-1}, solid (AlH$_3$)$_n$ in Nujol reveals a very strong, broad 1592 cm^{-1} band, and solid (AlD$_3$)$_n$ gives a corresponding 1163 cm^{-1} band [24, 25]. The present broad absorptions at 1720 and 1260 cm^{-1} are therefore due to solid (AlH$_3$)$_n$ and (AlD$_3$)$_n$, respectively, on the salt window. Thus, we have made solid alane through the reaction of the elements aluminum and hydrogen to give AlH, then AlH$_3$, Al$_2$H$_6$, and finally the solid (AlH$_3$)$_n$ polymeric material. The crystal structure for this solid material shows six-coordinate aluminum with equivalent hydrogen atoms in bridge-bonding arrangements between aluminum atoms [23]. The average Al-H distance is 1.72 Å. It is interesting to note that the solid Al-H distance is intermediate between the 1.74 Å bridge and the 1.58 Å terminal Al-H distances calculated [24, 25] for Al_2H_6 and that the solid frequency of 1720 cm^{-1} is likewise intermediate between the terminal (1932, 1915 cm^{-1}) and the bridged (1408, 1268 cm^{-1}) Al_2H_6 values observed here.

12.4
Titanium and Boron Trifluoride Give the Borylene FB=TiF$_2$

The boron trifluoride molecule is one of the least reactive we have investigated, and boron and fluorine are among the most difficult theoretical subject atoms. Three medium intensity absorptions were observed at 1414 cm^{-1} in the B–F stretching region and at 721 and 638 cm^{-1} in the Ti–F stretching region from the reaction of laser-ablated Ti and BF_3 in excess argon. Density functional calculations showed that the anticipated borylene molecule FB=TiF$_2$ was the lowest energy product [28], but the presence of low-energy singlet and triplet states in this difficult system called for higher levels of theory, and we asked Roos to perform CASSCF/CASPT2 calculations. At this higher level of theory, the singlet state was 8 kcal/mol lower than the triplet in energy [28].

Table 12.3 compares the observed and computed singlet-state frequencies for the singlet FB=TiF$_2$ borylene titanium difluoride product using CASPT2 and two density functional theoretical methods. The B3LYP frequencies are all slightly higher than the observed values, and the BPW91 values are slightly lower, as expected [19, 20]. The frequencies calculated by CASPT2 are slightly higher than the B3LYP values: a similar relationship has been found among computed frequencies for these different methods for the simple group 6 metal pnictides [29]. The B3LYP computed terminal

Table 12.3 Observed CASPT2 and density functional theory calculated fundamental frequencies for the FB=TiF$_2$ borylene molecule in the ground $^1A_1$ electronic states with C$_{2v}$ structures[a].

Approx. mode description	FB=TiF$_2$						
	Obs.	Calc. (CA)	Int.	Calc. (B3)	Int.	Calc. (BP)	Int.
B–F str., a$_1$	1404	1454	403	1444	424	1394	341
Ti–F str., b$_2$	721	755	284	742	243	736	219
Ti–F str., a$_1$	638	670	151	656	108	654	101
Ti–B str., a$_1$		421	8	398	0	405	0
TiBF def., b$_1$		318	19	325	23	318	29
TiBF def., h$_2$		314	4	321	4	311	2
TiF$_2$ bend, a$_1$		148	19	133	11	141	8
FBTiF def., b$_2$		103	54	79	0	83	0
BTiF$_2$ def., b$_1$		88	1	48	25	24	32

a) Frequencies and intensities are in cm^{-1} and km/mol. Observed in an argon matrix. Frequencies and intensities calculated with CASPT2/VTZP(6s5p3d2f1g for Ti) noted (CA), B3LYP/6-311 + G (3df)/SDD for Ti noted (B3), or BPW91/6-311 + G(3df)/SDD for Ti noted (BP)in the harmonic approximation. Mode symmetry notations are based on the C2v structure.

bond stretching frequencies are 2.8, 2.9, and 2.8% higher and again the BPW91 values are slightly lower. Overall, the agreement between calculated and observed frequencies is excellent, and this correlation for the rigorous CASPT2 wavefunction-based approach and two density functional methods confirms our identification of the FB=TiF$_2$ borylene complex [28]. Finally, occupation of the CASPT2 computed molecular orbitals gives an effective bond order of 1.81 for the FB=TiF$_2$ borylene complex [28].

12.5
Ti and CH$_3$F Form the Agostic Methylidene Product CH$_2$=TiHF

Methane activation is an important process, and we were initially interested in methane activation by early transition metal atoms, but we knew from earlier experiments that the reaction product yield is relatively low. Hence, we reasoned that transition metal atom reactions with the more electron-rich methyl halides would be more favorable, and the Ti and CH$_3$F reaction provided our first contribution to the methylidene complex literature [30].

Infrared spectra of Ti and CH$_3$F reaction products revealed three sets of absorptions that are grouped by their behavior on sample irradiation and annealing. The first group (noted **m** for methylidene) at 1602.8, 757.9, 698.6, and 652.8 cm^{-1} decreases on visible light irradiation but increases markedly on ultraviolet irradiation while the second set at 504.3, 646.3, and 1105.7 cm^{-1} (noted **i** for insertion product) decreases. These photochemical changes are reversible. In addition, new bands at 782.3 and 703.8 cm^{-1} (for dimethyl titanium difluoride) increase slightly upon UV irradiation and markedly upon annealing and with higher CH$_3$F reagent concentrations.

Density functional theory calculations found the lowest energy products as CH$_3$–TiF (triplet state) and CH$_2$=TiHF (singlet state) with the latter 22 kcal/mol higher energy than the former, but both structural isomers are trapped in the solid matrix and interconverted on light irradiation. The observed frequencies for the **i** group correlated with the calculated frequencies having the higher infrared intensities for CH$_3$–TiF and the **m** group absorptions corresponded to the strongest calculated infrared frequencies for CH$_2$=TiHF. The shifts with ^{13}CH$_3$F and CD$_3$F and computed isotopic frequencies agree well enough to confirm the vibrational assignments and the CH$_2$=TiHF molecular identification [30].

Table 12.4 compares the observed isotopic frequencies with those calculated using the large all-electron Gaussian basis 6-311++G(3df,3pd) and the B3LYP and BPW91 density functionals. Notice first that the frequencies computed by both functionals are higher than the observed values, with B3LYP slightly higher than BPW91, which is the typical relationship [19, 20]. Since the hybrid and pure density functionals have sufficiently different treatments of exchange and correlation, their general agreement shows that the system is reasonably well described.

Next, focus on the two frequencies computed in the 720–820 cm^{-1} region, which show Ti–C and Ti–F stretching character along with in-plane Ti–H and C–H (nonagostic) bending motion. For example, a pure Ti–C stretching mode at

Table 12.4 Observed and calculated fundamental frequencies of $CH_2=TiHF$ in the ground electronic state (^1A)[a].

Approx. mode description	$CH_2=TiHF$				$CD_2=TiDF$					$^{13}CH_2=TiHF$			
	Exp.	Calc.[b]	Int.[b]	Calc.[c]	Int.[c]	Exp.	Calc.[b]	Int.[b]	Calc.[c]	Int.[c]	Exp.	Calc.[b]	Int.[b]	Calc.[c]	Int.[c]
v_1 CH str.		3194.8	1	3135.1	1		2365.8	2	2321.6	2		3183.8	1	3124.3	0
v_2 CH str.		2861.3	5	2759.8	4		2081.7	3	2007.8	2		2854.8	5	2753.5	4
v_3 Ti–H str.	1602.8	1671.7	411	1630.5	327	1158.6	1196.4	220	1166.7	175	1602.8	1671.7	410	1630.5	327
v_4 CH_2 scis.		1328.2	24	1293.8	22		1048.9	31	1014.2	28		1318.4	24	1285.4	22
v_5 C–Ti str.[d]	757.8	816.6	128	818.7	127	644.9	693.9	58	686.0	77	748.8	803.9	148	804.9	141
v_6 Ti–F str.[d]	698.6	738.7	148	726.5	124	702.6	735.0	169	734.1	136	692.0	733.3	127	721.4	109
v_7 CH_2 wag	652.8	689.1	165	655.8	146	522.1	543.0	120	516.1	107	646.6	682.9	160	650.1	141
v_8 CTiH bend		608.9	15	610.8	16		472.6	10	482.0	8		607.7	14	609.2	15
v_9 CH_2 twist		495.8	25	493.2	30		353.5	9	351.4	11		495.8	25	493.1	31
v_{10} CH_2 rock		386.0	4	421.8	6		310.1	3	325.8	5		382.3	4	418.8	6
v_{11} CTiF bend		197.9	19	208.5	19		168.7	5	181.9	6		197.0	19	207.3	20
v_{12} TiH OOP bend		101.9	163	88.0	146		80.5	105	70.5	95		101.7	163	87.8	145

a) Frequencies and intensities are in cm^{-1} and km/mol. Intensities are all calculated values.
b) Calculated with B3LYP/6-311++G(3df,3pd) for all atoms.
c) Calculated with BPW91/6-311++G(3df,3pd) for all atoms.
d) These modes also involve in-plane Ti–H and C–H (nonagostic) bending motion.

757.8 cm^{-1} would shift 23.7 cm^{-1} upon carbon-13 substitution, and the present band shifts only 9.0 cm^{-1}, but a large 112.9 cm^{-1} deuterium shift is observed. The mostly Ti–F stretching mode shows an unusual blueshift upon deuteration, 4.0 cm^{-1} here, and a 6.6 cm^{-1} carbon-13 redshift, which are due to mixing with the above vibrational mode shifted to a lower position upon deuteration, and the frequency calculations will be examined for this behavior. Since the hybrid B3LYP functional predicts higher frequencies for most modes, as observed previously, than the pure BPW91 functional [19, 20], as a result the pure density functional values are closer to the observed frequencies, and the mode mixing is predicted more nearly correctly. The pure BPW91 functional gives a 7.6 cm^{-1} blueshift in the mostly Ti–F stretching mode upon deuteration and a 5.1 cm^{-1} carbon-13 redshift, which are within a few cm^{-1} of the observed values, but the B3LYP functional predicts instead a 3.7 cm^{-1} redshift and a 5.4 cm^{-1} redshift, respectively. The mostly Ti–C and Ti–F stretching modes clearly involve mode mixing with hydrogen motion. Finally, notice that the smaller Gaussian basis used earlier [30] gave frequencies within 7 cm^{-1} except for the very lowest modes and even using the SDD pseudopotential on Ti made little difference except in the Ti–F mode.

The CH$_2$=TiHF methylidene is distorted at the CH$_2$ group through the agostic (Greek for *hold on to oneself*) interaction between one hydrogen and the Ti center based on our original B3LYP calculations [30] and a subsequent more detailed theoretical analysis including compliance constants [21]. Grunenberg et al. found no α-agostic distortion with the HF method while MP2 overestimates the agostic interaction. The CCSD(T) method revealed an 86–88° agostic H–C–Ti angle depending on the basis set employed. These works underscore the necessity of polarization functions on carbon to characterize the agostic interaction [21, 30]. The B3LYP and BPW91 functionals with the large all-electron 6-311++G(3df,3pd) Gaussian basis gave 91.6 and 87.6° agostic H–C–Ti angles, respectively, and B3LYP with the small all-electron 6-311++G(2d,p) Gaussian basis resulted in a 91.5° agostic angle, but using the SDD pseudopotential for Ti gave a larger 94.0° agostic angle. Finally, Grunenberg et al. conclude that the strength of α-agostic bonding is in the range – though different in nature – of a typical hydrogen bond ($\leq$10 kcal/mol) [21].

The major 782.3 and 703.8 cm^{-1} absorptions for the third group of product bands reveal Ti isotopic splittings in natural abundance for the antisymmetric and symmetric stretching vibrations of a TiF$_2$ subunit. In addition, associated 1385.2 and 1378.4 cm^{-1} bands are characteristic of methyl bending modes and a 566.9 cm^{-1} band with the antisymmetric Ti–CH$_3$ stretching mode. These bands are in excellent agreement with frequencies computed for the (CH$_3$)$_2$TiF$_2$ molecule, which is chemically analogous to the (CH$_3$)$_2$TiCl$_2$ compound.

12.6
Zr and CH$_4$ Form the Agostic Methylidene Product CH$_2$=ZrH$_2$

The simple CH$_2$=ZrH$_2$ complex presents the best experimental case for agostic distortion in a methylidene transition metal dihydride complex [31]. The neon matrix spectrum for laser-ablated Zr codeposited with 2% CH_4 in neon at 4 K revealed weak bands at 1581.0 and 1546.2 cm^{-1} upon sample deposition. These absorptions increased on vis and UV irradiation and exhibited counterparts in the Zr–D stretching region, and the H/D frequency ratios, 1.3953 and 1.3901, characterize these vibrations as symmetric and antisymmetric Zr–H_2 stretching modes. Note that these new Zr–H_2 stretching modes appear between the 1530 cm^{-1} and the 1648 cm^{-1} antisymmetric stretching modes for ZrH_2 and ZrH_4 in solid neon [31]. Two absorptions at 757.0 and 634.5 cm^{-1}, which show carbon-13 shifts appropriate for C=Zr stretching and CH_2 wagging modes, are associated upon photolysis and annealing. In addition, the CH_2D_2 reaction provides diagnostic information for the identification of the $CH_2=ZrH_2$ species because of the low symmetry and non-equivalent C–H bonds brought about by agostic distortion [31].

The final confirmation for the identification of $CH_2=ZrH_2$ is the agreement between calculated and observed frequencies for four diagnostic vibrational modes. Table 12.5 lists the frequencies calculated (not scaled) for the C_1 symmetry $CH_2=ZrH_2$ structure at the B3LYP and CCSD levels of theory [31]. Notice that the four most intense calculated frequencies are 3.2, 3.6, 1.3, and 4.5% higher, as is appropriate for the B3LYP density functional [19, 20], and the more rigorous quantum mechanical CCSD method gives slightly higher frequencies, which correlate well with the experimental absorptions. The calculated and observed $^{13}CH_2=ZrH_2$ and $CD_2=ZrD_2$ frequencies are characteristic of their vibrational modes, as discussed in the original report [31]. Thus, we conclude that $CH_2=ZrH_2$ is distorted by agostic interaction, based on our neon (and argon) matrix spectra, and calculations at several levels of single reference theory. The small T1 diagnostic that we obtained (0.016) indicates that multireference character is not a problem for

Table 12.5 Vibrational frequencies (cm^{-1}) observed and calculated for C_1 ground-state $CH_2=ZrH_2$ at the CCSD/6-311++G(2d,p)/SDD and B3LYP/6-311++G(3df,3pd)/SDD levels of theory.

Mode	Calculated[a]	Intensity[a]	Calculated[b]	Intensity[b]	Observed[c]	Ratio[d]
CH_2 str.	3210	(1)	3179	(1)		
CH_2 str.	2842	(7)	2858	(5)		
ZrH_2 str.	1650	(384)	1634	(301)	1581	0.968
ZrH_2 str.	1597	(695)	1603	(544)	1546	0.964
CH_2 scis.	1384	(17)	1320	(16)		
C=Zr str.	787	(133)	767	(130)	757	0.987
CH_2 wag	713	(206)	665	(144)	635	0.955
ZrH_2 scis.	656	(55)	642	(85)		
ZrH_2 rock	531	(29)	515	(10)		
CH_2 twist	481	(3)	408	(23)		
CH_2 rock	385	(6)	310	(75)		
ZrH_2 wag	87	(457)	240	(131)		

a) CCSD, intensities in km/mol.
b) B3LYP.
c) Neon matrix.
d) Obs./B3LYP Calc. Ratio (i.e., scale factor).

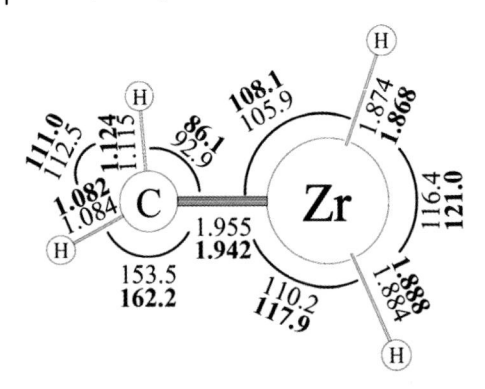

Figure 12.4 Structure and parameters calculated for C_1 ground-state $CH_2=ZrH_2$ using CCSD/6-311++G(2d,p)/SDD (in bold) and B3LYP/6-311++G(3df, 3pd)/SDD.

$CH_2=ZrH_2$. Although multireference calculations found a stable C_{2v} structure after imposing C_{2v} symmetry [32], our calculations find the lowest energy structure to be distorted. Figure 12.4 shows the structure, and the two methods give almost the same parameters save the agostic angle, which is smaller for CCSD, as found previously for $CH_2=TiHF$ [21] Finally, CCSD(T) calculations have also shown that the simple $CH_2=TiH_2$ methylidene is similarly distorted [3c].

12.7
Mo and $CHCl_3$ Form the Methylidyne $CH\equiv MoCl_3$

Methane activation with laser-ablated Mo atoms forms three products, the insertion CH_3-MoH, which is characterized by a Mo–H stretching mode at 1728.0 cm^{-1}. Visible irradiation promotes α-H transfer to give $CH_2=MoH_2$ with two Mo–H stretching modes at 1791.6 and 1759.6 cm^{-1}, and a second α-H transfer to form $CH\equiv MoH_3$ with a strong MoH_3 stretching mode at 1830 cm^{-1}. Next UV irradiation transfers H back to carbon. These photochemical processes are completely reversible. These products with comparable energies were identified by matching observed frequencies with those computed by density functional theory [33]. However, the reaction with $CHCl_3$ exclusively forms the methylidyne $CH\equiv MoCl_3$ as this product is 147 kcal/mol lower in energy than the reagents based on B3LYP calculations [34].

Four vibrational modes have been observed at 3058.2, 978.1, 438.7, and 658.9 cm^{-1} for the methylidyne $CH\equiv MoCl_3$, and these assignments are substantiated by the B3LYP frequency calculations summarized in Table 12.6 [34]. First, the highest band is characterized as a C–H stretching mode from its position and from its observed and calculated ^{13}C and D isotopic shifts. The 978.1 cm^{-1} band shows the large ^{13}C isotopic shift anticipated for the Mo–C stretching mode, which is confirmed by the resolved natural Mo isotopic shifts and the values computed using the Mo isotopic masses. The 438.7 cm^{-1} band is assigned to the strong degenerate Mo–Cl stretching mode even though the B3LYP calculation undershoots this by 13.7 cm^{-1} (3.1%), which is opposite from most of the present comparisons. The transition metal-

Table 12.6 Observed and calculated fundamental frequencies of isotopic HC≡MoCl$_3$ complexes in the ground $^1A_1$ electronic state with the C_{3v} structure[a].

Approximate Description	HC≡MoCl$_3$			H^{13}C≡MoCl$_3$			DC≡MoCl$_3$	
	Observed	Calculated	Intensity	Observed	Calculated	Observed	Calculated	Intensity
C−H str., a$_1$	3058.2	3212.2	35	3048.0	3200.4	2296.2	2387.9	24
HC≡Mo str., a$_1$	978.1[b]	1051.9	8	948.0	1019.1	932.4[b]	1005.8	6
Mo−X str., e	438.7	425.0	81 × 2	438.6	425.4	436.4	423.0	74 × 2
Mo−X str., a$_1$		380.8	9		380.7		380.7	9
H−C−Mo def., e	658.9	660.2	76 × 2	652.4	653.4	533.4	533.6	52 × 2
C−Mo−X def., e		237.7	7 × 2		232.5		212.3	7 × 2
Mo−X$_3$ umb., a$_1$		144.0	0		143.7		143.7	0
Mo−X$_2$ bend, e		100.4	0		100.4		100.4	0

a) Frequencies and intensities are in cm^{-1} and km/mol and computed with B3LYP/6-311++G(2d,p) in the harmonic approximation using the SDD core potential and basis set for Mo. The symmetry notations are based on the C3v structure.
b) Observed in an argon matrix. Band position for the major ^{98}Mo isotope.

chlorine bond is difficult to theoretically model, and a similar relationship is found for the Mo—F bond [34]. Finally, the 658.9 cm^{-1} band is assigned to the H—C—Mo deformation mode, and again the observed and calculated ^{13}C and D isotopic shifts verify this assignment.

12.8
Tungsten and Hydrogen Produce the WH$_4$(H$_2$)$_4$ Supercomplex

The first experimental observation of atomic W reactions with dihydrogen found evidence for WH$_4$ as a major reaction product, and annealing the solid neon matrix increased a group of six sharp new absorptions, which were assigned to WH$_6$, formed by the spontaneous reaction of WH$_4$ and H$_2$ [35]. These absorptions matched density functional calculated frequencies for the distorted C_{3v} prism WH$_6$ structure, which remains the highest neutral hydride and the only neutral metal hexahydride to be observed experimentally. This distorted structure is more stable than the octahedral form, which has been the subject of extensive theoretical calculations [36]. Later work in this laboratory showed that laser-ablated W atoms react with neat normal and *para*-hydrogen to form the largest possible physically stable tungsten hydride species, which is the tungsten tetrahydride tetradihydrogen [37].

Figure 12.5 illustrates infrared spectra for tungsten ablation and reaction with pure normal hydrogen during condensation at 4 K. The most prominent new absorptions are observed at 1859.1, 1830.3, and 437.2 cm^{-1} with weaker bands at 2500, 1781.6, 1007.6, and 551.5 cm^{-1} (marked with arrows). Annealing this sample to 6 K had little effect on the spectrum while full arc irradiation (>220 nm) reduced the absorptions and a subsequent annealing to 6.3 K only sharpened them. The seven absorptions marked with arrows track together upon annealing and UV irradiation, and they are thus associated with a common new product species. Note the absence of absorptions

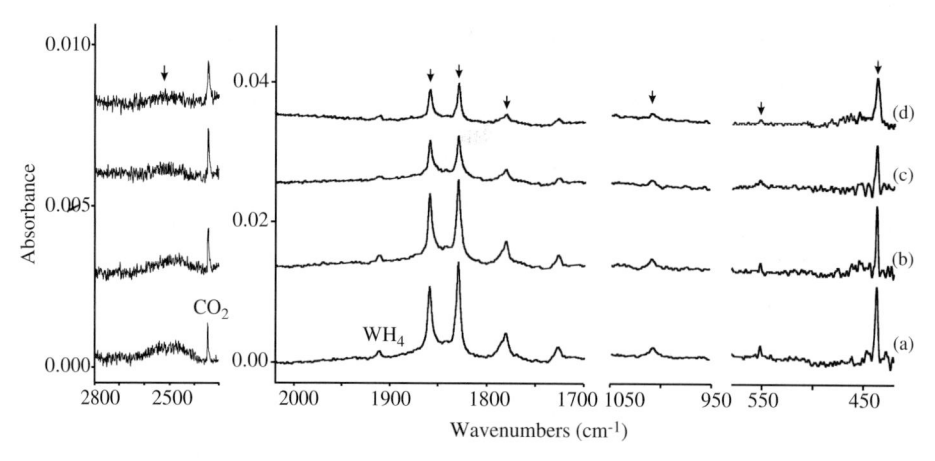

Figure 12.5 Infrared spectra for the laser-ablated W atom and pure hydrogen reaction product in solid normal hydrogen at 4 K. (a) Spectrum after reagent codeposition for 30 min, (b) after annealing at 6 K, (c) after >220 nm irradiation, and (d) after annealing at 6.3 K.

in the 1920–2020 cm^{-1} region where the strongest absorptions of WH$_6$ appeared in solid neon [35]. A final annealing at 7 K allowed the hydrogen matrix to evaporate, and the new product to aggregate leaving behind no infrared absorption. Deuterium counterparts for the stronger bands were also observed and are reported in our publication [37].

The observed infrared spectra reveal diagnostic absorptions due to molecular subunits that identify this new tungsten hydride–dihydrogen complex. First, the WH$_4$ molecule in tetrahedral symmetry has been characterized by the triply degenerate antisymmetric W–H stretching and bending modes at 1920.5 and 525.2 cm^{-1} in solid neon [35], and the weak new band at 1911.5 cm^{-1} in the solid hydrogen experiment can be assigned to WH$_4$ trapped on the surface where limited coordination can occur. The strong new absorptions at 1859.1 and 1830.3 cm^{-1} are slightly lower but still appropriate for W–H stretching modes as the 1.3951 and 1.3853 isotopic H/D frequency ratios indicate, and the new 551.5 cm^{-1} band is likewise due to an analogous H–W–H bending mode. Hence, the new product contains two or more W–H hydride bonds. Second, the new absorptions at 2500, 1781.6, 1007.6, and 437.2 cm^{-1} arise from the presence of side-bound dihydrogen molecules in this new product. The broad 2500 cm^{-1} band is characteristic of the H–H stretching mode for strongly complexed dihydrogen molecules as this mode was first observed at 2690 cm^{-1} in the important Kubas complex [38]. The 2500/1790 = 1.397 H/D isotopic ratio is in accord with that expected for an H–H stretching mode. The 1781.6 cm^{-1} band can be assigned to the antisymmetric W–(H$_2$) stretching mode on the basis of its 1.3892 isotopic H/D frequency ratio and its prediction from density functional calculations to fall about 80 cm^{-1} below the aforementioned highest antisymmetric W–H stretching mode at 1859.1 cm^{-1}. Such a mode was observed lower at 1570 cm^{-1} in the Kubas complex. The remaining two weak 1007.6 and strong 437.2 cm^{-1} bands are due to H$_2$–W–H$_2$ bending modes based on the prediction of such vibrational frequencies at 1065 and 414 cm^{-1} by density functional theory calculations [37]. The higher of these has the appropriate 1.3996 H/D isotopic frequency ratio, and the lower of these may be compared with the W–(H$_2$) deformation mode observed near 450 cm^{-1} for the Kubas complex. This identification of WH$_4$(H$_2$)$_4$ based on the above vibrational assignments is confirmed by the close correlation of the seven strongest observed and calculated frequencies (Table 12.7) [37].

Thus, we see that WH$_4$ is ligated by four dihydrogen molecules after preparation from the spontaneous reaction of W and neat H$_2$. Calculations show that the lowest energy complex that can be formed is the WH$_4$(H$_2$)$_4$ species [37]. In like fashion, laser-ablated Ru reacts with pure hydrogen to form an analogous RuH$_2$(H$_2$)$_4$ supercomplex [39].

12.9
Pt and CCl$_4$ Form the Carbene CCl$_2$=PtCl$_2$

Platinum metal is well known as a catalyst material for many chemical processes, and a number of platinum carbene complexes and their organometallic chemistry in

Table 12.7 Frequencies (cm^{-1}) calculated at the DFT/BP86/TZVPP level of theory for WH$_4$(H$_2$)$_4$. singlet state. D_{2d} symmetry.

Observed frequency[a]	Calculated frequency	Intensity[b]	Symmetry[c]	Mode description
	351	0	b$_1$	
437.2	414	171	b$_2$	Bending H$_2$–W–H$_2$
	428	6 × 2	e	Bending H$_2$–W–H$_2$
	499	0	a$_1$	
	539	0	a$_2$	
551.5	565	38 × 2	e	Bending H–W–H
	681	0	b$_1$	
	748	8 × 2	e	Bending H–W–H$_2$
	775	0	a$_2$	
	816	16 × 2	e	Bending H–W–H$_2$
	842	0	b$_2$	
	871	0	a$_1$	
	897	0	b$_1$	
1007.6	1065	172 × 2	e	Bending H$_2$–W–H$_2$
	1160	3	b$_2$	Bending H$_2$–W–H$_2$
	1284	0	a$_1$	
	1741	0	b$_1$	Asym. stretch W–H$_2$
	1767	0	a$_2$	Stretch W–H$_2$
1782.0	1790	40 × 2	e	Asym. stretch W–H$_2$
1830.6	1844	212	b$_2$	Sym. stretch W–H
1859.3	1868	53 × 2	e	Asym. stretch W–H
	1903	0	a$_1$	Totally sym. stretch W–H
2500	2657	208 × 2	e	Stretch H–H
	2683	11	b$_2$	Stretch H–H
	2740	0	a$_1$	Totally sym. stretch H–H

a) Observed here in solid hydrogen.
b) Calculated infrared intensity (km/mol).
c) Mode irreducible representation in D_{2d} symmetry.

catalytic reactions have been explored [40]. Carbon tetrachloride reacts extensively with metal atoms, and the simple methylidyne product ClC≡MoCl$_3$ is a testament to this point [34]. In the case of platinum, however, relative product energies terminate this reaction with the CCl$_2$=PtCl$_2$ carbene [41].

Laser-ablated Pt atoms react with carbon tetrachloride in excess argon during condensation at 8 K to give two strong new absorptions at 1008.3 and 886.5 cm^{-1} (884.6 cm^{-1} shoulder for chlorine isotopic splitting) (labeled **m** for methylidene) in Figure 12.6 along with bands at 1036.4 cm^{-1} (CCl$_3^+$), 1019.3 and 926.7 cm^{-1} (Cl$_2$CCl-Cl), and 898 cm^{-1} (CCl$_3$) produced by laser plume irradiation of the precursor. The latter bands are common to all laser-ablated metal experiments with CCl$_4$ [34, 40]. The new absorptions increased in concert by 10 and 20% upon irradiation in the visible ($\lambda > 420$ nm) and ultraviolet (240–380 nm) regions, respectively. Annealing at 28 K sharpened the bands and resolved the 884.6 cm^{-1} shoulder. A similar experiment with ^{13}CCl$_4$ (90% enriched) shifted the new absorptions to

12.9 Pt and CCl₄ Form the Carbene CCl₂=PtCl₂

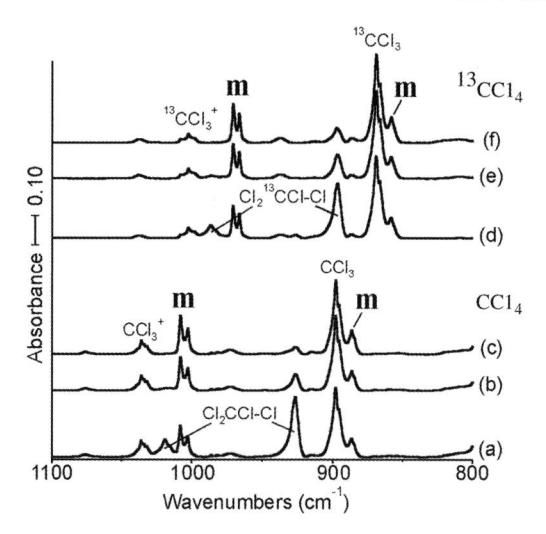

Figure 12.6 Infrared spectra observed in the 1100–800 cm^{-1} region for the laser-ablated platinum atom and carbon tetrachloride reaction products in excess argon at 10 K. (a) Pt and CCl$_4$ (0.5% in argon) codeposited for 1 h, (b) after visible ($\lambda > 420$ nm) irradiation for 20 min, and (c) after ultraviolet (240–380 nm) irradiation for 20 min. (d) Pt and ^{13}CCl$_4$ (0.5% in argon, 90% enriched) codeposited for 1 h, (e) after visible ($\lambda > 420$ nm) irradiation for 20 min, and (f) after ultraviolet (240–380 nm) irradiation for 20 min.

971.4 and to 858.5 cm^{-1} (856.1 cm^{-1} shoulder). The appearance of the ^{12}C product bands with about 1/10 of the ^{13}C product band absorbance confirms that a single carbon atom participates in these vibrational modes [41].

The two strong 1008.3 and 886.5 cm^{-1} product absorptions from the reaction of Pt and CCl$_4$ in solid argon are assigned to the CCl$_2$=PtCl$_2$ methylidene for the following reasons [36]. First, the observed bands are 1 and 2% higher than the harmonic frequencies calculated for the two strongest modes of this minimum energy product using the B3LYP density functional (Table 12.8). Furthermore, the strongest calculated (B3LYP) absorption for the 23 kcal/mol higher energy CCl$_3$-PtCl insertion product at 838 cm^{-1} is not detected. Second, the carbon-13 shifts, 36.9 and 28.0 cm^{-1}, observed for the two strong bands very closely match the computed values, 36.6 and 28.1 cm^{-1}, as the observed 12/13 isotopic frequency ratios 1.03 799 and 1.03 262 are almost the same as the calculated frequency ratios, 1.03 799 and 1.03 339. This means that the calculation is describing the same normal modes as observed for the reaction product, and it reinforces the match in calculated frequency position and high intensities. Third, the 886.5 cm^{-1} product band and 884.6 cm^{-1} shoulder have the appropriate 9/6 relative intensity for natural abundance chlorine isotopes (35,35/35,37 statistical population) for a vibration involving two equivalent chlorine atoms. The B3LYP density functional calculation predicts a 2.0 cm^{-1} shift, and we observe a 1.9 cm^{-1} shift for this isotopic effect. In the CCl$_2$=PtCl$_2$ methylidene case, the B3LYP density functional predicts slightly higher frequencies than observed, but the BPW91 pure density functional values are considerably lower than the observed values.

Table 12.8 Observed and calculated fundamental frequencies of the $CCl_2=PtCl_2$ methylidene in the ground $^1A_1$ electronic state with the C_{2v} structure[a].

Approximate description	$^{12}CCl_2=PtCl_2$					$^{13}CCl_2=PtCl_2$			
	Observed	B3LYP	Intensity	BPW91	Intensity	Observed	B3LYP	Intensity	BPW91	Intensity
$Cl_2-C=Pt$ str.,[b] a_1	1008.3	999.9	284	981.5	242	971.4	963.3	264	945.4	225
C--Cl str., b_2	886.5	869.7	198	838.9	190	858.5	841.6	185	811.8	178
CCl_2 wag, b_1		451.1	0	427.9	5		434.7	0	427.5	5
CCl_2 bend,[c] a_1		438.9	4	426.6	0		438.5	3	411.4	0
Pt--Cl str. b_1		365.3	69	367.9	60		365.1	68	367.5	60
Pt--Cl str., a_1		349.1	8	350.9	8		349.1	8	350.9	8
CCl_2 def., a_1		227.7	0	222.0	0		227.7	0	222.0	0
CCl_2 rock, b_2		213.2	0	207.2	0		212.3	0	206.4	0
$PtCl_2$ wag, b_2		102.2[d]	1	101.4	1		102.2	1	101.4	1

a) Frequencies and intensities are in cm^{-1} and km/mol. Observed in an argon matrix. Frequencies and intensities computed with B3LYP/6-311++G(3df,3pd) in the harmonic approximation using the SDD core potential and basis set for Pt and using the BPW91 functional. Symmetry notations are based on the C_{2v} structure.
b) Mode has antisymmetric $Cl_2-C=Pt$ stretching character.
c) Mode has some symmetric $Cl_2-C=Pt$ stretching character.
d) Three real 96, 65, and 23 cm^{-1} frequencies are not listed.

Interestingly, the 1008.3 cm^{-1} vibrational coordinate involves mostly Pt=C stretching mixed with symmetric C–Cl$_2$ displacement. For a pure symmetric Cl–C–Cl stretching mode with the 115.3° valence angle calculated here, the 12/13 isotopic frequency ratio would be 1.02 503, which is substantially less than the observed 1.03 799 value. This underscores the description of this normal mode as C vibrating back and forth between Pt and two Cl atoms. The calculated 438.9 cm^{-1} mode with only 0.4 cm^{-1} carbon-13 shift is the symmetric Cl$_2$–C=Pt stretching counterpart, which involves little carbon motion. The matrix infrared spectrum with the two strongest modes calculated for CCl$_2$=PtCl$_2$ facilitates the identification of this simple platinum carbene complex.

12.10
Th and CH$_4$ Yield the Agostic Methylidene Product CH$_2$=ThH$_2$

Thorium and uranium chemistries are centered on IV and VI oxidation state compounds [22], respectively, so we thought it would be possible to prepare some of the above group 4 and 6 complexes using the same reactions of these two least radioactive early actinide metals. Following Ti, Zr, and Hf reactions, Th experiments were conducted first with methane, and a group of five bands (1435.7, 1397.1 cm^{-1} ThH$_2$ stretch, 670.8 cm^{-1} C=Th stretch, 634.6 cm^{-1} CH$_2$ wag, and 458.7 cm^{-1} ThH$_2$ bend) tracked together upon photolysis and annealing and were assigned to CH$_2$=ThH$_2$ based on the mode characterizations and excellent agreement with calculated frequencies, as shown in Table 12.9 [42]. Notice that the isotopic shifts for the diagnostic Th=C bond stretching mode are nicely matched by the calculations. The computed structure reveals comparable agostic distortion to that calculated for CH$_2$=ZrH$_2$. Following the results with Ti, thorium reactions with methyl fluoride gave a higher product yield, and the Th–H, Th–F, and Th=C stretching modes and the CH$_2$ wagging mode correlate well with B3LYP computed values for the CH$_2$=ThHF product [43].

12.11
U and CHF$_3$ Produce the Methylidyne CH≡UF$_3$

Following the logic described above for Mo, it appeared that higher oxidation-state uranium products would be favored with more heavily halogenated reagents. The analogous reaction with uranium and CHF$_3$ produced similar product spectra. New absorptions were observed at 576.2, 540.2, and 527.5 cm^{-1} in the infrared spectrum recorded after the initial reaction of U and CHF$_3$. These bands increase by 30% upon ultraviolet irradiation ($\lambda > 290$ nm) and another 20% upon further ultraviolet irradiation ($\lambda > 220$ nm), and they are due to the major reaction product. The reaction with CDF$_3$ gave the same upper band, the strong band shifted to 535.9 cm^{-1}, and the lower band shifted below our region of observation.

On the basis of the agreement with density functional computed frequencies, Table 12.10, the higher bands are assigned to the symmetric stretching mode of a

Table 12.9 Observed and calculated fundamental frequencies of $CH_2=ThH_2$ [a].

Approximate mode description	$CH_2=ThH_2$			$^{13}CH_2=ThH_2$			
	Observed	Calculated	Intensity	Observed	Calculated	Intensity	Observed	Calculated	Intensity

Approximate mode description	Observed	Calculated	Intensity	Observed	Calculated	Intensity	Observed	Calculated	Intensity
CH_2 stretch		3142.6	2		3132.2	2		2321.7	2
CH_2 stretch		2861.4	11		2854.9	11		2084.7	2
ThH_2 stretch	1435.7	1434.9	350	1435.7	1434.8	350	[b]	1023.5	110
ThH_2 stretch	1397.1	1394.2	698	1391.7	1394.2	698	[b]	1005.7	98
CH_2 bend		1327.5	11		1320.5	11		989.0	340
$C=Th$ stretch	670.8	679.6	178	651.5	659.7	173	602.9	614.8	127
CH_2 wag	634.6	633.0	161	629.2	627.5	157	499.2	495.1	109
ThH_2 bend	458.7	492.8	110		492.3	108		355.8	34
ThH_2 rock		460.8	5		458.2	4		344.4	29
CH_2 twist		343.0	30		342.5	30		245.3	18
ThH_2 wag		321.9	65		321.6	66		230.2	30
CH_2 rock		248.4	62		248.1	62		177.5	30

a) B3LYP/6-311++G(3df,3pd)/SDD level of theory. Frequencies and infrared intensities are in cm^{-1} and km/mol. Observed are in argon matrix. Intensities are calculated values.
b) Region covered by CD_4 precursor absorption.

12.11 U and CHF₃ Produce the Methylidyne CH≡UF₃

Table 12.10 Observed and calculated fundamental vibrational frequencies for the C_{3v} CH≡UF₃ (X = H, D, F) molecules.

Mode description	¹³CH≡UF₃ Observed[a]	¹³CH≡UF₃ Calculated[b]	¹³CH≡UF₃ Observed[a]	¹³CH≡UF₃ Calculated[b]	CD≡UF₃ Observed[a]	CD≡UF₃ Calculated[b]	CH≡UF₃···CD≡UF₃ Calculated[c]	CH≡UF₃···CD≡UF₃ Calculated[c]
C—X str., a₁	—	2979(2)	[d]	2969(2)	—	2200(1)	3069(1)	2267(0)
U≡CX str., a₁	—	747(46)	[d]	721(42)	—	717(41)	715(20)	687(16)
U—F sym. str., a₁	576.2	585(122)	[d]	585(123)	576.2	586(123)	586(142)	586(141)
U—F antisym. str., e	540.2	561(284)	539.2	559(280)	535.9	541(207)	575(560)	545(478)
U≡C—X bend, e	527.5	508(34)	[d]	506(24)	—	412(49)	520(43)	436(71)

a) Absorptions observed in argon matrix.
b) Vibrational frequencies (cm⁻¹) and intensities (km/mol, in parenthesis) are calculated using PW91/TZ2P. Three real lower frequency bending modes (a⁻¹, e, e) are not listed.
c) B3LYP/6-311++G(3df,3pd)/SDD calculation.
d) Sample too dilute to observe weaker bands.

trigonal UF$_3$ group and the strongest band to the degenerate antisymmetric counterpart. The lowest band is due to the degenerate H−U−F bending mode, which is blueshifted in the chloroform product. The B3LYP values are higher than the PW91 values, and in the case of the degenerate H−U−F bending mode, closer to the observed spectral bands. Theoretical analysis shows that a triple bond is formed between carbon and uranium in these methylidyne HC≡UF$_3$ molecules [44].

Expanding the extensive methylidene and methylidyne organometallic chemistry of the transition metals [45] to include actinide metals is a difficult challenge. This may be due in part to the fact that actinide 6d and 5f valence orbitals do not behave like transition metal nd orbitals, but we find some transition metal-like behavior in the early actinides. These investigations into Th and U methylidenes and methylidynes [42–44, 46] provide information on the existence and the nature of novel actinide-carbon multiple bonded species.

Acknowledgment

The author gratefully acknowledges support for this research from the National Science Foundation and the Department of Energy (US), as well as the contributions of the many coauthors whose names are given in the references below. The author would also like to thank three distinguished quantum chemists, namely, C. Bauschlicher, P. Pyykkö, and B. Roos, who have greatly contributed to his quantum chemical education through many e-mail messages.

References

1 Jacox, M.E. (1994) *Chem. Phys.*, **189**, 149.
2 Bondybey, V.E., Smith, A.M., and Agreiter, J. (1996) *Chem. Rev.*, **96**, 2113.
3 (a) Zhou, M., Andrews, L., and Bauschlicher, C.W., Jr. (2001) *Chem. Rev.*, **101**, 1931 and references therein;
(b) Andrews, L. and Citra, A. (2002) *Chem. Rev.*, **102**, 885 and references therein;
(c) Andrews, L. (2004) *Chem. Soc. Rev.*, **33**, 123 and references therein;(d) Andrews, L. and Cho, H.-G. (2006) *Organometallics*, **25**, 4040 and references therein.
4 Andrews, W.L.S. and Pimentel, G.C. (1966) *J. Chem. Phys.*, **44**, 2361.
5 Billups, W.E., Chang, S.-C., Hauge, R.H., and Margrave, J.L. (1995) *J. Am. Chem. Soc.*, **117**, 1387.
6 Knight, L.B., Jr., Gregory, B.W., Corbranchi, S.T., Feller, D., and Davidson, E.R. (1987) *J. Am. Chem. Soc.*, **109**, 3521.
7 Burkholder, T.R. and Andrews, L. (1991) *J. Chem. Phys.*, **95**, 8697.
8 Li, S., Weimer, H.A., Van Zee, R.J., and Weltner, W. (1997) *J. Chem. Phys.*, **106**, 2583.
9 Tam, S., Macler, M., DeRose, M.E., and Fajardo, M.E. (2000) *J. Chem. Phys.*, **113**, 9067.
10 Frisch, M.J. et al. (2004) Gaussian 03, Revision E.01, Gaussian, Inc., Wallingford, CT.
11 (a) Becke, A.D. (1993) *J. Chem. Phys.*, **98**, 5648; (b) Lee, C., Yang, Y., and Parr, R.G. (1988) *Phys. Rev. B*, **37**, 785; (c) Stephens, P.J., Devlin, F.J., Chabalowski, C.F., and Frisch, M.J. (1994) *J. Phys. Chem.*, **98**, 11623 and references therein.
12 (a) Becke, A.D. (1988) *Phys. Rev. A*, **38**, 3098; (b) Perdew, J.P., Burke, K., and Wang, Y. (1996) *Phys. Rev. B*, **54**, 16533 and references therein.

13 (a) Frisch, M.J., Pople, J.A., and Binkley, J.S. (1984) *J. Chem. Phys.*, **80**, 3265; (b) Andrae, D., Haeussermann, U., Dolg, M., Stoll, H., and Preuss, H. (1990) *Theor. Chim. Acta*, **77**, 123 (transition metals); (c) Dolg, M., Stoll, H., and Preuss, H. (1989) *J. Chem. Phys.*, **90**, 1730 (actinide metals).

14 (a) Purvis, G.D. and Bartlett, R.J. (1982) *J. Chem. Phys.*, **76**, 1910; (b) Pople, J.A., Head-Gordon, M., and Raghavachari, K. (1987) *J. Chem. Phys.*, **87**, 5968.

15 (a) Andrews, L. (1968) *J. Am. Chem. Soc.*, **90**, 7368; (b) Andrews, L. (1969) *J. Chem. Phys.*, **50**, 4288.

16 (a) Hatzenbuhler, D.A. and Andrews, L. (1972) *J. Chem. Phys.*, **56**, 3398; (b) Andrews, L. and Smardzewski, R.R. (1973) *J. Chem. Phys.*, **58**, 2258.

17 Allen, W.D., Horner, D.A., Dekock, R.L., Remington, R.B., and Schaefer, H.F., III (1989) *Chem. Phys.*, **133**, 11 and references therein.

18 Wang, X. and Andrews, L. (2009) *Mol. Phys.*, **107**, 739.

19 Scott, A.P. and Radom, L. (1996) *J. Phys. Chem.*, **100**, 16502.

20 Andersson, M.P. and Uvdal, P.L. (2005) *J. Phys. Chem. A*, **109**, 3937.

21 von Frantzius, G., Streubel, R., Brandhorst, K., and Grunenberg, J. (2006) *Organometallics*, **25**, 118.

22 Cotton, F.A., Wilkinson, G., Murillo, C.A., and Bochmann, M. (1999) *Advanced Inorganic Chemistry*, 6th edn, John Wiley & Sons. Inc., New York.

23 Turley, J.W. and Rinn, H.W. (1969) *Inorg. Chem.*, **8**, 18.

24 Andrews, L. and Wang, X. (2003) *Science*, **299**, 2049.

25 Wang, X., Andrews, L., DeRose, M.E., Tam, S., and Fajardo, M.E. (2003) *J. Am. Chem. Soc.*, **125**, 9218.

26 Downs, A.J. and Pulham, C.R. (1994) *Chem. Soc. Rev.*, **1994**, 175.

27 (a) Chertihin, G.V. and Andrews, L. (1993) *J. Phys. Chem.*, **97**, 10295; (b) Kurth, F.A., Eberlein, R.A., Schnöckel, H., Downs, A.J., and Pulham, C.R. (1993) *J. Chem. Soc., Chem. Commun.*, 1302; (c) Pullumbi, P., Bouteiller, Y., Manceron, L., and Mijoule, C. (1994) *Chem. Phys.*, **185**, 25.

28 Wang, X., Roos, B.O., and Andrews, L. (2010) *Angew. Chem. Intl. Ed.*, **49**, 157.

29 Wang, X., Andrews, L., Lindh, R., Veryazov, V., and Roos, B.O. (2008) *J. Phys. Chem. A*, **112**, 8030.

30 Cho, H.-G. and Andrews, L. (2004) *J. Phys. Chem. A*, **108**, 6294 (Ti + CH_3F).

31 Cho, H.-G., Wang, X., and Andrews, L. (2005) *J. Am. Chem. Soc.*, **127**, 465 (Zr + CH_4).

32 Cundari, T.R. and Gordon, M.S. (1992) *J. Am. Chem. Soc.*, **114**, 539.

33 Cho, H.-G. and Andrews, L. (2005) *J. Am. Chem. Soc.*, **127**, 8226 (Mo + CH_4).

34 Lyon, J.T., Cho, H.-G., and Andrews, L. (2008) *Organometallics*, **27**, 6373 (Mo + $CHCl_3$).

35 Wang, X. and Andrews, L. (2002) *J. Phys. Chem. A*, **106**, 6720 (W + H_2).

36 Kaupp, M. (1996) *J. Am. Chem. Soc.*, **118**, 3018.

37 Wang, X., Andrews, L., Infante, I., and Gagliardi, L. (2008) *J. Am. Chem. Soc.*, **130**, 1972 (W + H_2).

38 Kubas, G.J., Ryan, R.R., Swanson, B.I., Vergamini, P.J., and Wasserman, H.J. (1984) *J. Am. Chem. Soc.*, **106**, 451.

39 Wang, X. and Andrews, L. (2008) *Organometallics*, **27**, 4273 (Ru + H_2).

40 Herndon, J.W. (2007) *Coord. Chem. Rev.*, **251**, 1158 and earlier review articles in this series.

41 Cho, H.-G. and Andrews, L. (2008) *J. Am. Chem. Soc.*, **130**, 15836 (Pt + CCl_4).

42 Andrews, L. and Cho, H.-G. (2005) *J. Phys. Chem. A*, **109**, 6796 (Th + CH_4).

43 Lyon, J.T. and Andrews, L. (2005) *Inorg. Chem.*, **44**, 8610 (Th + CH_3X).

44 Lyon, J.T., Hu, H.-S., Andrews, L., and Li, J. (2007) *Proc. Natl. Acad. Sci.*, **104**, 18919 (U + CHF_3).

45 Schrock, R.R. (2002) *Chem. Rev.*, **102**, 145.

46 Lyon, J.T. and Andrews, L. (2006) *Inorg. Chem.*, **45**, 1847 (U + CH_3X).

13
Astronomical Molecular Spectroscopy
Timothy W. Schmidt

13.1
The Giants' Shoulders

Stargazing, in its most primitive form, has entertained our species for hundreds of thousands of years. The ancients used knowledge of the positions of fixed stars as navigational aids and also recognized wandering stars (planets), comets, and novae. The astrolabe, an instrument for fixing the positions of stars, was invented in Greece some 2200 years ago. In medieval times, the positions of stars were of great importance to the Islamic world to indicate the correct times for prayer and the direction of Mecca, and the astrolabe was an important tool. Indeed, one of the greatest observational astronomers, Tycho Brahe,[1] performed his measurements without a telescope. He used a "great mural quadrant," whereby stars were sighted through a slot in a wall with a tool attached to a 90° arc. Accurate measurements of the positions of stars and planets were made in this way, and they remained the most accurate until the introduction of well-built telescopes. Brahe died in 1601, passing his legacy to Johannes Kepler,[2] who had joined him as an assistant only a year before. Kepler continued to work with Brahe's observations for a decade, during which he developed the laws of planetary motion for which he is most famous, and advanced theories on geometric optics. Kepler was a contemporary of Galileo,[3] who is credited with the development and improvement of the refractive telescope. Galileo observed the four largest moons of Jupiter and described the phases of Venus as being similar to the Earth's moon. Hans Lipperhey,[4] a German–Dutch lensmaker, is widely considered the inventor of the telescope. However, lenses themselves had been known for at least 2000 years, and are the subject of many writings from ancient and medieval times. Indeed, Ibn al-Haytham,[5] considered by many the first scientist, wrote extensively on optical properties of light and vision in his *Kitáb al-Manázir*

---

1) Tyge Ottesen Brahe, 1546–1601, Danish astronomer.
2) 1571–1630, German astronomer and physicist.
3) Galileo Galilei, 1564–1642, Italian physicist, astronomer, and mathematician.
4) 1570–1619.
5) Abu Ali Hasan Ibn al-Haytham (called Alhazen in Europe), 965–1039, Persian polymath.

Computational Spectroscopy: Methods, Experiments and Applications. Edited by Jörg Grunenberg
Copyright © 2010 WILEY-VCH Verlag GmbH & Co. KGaA, Weinheim
ISBN: 978-3-527-32649-5

(*Book of Optics*, 1021), summarized by Roger Bacon over 200 years later in his *Perspectiva* (1267). Al-Haytham's work was translated into Latin in 1270 as *Opticae thesaurus Alhazeni*. He described many studies on parabolic and spherical mirrors, the foundation for all modern optical telescopes, and performed the first recorded experiments on the separation of white light into its constituent colors.

Around 1650, in Bologna, Grimaldi[6] observed that small apertures cause what he referred to as *diffringere*. Moreover, he also observed that diffraction also resulted in coloration. As an astronomer, Grimaldi made accurate maps of the visible geographical features of the Earth's moon and named them after scientists. Grimaldi's diffraction was explained by the wave theory of light put forward by Christiaan Huygens[7] in his *Traité de la lumière* (1690), though the reason of the appearance of different colors eluded him. In addition to studying light, Huygens described the rings of Saturn and the largest of its moons, Titan. The atmosphere of Titan is today the subject of intense chemical research following the successful deployment of the Huygens probe by the Cassini orbiter in 2005.

Isaac Newton[8] wrote to the Royal Society in 1672 a letter enclosing his New Theory about Light and Colors:

Where Light is declared to be not Similar or Homogeneal, but consisting of difform rays, some of which are more refrangible than others: And Colors are affirm'd to be not Qualifications of Light, deriv'd from Refractions of natural Bodies, (as 'tis generally believed;) but Original and Connate properties, which in divers rays are divers: Where several Observations and Experiments are alledged to prove the said Theory.

Newton's letter begins, "To perform my late promise to you, I shall without further ceremony acquaint you, that in the beginning of the Year 1666 (at which time I applied my self to the grinding of Optick glasses of other figures than Spherical), I procured me a Triangular glass-Prisme, to try therewith the celebrated Phænomena of Colours. And in order thereto having darkened my chamber, and made a small hole in my window-shuts, to let in a convenient quantity of the Suns light, I placed my Prisme at his entrance, that it might be thereby refracted to the opposite wall. It was at first a very pleasing divertisement, to view the vivid and intense colours produced thereby."

Newton explained refraction as a property of a corpuscular light, in which the particles of different colors were refracted to different extents. This communication marks the birth of spectroscopy, with the description of color arising from the selective absorption and transmission of various components of the visible spectrum. Newton communicated his studies on light to the general public in his *Opticks: or, a Treatise of the Relexions, Refractions, Inflexions and Colours of Light*, released in 1704. The treatise was written in contemporary English, rather than Latin, and dealt with

---

6) Francesco Maria Grimaldi, 1618–1663, Italian mathematician, physicist, and Jesuit priest.
7) 1629–1695, Dutch astronomer, physicist, and mathematician.
8) 1643–1727, English physicist, astronomer, mathematician, and alchemist.

everyday phenomena in the language of everyday people. Importantly, just like Al-Haytham, Kepler, Galileo, Grimaldi, and Huygens, Newton contributed to the development of astronomy and designed a reflecting telescope that now bears his name. With the telecope and the prism, we have the tools of astronomical spectroscopy. Newton, *standing on the shoulders of giants*,[9] provided the tools with which to investigate phenomena in the greatest laboratory of all: Outer Space.

13.2
The First Spectroscopists and Seeds of Quantum Theory

The wave theory of Huygens and the corpuscular theory of Newton were both hugely successful, yet the wave theory of light gained the upper hand during the nineteenth century. Ironically, Newton's reputation and achievements served as a barrier to further developments in fundamental understanding of light until the late eighteenth century. In November 1801, Thomas Young[10] delivered the Bakerian Lecture to the Royal Society entitled "On the Theory of Light and Colours" [1]. In this lecture, he begins carefully by deferring to Newton, going so far as to claim that his theories are mostly a reinterpretation of Newton's own writings: "A more extensive examination of Newton's various writings has shown me, that he was in reality the first that suggested such a theory as I shall endeavour to maintain." Yet, by the end of the lecture, Young has all but destroyed the corpuscular theory of light by demonstrating, quantitatively, that the colors of thin plates, oil films, and diffraction gratings ("Mr Coventry's Exquisite Micrometers") can all be explained by an *undulatory* theory of light. He postulates that "The Sensation of different Colours depends on the different frequency of Vibrations excited by Light in the Retina," having himself dissected many [2]. Moreover, he accurately calculates the wave number of the extrema of the visible spectrum as $37\,640\,\text{in.}^{-1}$ ($14\,820\,\text{cm}^{-1}$, 675 nm) and $59\,750\,\text{in.}^{-1}$ ($23\,520\,\text{cm}^{-1}$, 425 nm), with yellow estimated at $44\,000\,\text{in.}^{-1}$ ($17\,320\,\text{cm}^{-1}$, 577 nm) (Figure 13.1). William Herschel[11] had demonstrated, in 1800, that the sun's radiation extended into the infrared, and Ritter made similar observations in the ultraviolet the following year. In the same volume of *Philosophical Transactions*, as Young's lecture is published, Herschel publishes estimates of the sizes of the asteroids Ceres and Pallas [3] and Wollaston[12] remarks on dark lines within the solar spectrum [4]. Analyzing the light of a flame, he says that the blue part of the flame exhibits a series of four bands from the red to the blue, and a fifth band at the boundary of the blue and violet regions. The red band was terminated with a

---

9) Newton used this ancient metaphor in a letter to Robert Hooke, which some have taken as a sarcastic reference to Hooke's stature.
10) 1773–1829, English physician, physicist, and Egyptologist.
11) Friedrich Wilhelm Herschel, 1738–1822, German-born British astronomer. Discovered Uranus and named it George, for the king of England.
12) William Hyde Wollaston, 1766–1828, English chemist and physicist. Discovered palladium and rhodium.

Colors.	Length of an undulation in parts of an inch, in air.	Number of undulations in an inch.	Number of undulations in a second.
Extreme	.0000266	37640	463 millions of millions
Red	.0000256	39180	482
Intermediate	.0000246	40720	501
Orange	.0000240	41610	512
Intermediate	.0000235	42510	523
Yellow	.0000227	44000	542
Intermediate	.0000219	45600	561 ($= 2^{48}$ nearly)
Green	.0000211	47460	584
Intermediate	.0000203	49320	607
Blue	.0000196	51110	629
Intermediate	.0000189	52910	652
Indigo	.0000185	54070	665
Intermediate	.0000181	55240	680
Violet	.0000174	57490	707
Extreme	.0000167	59750	735

Figure 13.1 Young's tabulated wavelengths and wave numbers for visible light, from his 1801 Bakerian Lecture. The frequencies suppose the speed of light to be 500 000 000 000 ft. in 8.5 min, 2.99×10^8 ms^{-1}, from astronomical measurements.

bright line of yellow. On the spectrum of an electrical discharge in air, he says, "I cannot undertake to explain."

Fraunhofer,[13] who had become a skilled glassmaker and instrument maker, invented the spectroscope in 1814. He located 574 dark lines appearing in the solar spectrum, and still denoted Fraunhofer lines in his honor. The most prominent of these dark lines were labeled with letters A–K, and Fraunhofer noted the coincidence between the D-lines and a yellow emission feature known to occur in flames, which Wollaston had described. In the early 1820s, he affixed his spectroscope to a telescope in order to study the spectrum of stars other than the sun. The brightest star, *Sirius*, was found to exhibit a spectrum quite different from and simpler than that of the sun, while that of the Venus was virtually identical, it being seen in reflected sunlight.

Slowly, many of the dark lines exhibited in the solar spectrum were observed in emission from laboratory flames, with various lines becoming associated with particular elements. Indeed, the D-lines of sodium observed by Wollaston and Fraunhofer are still referred to as such. The exact explanation for the appearance of lines in absorption and emission came from Kirchhoff and Bunsen[14] who published, in 1860, a list of laboratory lines and their correspondences with the absorption features in the solar spectrum (Figure 13.2) [5]. Their work conclusively proved that the sun contained the same elements as appeared on earth, among them calcium and sodium. A few years earlier, William Swan[15] described the emission

---

13) Joseph (von) Fraunhofer, 1787–1826, German.
14) Robert Wilhelm Eberhard Bunsen, 1811–1899, and Gustav Robert Kirchhoff, 1824–1887, German physicists and chemists. Discovered rubidium and caesium.
15) 1818–1894, Scottish.

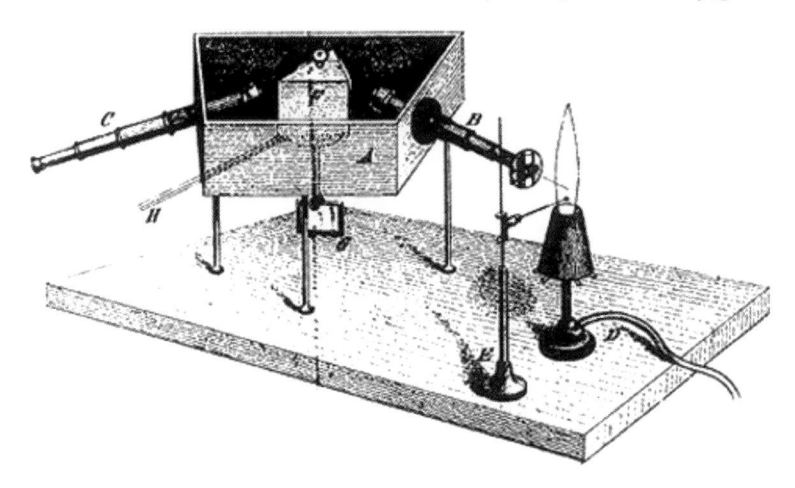

Figure 13.2 Bunsen and Kirchhoff's spectroscope from *Annalen der Physik* (1860), showing Bunsen's famous burner. With this apparatus, one could observe the Swan bands of C_2 and the hydrocarbon bands due to the CH radical.

bands that bear his name to the Royal Society of Edinburgh [6]. These had been observed by Wollaston as the dominant emission from the blue part of a flame.

William Huggins[16] pounced on the scientific opportunities facilitated by the work of Kirchhoff and Bunsen and published the spectra of many elements in 1864 [7]. He built a telescope and spectrograph in his garden and collected the spectra of various stars, nebulae, and comets. This work was published almost simultaneously with that of the American L.M. Rutherfurd, who also collected and classified various stellar spectra. The first cometary spectrum, of Comet Tempel, was reported in 1864 by Donati,[17] who reported the appearance of the Swan bands [8]. The appearance of Swan bands in cometary spectra was confirmed by Huggins in 1868 in his report of the spectrum of Comet Winnecke and also in Coggia's Comet of 1874 [9, 10]. Remarkably, by comparison with the positions of Swan bands obtained from a terrestrial flame, Huggins demonstrated that the relative velocity of the comet was some 40 miles per second. The bright band near Fraunhofer's G-line at 4300 Å was conspicuously absent from the comet's spectrum, being normally associated with the Swan bands.

The Great Comet of 1881 was first observed by Tebbutt at Windsor, New South Wales, now on the outskirts of Sydney,[18] and its spectrum was obtained by Huggins and others (Figure 13.3) [11]. Huggins correctly concluded that the bright lines appearing at 3883 and 3870 Å were due to carbon, in combination with nitrogen, noting Liveing and Dewar's demonstration that the spectrum was obtained from cyanogen [12]. Huggin's 1881 spectrum also revealed a band near 4050 Å, which would not be identified for another 70 years.

16) 1824–1910, English astronomer.
17) Giovanni Battista Donati, 1826–1873, Italian astronomer.
18) John Tebbutt, 1834–1916, Australian astronomer.

Figure 13.3 *Top*: The spectrum of the Great Comet of 1881, discovered by Tebbutt, as recorded and published by Huggins. The Fraunhofer lines indicated are due to Ca^+ (H, K), Fe (G), and H $n = 5 \to 2$ (h). Two molecular species are responsible for the emissions, CN and C_3. The latter would not be identified until 1951. *Bottom*: Tebbutt and his observatories as depicted in Australia's original $100 note.

By the late nineteenth century, spectroscopy was established as an astronomical tool of great utility. The inherent curiosity of our species had spurred fundamental studies of nature that had by now revealed the spectra of elements and molecules, though these were not yet identified as such. The comets had revealed the great ubiquity of the carbonaceous Swan bands, a hydrocarbon-associated emission of about 4300 Å, and bright emissions from a cyanogen-related form of carbon. These phenomena were revealed to fundamental physicists by astronomical spectroscopists, but a proper understanding of them would have to wait until the advent of quantum mechanics.

This chapter is about the relationship between astronomical molecular spectroscopy and computational spectrometry. A brief account of how the former gave rise to the discovery of the quantum mechanical phenomena that necessitated the latter has already been given. Happily, in 2010, we are approaching the age whereby quantum mechanics can "give back" to astronomy and assist in solving problems, some as old as 90 years. Between these periods, molecular astronomy and computational spectrometry have walked hand in hand, one aiding the other in development and interpretation. In the following sections, we will deal with molecular astronomy across four orders of magnitude of the electromagnetic spectrum, using case studies to illustrate how computational spectrometry has advanced since the beginning of the quantum theory.

13.3
Small Molecules

The spectra of many small molecules were obtained astronomically in the late nineteenth and early twentieth centuries. However, in many instances, their identification depended on the detailed understanding of spectroscopy afforded by the developments in quantum theory in the first few decades of the twentieth century. As of 2010, spectroscopic constants of small astronomical molecules can be calculated to within 0.1%, but it has not always been so. Nevertheless, computational spectrometry has played its part in interpreting astronomical spectra throughout the twentieth century, and the exquisite data afforded by molecular spectroscopy have provided quantum chemists with the sparring partners necessary to hone their code.

13.3.1
CH, CN, CO, CO$^+$

By the end of the nineteenth century, the hydrocarbon spectrum observed by Wollaston had been extensively reproduced in the laboratory. Although it was not possible to establish with certainty whether hydrogen was required for production of the Swan bands at 5635, 5165, and 4737 Å, it appeared that it *was* required for the 4315 Å band, and as such this and the other bands associated with this feature came to be known as the "hydrocarbon bands." Similarly, the "cyanogen bands," identified first in comets, were known to involve both carbon and nitrogen. Liveing and Dewar referred to these as the nitrocarbon bands in 1880 [12], and they were observed brightly in the Great Comet of 1881 [11]. Other band systems known to be associated with carbon were the Ångström bands [13] and the "comet tail bands" [14].

The comet tail bands were first observed in the spectra of comets Daniel (1907d) and Morehouse (1908c), especially bright in the tails. They were reproduced by Fowler and shown most likely to be due to CO [14, 15]. As they were always produced at very low pressures, some referred to these spectra as the low-pressure carbon bands. While CO is a major component of cometary gas, the lowest transition that can be excited from the ground state is already in the vacuum ultraviolet and inaccessible to early astronomers. As such, the spectrum of neutral CO was not known at first by cometary spectra but by experimentation with discharge tubes.

Birge noted in 1926, "Until very recently it was impossible to give the exact chemical origin of practically any band system, even of those most exhaustively investigated. The long controversy concerning the Swan bands is a case in point." Indeed, the origin of the Swan bands in C_2 would not be accepted for a few years, and Birge thought "it is doubtful such a molecule (diatomic) exists" [16]. In 1926,

The National Research Council (Washington) published "Molecular Spectra in Gases," the report of the National Research Council Committee on Radiation in Gases. A bound copy would set you back $4.50, but it would have been worth it just to read Chapter 2: "a thirty page survey of 'quantum dynamics and the correspondence principle', ... entirely on the basis of the old quantum theory rather than the Heisenberg–Schroedinger mechanics, which were developed too late for incorporation" [17]. Notwithstanding, the theories were advanced sufficiently that moments of inertia extracted from the quantized rotational structure were sufficient to identify the carriers of the hydrocarbon and cyanogen bands as CH and CN, respectively. The identification of the comet tail bands with CO^+ stems from a combination of the chemical evidence, the measured spectroscopic constants (now more or less understood), and the fact that they occur in the region of the discharge tube associated with cations [16]. These three diatomic molecules were among the first radicals to be identified. They were discovered in space and contributed to the development of the spectroscopy, which in the mid-1920s was developing into a quantitative science.

In 1940, MacKellar reported on "Evidence for the molecular origin of some hitherto unidentified interstellar lines" [18]. Coincidences between the lowest J-lines of CH and CN demonstrated these species to be interstellar. Indeed, from these spectra, the temperature of interstellar space is calculated to not to exceed 2.7 K. In fact, this is also the first evidence for the cosmic microwave background (CMB), for the discovery of which, in 1965, Penzias and Wilson were awarded the 1978 Nobel Prize in Physics. The dipolar CN molecule has since been used as a molecular thermometer of the CMB [19]. In order to calculate column densities of CN, the oscillator strengths must be known. The oscillator strengths in use are those of Roth, Meyer, and Hawkins [20], who calculated values from a combination of total electronic oscillator strength (from Jackson's laser excitation studies [21]), Franck–Condon and Hönl–London factors.

Cartwright and Hay, at Los Alamos, made an attempt to calculate oscillator strengths for the CN violet and red systems in 1982 [22]. Their CI study utilized a set of orbitals derived from an MCSCF calculation of the anion, which is isoelectronic with N_2. With a [4s 3p 1d] basis, a decent agreement was found between the measured and the calculated fluorescence lifetimes for the $B^2\Sigma^+$ state (violet system), with the calculated values about 6% high. However, the calculations for the $A^2\Pi \rightarrow X^2\Sigma^+$ red system were not in agreement with the experiment. Sumner Davis et al. at Berkeley sought to clarify the relative oscillator strengths of the violet and red systems in 1986, after discussions with Cartwright and Hay [23]. They directly compared the oscillator strengths of the red and violet systems, and calibrated their results with a check of the fluorescence lifetime of the $v' = 0$ level of the $B^2\Sigma^+$ state. The results confirmed the validity of the experimental findings, and prompted theorists at NASA to reattempt the calculation of the CN red system. In 1988, Bauschlicher, Langhoff, and Taylor calculated oscillator strengths and emission lifetimes for the red and violet CN systems [24]. They performed MRCI calculations with a [5s 4p 3d 2f 1g] basis, calculating the radiative lifetime of the $v' = 0$ level of the violet system in accord with that measured by Davis, but a little lower than some earlier investigations. Moreover, their calculated lifetimes for the violet system were nearly identical with those of Cartwright and Hay. Again, the calculated lifetimes for the red system were about twice as long as the experimental values. Lu, Huang, and Halpern remeasured the vibrational levels of the red system in 1992, confirming the error in the calculated lifetimes [25]. They suggested that the calculations had omitted some effects assumed to be small, such as rotational interactions or quadrupole terms. So, while astronomers can be satisfied that the experimental oscillator strengths are reliable, it remains for computational spectrometry to predict the correct lifetimes for the CN red system.

13.3.2
Dicarbon: C_2

The Swan bands were known to be associated with carbon, but there persisted uncertainty as to the carrier as late as 1927 [26]. Johnson concludes, "The evidence both of direct experiment and of analysis is conclusive in assigning the Swan bands to a HC-CH molecule." The same year Mulliken argued that the Swan bands are "probably C_2" [27], and by 1929 Johnson had come around to this way of thinking, influenced, in part, by "more especially the rapid theoretical developments due to the work of Mulliken and others" [28]. This paper also revealed more information on Fowler's "high-pressure bands of carbon" [14, 15], which were shown to share the lower state with the Swan bands.[19] In the same paper, Fowler described reproduction of the "tail bands" of comets in low-pressure CO, alluding to conversations with Huggins that suggested that it was these bands that were seen in the 1868 comet (Bronson). In 1939, Mulliken predicted 33 electronic states of C_2, and calculated bond lengths with remarkable accuracy using semiempirical formulas honed on the knowledge of CN, N_2^+, N_2, and O_2 [29]. It was unknown then whether the triplet $^3\Pi_u$ or singlet $^1\Sigma_g^+$ state should be the "normal" state of C_2, with Mulliken leaning toward the triplet, without ruling out other possibilities, including $^3\Sigma_g^-$. These issues were resolved by the discovery of a band system by Ballik and Ramsay in 1958 that shared the lower state with the Swan and Fox–Herzberg systems [30]. This had, as its upper level, the $^3\Sigma_g^-$ state that Mulliken had entertained as a possible ground state. Moreover, perturbations in the $^3\Sigma_g^-$ levels were found to match equal and opposite perturbations in the $^1\Sigma_g^+$ state, revealing the singlet state as the ground state of C_2. In a conversation between Ramsay and the author,[20] Ramsay described that Herzberg took as many as eight matching perturbations as sufficient evidence as to this now established fact. The band systems of greatest astrophysical importance then known were the $d^3\Pi_g \to a^3\Pi_u$ Swan, $e^3\Pi_g \to a^3\Pi_u$ Fox–Herzberg, $b^3\Sigma_g^- \to a^3\Pi_u$ Ballik–Ramsay, $D^1\Sigma_u^- \to X^1\Sigma_g^+$ Mulliken, and $A^1\Pi_u \to X^1\Sigma_g^+$ Phillips systems. The $c^3\Sigma_u^+$ state predicted by Mulliken could not be identified in spectra until the next century (Figure 13.4).

---

19) These bands were, in fact, shown by Herzberg to be Swan bands with $v' = 6$, this state being excited selectively from the recombination of two carbon atoms in a three-body collision.
20) International Symposium on Free Radicals, Big Sky, Montana, 2007.

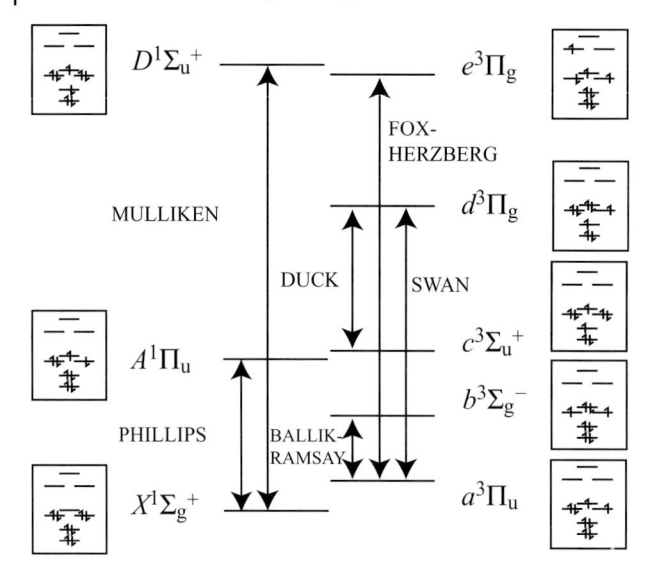

Figure 13.4 Some of the valence states of C_2, with notable band systems indicated. The "Duck" bands were discovered in the author's laboratory in 2006, the name being coined by Klaas Nauta for $d-c$, and the proximity of a pub called the *Duck and Swan* (not to scale).

With the ground state of C_2 fixed, and the positions of several states known, Clementi proceeded to calculate the oscillator strengths of several band systems of C_2 with impressive accuracy [31], using the "magic formula" of Mulliken to calculate the hybridization of the molecular orbitals [32], checked against LCAO-MO calculations. Clementi used a compromise between his calculated values and the contemporary experimental values for the oscillator strength of the Swan system to estimate the abundance of dicarbon in the solar reversing layer, the Swan bands having been observed in absorption in the spectrum of the sun [33].

In 1977, Souza and Lutz confirmed the dicarbon molecule as a component of the interstellar medium, as evidenced by Phillips band absorption in the line of the sight toward Cyg OB2 No. 12 [34]. This was the first new molecule identified in the interstellar medium by optical spectroscopy since the late 1930s. In arriving at a total abundance in this line of sight of 1×10^{14} cm^{-2}, they used the experimentally determined oscillator strength for the 1-0 band of 2.4×10^{-3} from Roux [35], which did not differ so much from the earlier calculated result of Clementi. The following year Chaffee and Lutz observed dicarbon toward ζ-Ophiuchi, reporting a total column density of 7.9×10^{12} cm^{-2} [36].

At the same time, with these astronomical developments, there were several attempts to conquer C_2 by theory. Arnold and Langhoff published their CI study of the Swan system in 1978 [37], and the same year a similar study by Zeitz *et al.* was also published [38]. Chabalowski *et al.* improved upon these calculations in 1981 [39], reporting accurate oscillator strengths for the Swan and Fox–Herzberg systems, and reported the results of similar calculations on the Mulliken, Phillips, and Ballik–Ramsay systems in 1983 [40]. These calculations are in very good agreement with the most recent ones from the author's research group in Sydney. The motivation for undertaking these calculations in Sydney is an interesting story and highlights how computational spectrometry is now so advanced as to precede experiment.

In 2004, Robert Sharp (Anglo-Australian Observatory) and the author undertook observations on the Red Rectangle (q.v.). It was pointed out to us by Robert Glinski (Tennessee Tech.) that our spectra of the 5800 Å Red Rectangle bands (RRBs) had also captured the (0,1) Swan band of C_2. The Swan bands are dwarfed by the RRBs for which the carrier is unknown. Nevertheless, we undertook to model the C_2 photophysically, on the initial assumption that C_2 was a "solved problem." This sort of kinetic model had been built before by various groups [41]. However, there appeared to be some uncertainty in the transition moments of some of the band systems invoked in the models, which normally included the $a^3\Pi_u, b^3\Sigma_g^-, c^3\Sigma_u^+, d^3\Pi_g, X^1\Sigma_g^+$, and $A^1\Pi_u$ states. I approached my friend and colleague, George Bacskay, to calculate these states using the highest level possible in 2006. Partitioning the work between us and a graduate student, Damian Kokkin, we obtained, eventually, excellent results for the primary spectroscopic constants, generally to within 0.1%. This triumph was confused only by the terrible agreement with the literature for the $c^3\Sigma_u^+$ state, where ω_e was calculated at 2061 cm^{-1} with the experiment reporting 2085 cm^{-1} [42]. Such an agreement might have been satisfactory 20 years earlier, but since all other states agreed with the reported spectroscopy to within 0.1%, we were confident that the calculations for the $c^3\Sigma_u^+$ state were indeed accurate, and searched the literature thoroughly. As it turned out, the $c^3\Sigma_u^+$ state had actually never been observed, except in a photodetachment spectrum of the anion [43]. The reported constants were the result of fitting five spectroscopic constants to five perturbations of the $c^3\Sigma_u^+$ state with the $A^1\Pi_u$ state [44].

Being ambitious spectroscopists, and with the tools at our disposal, we undertook a laboratory search for the predicted $d^3\Pi_g \leftarrow c^3\Sigma_u^+$ fluorescence excitation spectrum, monitoring Swan emission, which theory had shown would dominate the Einstein A coefficient. Due to the metastable nature of the $c^3\Sigma_u^+$ state, it having no symmetry or spin allowed downward transitions, it is formed easily enough (as it turned out) in a pulsed electric discharge through argon containing about 1% acetylene. We obtained excellent signal on the first day of trying and after some weeks had extracted spectra of sufficient quality to extract the spectroscopic constants predicted by our calculations [45, 46]. The experiments confirmed the validity of the computational results, with ω_e measured at 2061.940 cm^{-1}, only 0.03% above the calculated value. This example of computational spectrometry taking priority of experiment, while seemingly minor, gives hope for the future, when age-old problems of astronomical spectroscopy may be solved using computers.

13.3.3
The Carbon Trimer: C_3

What appear to be four lines centered around 4050 Å in Huggin's spectrum of the Great Comet of 1881 would prove a mystery for 70 years [11]. The bands appeared in many comets, though not all, sometimes when the CH spectrum around 4300 Å was absent. In her report of the study of the spectrum of Comet Mellish (1915), Glancy notes the appearance of the 4050 Å group, which had also been observed in Comets Zlatinsky (1914) and Brooks (1910) [47]. Importantly, she connects these bands with the 1916 observations of Raffety in his study on discharges with carbon electrodes [48]. Indeed, Raffety's work appears to be the first laboratory detection of the 4050 Å group, and these bands were denoted Raffety's Bands by Bobrovnikoff in his 1931 review, in which he suggests CN as the carrier of the bands [49]. Swings et al. discussing the appearance of the bands in Comet Cunningham (1940), regard the use of the term "Raffety's Bands" inadequate and suggest the carrier to be an as yet uninvestigated polyatomic molecule [50]. In 2010, the bands were commonly referred to as the "comet bands" for obvious reasons. The comet bands were reproduced by Herzberg in the laboratory in 1942 [51]. He suggested CH_2 as the carrier, but it was shown by Monfils and Rosen, using a deuterated precursor, that the carrier possessed no hydrogens [52]. However, suggesting nothing else, the identification remained a mystery until 1951, with a landmark paper by Douglas [53]. Comparing laboratory spectra obtained with natural and ^{13}C-enriched precursors, Douglas concluded that the carrier was most likely C_3, which was confirmed by a follow-up study in 1954 [54]. The spectroscopy of C_3 was analyzed in detail in 1965 by Gausset et al., a paper which, at the time of writing, has been cited nearly 250 times [55].

The carbon trimer poses a tricky problem for computational spectrometry, the upper state being plagued by the complications of Renner–Teller coupling. The ground state is also rather unusual, having a vibrational frequency of only 63.4 cm^{-1}. Liskow et al. were able to calculate this at 69 cm^{-1} in 1972 [56], but this was perhaps fortuitous as very high-quality calculations by Saha and Western in 2006 yielded 85 cm^{-1} for this mode [57]. In 1977, Perić-Radić et al. in Bonn treated both the ground and the excited states of C_3 with configuration interaction using a polarized double-zeta basis [58]. They obtained excellent agreement with the excited state position, predicting a value only 0.03 eV too low, or 4 nm to the red. However, the calculated ground-state bending frequency is quite poor. They calculated the total oscillator strength for the $A^1\Pi_u \leftarrow X^1\Sigma_g^+$ transition to be about $f = 0.061$. Another result reported is the theoretical position of the lowest triplet state of C_3, which Perić-Radić and coworkers put at 2.04 eV seemingly confirming the 2.10 eV phosphorescence observed in matrix isolation spectra as originating in the $a^3\Pi_u$ state. The precise position of this state is not known from gas-phase spectroscopy, and may be astronomically interesting. Chabalowski et al. improved on the previous results of the Bonn group in 1986 with a study employing a much larger basis set, revising the oscillator strength to $f = 0.052$ [59]. This would correspond to an observable emission lifetime of about 95 ns for the $A^1\Pi_u \rightarrow X^1\Sigma_g^+$ transition.

In 2001, Maier and coworkers succeeded in observing C_3 in absorption toward a number of astronomical sources, namely, ζ-Ophiuchi, 20 Aquilae, and ζ-Persei (Figure 13.5) [60]. In calculating the column densities of C_3 toward these sources, Maier et al. opted for a value of $f = 0.016$, some three times smaller than the ab initio result of Chabalowski et al. Now, the theoretical result is a total oscillator strength, assuming a Franck–Condon factor of unity. Nevertheless, the predicted emission

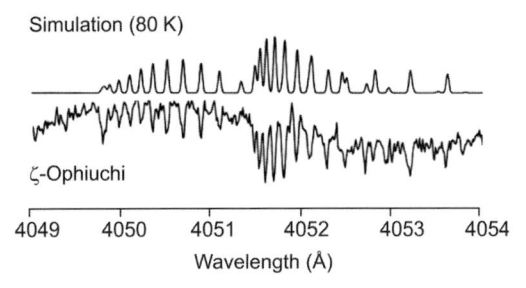

Figure 13.5 The absorption spectrum of C_3 seen toward ζ-Ophiuchi by Maier et al. [60] compared with a calculated spectrum at 80 K.

lifetime is directly comparable with observations. Indeed, the oscillator strength of C_3 is calculated from emission lifetimes on the order of 200 ns observed by Becker et al. in 1979 [61]. A recent reinvestigation by Zhang et al. of the lifetimes of rovibrational lines of C_3, in 2005, shows Becker and coworkers' experimental results to be essentially correct [62]. Their reported emission lifetimes span 192–227 ns for low J-levels of the $A^1\Pi_u(000)$ state. The discrepancy between these values and those predicted *ab initio* is possibly due to the variation of the transition moment with geometry. The theoretical value of Chabalowski et al. is a f_e value, evaluated only at the equilibrium geometry. However, a 2002 MR-AQCC calculation quotes a value of $f_e = 0.02$, in accord with the derived experimental total oscillator strength of 0.0246 [63]. Although the 2006 results of Saha and Western are generally excellent, they do not report oscillator strengths [57]. It would appear that the defining computational spectrometry for C_3 is yet to be reported. C_3 has now been observed in many translucent sight lines toward reddened stars [64] and in the infrared spectrum of the molecule factory, IRC + 10 216, and Saggitarius B2 [65, 66]. As a fundamental component of the interstellar medium, accurate calculations on the spectrosocpy of C_3 will certainly illuminate its role in the chemistry of space.

13.3.4
Radioastronomy

In the mid-1930s, it was shown by Cleeton and Williams that gases could absorb microwaves in the centimeter region [67]. Following the World War II, scientists who had diverted their attention to the development of radar could redirect their efforts to fundamental science. In 1946, there was an explosion of spectroscopy of water and ammonia in the centimeter region [68, 69]. However, detection of ammonia and then water emission from extraterrestrial sources did not come until the reports of Cheung et al. in the late 1960s [70, 71]. Cheung's coauthors included Charles Townes, who had won the Nobel Prize in Physics in 1964 "for fundamental work in the field of quantum electronics, which has led to the construction of oscillators and amplifiers based on the maser-laser principle." These reports were the first of the existence of polyatomic molecules in the interstellar regions, as opposed to transient species in comet tails. While the first molecules in space were identified through electronic spectroscopy, it was the radioregions that yielded, by far, the majority of interstellar molecular identifications. As of July 2000, there were 123 interstellar molecules known, with new discoveries being added at a rate of about 4 per year [72]. In molecular radioastronomy, line positions can be measured to a precision of 1 part in 10^7, with laboratory measurements matching this precision [73]. Since rotational spectroscopy is a ground-state phenomenon, it poses excellent challenges and opportunities to various computational methods. As radioastronomy progressed, many new unidentified lines, or "U" lines, were discovered. The discovery of the identity of these interstellar molecules is greatly enhanced by theoretical predictions of structure and centrifugal distortion. An excellent example of this is the first detection of an interstellar anion. In 1995, Kawaguchi and coworkers reported a series of harmonically related U-lines, a signature of a linear molecule, in a survey toward IRC + 10 216, a source often called the "molecule factory" [74]. Seven lines were identified consistent with a linear molecule having a rotational constant of 1376.8641(4) MHz. From this, the unidentified species was designated B1377. Aoki performed *ab initio* calculations of various candidates [75]. He noted, in 2000, that the rotational constant was similar to C_6H and C_5N, and while noting that no anion had yet been discovered in interstellar space, recommended that "the U-lines with the rotational constant of 1377 MHz may originate from the C_6H^- anion. The B1377 spectrum was reproduced in the laboratory in 2006 by McCarthy *et al.* at the Harvard-Smithsonian Center for Astrophysics [76]. The laboratory B-constant was measured at 1376.86298(7) MHz, compared to the astronomical measurement, 1376.86248 (294) MHz. John Stanton performed a CCSD(T)/cc-pVTZ calculation to predict a rotational constant of 1376.9 MHz, confirming without doubt the chemical origin of the B1377 spectrum. Consequently, C_4H^- was discovered in the envelope of the carbon star IRC + 10216 [77]. The anion had been previously identified, along with C_8H^-, in the laboratory by the Harvard-Smithsonian group [78]. They compared measured constants to Stanton's CCSD(T)/cc-pVTZ calculations with the vibration–rotation correction calculated at the CCSD(T)/cc-pVDZ level of theory. For both anions, the error is only 0.02%, highlighting the importance and success of computational spectrometry in aiding the identification of laboratory and interstellar spectra. C_8H^- has since been detected in the Galactic molecular source TMC-1 [79].

13.4
The Diffuse Interstellar Bands

The first descriptions, in 1922, of unidentified absorption features seen toward reddened stars were due to Mary Lea Heger [80]. Merrill's studies had demonstrated that these features did not follow the oscillatory motion when observed toward binary systems and thus were due to the interstellar medium [81]. Merrill knew of four features, at 5780.4, 5796.9, 6283.9, and 6613.9 Å with a vague feature at 4427 Å being "suspected." There are now hundreds of confirmed *diffuse interstellar bands*, but not a single one has been assigned to a carrier, despite decades of detailed investigation. In the late 1970s, it was generally assumed that the carriers arose from impurities in grains acting as color centers. However, Smith *et al.* renewed interest in molecular carriers [82]. Also in 1977, Douglas, who had years earlier identified C_3 in the laboratory, suggested carbon chains as carriers of the diffuse interstellar bands and that the observed widths of the lines arose from the rapid internal conversion [83]. It was known by then from radioastronomy that carbon chains existed in the interstellar medium.

Computational results aid in interpreting and understanding laboratory spectra. While calculations may not always be able to reproduce oscillator strengths and band positions quantitatively, there is still value in qualitative results. Coulson would have agreed. At the molecular quantum mechanics conference in 1960 at Boulder, Colorado, he is said to have pleaded in his after dinner speech "give us insight, not numbers!" [84]. An example of this with regard to carbon chain molecules and DIBs is in the strengths of electronic transitions of odd, hydrogen-terminated carbon chains, $HC_{2n+1}H$ [85]. These molecules possess one π_x and one π_y electron for each carbon in the chain, ensuring that the ground state has odd occupancies of each of the two perpendicular π-systems. The ground state is thus a triplet, and electronic excitations take place by promoting electrons from the nth $\pi_{x/y}$ orbital to the $n+1$th, or from the singly occupied $n+1$th to the $n+2$th (LUMO). At the Hückel level, these transitions give rise to a pair of degenerate excited states, due to the symmetry of the energy-level spacings in the π-system. Introducing configuration interaction acts to split these states into even and odd combinations of the one-electron excitations. As it turns out, the lower energy excited state, which had been the focus of laboratory investigations, is the odd combination [86]. This has the consequence that the transition moments of each one-electron excitation cancel, and the $A^3\Sigma_u^- \leftarrow X^3\Sigma_g^-$ carries very little oscillator strength and is therefore not likely to be a good candidate for astronomical detection (a match to a DIB would have been self-evident). The oscillator strength is carried by the $B^3\Sigma_u^- \leftarrow X^3\Sigma_g^-$ transition, which we calculated by CASSCF to have oscillator strengths greater than unity and increasing with chain length. For $HC_{19}H$, which we detected using R2C2PI spectroscopy, the CASSCF $B^3\Sigma_u^- \leftarrow X^3\Sigma_g^-$ oscillator strength is calculated to be about 10. Higher level calculations by Mühlhäuser et al. using MRCI for smaller members of the series predict an f-value about two-thirds this value [87]. Indeed, these consequences are predicted by the pairing theorem put forward by Coulson and Rushbrooke in 1940 [88].

Observations on infrared emissions at characteristic wavelengths in the 1970s led to polycyclic aromatic hydrocarbons (PAHs) being suggested as the carrier [89]. PAHs are transiently heated by absorption of ultraviolet photons and then fluoresce in the infrared through vibrational transitions on the ground state following internal conversion. Léger and d'Hendecourt dedicated their 1985 paper on the DIBs to the memory of Douglas, in which they hypothesize that the DIBs are due to polycyclic aromatic hydrocarbons [90]. A similar hypothesis was put forward by van der Zwet and Allamandola [91]. Crawford, Tielens, and Allamandola, realizing that even small PAH cations exhibited transitions in the DIB range, proposed this class of molecular ion as a set of candidates [92]. With carbon chains and PAHs, neutral or ionized, we have the most seriously considered candidates for the DIB carriers. Computational spectroscopy may be of great use to experimentalists in identifying candidate carriers for investigation in the laboratory. For instance, for PAHs comprising between 4 and

10 fused benzene rings, there are over 20 000 possibilities! It is clearly impossible to synthesize every possible species and measure its excitation under simulated astrophysical conditions.

Weisman et al. applied TDDFT with the 6-31G* basis to show that closed shell PAH cations exhibit spectra dominated by a strong absorption in the DIB range, with oscillator strengths in the $f \approx 0.2$ range [93]. Since few DIBs exhibit any perfect correlation in equivalent width, it is generally assumed that each DIB arises from a different species, hence the need for a single strong transition to dominate the spectroscopy. Recently, Hammonds et al. performed a similar study on hydrogenated and protonated PAHs [94]. For more standard PAHs and their cations, a database is maintained by Malloci et al. at http://astrochemistry.ca.astro.it/database/. At the time of writing, it contained calculated spectra of 40 PAHs in various charge states.

In 2010, with the DIB problem still unresolved, the problem lies more with experimental techniques. It is rather simple these days to select a candidate molecule, optimize the geometry at the B3LYP/6-31G* or a similar level, and run a TDDFT calculation. The positions of excited states may be within 0.3 eV of the eventual answer, but matches to DIBs must be spectroscopically exact. In the author's opinion, a real contribution to solving the problem may come from finding, using theoretical methods, a chromophore with large oscillator strengths and near-vertical excitations. In doing so, laboratory measurements may be guided toward more likely DIB carriers and effort will be more fruitfully spent.

13.4.1
The Hump

The most prominent feature on the interstellar extinction curve, and thus the biggest DIB of all, is the so-called 2175 Å "hump," discovered in 1965 [95]. It was immediately realized that this could be explained by graphitic particles in the interstellar medium, an idea that fits well with the later proposed carriers of the DIBs and the AIBs. In 1991, using the CNDO/S method, Braga et al. suggested that fullerenes could be responsible for the hump [96]. Sitting between fullerenes and graphitic particles, carbon onions were proposed in 2003, from comparisons with actual spectra [97]. Recent calculations at the B3LYP/4-31G level have shown that a mixture of dehydrogenated PAHs could give rise to this feature [98, 99]. So, it would appear that there is a range of materials that could conspire together to give rise to the 2175 Å hump. Unfortunately, computational spectrometry is not yet at a stage to rule out candidates, but rather shows that a great variety of aromatic carbonaceous materials may be responsible.

13.5
The Red Rectangle, HD44179

The Red Rectangle is one of the brightest objects in the sky when viewed at a wavelength of 3.3 μm (Figure 13.6) [100]. It has a peculiar shape, which at low resolution

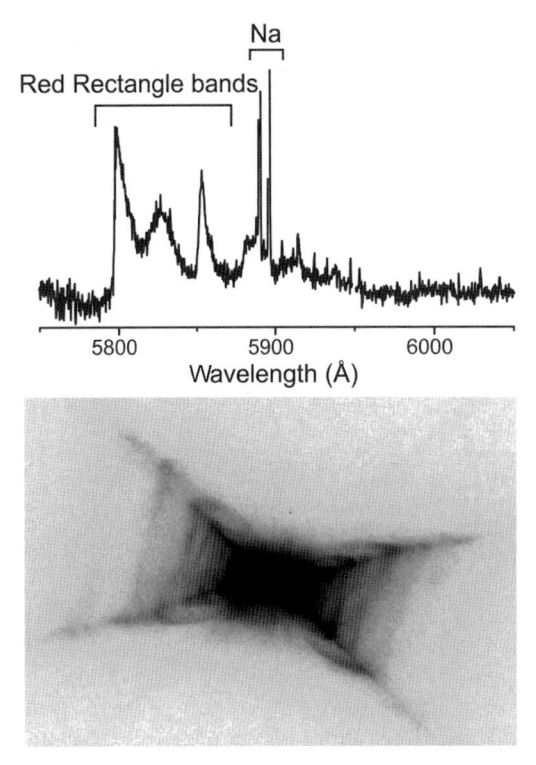

Figure 13.6 The emission spectrum from the biconial Red Rectangle nebula (below). The Red Rectangle bands are presumed to be due to as yet unidentified molecules. Spectrum courtesy of Rob Sharp, Anglo-Australian Observatory. Photograph from Hubble Heritage archive.

appears to be a red rectangle, but Hubble Space Telescope images reveal to be a beautiful biconical nebula, a view almost exactly side-on, so as to appear like a bow-tie [101]. In 1980, Gary Schmidt et al. reported molecular emission in the nebula, the most prominent series of emission being on the red side of 5800 Å [102]. In 1995, Sarre noted a tantalizing coincidence between the blue edge of these emissions and several DIBs, raising the possibility that the carriers may be the same species and opening a new front in the experimental assault to uncover the carriers [103]. A crucial part of the hypothesis is that the Red Rectangle bands, being degraded to the red, are due to hotter versions of the same species as the DIB absorber. We showed, using a combination of DFT and molecular mechanics, that the broadening observed in the RRBs is consistent with large organic molecules heated at moderate temperatures of a few 10s of Kelvin, getting colder toward the edges of the nebula [104]. A suggestion by Glinski that the carrier could be phosphorescence from C_3 is unlikely, yet the precise position of the lowest triplet state of C_3 remains unknown [105]. It also remains to be proven whether the RRBs actually do converge upon the DIB positions, but nevertheless, the carriers of the RRBs remain undiscovered as of 2010 and still pose a very interesting problem for astronomical spectroscopy [106]. As with the carriers of the DIBs, computation can aid in reducing the size of the field of candidates for experimental investigation.

13.6
The Aromatic Infrared Bands

As mentioned above, a set of infrared emission features observed since the 1970s [108] were suggested to be due to polycyclic aromatic hydrocarbons [89]. These bands were referred to, for a long time, as the "unidentified infrared bands," though more commonly in 2010 it is accepted to call them the "aromatic infrared bands" (AIBs). The strongest AIBs lie at 3.3, 6.2, 7.7, 8.6, 11.3, and 12.4 µm, which had been noted by Duley and Williams to coincide with typical vibrational frequencies of organic molecules [109]. There has been much effort dedicated to understanding the AIBs and what these emission features tell us about interstellar and circumstellar chemistry. Much of the computational work has been carried out by Charles W. Bauschlicher, Jr., at NASA, with over 40 papers on the topic. In 2008, the NASA group reported DFT calculations on PAHs as large as $C_{130}H_{26}$ [110]. The largest PAH ever studied spectroscopically as an isolated gas phase molecule is hexa-*peri*-hexabenzocoronene, $C_{42}H_{18}$, also reported in 2008 [107]. Bauschlicher and coworkers, content to use the 4–31G basis set with the B3LYP functional, have provided the "insight" that Coulson would have requested. They suggest that the AIBs are due to a mixture of large and small PAHs, which are symmetrical and compact. Substitutions with nitrogen may improve the agreement between astronomical and synthetic spectra further still. These conclusions were supported by a 2009 report by the same authors, where irregularly shaped PAHs were studied at sizes up to $C_{120}H_{36}$ [111]. The spectroscopy of nitrogen containing PAHs (PANHs) is largely unexplored, especially for larger members. If these are indeed the carriers of the AIBs, it is also possible that they enter the diffuse interstellar regions and present themselves as the carriers of the DIBs. If this turns out to be the case, then it is clear that computational spectrometry would have played a significant role in guiding experimentalists toward these species.

13.7
The Holy Grail

One chapter is insufficient to mention every interaction between computational spectrometry and astronomical spectroscopy. However, what I have attempted to communicate above is the complexity and bilaterality of the relationship that these two fundamental fields of endeavor enjoy. It was astronomers and their laboratory-bound companions who first studied the spectra of atoms and small molecules, as observed in astronomical sources such as the sun, other stars, nebulae, and comets. It was in part due to these data that led Bohr and others to propose quantized energy-level structures for atoms, but Mulliken more than others advanced our understanding of molecular structure by seeking to explain the band spectra presented by astronomers and molecular spectroscopists. His Nobel Prize in Chemistry, awarded "for his fundamental work concerning chemical bonds and the electronic structure of molecules by the molecular orbital method," in 1966, was overdue. By then computational spectrometry was beginning to take off as a subject with the possibilities afforded by the developments in computing. In 2010, computational results are used regularly for large molecules as a qualitative guide for experimental investigation, and there are some results, such as the spectroscopy of the $c^3\Sigma_u^+$ state of C_2, where computation was the first to arrive at the correct answer.[21] Our work on C_2 showed the possibility to calculate vibrational frequencies of second-row diatomic molecules to within 0.1%. However, the absolute positions of electronic transitions were generally calculated to within about 0.5% of the true values. Coulson may have wanted insight rather than numbers, but in the coming years it need not be "either–or." What I will put forward is the "Holy Grail" of computational spectrometry as it relates to astronomical spectroscopy, to calculate the electronic spectra of a large organic molecule to spectroscopic accuracy. This may at first sound like an unreasonable request – because it is! Let us lay down this challenge in 2010 to calculate electronic spectra of molecules with accuracies of $1\ cm^{-1}$ or better on each band. We have shown that this is still not possible for C_2 in a single calculation, but a series of calculations that isolate the effects of basis set size and level of correlation may be found to achieve the task. Just as the list of known molecules in space had zero members in 1940, so is our list of accurately calculated electronic spectra in 2010, some 70 years later. The author sincerely hopes that in the next 70 years of his life (by which time he will be 105!) he will see computational spectrometry develop to the stage where calculations of electronic spectra make redundant the careful collection of laboratory spectra. The only pity would be that playing with lasers, vacuums, and discharges is a lot of fun. Let us begin.

References

1 Young, T. (1802) *Philos. Trans. R. Soc. Lond.*, **92**, 12.
2 Young, T. (1801) *Philos. Trans. R. Soc. Lond.*, **91**, 23.
3 Herschel, W. (1802) *Philos. Trans. R. Soc. Lond.*, **92**, 213.
4 Wollaston, W.H. (1802) *Philos. Trans. R. Soc. Lond.*, **92**, 365.
5 Kirchhoff, G. and Bunsen, R. (1860) *Annalen der Physik und der Chemie*, **110**, 161.
6 Swan, W. (1857) *Trans. R. Soc. Edinb Earth*, **21**, 411.
7 Huggins, W. (1864) *Philos. Trans. R. Soc. Lond.*, **154**, 139.
8 Donati, G.B. (1864) *Astron. Nachr.*, **62**, 375.
9 Huggins, W. (1868) *Philos. Trans. R. Soc. Lond.*, **158**, 529.
10 Huggins, W. (1874) *Proc. R. Soc. Lond.*, **23**, 154.
11 Huggins, W. (1881) *Proc. R. Soc. Lond.*, **33**, 1.
12 Liveing, G.D. and Dewar, J. (1880) *Proc. R. Soc. Lond.*, **30**, 494.
13 Ångström, K. and Thalén, T.-R. (1875) *Nova Acta Reg. Soc. Sc. Upsal.*, **3**, 9.

21) Both the calculations and the experiments were undertaken in my group. Although the calculations were essentially finished before the experiments had begun, they were in press at a later date.

14 Fowler, A. (1910) *Mon. Not. R. Astron. Soc.*, **70**, 176.

15 Fowler, A. (1910) *Mon. Not. R. Astron. Soc.*, **70**, 484.

16 Birge, R.T. (1926) *Phys. Rev.*, **28**, 1157.

17 van Vleck, J.H. (1927) *J. Opt. Soc. Am.*, **15**, 201.

18 MacKellar, A. (1940) *Proc. Astron. Soc. Pacific*, **57**, 187.

19 Słyk, K., Bondar, A.V., Galazutdinov, G.A., and Krełowski, J. (2008) *Mon. Not. R. Astron. Soc.*, **390**, 1733.

20 Roth, K.C., Meyer, D.M., and Hawkins, I. (1993) *Astrophys. J.*, **413**, L67.

21 Jackson, W.M. (1974) *J. Chem. Phys.*, **61**, 4177.

22 Cartwright, D.C. and Hay, P.J. (1982) *Astrophys. J.*, **257**, 383.

23 Davis, S.P., Shortenhaus, D., Stark, G., Engelman, R., Jr., Phillips, J.G., and Hubbard, R.P. (1986) *Astrophys. J.*, **303**, 892.

24 Bauschlicher, C.W., Jr., Langhoff, S.R., and Taylor, P.R. (1988) *Astrophys. J.*, **332**, 531.

25 Lu, R., Huang, Y., and Halpern, J.A. (1992) *Astrophys. J.*, **395**, 710.

26 Johnson, R.C. (1927) *Philos. Trans. R. Soc. Lond. A*, **226**, 157.

27 Mulliken, R.S. (1927) *Phys. Rev.*, **29**, 637.

28 Johnson, R.C. and Asundi, R.K. (1929) *Proc. R. Soc. Lond. A*, **124**, 668.

29 Mulliken, R.S. (1939) *Phys. Rev.*, **56**, 778.

30 Ballik, E.A. and Ramsay, D.A. (1959) *J. Chem. Phys.*, **31**, 1128.

31 Clementi, E. (1960) *Astrophys. J.*, **132**, 898.

32 Mulliken, R.S. (1953) *J. Phys. Chem.*, **56**, 295.

33 Clementi, E. (1960) *Astrophys. J.*, **133**, 303.

34 Souza, S.P. and Lutz, B.L. (1977) *Astrophys. J.*, **216**, L49.

35 Roux, F., Cerny, D., and d'Incan, J. (1976) *Astrophys. J.*, **204**, 940.

36 Chaffee, F.H., Jr. and Lutz, B.L. (1978) *Astrophys. J.*, **221**, L91.

37 Arnold, J.O. and Langhoff, S.R. (1978) *J. Quant. Spectrosc. Radiat. Transf.*, **19**, 461.

38 Zeitz, M., Peyerimhoff, S.D., and Buenker, R.J. (1978) *Chem. Phys. Lett.*, **58**, 487.

39 Chabalowski, C.F., Buenker, R.J., and Peyerimhoff, S.D. (1981) *Chem. Phys. Lett.*, **83**, 441.

40 Chabalowski, C.F., Peyerimhoff, S.D., and Buenker, R.J. (1983) *Chem. Phys.*, **81**, 57.

41 van Dishoeck, E.F. and Black, J.H. (1982) *Astrophys. J.*, **258**, 533.

42 Kokkin, D.L., Bacskay, G.B., and Schmidt, T.W. (2007) *J. Chem. Phys.*, **126**, 084302.

43 Bragg, A.E., Wester, R., Davis, A.V., Kammrath, A., and Neumark, D.M. (2003) *Chem. Phys. Lett.*, **376**, 767.

44 Davis, S.P., Abrams, M.C., Phillips, J.G., and Rao, M.L.P. (1988) *J. Opt. Soc. Am. B*, **5**, 2280.

45 Kokkin, D.L., Reilly, N.J., Morris, C.W., Nakajima, M., Nauta, K., Kable, S.H., and Schmidt, T.W. (2006) *J. Chem. Phys.*, **125**, 231101.

46 Joester, J.A., Nakajima, M., Reilly, N.J., Kokkin, D.L., Nauta, K., Kable, S.H., and Schmidt, T.W. (2007) *J. Chem. Phys.*, **127**, 214303.

47 Glancy, A.E. (1919) *Astrophys. J.*, **49**, 196.

48 Raffety, C.W. (1916) *Philos. Mag. Ser.*, **6** (32), 546.

49 Bobrovnikoff, N.T. (1931) *Astrophys. J.*, **73**, 61.

50 Swings, P., Elvey, C.T., and Babcock, H.W. (1940) *Astrophys. J.*, **94**, 320.

51 Herzberg, G. (1942) *Astrophys. J.*, **96**, 314.

52 Monfils, A. and Rosen, B. (1949) *Nature*, **164**, 713.

53 Douglas, A.E. (1951) *Astrophys. J.*, **114**, 446.

54 Clusius, K. and Douglas, A.E. (1954) *Can. J. Phys.*, **32**, 319.

55 Gausset, L., Herzberg, G., Lagerqvist, A., and Rosen, P. (1965) *Astrophys. J.*, **142**, 45.

56 Liskow, D.H., Bender, C.F., and Schaefer, H.F., III (1972) *J. Chem. Phys.*, **56**, 5075.

57 Saha, S. and Western, C.M. (2006) *J. Chem. Phys.*, **125**, 224307.

58 Perić-Radić, J., Römelt, J., Peyerimhoff, S.D., and Buenker, R.J. (1977) *Chem. Phys. Lett.*, **50**, 344.

59 Chabalowski, C.F., Buenker, R.J., and Peyerimhoff, S.D. (1986) *J. Chem. Phys.*, **84**, 268.

60 Maier, J.P., Lakin, N.M., Walker, G.H., and Bohlender, D.A. (2001) *Astrophys. J.*, **553**, 267.
61 Becker, K.H., Tatarczyk, T.A., and Perić-Radić, J. (1979) *Chem. Phys. Lett.*, **60**, 502.
62 Zhang, G., Chen, K.-S., Merer, A.J., Hsu, Y.-C., Chen, W.-J., Shaji, S., and Liao, Y.-A. (2005) *J. Chem. Phys.*, **122**, 244308.
63 Monninger, G., Förderer, M., Gürtler, P., Kalhofer, S., Petersen, S., Nemes, L., Szalay, P.G., and Krätschmer, W. (2002) *J. Phys. Chem. A*, **106**, 5779.
64 Oka, T., Thorburn, J.A., McCall, B.J., Friedman, S.D., Hobbs, L.M., Sonnentrucker, P., Welty, D.E., and York, D.G. (2003) *Astrophys. J.*, **582**, 823.
65 Cernicharo, J., Barlow, M.J., and Gonzalez-Alfonso, E. *et al.* (1996) *Astron. Astrophys.*, **315**, L201.
66 Cernicharo, J., Goicoechea, J.R., and Caux, E. (2001) *Astrophys. J.*, **534**, L199.
67 Cleeton, C.E. and Williams, N.H. (1934) *Phys. Rev.*, **45**, 234.
68 Herschberger, W.D. (1945) *J. Appl. Phys.*, **17**, 495.
69 Becker, G.E. and Autler, S.H. (1946) *Phys. Rev.*, **70**, 300.
70 Cheung, A.C., Rank, D.M., Townes, C.H., Thornton, D.D., and Welch, W.J. (1969) *Nature*, **221**, 626.
71 Cheung, A.C., Rank, D.M., Townes, C.H., Thornton, D.D., and Welch, W.J. (1968) *Phys. Rev. Lett*, **21**, 1701.
72 McCarthy, M.C. and Thaddeus, P. (2001) *Chem. Soc. Rev.*, **30**, 177.
73 Thaddeus, P. and McCarthy, M.C. (2001) *Spectrochim. Acta A*, **57**, 757.
74 Kawaguchi, K., Kasai, Y., Ishikawa, S., and Kaifu, N. (1995) *Publ. Astron. Soc. Jpn.*, **47**, 853.
75 Aoki, K. (2000) *Chem. Phys. Lett.*, **323**, 55.
76 McCarthy, M.C., Gottlieb, C.A., Gupta, H., and Thaddeus, P. (2006) *Astrophys. J.*, **652**, L141.
77 Cernicharo, J., Guélin, M., Agúndez, M., Kawaguchi, K., McCarthy, M., and Thaddeus, P. (2007) *Astron. Astrophys.*, **467**, L37.
78 Gupta, H., Brünken, S., Tamassia, F., Gottlieb, C.A., McCarthy, M.C., and Thaddeus, P. (2007) *Astrophys. J.*, **655**, L57.
79 Brünken, S., Gupta, H., Gottlieb, C.A., McCarthy, M.C., and Thaddeus, P. (2007) *Astrophys. J.*, **664**, L43.
80 Heger, M.L. (1922) *Lick Observatory Bull.*, **10**, 146.
81 Merrill, P.W. (1936) *Astrophys. J.*, **83**, 126.
82 Smith, W.H., Snow, T.P., Jr., and York, D.G. (1977) *Astrophys. J.*, **218**, 124.
83 Douglas, A.E. (1977) *Nature*, **269**, 130.
84 Frenking G., Kim, K.S., and Scuseria G.E. (eds) (2005) *Theory and Applications of Computational Chemistry: The First Forty Years*, Elsevier.
85 Ding, H., Schmidt, T.W., Pino, T., Boguslavskiy, A.E., Güthe, F., and Maier, J.P. (2003) *J. Chem. Phys.*, **119**, 814.
86 Ball, C., McCarthy, M.C., and Thaddeus, P. (2000) *Astrophys. J.*, **525**, L61.
87 Mühlhäuser, M., Haubrich, J., and Peyerimhoff, S.D. (2002) *Chem. Phys.*, **280**, 205.
88 Coulson, C.A. and Rushbrooke, G.S. (1940) *Proc. Camb. Philol. Soc.*, **36**, 193.
89 Léger, A. and Puget, J.L. (1984) *Astron. Astrophys.*, **137**, L5.
90 Léger, A. and d'Hendecourt, L. (1985) *Astron. Astrophys.*, **146**, 81.
91 van der Zwet, G.P. and Allamandola, L.J. (1985) *Astron. Astrophys.*, **146**, 76.
92 Crawford, M.K., Tielens, A.G.G.M., and Allamandola, L.J. (1985) *Astrophys. J.*, **293**, L45.
93 Weisman, J.L., Lee, T.J., and Head-Gordon, M. (2001) *Spectrochim. Acta A*, **57**, 931.
94 Hammonds, M., Pathak, A., and Sarre, P.J. (2009) *Phys. Chem. Chem. Phys.*, **11**, 4458.
95 Stecher, T.P. (1965) *Astrophys. J.*, **142**, 1683.
96 Braga, M., Larsson, S., Rosen, A., and Volosov, A. (1991) *Astron. Astrophys.*, **245**, 232.
97 Chhowalla, M., Wang, H., Sano, N., Teo, K.B.B., Lee, S.B., and Amaratunga, G.A.J. (2003) *Phys. Rev. Lett.*, **90**, 155504.
98 Malloci, G., Mulas, G., Cecchi-Pestellini, C., and Joblin, C. (2008) *Astron. Astrophys.*, **489**, 1183.

99 Cecchi-Pestellini, C., Malloci, G., Mulas, G., Cecchi-Pestellini, C., Joblin, C., and Williams, D.A. (2008) *Astron. Astrophys.*, **486**, L25.

100 Cohen, M., Anderson, C.M., Cowley, A., Coyne, G.V., Fawley, W.M., Gull, T.R., Harlan, E.A., Herbig, G.H., Holden, F., Hudson, H.S., Jakoubek, R.O., Johnson, H.M., Merrill, K.M., Schiffer, F.H., Soifer, B.T., and Zuckerman, B. (1975) *Astrophys. J.*, **196**, 179.

101 Cohen, M., van Winckel, H., Bond, H.E., and Gull, T.R. (2004) *Astron. J.*, **127**, 2362.

102 Schmidt, G.D., Cohen, M., and Margon, B. (1980) *Astrophys. J.*, **239**, L133.

103 Sarre, P.J., Miles, J.R., and Scarrott, S.M. (1995) *Science*, **269**, 674.

104 Sharp, R.G., Reilly, N.J., Kable, S.H., and Schmidt, T.W. (2006) *Astrophys. J.*, **639**, 194.

105 Glinski, R.J. and Nuth, J.A. (1997) *Astrophys. Space Sci.*, **249**, 143.

106 Glinski, R.J., Michaels, P.D., Anderson, C.M., Schmidt, T.W., Sharp, R.G., Sitko, M.L., Bernstein, L.S., and van Winckel, H. (2009) *Astrophys. Space Sci.*, **323**, 337.

107 Kokkin, D.L., Troy, T.P., Nakajima, M., Nauta, K., Varberg, T.D., Metha, G.F., Lucas, N.T., and Schmidt, T.W. (2008) *Astrophys. J. Lett.*, **681**, L49.

108 Gillett, F.C., Forrest, W.J., and Merrill, K.M. (1973) *Astrophys. J.*, **183**, 87.

109 Duley, W.W. and Williams, D.A. (1981) *Mon. Not. R. Astron. Soc.*, **196**, 269.

110 Bauschlicher, C.W., Jr., Peeters, E., and Allamandola, L.J. (2008) *Astrophys. J.*, **678**, 316.

111 Bauschlicher, C.W., Jr., Peeters, E., and Allamandola, L.J. (2009) *Astrophys. J.*, **697**, 311.

Index

a

ab initio methods, wavefunction-based 152
ablation, laser 353–354
absolute configuration (AC) 241, 261–270
absolute deviation, mean 83–84
absolute shieldings 37–38
absorption
– energy 166
– X-ray 329–330
acetamide, N-methyl- 165–166
acetonitrile 101
N-acyl azoles 38
adiabatic internal coordinate modes (AICoM) 124–131
adiabatic local density approximation (ALDA) 155
adiabatic vibrational modes 105–149
adsorbates, chiral 262
agostic methylidene product 360–366, 371
Ahlrich-like hybrid bases 46
alanine subunits 304
alcohols, tempo–alcohol complexes 89
ALDA (adiabatic local density approximation) 155
aldehydes 123
alkines 134
alumina, NMR spectra 337–341
aluminum 341, 356–359
amines, cyclic 39
amino acids 38
– helicogenic 100
amplitude function 9
analysis
– PED 128–129
– population 229
– quantitative 1
– spectral 23
anapole moment, nuclear 209

anatomy, molecular 243–246
angular degrees of freedom 192
angular momentum, molecular 178
anharmonic oscillator, diatomic molecule 20–23
anharmonicity constant 117
anisotropy, chemical shift 329
anisotropy ratio 227
annealing 366–367
anticommutation rules 204
antiinflammatory activity 263
antimatter 201
antisymmetric stretching modes 135
apo-calbindin D_{9K} 307–308
apparent surface charges (ASC) approach 159
applications
– computational dielectric spectroscopy 299–316
– DeVoe method 252–256
approximation
– adiabatic local density 155
– Born–Oppenheimer 65, 180–182, 185–186
– independent systems 247
– NDO 259
– point dipole 76
– random-phase 257
– Tamm–Dancoff 155
aromatic heterocycles 37
aromatic hydrocarbons, polycyclic 325, 391
aromatic IR bands 394
aromatic rings, conjugated 97
aromaticity 41–45
aryl–aryl torsions 263
astronomical molecular spectroscopy 377–398
asymmetry, matter/antimatter 201
asymmetry parameter 82

Computational Spectroscopy: Methods, Experiments and Applications. Edited by Jörg Grunenberg
Copyright © 2010 WILEY-VCH Verlag GmbH & Co. KGaA, Weinheim
ISBN: 978-3-527-32649-5

atom–molecule reactions, simple 353
atomic orbitals (GIAOs), gauge including 80, 328
atropisomers 242
augmented Liouville equation 287
autocorrelation 288
averaging of frequencies 135–139
axial chirality 215
azapentalenes 44
azines 39
azoles 38, 41
azolides 38

b

β-sheet conformers 230, 303
background, cosmic microwave 384
"background-corrected" spectra 327
Badger–Herschbach–Laurie equations 111
Badger rule 105–149
– bond-specific 118
Bakerian Lecture 379
bands
– aromatic 394
– comet tail 383–384
– diffuse interstellar 390–392
– Swan 382
bandwidth, frequency-dependent 261
baryogenesis 202
base pairs, Watson–Crick 140
bases
– Ahlrich-like hybrid 46
– Tröger's 38
basis sets 82–84
bath, system–bath decomposition 86
BEC (Bose–Einstein condensation) 210
benchmarks, QM 163
bending modes 123
benzene
– chromophore 270
– ICSS view 42
bicyclobutane 128–129
bidentate bridging 338
3,8''-biflavonoid morelloflavone 264
bimanes 45
binding protein, rat fatty acid 305
BINOL (1,1'-bi(2-naphthol)) 252, 254
biological cells 310–311
biomolecules, solvated 303–311
biopolymers 252, 271
biphenyls, polychlorinated 325
birefringence, circular 223–239
BMIM$^+$ (1-butyl-3-methyl-imidazolium) 284, 314
BO, see Born–Oppenheimer ...

Bohr magneton 67
Bohr's theory 5
Boltzmann weights 246, 264, 268
bond lengths 105
– effective 142
– equilibrium 141
– experimental 114–116
bond-specific Badger rules 118
bonds
– delocalized 134
– hydrogen 96
– intramolecular hydrogen 51
– metal–metal 111
– polarity 107
– spectroscopic and geometrical constants 106–108
– stretching 113–116
borazine 51
Born–Oppenheimer (BO) approximation 65
– dipole moments 185–186
– discussion 180–182
Born–Oppenheimer (BO) Hamiltonian 67
boron trifluoride 359–360
borylene 359–360
Bose–Einstein condensation (BEC) 210
bosons 19, 189
Bouchiat–Hamiltonian 211
boundary conditions
– conducting 287, 305
– toroidal 295
Brahe, Tycho 377
bridging 338
Bunsen, Robert Wilhelm Eberhard 380–381
buta-1,3-diene 32
1-butyl-3-methyl-imidazolium (BMIM$^+$) 284, 314

c

calculations
– coupled-cluster 213
– coupling constants 54
– direct time-domain 231–238
– errors 55
– generalized dielectric constant 286–293
– harmonic frequencies 355
– Hückel 208
– magnetic tensors 63–104
– multireference 364
– nonadiabatic 173–199
– SSCC 48–50
– vibrational frequencies of new molecules 353–375
canonical linear harmonic oscillator 11–12, 18–19

Car–Parrinello molecular dynamics (CPMD) scheme 162
carbenes 367–371
carbon monoxide ion (CO^+) 383–385
carbon tetrachloride (CCl_4) 367–371
carbon trimer 387-389
carbonaceous Swan bands 382
Cartesian vectors, orthonormal 193
CASSCF (complete active space self-consistent field) 153
cassiterite 344
catalysts, enantioselective 254
cations
- organic 284
- trihydrogen 25
CC (coupled cluster) theory 153
CCl_4 (carbon tetrachloride) 367–371
CD (circular dichroism), vibrational 223–239
CE (Cotton effect) 242, 256
cells, biological 310–311
center of mass motion 24, 283
center of mass operator 176
charge
- "charge arm" 284
- charged dipolar systems 279–321
- Mulliken 356
- nuclear charge distribution model 213
- surface 159
chemical reactions
- simple atom–molecule 353
- surface 324–325
chemical shift 37–41
- CSA tensor 329
chemistry
- environmental 323–351
- organometallic 374
- quantum 1–36
CHF_3 (fluoroform) 371–374
CH_3F (fluoromethane) 360–362
CHFClBr 206–207, 212–214
chiral adsorbates 262
chiral molecules 201–221
- randomly oriented 233
chiral phosphorus compounds 214
chirality
- axial 215
- intrinsic 244
- operator 203
- rule 251
chiroptical spectroscopic technique 224
chirospecific response 232
chloroform 101, 305, 364–366
chromophores 243–244, 248–252
- benzene 270

CI (configuration interaction), truncated 259
circular birefringence (CB) 223–239
circular dichroism (CD)
- electronic 241–277
- vibrational 223–239
circular polarized γ-radiation 207
CIS (configuration interaction singles) 153–155, 161–162
Clausius–Mosotti equation 184
closed shell systems, neutral 140
CMB (cosmic microwave background) 384
CO^+ (carbon monoxide ion) 383–385
"coarse-grained" models 66
codeposition 357, 369
cold molecules 206
collective dipole moment, total 282–286
color, "Theory of Light and Colours" 379
comet rule, empirical 270
comet tail bands 383–384
commuting operators 178–179
complete active space self-consistent field (CASSCF) 153
complex electric field 235
complex numbers 15, 26–27
- hyper- 18
complexes
- Al–organic 337–339
- Kubas 367
- metal dihydride 362
- super- 366–367
- tempo–alcohol 89
compliance constants 139–140, 362
computation, symbolic 34
computational conformational optimization 253
computational dielectric spectroscopy
- applications 299–316
- charged dipolar systems 279–321
computational NMR spectroscopy 37–61
computational simulation, see simulation
computational spectrometry
- carbon trimer 388
- concepts 1–36
- environmental chemistry 323–351
condensation, Bose–Einstein 210
Condon
- Franck–Condon factors 384
- Franck–Condon-like response 160
conducting boundary conditions 287, 305
conductivity
- dielectric 312–314
- frequency-dependent 313
- parabola 316
- static 289, 312

configuration, absolute/relative   241, 261–270
configuration interaction (CI), truncated   259
configuration interaction singles (CIS)   153–155, 161–162
conformation   241
– peptides   100
conformational manifolds   246–247
conformational optimization, computational   253
conformational search, Monte Carlo-based   267
conformers, β-sheet   230, 303
conjugated aromatic ring   97
conservation of orbital symmetry   33
constants of integration   10
contaminants, organic   323
continuum models   53
– PCM, see polarizable continuum model
coordinate matrix   7
coordinate transformation   175, 193
coordinates
– electronic   180, 185–186
– mass-weighted   119
– molecule-fixed system   192–195
– redundant sets   137
– translation-free internal   177–178
correlated Gaussian functions   189
correlation, auto-   288
correlation coefficients   142
correlation functions, time   286
cosine functions, phase-shifted   302
cosmic microwave background (CMB)   384
Cotton effect (CE)   242, 256
Coulomb, DC–HF theory   206
Coulomb interactions   174, 180, 228, 295
Coulomb's law   294
counterions, iminio   267
coupled-cluster calculations   213
coupled-cluster (CC) theory   153
coupled oscillators   248–251
coupling
– degenerate   251
– hyperfine   71, 81, 91
– mass   123
– quadrupole   71–72
– Renner–Teller   388
– spin–orbit   70–71
– vibronic   261
coupling constants
– Fermi   203
– hexafluorocyclotriphosphazene   54
– hyperfine   66–84
– spin–spin   45–52

CPT symmetry   201
creation operator   15–16
cross-correlation function   227
cross-polarization detection configuration   232–234
crystal field, lattice   209
crystal truncation rod (CTR)   329–330
crystals, macroscopic   3
CSA (chemical shift anisotropy) tensor   329
CTR (crystal truncation rod)   329–330
cuprite   29
cyanide   383–385
cybotactic region   101
cyclic amines   39
cyclobutene   32
cyclohexa-1,3-diene, 5-methyl   245
cyclohexane   158
cyclopentadienyl anion   267
D-lines of sodium   380

d

d-orbital hole   29
dAMP (2′-deoxyadenosine 5′-monophosphate)   339
Daphniphyllum macropodum   267
Davydoff splitting   253
DC–HF (Dirac-Coulomb–Hartree-Fock) theory   206
Debye equation   184
Debye processes   300
decomposition
– singular value   269
– system–bath   86
defects, "free from"   30
deformation mode   366–367
degenerate coupling   251
degrees of freedom, angular   192
delocalization, spin   97
delocalized bonds   134
delta function, Kronecker   5
DEMO (dynamic extended molecular orbitals)   187
denatured proteins   229
density functional theory (DFT)   131, 163, 212–214
– astronomical molecular spectroscopy   392–394
– environmental chemistry   330–333, 342–345
– magnetic parameters   64, 72–73, 79, 82
– NMR spectroscopy   46–50
– periodic DFT MD simulations   336
– time-dependent, see time-dependent density functional theory

density matrix   65
2′-deoxyadenosine 5′-monophosphate (dAMP)   339
destruction operator   15–16
detection, cross-polarization   232–234
determinant
– "excited"   152
– Slater   68, 152
deviation, mean absolute   83–84
DeVoe method   248–251
DFT, see density functional theory
di-(tert-butyl nitroxide (dtbn)   85, 95
diagnostic interionic modes   355
dialane, dibridged   356–359
diastereomeric compounds   266
diatomic molecules
– as anharmonic oscillator   17–20
– Badger-type relationships   112–118
– dipole moments   190
– heteronuclear   195
3a,6a-diazapentalene   45
dibridged dialane   356–359
dicarbon   385–387
– valence states   386
dichroism
– linear   248
– vibrational circular   223–239
dicyanoamide, 1-ethyl-3-methyl-imidazolium   290, 301
dielectric conductivity   312–314
dielectric constant
– generalized   286–293, 315–316
– imaginary part   226
– local   293
– solvents   88
dielectric field equation   279–282
dielectric permittivity   314
dielectric polarization, total   279
dielectric relaxation time   300
dielectric solvent effects (DSE)   53
dielectric spectroscopy, computational   279–321
differential operators   9
differential overlap   259
diffuse interstellar bands (DIBs)   390–392
diffusion tensor, generalized   98
diffusive operators   66
dihedral angle   212, 254
dihydride complex, metal   362
1,2-dihydro-1,2-azaborine   52
dimer model   335
dimers   253
dimethyl ether   24
dipolar displacement   288, 295

dipolar moment, molecular electric   32
dipolar spin–spin interaction, magnetic   71
dipolar systems, charged   279–321
dipole densities, translational/rotational   280
dipole–dipole tensor   294
dipole moment
– diatomic molecules   190
– EEDM   202
– measurements   184–185
– nonadiabatic calculation   173–199
– total collective   282–286
dipoles
– oscillating electric   233
– transition   250
Dirac constant   4
Dirac-Coulomb–Hartree-Fock (DC–HF) theory   206
Dirac matrices   204
Dirac operators
– creation and destruction   15–16
– parity-conserving   208
Dirac pseudoscalar   203
Dirac–van Vleck vector model   46
direct magnetic dipolar spin–spin interaction   71
direct time-domain calculation   227–231
direct time-domain measurement, VOA free induction decay field   231–238
discreteness, laws of   3–5
dispersion, optical rotatory   223
dispersion–repulsion contribution   94
dissection
– molecular   244
– polyatomic molecules   118–131
dissolved organic matter (DOM)   323
distance, "effective"   77
DNA, computational dielectric spectroscopy   309–310
DOM (dissolved organic matter)   323
donor ions, proton   52
double bonds   143
double-hybrid functionals   260
double perturbation theory   72
ds-DNA   310
DSE (dielectric solvent effects)   53
dtbn (di-tert-butyl nitroxide)   85, 95
dynamic criterion   127
dynamic extended molecular orbitals (DEMO)   187
dynamical effects   84–98
dynamical variables   18
dynamics
– molecular, see molecular dynamics
– water   343–345

e

ECD (electronic circular dichroism)   241–277
– hybrid approaches   247–256
– solid-state   266–268
– spectra   254–255, 268–270
– TDDFT   246, 257–272
EEDM (electron–electric dipole moment)   202
effective bond lengths   142
effective distance,   77
effective force constants   131
effective Green function   298
effects
– Cotton   242, 256
– dielectric solvent   53
– dynamical   84–98
– electron correlation   212
– environmental   86–89
– Mills–Nixon   41
– modeling of solvent effects   157–161
– polarity   158
– short-/long-range solvation   167
– solvent   53–54
– stereoelectronic   84–98
– Zeeman   22
Einstein
– BEC   210
– Einstein's relation   99
– Nernst–Einstein relation   312–313
electric dipolar moment, molecular   32
electric dipoles, oscillating   233
electric field, complex   235
electric field gradient   81–82
electric Hessian matrix   75
electric resonance experiments, molecular beam   185
electric transition dipole   250
electrocyclic process, stereospecific   32
electron correlation effects   212
electron–electric dipole moment (EEDM)   202
electron paramagnetic resonance (EPR)   63–104
electron-rich methyl halides   360
electron spin resonance (ESR)   101, 213, 333
– observables   64
electronic Bohr magneton   67
electronic circular dichroism (ECD)   241–277
– hybrid approaches   247–256
– solid-state   266–268
– spectra   254–255, 268–270
electronic coordinates   180, 185–186
electronic effects, stereo-   84–98
electronic spectroscopy, parity violation   209
electronic structure theory   67–69
electronic Zeeman interaction   72
electrostatic interactions   20
electrostatic potential, molecular   44
electrostatics, finite system   294–299
empirical comet rule   270
S-enantiomer   205
enantioselective catalysts   254
enantioselective HPLC   263
Encyclopedia of Computational Chemistry   333
energy
– absorption   166
– excitation energy errors   164–165
– gas-phase   92–93
– kinetic   80
– PED analysis   128–129
– potential energy surface   118
– residual   8
– Stark rotational   184
– vibrational–rotational   21–22
– zero-point   8
energy matrix   7
enthalpy, reaction   332
environmental chemistry   323–351
environmental effects   86–89
EPR spectra, free radicals   63–104
equation of motion (EOM) formalism   154
equations, see laws and equations
equilibrium
– bond lengths   141
– generalized dielectric constant   286–293
– geometry   90
errors
– excitation energy   164–165
– in theoretical calculations   55
– self-interaction   258
ESR (electron spin resonance)   101, 213, 333
– observables   64
ether, dimethyl   24
1-ethyl-3-methyl-imidazolium dicyanoamide   290, 301
ethylene, twisted   211
Euler–Lagrange equations   129
Ewald scheme   298
exchange correlation functional, hybrid   155
exchange potential, Kohn–Sham   156
excitation energy errors   164–165
excitation spectra, fluorescence   387
"excited" determinants   152
exciton chirality rule   251
exciton-coupled ECD   245
experimental bond lengths   114–116
experimental coupling constants   54

experimental force constants, bond stretching   113–116
extended X-ray absorption fine structure (EXAFS) spectroscopy   329–330
extinction coefficients, molar   157, 249

f

F-substituted borazines   51
FC (Fermi contact)   46
Fe-hydroxide dimer model   335
femtosecond spectral interferometric approach   232–238
α-FeOOH (goethite)   324, 334–336
Fermi contact (FC)   46
Fermi coupling constant   203
Fermi resonances   132–133
fermions   189
Feynman theorem, Hellmann–   179, 186, 190–191
FID (free induction decay) field   231–238
field
– CASSCF   153
– complex electric   235
– dielectric   279–282
– FID   231–238
– force field parameterization   345
– interaction with molecules   182–184
– lattice crystal   209
– Lorentz   296
– Maxwell   279, 294
– reaction field method   296–297
– reactive force   333
– self-consistent   28, 73
field gradient, electric   81–82
field-reduced splitting   195
fine structure, EXAFS spectroscopy   329–330
finite system electrostatics   294–299
first hydration layers   317
fluorescence excitation spectra   387
2-fluorobenzamide   38
fluoroform (CHF_3)   371–374
fluoromethane (CH_3F)   360–362
Fokker–Planck operators   66
folding motion   128
force
– long-range intermolecular   183
– reactive force field   333
force constant matrix   139
force constants
– compliance   139–140
– effective   131
– experimental   113–116
– intrinsic   138
– quadratic   131

force field parameterization   345
four-particle molecule   188
Fourier–Laplace transform   287, 293, 300–302
Fourier transform spectral interferometry (FTSI)   234–237
Fourier transformation   14
Fox–Herzberg system   385–386
fractionation factors, isotopic   328
Franck–Condon factors   384
Franck–Condon-like response   160–161
Fraunhofer, Joseph (von)   380
"free from defects"   30
free induction decay (FID) field   231–238
free radicals, EPR spectra   63–104
Free–Wilson matrices   43
frequencies
– fundamental   359–361, 365, 370–373
– harmonic   355
– intrinsic   135–139
– IR active   358
– local modes   134–135
– overtone   136
– vibrational   326, 353–375
frequency-dependent bandwidth   261
frequency-dependent conductivity   313
friction tensor   99
FTSI (Fourier transform spectral interferometry)   234–237
functionals   82–84
– DFT, see density functional theory
– double-hybrid   260
– hybrid   69
– hybrid exchange correlation   155
functions
– correlated Gaussian   189
– cross-correlation   227
– effective Green   298
– Gaussian   13
– Green   279
– Havriliak–Negami   301
– Kronecker delta   5
– one-center Gaussian   191
– phase-shifted cosine   302
– polynomial   190
– SCF wavefunction   187
– time correlation   286
– time-correlation function theory   224–227
– variational   179, 189
– wavefunction   9
– Whittaker M/W   10–13
fundamental frequencies   359–361, 365, 370–373

g-factor 21–22
γ-radiation, circular polarized 207

g

g-tensor 66–84
– isotropic 85
Galilei, Galileo 377
gas-phase energy 92–93
gauge including atomic orbitals (GIAOs) 80, 328
Gaussian functions 13
– correlated 189
Gaussian theorem 281
generalized dielectric constant (GDC) 286–293, 315–316
generalized diffusion tensor 98
geometry
– molecular 106–108, 182
– optimization 109
Glashow–Salam theory, Weinberg– 202
glass, surfaces 341
GLOB model 93
goethite (α-FeOOH) 324, 334–336
gold nanoclusters 262
gradient, electric field 81–82
gradient vector 139
great mural quadrant 377
Green function 279
– effective 298
Green–Kubo approach 289
group theory 201
groups
– Schleyer's 44
– symmetry-breaking 244

h

halides, electron-rich methyl 360
Hamiltonian 7
– Born–Oppenheimer 67
– Bouchiat– 211
– molecular 23, 174–178, 192
– nonrelativistic 16
– radiation–matter interaction 224–225
– spin 66–84
harmonic frequencies 355
harmonic oscillator
– canonical linear 11–12, 18–19
– quantum theories 5–20
harmonic potential 120
harmonic wave number 26
Hartree–Fock (HF) theory 69
– DC–HF 206
– time-dependent 257
Havriliak–Negami function 301

Heisenberg's principle of indeterminacy 5
helical systems 215
helicity rules 269
helicogenic amino acids 100
α-helix, VCD spectra 231
Hellmann–Feynman theorem 179, 186, 190–191
hen egg white lysozyme 305
Hermite polynomial 11–13
Herschbach–Laurie equations, Badger– 111
Herzberg system, Fox– 385–386
Hessian matrix 105
– electric 75
heterocycles, aromatic 37
heterodyned spectral interferograms 237
heteronuclear diatomic molecules 195
heteropentalenes 42–43
hexafluorocyclotriphosphazene 54
HF (Hartree–Fock) theory 69
– DC–HF 206
– time-dependent 257
high harmonic spectra 31
high-resolution spectroscopy 206
Hohenberg–Kohn theorems 68
hohlraum radiation 19
Hönl–London factors 384
Hooke's law 6
Hückel calculations 208
"Hump", DIB 392
Huygens, Christiaan 378
hybrid bases, Ahlrich-like 46
hybrid ECD approaches 247–256
hybrid functionals 69, 155
– double- 260
hybrid QM/molecular mechanics (QM/MM) approach 159–161
hybridization 30
hydrated ionic liquids 318
hydration layers, first and second 317
hydrazones 39
hydrocarbons, polycyclic aromatic 325, 391
hydrodynamic approach, mesoscopic 99
hydrogen
– dibridged dialane 356–359
– solid 358
– supercomplex 366–367
hydrogen atoms, data analysis 82–83
hydrogen bonds
– intramolecular 51
– solute–solvent 96
hydrolysis 342
hypercomplex numbers 18
hyperfine coupling 71, 81
– constants 66–84

– vinyl   91
hyperfine structure   203
hypersurface   25

i

ICSS (isochemical shielding surfaces)   41–42
identical particles, permutation   178
imaging, tomographic   30
iminio counterion   267
INDCO   96
independent systems approximation (ISA)   247
indeterminacy, principle of   5
induction decay, free   231–238
inelastic neutron scattering (INS)   330–331
infrared, see IR
integration, constants of   10
intensities, Raman   327
intensity of transition   14
intercept   40
interface, mineral–water   342–343
interferograms, heterodyned spectral   236
interferometer, Mach–Zehnder   235
interferometry, spectral   232–238
interionic modes   355
intermolecular forces   183
internal coordinate modes, adiabatic   124–131
internal coordinates, translation-free   177–178
internal displacement coordinates   121
internuclear axis   194
interstellar bands, diffuse   390–392
interstellar regions, polyatomic molecules   389
intramolecular hydrogen bonds   51
intrinsic chirality   244
intrinsic force constants   138
intrinsic frequencies   135–139
inverse Fourier transform   14
ionic liquids
– hydrated   318
– molecular   311–316
ionic lithium dioxide molecule   354–356
ionization   353
ions
– CO^+ 383–385
– isotropic monoatomic   311
– proton donor   52
IR active frequencies   358
IR spectra
– aromatic bands   394
– laser ablation   357, 366
IR/Raman spectra   325–328, 334–336

iron, see Fe
ISA (independent systems approximation)   247
isochemical shielding surfaces (ICSS)   41–42
isolated peptides, VCD spectra   228
isolated stretching modes   132–134
isolation, matrix   353–354
isotope shift   203
isotope substitution   132–133
isotopic fractionation factors   328
isotopic variants   21–23
isotropic g-tensor   85
isotropic monoatomic ions   311
isotropic polarizability   191

j

Jones interaction, Lennard–   296

k

Kepler, Johannes   377
ketones   123
kinetic energy operator   80
Kirchhoff, Gustav Robert   380–381
Kobayashi–Maskawa mechanism   202
Kohn–Sham exchange potential   156
Kohn–Sham procedure   68–69
Kramers–Kronig relation   249
Kramers–Kronig transformation   224
Kramers operator   208
Kronecker delta function   5
Kubas complex   367
Kubo approach, Green–   289
α-lactalbumin   305

l

Lagrange multiplier   124–126
Laplace, Fourier–Laplace transform   287, 293, 300–302
laser ablation   353–354
lattice crystal field   209
lattice sum techniques   298
laws and equations
– anticommutation rules   204
– asymmetry   82
– augmented Liouville equation   287
– Badger–Herschbach–Laurie equations   111
– Badger rule   105–149
– Clausius–Mosotti equation   184
– Coulomb's law   294
– Debye equation   184
– dielectric field equation   279–282
– dipolar displacement   288
– Einstein's relation   99
– empirical comet rule   270

- Euler–Lagrange equations 129
- exciton chirality rule 251
- Gaussian theorem 281
- helicity rules 269
- Hellmann–Feynman theorem 179, 186, 190–191
- Hohenberg–Kohn theorems 68
- Hooke's law 6
- Kramers–Kronig relation 249
- laws of discreteness 3–5
- linear polarization 225–226
- linear response equations 76–82
- Liouville equation 65
- Nernst–Einstein relation 312–313
- Newton's second law 6, 119
- Poisson equation 294, 310
- quantum laws 3–5
- Raman intensities 327
- Rosenfeld equation 242–243, 246
- Schrodinger's equation 9–12, 24, 67
- Schwinger–Lüders–Pauli theorem 201
- vibrational secular equation 131
laws of discreteness 3–5
LD (linear dichroism) measurements 248
leading parameter principle 129
Lecture, Bakerian 379
Lennard–Jones interaction 296
light, "Theory of Light and Colours" 379
line shapes 98–101
- VCD 227
linear dichroism (LD) measurements 248
linear harmonic oscillator, canonical 11–12, 18–19
linear optical activity susceptibility 227, 234
linear polarization 225
linear response equations 76–82
linear response theory 72–76
D-lines of sodium 380
Liouville equation 65
- augmented 287
Liouville transformation 27
liquid-state NMR 328
liquids, molecular ionic 311–316
lithium dioxide molecule 354–356
llama antibody heavy-chain variable domain 305
local density approximation, adiabatic 155
local dielectric constant 293
local excitations 164–165
local mode frequencies 134–135
localized vibrational modes 122–124
London factors, Hönl– 384
long-range intermolecular forces 183
long-range solvation effects 167

Lorentz field 296
Lorentzian curve 127
low-frequency motions 92
Lüders–Pauli theorem, Schwinger– 201
lysozyme, hen egg white 305

m

Mach–Zehnder interferometer 235
macropodumine C 267
macroscopic crystals 3
MAD (mean absolute deviation) 83–84
magic-angle spinning (MAS), liquid-state 329, 339–341
magnetic dipolar spin–spin interaction 71
magnetic parameters 84–86
magnetic shielding, nuclear 208
magnetic tensor 63–104
magnetic transition dipole 250
magnetogyric ratios 37
magneton, Bohr 67
manifolds, conformational 246–247
maser 389
Maskawa mechanism, Kobayashi– 202
mass, reduced 106
mass coupling 123
mass-weighted coordinates 119
matrix
- density 65
- Dirac matrices 204
- force constant 139
- Free–Wilson 43
- Hessian 75, 105
- orthogonal 193
- Wilson 121
matrix isolation laser ablation 353–354
matrix mechanics 6–9
matrix method 251–252
matter/antimatter asymmetry 201
Maxwell field 279, 294
MC (Monte Carlo)-based conformational search 267
MC (Monte Carlo) simulations 334
MD, see molecular dynamics
mean absolute deviation (MAD) 83–84
measurements
- dipole moments 184–185
- direct time-domain 231–238
- LD 248
- vibrational OA-FID 237–238
mechanics
- matrix 6–9
- molecular, see molecular mechanics
- quantum, see quantum mechanics
- wave 9–15

memory, temporal 286
mesoscopic hydrodynamic approach 99
mesoscopic level 293–294
metals
– dihydride complex 362
– metal–metal bonds 111
– transition 374
methane 362–364, 371
– AICoM frequencies 138
methanol 101
N-methyl-acetamide (NMA) 165–166
5-methyl cyclohexa-1,3-diene 245
methyl halides, electron-rich 360
methyl radical 90
methylidene 360–366, 371
methylidyne 383–385
microscopic reversibility 8
microwave background, cosmic 384
Millikan's measurement 4
Mills–Nixon effect 41
mineral surfaces 341
mineral–water interface 342–343
Möbius rings 41
modeling of solvent effects 157–161
models, see theories and models
modes
– deformation 366–367
– interionic 355
– normal 127–129
– stretching 132–134
– vibrational 105–149
– XH stretching 132–134
molar extinction coefficient 157, 249
molecular anatomy 243–246
molecular angular momentum 178
molecular beam electric resonance experiments 185
molecular dissection 244
molecular dynamics (MD) 63, 95–98, 165–167, 262, 342, 345
– Car–Parrinello 162
– computational dielectric spectroscopy 282, 311–313, 317–318
– periodic DFT MD simulations 336
– QM/MM MD 227–231
– vibrational circular dichroism 224, 228, 238
molecular electric dipolar moment 32
molecular electrostatic potential 44
molecular geometry 182
molecular Hamiltonian 23, 174–178, 192
molecular ionic liquids 311–316
molecular mechanics 18
– QM/MM approach 159–161
– QM/MM MD 227–231

molecular orbitals (MO) 68
– dynamic extended 187
molecular resolution 282–286
molecular spectroscopy, astronomical 377–398
molecular structure
– and quantum mechanics 23–33
– conformational manifolds 246–247
molecule-fixed coordinate systems 192–195
molecules
– bio- 303–311
– chiral 201–221, 233
– cold 206
– diatomic 20–23, 112–118, 190
– four-particle 188
– heteronuclear diatomic 195
– interaction with external fields 182–184
– ionic lithium dioxide 354–356
– neutral 177
– overtone frequencies 136
– physical properties 4
– polyatomic 23, 118–131, 140–143
– quasi-diatomic 122–124
– small 383–390
– spectroscopic and geometrical constants 106–108
– super- 53, 151
molybdenum 364–366
monoatomic ions, isotropic 311
monodentate bridging 338
mononucleotides 339
Monte Carlo-based conformational search 267
Monte Carlo (MC) simulations 334
morelloflavone, 3,8''-biflavonoid 264
Morse potential 107, 109
Mosotti equation, Clausius– 184
Mössbauer spectroscopy 207–208
motion
– center of mass 24, 283
– EOM formalism 154
Mulliken charges 356
MUlti-SIte Complexation (MUSIC) model 325
multiexponential fit 302
multiplier, Lagrange 124–126
multireference calculations 364
mural quadrant, great 377

n
Na, see sodium
NALMA/NAGMA 305
nanoclusters, gold 262

natural organic matter (NOM) 337
nebula, "Red Rectangle" 392
Negami function, Havriliak– 301
neglect of differential overlap approximation (NDO) 259
Nernst–Einstein relation 312–313
neutral closed shell systems 140
neutral molecules 177
neutron scattering, spectra 330–331
new molecules, vibrational frequencies 353–375
Newton's second law 6, 119
NICS (nucleus-independent chemical shifts) 37, 41–45
nitrogen, astronomical spectroscopy 394
nitroxide radical 88
Nixon effect, Mills– 41
NMR
– liquid-state 328
– MAS 339–341
– solid-state 341
– solution-state 337–340
– spectra 328–329
– triplet wavefunction model (NMRTWM) 46
NMR properties 37–41
NMR spectroscopy
– computational 37–61
– parity violation 208–209
NOM (natural organic matter) 337
non-BO approaches 180
nonadiabatic calculation of dipole moments 173–199
noncommuting operators 15
noninteracting reference system 68
nonrelativistic Hamiltonian 16
A-nor-2-thiacholestane 211
normal mode vectors 121
normal modes, characterization 127–129
normalization 10–13
nuclear anapole moment 209
nuclear Bohr magnetons 67
nuclear charge distribution model 213
nuclear magnetic resonance, see NMR
nuclear magnetic shielding 208
nuclear–orbit interaction 71
nucleotides 309
nucleus-independent chemical shifts (NICS) 37, 41–45
number
– complex 15, 26–27
– harmonic wave 26
– hypercomplex 18
– operator 15

O

OA (optical activity)
– linear susceptibility 234
– vibrational 224
observables 17
– spectroscopic 65
observed orbitals 29
octahedral coordination 341
one-center Gaussian functions 191
ONIOM scheme 93
operators
– center of mass 176
– chirality 203
– commuting 178–179
– differential 9
– Dirac's 15–16
– kinetic energy 80
– Kramers 208
– noncommuting 15
– number 15
– parity 204
– parity-conserving Dirac 208
– stochastic 66
– vector 78
Oppenheimer . . ., see Born–Oppenheimer . . .
optical activity (OA)
– linear susceptibility 234
– Raman 241
– vibrational 224
optical rotation (OR) 241
optical rotatory dispersion (ORD) 223
optimization
– computational conformational 253
– geometry 109
orbit, spin–/nuclear–orbit interaction 70–71
orbitals 3
– conservation symmetry 33
– d-orbital hole 29
– gauge including atomic 80, 328
– molecular 68
– observed 29
– perturbed 74
– "quasi-restricted" 80
organic cations 284
organic complexes, Al– 337–339
organic contaminants 323
organic matter
– dissolved 323
– natural 337
organic molecules, overtone frequencies 136
organometallic chemistry 374
orthogonal matrix 193
orthonormal Cartesian vectors 193
oscillating electric dipoles 233

oscillator strength, DIBs 391
oscillators
– anharmonic 20–23
– coupled 248–251
– harmonic 5–20
out-of plane bendings 90
overlap factors, spatial 164
overtone frequencies, organic molecules 136
overtone spectroscopy 134–135

p

PAHs (polycyclic aromatic hydrocarbons) 325
– interstellar regions 391
parabola, conductivity 316
paramagnetic resonance, electron, see EPR
parity-conserving Dirac operator 208
parity operator 204
parity violation in chiral molecules 201–221
Parrinello MD scheme, Car– 162
particle in a box 2
particle mesh–Ewald (PME) method 299
Pauli theorem, Schwinger–Lüders– 201
PCBs (polychlorinated biphenyls) 325
pentalene dianion 42
peptides
– computational dielectric spectroscopy 304–305
– conformations 100
– spin-labeled 101
– VCD spectra 228
periodic DFT MD simulations 336
permanent dipole/quadrupole moments 183
permittivity, dielectric 314
permutation of identical particles 178
perturbation theory
– double 72
– orbitals 74
– Stark shift 195–196
phase-shifted cosine functions 302
phenomenological spin Hamiltonian 63
phenyl radical 86
philosophical point of view 2
phosphate
– binding on alumina 339–340
– IR/Raman spectra 334–336
phospholipase C-$\gamma 1$ 307
phosphorus compounds, chiral 214
photolysis 357
photons
– physical properties 4
– two-photon Ramsey-fringe spectroscopy 207
photosynthetic reaction center 262

physical properties of molecules and photons 4
(1S)-β-pinene 237
pioneer quantum mechanics 17
Planck constant 4
plane bendings, out-of 90
platinum 367–371
plumbagin 38
PME (particle mesh–Ewald) method 299
point dipole approximation 76
Poisson equation 294, 310
polarity, bonds 107
polarity effects 158
polarizability, isotropic 191
polarizability tensor 257
polarizable continuum model (PCM) 87, 159–160, 165–166
– environmental chemistry 331
polarization
– cross- 232–234
– Debye equation 184
– linear 225–226
– total dielectric 279
polarized γ-radiation, circular 207
polyatomic molecules 23
– Badger-type relationships 140–143
– dissection 118–131
– interstellar regions 389
polychlorinated biphenyls (PCBs) 325
polycyclic aromatic hydrocarbons (PAHs) 325
– interstellar regions 391
polymers 248
polynomial, Hermite 11–13
polynomial function 190
population analysis 229
porphyrin 254–256
post-Hartree–Fock (HF) method 257
potential
– harmonic 120
– Kohn–Sham exchange 156
– molecular electrostatic 44
– Morse 107, 109
potential energy distribution (PED) analysis 128–129
potential energy surface (PES) 118
principle of indeterminacy, Heisenberg's 5
probability, transition 14
probe–solvent interactions 65
property surface 89
proteins
– computational dielectric spectroscopy 305–309
– structure 229

proton donor ions 52
proxyl radical 85
pseudoeigenvalue problem 120
pseudoscalar, Dirac 203
pyramidal structures 91

q

quadrant, great mural 377
quadratic force constants 131
quadrupole coupling 71–72
quadrupole moment 183
quantitative analysis 1
quantum chemistry 1–36
quantum laws 3–5
quantum mechanics
– and molecular structure 23–33
– approach to ECD 256–271
– QM benchmarks 163
– QM/MM approach 159–161
– QM/MM MD 227–231
– simulation methods 152–157
quantum theory
– harmonic oscillator 5–20
– history 17, 377–383
quasi-diatomic molecules 118–131
quasi-elastic neutron scattering (QENS) 330–331
"quasi-restricted" orbitals (QROs) 80

r

radiation, hohlraum 19
radiation–matter interaction Hamiltonian 224–225
σ-radical (vinyl) 84
radicals
– free 63–104
– tempo 85, 88–89
– vinyl 91
– vinyl 84–86
radioastronomy 389–390
RAHB (resonance-assisted hydrogen bond) 51
Raman intensities 327
Raman optical activity (ROA) 241
Raman spectra 325–328, 334–336
Ramsey-fringe spectroscopy, two-photon 207
random-phase approximation (RPA) 257
randomly oriented chiral molecules 233
rare earth nuclei 209
rat fatty acid binding protein 305
RC (relative configuration) 241
reaction field (RF) method 296–297
reactions
– enthalpies 332
– simple atom–molecule 353
– surface 324–325
reactive force field 333
"Red Rectangle" 392–394
Redfield limit 63
reduced mass 106
redundant coordinate sets 137
reference compound 40
reference system, noninteracting 68
refractive index 250
relative configuration (RC) 241
relaxation time, dielectric 300
Renner–Teller coupling 388
representation theory 201
residual energy 8
resonances, Fermi 132–133
response
– chirospecific 232
– Franck–Condon-like 160
– linear 72–76
retinol 246
reversibility, microscopic 8
rhodopsin 246
rigid rotator model 195
rings
– conjugated aromatic 97
– Möbius 41
ROA (Raman optical activity) 241
rod, crystal truncation 329–330
Rosenfeld equation 243, 246
rotation, optical 241
rotational dipole densities 280
rotational energy, Stark 184
rotational spectra 173
rotational strength 258
rotation(al) ...see also vibration(al)–rotation(al) ...
rotator, rigid 195
rotatory dispersion, optical 223
Royal Society 378–379
RPA (random-phase approximation) 257
rutile 344
Rydberg excitations 164–165

s

Salam theory, Weinberg–Glashow– 202
salicylic acid 337
scaling parameters 73
scattering, neutron 330–331
Schleyer's group 44
Schrodinger's equation 9–12
– Born–Oppenheimer 67
– polyatomic molecules 23
Schwinger–Lüders–Pauli theorem 201

second hydration layers   317
secular equation, vibrational   131
self-consistent field (SCF)   28, 73
– complete active space   153
– wavefunction   187
self-interaction error   258
self-term, water–water   306
semiempirical methods   152
Sham..., see Kohn–Sham...
β-sheet conformers   230, 303
shielding
– absolute   37–38
– isochemical   41
– nuclear magnetic   208
– through-space NMR   41
shift
– chemical   37–41, 329
– isotope   203
– Stark   195–196
– toroidal   283
short-range solvation effects   167
short-time dynamical effects   89–98
simulations
– Monte Carlo   334
– periodic DFT MD   336
– quantum mechanical methods   152–157
– UV-Vis spectroscopy   151–171
single bonds   143
singular value decomposition (SVD)   269
Slater, Born–Oppenheimer
    approximation   180
Slater determinant   68, 152
slope   40
small molecules, astronomical
    spectroscopy   383–390
Sn, see tin
sodium, D-lines   380
solid hydrogen   358
solid-state ECD   266–268
solid-state NMR   341
solid surfaces   324
solute–solvent hydrogen bonds   96
solute–solvent interactions   53
solution-state NMR   337–340
solvated biomolecules   303–311
solvation effects, short-/long-range   167
solvatochromism   157–158
solvent effects   53–54
– modeling   157–161
solvent–solvent interactions   65
solvents, dielectric constant   88
Soret transition   255
spatial overlap factors   164
speciation   323–324

spectra
– "background-corrected"   327
– ECD   254–255, 268–270
– fluorescence excitation   387
– high harmonic   31
– IR/Raman   325–328, 334–336
– neutron scattering   330–331
– NMR   328–329
– rotational   173
– types   325
– VCD   229–231
spectral analysis   23
spectral interferograms, heterodyned   236
spectral interferometry, femtosecond
    232–238
spectrometry, computational, see
    computational spectrometry
spectroscopic constants, molecules   106–108
spectroscopic observables   65
spectroscopy   1
– astronomical   377–398
– chiroptical techniques   224
– computational dielectric   279–321
– computational NMR   37–61
– electronic   209
– EXAFS   329–330
– high-resolution   206
– history   377–383
– interstellar nitrogen   394
– Mössbauer   207–208
– NMR   208–209
– overtone   134–135
– Ramsey-fringe   207
– UV-Vis   151–171
– vibration–rotation   206–207
spherical vector components   196
spin, total   78
spin delocalization   97
spin Hamiltonian   66–84
spin-labeled peptides   101
spin–orbit coupling   70–71
spin resonance, electron, see ESR
spin–spin coupling constants (SSCC)   45–52
– calculations   48–50
spin–spin interaction, magnetic dipolar   71
spinning, magic-angle   329, 339–341
splitting
– Davydoff   253
– field-reduced   195
– zero-field   66–84
stability criterion   127
standing wave, X-ray   329–330
Stark rotational energy   184
Stark shift, perturbation theory   195–196

states
- transition 3
- vibrational–rotational 196
static conductivity 289, 312
stereochemistry, DeVoe method 252–256
stereoelectronic effects 84–98
stereospecific electrocyclic process 32
stochastic operators 66
stretching, bonds 113–116
stretching modes
- antisymmetric 135
- isolated 132–134
structure, molecular 246–247
structure theory, electronic 67–69
subazaporphyrins 38
substitution, isotope 132–133
subunits, alanine 304
sum techniques, lattice 298
supercomplex 366–367
supermolecules 53, 151
surface charges, apparent 159
surface reactions 324–325
surfaces
- isochemical shielding 41
- mineral/glass 341
- property 89
- solid 324
susceptibility, linear OA 227, 234
SVD (singular value decomposition) 269
Swan bands, carbonaceous 382
Swan system 385–386
symbolic computation 34
symmetric top 186
symmetry
- CPT 201
- molecular systems 178–179
- orbital 33
symmetry-breaking groups 244
symmetry criterion 127
system–bath decomposition 86

t

Tamm–Dancoff approximation (TDA) 155
"task-specific" ionic liquids 311
tautomerism 43–44
Taylor expansion 119, 182
Teller coupling, Renner– 388
tempo radical 85
- tempo–alcohol complexes 89
- tempo-choline 88
temporal memory 286
tensor
- CSA 329
- diffusion 98

- dipole–dipole 294
- friction 99
- g- 66–85
- magnetic 63–104
- polarizability 257
tesserae 87, 93
tetrahedral coordination 341
theoretical calculations, errors 55
theories and models
- Bohr's theory 5
- "coarse-grained" models 66
- continuum models 53
- coupled cluster theory 153
- Dirac–van Vleck vector model 46
- double perturbation theory 72
- electronic structure theory 67–69
- environmental chemistry 331–333
- EPR general model 64–66
- Fe-hydroxide dimer model 335
- GLOB model 93
- group theory 201
- harmonic oscillator 5–20
- Hartree–Fock 69
- history of quantum theory 17, 377–383
- linear response theory 72–76
- MUSIC 325
- NMRTWM 46
- nuclear charge distribution model 213
- PCM, see polarizable continuum model
- PES concept 118
- QM/MM approach 159–161
- QM/MM MD approach 227–231
- representation theory 201
- rigid rotator model 195
- Stark shift 195–196
- "Theory of Light and Colours" 379
- time-correlation function theory 224–227
- Weinberg–Glashow–Salam theory 202
thorium 371
through-space NMR shieldings 41
time-correlation function theory 224–227
time-correlation functions 286
time-dependent density functional theory
   (TDDFT) 151, 154–157, 162–165
- ECD 246, 257–272
time-dependent HF 257
time-dependent spectroscopy 151
time-domain approaches 223–239
time-domain calculations/measurement,
   direct 231–238
tin dioxide 343–345
titania, environmental chemistry 341–345
titanium 359–362
toluene 101

tomographic imaging 30
toroidal boundary conditions 295
toroidal shifts 283
torsion 92
– aryl–aryl 263
torsional modes 123
total collective dipole moment 282–286
total dielectric polarization 279
total spin 78
toxicology 323
transformation
– coordinate 175, 193
– Fourier 14, 234–237
– Fourier–Laplace transform 287, 293, 300–302
– Kramers–Kronig 224
– Liouville 27
transition dipole 250
transition metals 374
transition probability 14
transition states 3
translation-free internal coordinates 177–178
translational dipole densities 280
trifluoride, boron 359–360
trihydrogen cations 25
trindoles 38
triple bonds 143
Tröger's bases 38
truncated configuration interaction 259
truncation rod, crystal 329–330
tungsten 366–367
twisted ethylene 211
two-photon Ramsey-fringe spectroscopy 207
tyrosine 210

u

ubiquitin 307–308
unfolded proteins 229
uranium 371–374
UV-Vis spectroscopy, simulation 151–171

v

valence states, dicarbon 386
van der Waals interaction 228
vaporization 353
variational function 179, 189
vector model, Dirac–van Vleck 46
vector operators 78
vectors
– gradient 139
– normal mode 121
– orthonormal Cartesian 193

– spherical components 196
vibrational circular dichroism (VCD) 223–239
vibrational correction 98
vibrational frequencies
– IR/Raman spectra 326
– new molecules 353–375
vibrational modes
– adiabatic 105–149
– localized 122–124
vibrational OA-FID measurements 237–238
vibrational optical activity (VOA) 224, 231–238
vibrational–rotational energy 21–22
vibrational–rotational spectroscopy 206–207
vibrational–rotational states 196
vibrational secular equation 131
vibronic coupling 261
vinyl, hyperfine coupling constants 91
vinyl radical 84–86
VOA (vibrational optical activity) 224

w

W, *see* tungsten
Waals, van der, *see* van der Waals
wagging 92
water
– dynamics 343–345
– environmental chemistry 341–345
– water–water self-term 306
Watson–Crick base pairs 140
wave, standing 329–330
wave mechanics 9–15
wavefunction 9
– SCF 187
– wavefunction-based *ab initio* methods 152
Weinberg–Glashow–Salam theory 202
Whittaker M function 10
Wilson matrix 121
– Free– 43

x

X-ray standing wave (XSW) techniques 329–330
XC kernel 75
XH stretching modes 132–134

y

Young, Thomas 379–380

z

Zeeman effect 22
Zeeman interaction, electronic 72
Zehnder interferometer, Mach– 235

zero-field splitting 66–84
zero-frequency limit 292
zero-point energy 8
zinc(II), environmental chemistry 341–343

zinc finger 303
ZINDO method 154
zirconium 362–364
ZORA 212